Materials for Energy

Advances in Materials Science and Engineering

Series Editor:
Sam Zhang

Semiconductor Nanocrystals and Metal Nanoparticles
Physical Properties and Device Applications
Tupei Chen and Yang Liu

Advances in Magnetic Materials
Processing, Properties, and Performance
Sam Zhang and Dongliang Zhao

Micro- and Macromechanical Properties of Materials
Yichun Zhou, Li Yang, and Yongli Huang

Nanobiomaterials
Development and Applications
Dong Kee Yi and Georgia C. Papaefthymiou

Biological and Biomedical Coatings Handbook
Applications
Sam Zhang

Hierarchical Micro/Nanostructured Materials
Fabrication, Properties, and Applications
Weiping Cai, Guotao Duan, and Yue Li

Biological and Biomedical Coatings Handbook, Two-Volume Set
Sam Zhang

Nanostructured and Advanced Materials for Fuel Cells
San Ping Jiang and Pei Kang Shen

Hydroxyapatite Coatings for Biomedical Applications
Sam Zhang

Carbon Nanomaterials
Modeling, Design, and Applications
Kun Zhou

Materials for Energy
Sam Zhang

For more information about this series, please visit: www.crcpress.com

Materials for Energy

Edited by
Sam Zhang

CRC Press
Taylor & Francis Group
Boca Raton London New York

CRC Press is an imprint of the
Taylor & Francis Group, an **Informa** business

First edition published 2021
by CRC Press
6000 Broken Sound Parkway NW, Suite 300, Boca Raton, FL 33487-2742

and by CRC Press
2 Park Square, Milton Park, Abingdon, Oxon, OX14 4RN

© 2021 Taylor & Francis Group, LLC

First edition published by Willan 2008
Sixth edition published by Routledge 2009

CRC Press is an imprint of Taylor & Francis Group, LLC

ISBN: 978-0-367-35021-5 (hbk)
ISBN: 978-0-367-54527-7 (pbk)
ISBN: 978-0-429-35140-2 (ebk)

Typeset in Times
by codeMantra

Contents

Preface..vii
Editor ix
Contributors xi

Chapter 1   Halide Perovskite Photovoltaics.. 1

   Guifang Han and Sam Zhang

Chapter 2   Carbon Nanomaterials for Flexible Energy Storage Devices 45

   Huijuan Lin, Hui Li, and Chenjun Zhang

Chapter 3   Triboelectric Materials for Nanoenergy... 77

   Xiude Yang, Jun Dong, Juanjuan Han, and Qunliang Song

Chapter 4   III-N Ultraviolet Light Emitters for Energy-Saving Applications..... 125

   Dong-Sing Wuu

Chapter 5   *In-situ* Growth of Spherical Graphene Films on Cemented
   Carbide for Spatial Sensor Matrix 167

   *Xiang Yu, Zhen Zhang, Jing-xuan Pei, Jian-kang Huang, and
   Xiao-yong Tian*

Chapter 6   Membrane Materials for Vanadium Redox Flow Battery................ 199

   Jiaye Ye and Lidong Sun

Chapter 7   Thin-Film Solid Oxide Fuel Cells.. 239

   Jong Dae Baek, Ikwhang Chang, and Pei-Chen Su

Chapter 8   *In-Situ* Mechanistic Study of Two-Dimensional Energy
   Materials by Well-Defined Electrochemical On-Chip Approach.... 285

   Yu Zhou, Shuang Yang, and Fangping Ouyang

Chapter 9   Phase Change Materials for Thermal Energy Storage..................... 317

   Baris Burak Kanbur, Zhen Qin, Chenzhen Ji, and Fei Duan

Chapter 10 Strategies for Performance Improvement of Organic Solar Cells.... 373

Wei-Long Xu and Xiao-Tao Hao

Chapter 11 Surface Passivation Materials for High-Efficiency Silicon
Solar Cells 413

*Shui-Yang Lien, Chia-Hsun Hsu, Xiao-Ying Zhang, and
Pao-Hsun Huang*

Chapter 12 Organic Solar Cell 443

Shaohui Zheng

Chapter 13 High-Performance Electrolytes for Batteries 479

Yixiang Ou

Index 507

Preface

In the contemporary society, energy usage is far more advanced, even "outraged" in a sense, than ever before. Handphone batteries, solar cells, and electrical cars are common concept and application examples. Earth ridden energies such as coal and oil are depleting. Therefore, people turn to other sources of energies, such as the sun and the tide. Thus, energy conversion has never been so important. Creation of an energy application/conversion relies on and can be realized by creation of certain "materials". Without the right material, however great a conversion idea is, it stays forever as an idea. The job is, therefore, the creation of the right material for the right energy conversion or storage. Toward this purpose, I organized a group of material experts specializing in various energy conversion and storage applications, and this has led to the formation of the current book. This book highlights current developments in energy conversion, storage, and applications from materialists' angle.

Chapter 1, "Halide Perovskite Photovoltaics," discusses halide perovskite materials for photovoltaic applications with regard to optical properties, along with the feasibility of bandgap tuning in ABX_3 perovskites. Chapter 2, "Carbon Nanomaterials, for Flexible Energy Storage Devices," details the necessities to develop flexible energy storage devices. Chapter 3, "Triboelectric Materials, for Nanoenergy," elaborates various paired materials used for triboelectric nanogenerator deemed as an ideal source to achieve nanoenergy for power supply to the worldwide sensor network in the near future. Chapter 4, "III-N Ultraviolet Light Emitters for Energy-Saving Applications," elaborates the growth and device fabrication of group III nitride-based LEDs materials for ultraviolet light-emitting diodes. Chapter 5, "In-Situ Growth of Spherical Graphene Films on Cemented Carbide for Spatial Sensor Matrix," reviews the preparation of spherical graphene films and their potential applications in the sensor field and introduces a new method for transfer-free graphene with controllable structure growth directly on the cemented carbide. Chapter 6, "Membrane Materials for Vanadium Redox Flow Battery," discusses the membrane materials for vanadium redox flow battery, a large-scale energy storage technique toward the power grid. Chapter 7, "Thin Film Solid Oxide Fuel Cells," focuses on current development of TF-SOFC architectures in terms of a supporting substrate and state-of-the art methodology to improve stability against mechanical and thermal stress. Chapter 8, "In-Situ Mechanistic Study of Two-Dimensional Energy Materials by Well-Defined Electrochemical On-Chip Approach," summarizes progress for in-situ technology and mechanism on 2D materials electrochemical devices. Chapter 9, "Phase Change Materials for Thermal Energy Storage," explains the theoretical, numerical, and experimental procedures of the phase change material analysis with real-scale unit setup and feasibility assessments. Chapter 10, "Strategies for Organic Solar Cells Performance Improvement," describes the strategies to improve the performance of organic solar cells and the challenges and opportunities for organic solar cells in the future. Chapter 11, "Surface Passivation Materials for High-Efficiency Silicon Solar Cells," describes the surface passivation materials and techniques to improve performance of passivated emitter and rear contact solar cells. Chapter 12, "Organic

Solar Cell," reviews organic solar cells and the materials to create them, Chapter 13, "High-Performance Electrolytes for Batteries," introduces the development of high-performance electrolyte materials and preparations used for batteries.

Researchers, technologists, graduates, and senior students who are interested in energy should find this book useful.

I'd like to take this opportunity to thank all the chapter authors for their support and hardwork. I'd like to thank Ms Allison Shatkin and the management team of the publishing company for their continued support in getting this book to its current form.

Thank you all!

This work was supported by Fundamental Research Funds for the Central Universities: SWU118105.

Sam Zhang
March 4, 2020

Editor

Sam Zhang (張善勇), Fellow of the Royal Society of Chemistry, was born and brought up in the famous "City of Mountains," Chongqing, China. He earned a bachelor of engineering in materials in 1982 at Northeastern University (Shenyang, China), a master of engineering in materials in 1984 at Iron and Steel Research Institute (Beijing, China), and a PhD in ceramics in 1991 at the University of Wisconsin-Madison, USA. He was a tenured Full Professor (from 2006 to 2018) at the School of Mechanical and Aerospace Engineering, Nanyang Technological University, Singapore. In January 2018, he joined the School of Materials and Energy, Southwest University, China, and assumed the duty as Director of the Centre for Advanced Thin Film Materials and Devices of the university.

Professor Zhang was the founding Editor-in-Chief for *Nanoscience and Nanotechnology Letters* (USA) (2008–December 2015) and has been the Principal Editor for *Journal of Materials Research* (USA) responsible for the field of thin films and coatings (since 2003). He has also been an Editorial Board Member of *Surface and Coatings Technology* since June 2012.

Professor Zhang has been serving the world's first "Thin Films Society" (www. hinfilms.sg) as its founding and current president since 2009. He has authored or edited 13 books, of which 12 were published with CRC Press/Taylor & Francis. Of these books, *Materials Characterization Techniques* has been adopted as a textbook by 30 American and European universities as of October 2015. This book was also translated into Chinese and published by China Science Publishing Company in October 2010 and distributed nationwide in China. All of Sam's books are a click away on Amazon.com. His books are also available in university and/or national libraries all over the world. Since August 2010, Professor Zhang has been serving as the Series Editor for Advances in Materials Science and Engineering Book Series published by CRC Press/Taylor & Francis.

Professor Zhang was elected as Fellow of Royal Society of Chemistry (FRSC) and Fellow of Thin Films Society (FTFS) in 2018 and Fellow of Institute of Materials, Minerals and Mining (FIoMMM) in 2007. His current research centers in areas of energy films and coatings for solar cells, hard yet tough nanocomposite coatings for tribological applications by vapor deposition, measurement of fracture toughness of ceramic films and coatings, and electronic/optical thin films. Over the years, he has authored or coauthored over 300 peer-reviewed international journal papers. As of March 10, 2020, from Web of Science (http://www.researcherid.com/rid/A-3867-2011), his articles with citation data: 317, the number of the times cited: 9,510, average citations per article: 30.5, and h-index: 51.

Professor Zhang is featured in *Who's Who in Engineering Singapore* (published October 2007) and in the 26th, 27th, and 28th editions of *Who's Who in the World*. His other honorary appointments include Honorary Professor of the Institute of Solid State Physics, Chinese Academy of Sciences (since November 2004); Guest Professor of Zhejiang University (since November 2006); Guest Professor of Harbin Institute of Technology (since September 2007); Guest Professor of Southwest Jiaotong University (since March 2018); and Guest Professor of Khashi University (since June 2018).

Contributors

Jong Dae Baek
Department of Automotive
    Engineering
Yeungnam University
Gyeongsangbuk-do, South Korea

Ikwhang Chang
Department of Automotive
    Engineering
Wonkwang University
Iksan, South Korean

Jun Dong
Institute for Clean Energy and
    Advanced Materials (ICEAM)
Southwest University
and
School of Materials Science and
    Engineering
Yangtze Normal University
Chongqing, China

Fei Duan
School of Mechanical and Aerospace
    Engineering
Nanyang Technological University
Jurong West, Singapore

Guifang Han
School of Materials Science and
    Engineering
Shandong University
Jinan, China

Juanjuan Han
Institute for Clean Energy and
    Advanced Materials (ICEAM)
Southwest University
Chongqing, China

Xiao-Tao Hao
School of Physics
State Key Laboratory of Crystal
    Materials
Shandong University
Shandong, China

Chia-Hsun Hsu
School of Opto-Electronic and
    Communication Engineering
Xiamen University of Technology
Xiamen, China

Jian-kang Huang
School of Materials Science and
    Technology
China University of Geosciences
Beijing, China

Pao-Hsun Huang
School of Information Engineering
Jimei University
Xiamen, China

Chenzhen Ji
Nanyang Technological University
Jurong West, Singapore

Baris Burak Kanbur
Nanyang Technological University
Jurong West, Singapore

Hui Li
Key Laboratory of Flexible Electronics
    (KLOFE)
Institute of Advanced Materials (IAM)
Jiangsu National Synergetic Innovation
    Center for Advanced Materials
    (SICAM)
Nanjing Tech University
Nanjing, China

Shui-Yang Lien
School of Opto-Electronic and
    Communication Engineering
Xiamen University of Technology
Xiamen, China

Huijuan Lin
College of Materials Science and
    Engineering
Nanjing Tech University
Nanjing, China

Yixiang Ou
Beijing Radiation Center
Beijing Academy of Science and
    Technology
Beijing, China

Fangping Ouyang
School of Physics and Electronics
Central South University
Changsha, China

Jing-xuan Pei
School of Materials Science and
    Technology
China University of Geosciences
Beijing, China

Zhen Qin
Nanyang Technological University
Jurong West, Singapore

Qunliang Song
Institute for Clean Energy and
    Advanced Materials (ICEAM)
Southwest University
Chongqing, China

Pei-Chen Su
School of Mechanical and Aerospace
    Engineering
Nanyang Technological University
Jurong West, Singapore

Lidong Sun
State Key Laboratory of Mechanical
    Transmission
School of Materials Science and
    Engineering
Chongqing University
Chongqing, China

Xiao-yong Tian
School of Materials Science and
    Technology
China University of Geosciences
Beijing, China

Dong-Sing Wuu
Department of Materials Science and
    Engineering
National Chung Hsing University
Taichung, Taiwan

Wei-Long Xu
School of Physics
State Key Laboratory of Crystal
    Materials
Shandong University
Shandong, China

and

School of Photoelectric Engineering
Changzhou Institute of Technology
Jiangsu, China

Shuang Yang
Key Laboratory for Ultrafine Materials
    of Ministry of Education
School of Materials Science and
    Engineering
East China University of Science and
    Technology
Shanghai, China

Xiude Yang
Institute for Clean Energy and
    Advanced Materials (ICEAM)
Southwest University
Chongqing, China

and

School of Physics and Electronic
    Science
Zunyi Normal College
Zunyi, China

Jiaye Ye
State Key Laboratory of Mechanical
    Transmission
School of Materials Science and
    Engineering
Chongqing University
Chongqing, China

Xiang Yu
Beijing Key Laboratory of Materials
    Utilisation of Nonmetallic Minerals
    and Solid Wastes
National Laboratory of Mineral
    Materials
School of Materials Science and
    Technology
China University of Geosciences
    (Beijing)
Beijing, China

Chenjun Zhang
Key Laboratory of Flexible Electronics
    (KLOFE)
Institute of Advanced Materials (IAM)
Nanjing Tech University
Nanjing, China

Sam Zhang
Center for Advanced Thin Films and
    Devices
School of Materials and Energy
Southwest University
Chongqing, China

Xiao-Ying Zhang
School of Opto-Electronic and
    Communication Engineering
Xiamen University of Technology
Xiamen, China

Zhen Zhang
School of Materials Science and
    Technology
China University of Geosciences
Beijing, China

Shaohui Zheng
School of Materials and Energy
Southwest University
Chongqing, China

Yu Zhou
School of Physics and Electronics
Central South University
Changsha, China

1 Halide Perovskite Photovoltaics

Guifang Han
Shandong University

Sam Zhang
Southwest University

CONTENTS

1.1 Solar Energy and Photovoltaics 1
1.2 Halide Perovskite Materials 4
    1.2.1 Structure of Halide Perovskite Materials 5
    1.2.2 Optical Property and Bandgap Tunability of Halide
            Perovskite Materials 7
    1.2.3 Optoelectronic Property of Halide Perovskite Materials 12
1.3 Stability of Halide Perovskites 15
    1.3.1 Intrinsic Thermal Stability of Halide Perovskite Materials 15
    1.3.2 Phase Stability of FAPbI$_3$ Perovskite 18
    1.3.3 Phase Stability of All-Inorganic CsPbI$_3$ Perovskite 23
1.4 Summary 35
References 36

1.1 SOLAR ENERGY AND PHOTOVOLTAICS

The sun not only gives us light and happy mood, but is also a natural power source for our Earth. It drives the circulation of global wind and ocean currents, the evaporation and condensation of water cycle that creates rivers and lakes, and the biological cycle of photosynthesis that causes the diversity of nature and life (Lewis and Crabtree 2005). It provides us clean and abundant energy. The energy from sunlight to our Earth in 1 hour is more than the total energy consumption by humans in an entire year (Lewis and Crabtree 2005, Lewis and Nocera 2006, Cook et al. 2010). The energy released by the earthquake in San Francisco 1906 with a magnitude of 7.8 is equal to the amount of energy the sun delivers to the Earth in 1 second (Crabtree and Lewis 2007). Solar energy is the largest resource among various renewable energy sources by far. Solar energy can be converted to electricity through photovoltaic cells, fuel through natural or artificial photosynthesis, and thermal energy by heat engines or other techniques (Crabtree and Lewis 2007). Here in this chapter, we focus on how to convert solar energy into electricity through photovoltaic cells, emphasizing the principle of photovoltaics, perovskite materials, and the structure and properties of these materials.

Photovoltaic or solar cell is a device that absorbs light and converts it into electricity. Normally, for a semiconductor with a bandgap of E_g, it absorbs light with energy ($h\nu$) higher than its bandgap, and an electron is excited into the conduction band leaving a hole behind in the valence band (Figure 1.1a). If the excited electron could be collected and passed through an outer circuit, electricity is "generated". Figure 1.1b depicts a typical current–voltage (I–V) curve of a solar cell. The open circuit voltage (V_{oc}) is the maximum voltage that a device could obtain. The short circuit current (J_{sc}) is the maximum current that a device could achieve. The power conversion efficiency (η) of one solar cell is the ratio of the output electricity to the input energy of sunlight. In practice, the efficiency η is determined as the ratio of the maximum power output, P_{max}, generated by the solar cell to the power input, P_{in}, based on the measurement of I–V curve: (Würfel 2007)

$$\eta = \frac{P_{max}}{P_{in}} = \frac{J_{mp} \cdot V_{mp}}{P_{in}} = \frac{J_{sc} \cdot V_{oc} \cdot FF}{P_{in}} \tag{1.1}$$

where J_{mp} and V_{mp} are the current density and voltage at the maximum power point (Figure 1.1b). To simplify the calculation and relate the efficiency with practically measurable parameters, fill factor (FF) is introduced, which is defined as the ratio of the areas of two rectangles determined by J_{mp} and V_{mp} (blue in Figure 1.1b) and by V_{oc} and J_{sc} (green in Figure 1.1b), respectively. Accordingly, the three parameters of V_{oc}, J_{sc}, and FF combine to determine the efficiency of a device as shown in Eq. (1.1). The input energy of sunlight (P_{in}) is 100 (mW cm^{-2}) based on one sun condition as a standard level for comparison of efficiency of devices fabricated in different labs and geographic conditions.

Before going into details of how to improve the efficiency of solar cell, we first look at the spectrum and energy distribution of standard sunlight shown in Figure 1.2 (based on solar cell application). Sunlight is actually electromagnetic radiation.

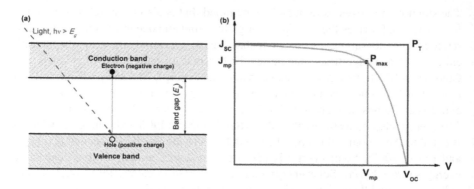

FIGURE 1.1 (a) Illustration of band structure and photoelectric effect in a bulk semiconductor and (b) a typical current–voltage (I–V) curve of a solar cell device for efficiency calculation (Han et al. 2017). (Reprinted with permission from Elsevier.)

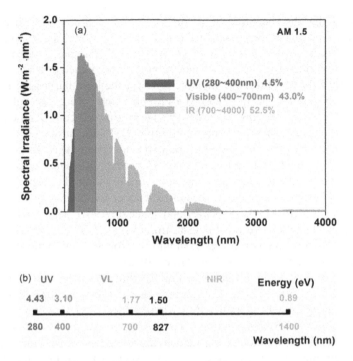

FIGURE 1.2 (a) Energy distribution of solar spectrum and (b) the corresponding energy level and wavelength. (Drawn based on the data of AM 1.5G according to ASTM G173) (Han et al. 2017). (Reprinted with permission from Elsevier.)

According to the wavelength from lower to higher, people divide the spectrum of sunlight into three regions: ultraviolet (UV, wavelength of 300–400 nm), visible (VL, wavelength of 400–700 nm), and infrared (IR, wavelength of 700–4,000 nm) as presented in Figure 1.2a. The total power of the solar spectrum is integrated to 100 mW cm^{-2} at AM 1.5 G, which is usually named as standard one sun condition. Currently, most of solar cells utilize UV and visible regions of the solar spectrum. Near infrared (NIR, wavelength of 700–1,400 nm) is also used by some types of solar cells. The rest of the energy in the long-wavelength region is lost as heat. The energy level of photons, as shown in Figure 1.2b, can be calculated by the Planck–Einstein relation (French and Taylor 1978):

$$E = h\nu = \frac{hc}{\lambda} \tag{1.2}$$

where h is the Planck constant, ν is the light frequency, c is the speed of light, and λ is the wavelength of incident light. This relation accounts for the quantum nature of light. The short-wavelength UV light, i.e. highest energy, only occupies <5% of the total solar energy. While the long-wavelength, low-energy NIR accounts for 52.5% of the solar energy. The visible light located in the middle covers around 43% of the solar energy.

The E_g of the absorber used limits the maximum value of V_{oc}. Therefore, the higher the E_g, the higher the possible value of output V_{oc}. The J_{sc} is a product of light harvesting efficiency, charge separation efficiency, and charge collection efficiency. The light harvesting efficiency is dependent on the absorbance of a semiconductor. The more photons are absorbed, the more efficient the light harvesting is. In principle, a semiconductor can only absorb a photon whose energy is higher than its E_g. However, photons bearing much higher energy than E_g can excite electrons to energy levels above conduction band minimum (CBM), and subsequently electrons rapidly relax to the CBM by releasing the extra energy as heat. To harvest more photons, the E_g should be as low as possible. Consequently, there exists an optimal bandgap energy for photovoltaic application in consideration of the spectrum losses. Shockley and Queisser calculated a theoretical conversion efficiency of around 31% for single junction solar cells (Shockley and Queisser 1961). M. Green proposed a simple empirical relation to estimate the minimal value of the reverse saturation current density. As such, the optimized E_g should be around 1.5 eV, which agrees well with the experimental data (Shah 2010, Green 1982).

Halide perovskite solar cell is a new type of thin film solar cell. Since its first report in 2009, the efficiency in 2019 already reached to more than 25%, which is the highest among all polycrystalline thin film solar cells. The low-cost solution process and feasibility of bandgap tuning make this material a promising candidate for novel thin film photovoltaics. Section 1.2 elaborates the feature structure and optical and optoelectronic properties of these materials. Section 1.3 explores the stability of halide perovskites including intrinsic thermal stability and strategies developed to stabilize these perovskite structures. A summary follows in Section 1.4.

1.2 HALIDE PEROVSKITE MATERIALS

The term "perovskite" originates from the mineral structure of $CaTiO_3$, named after a Russian geologist Count Lev Aleksevich von Perovski (Tanaka and Misono 2001). Nowadays, perovskite refers to one big group of compounds, ABO_3, which has a crystal structure similar to that of $CaTiO_3$. It is one of the very important and most frequently encountered structures in solid-state inorganic compounds (Roy and Muller 1974). Perovskite covers most of the metallic ions in the periodic table with a significant number of anions. Perovskite is a very important crystal structure due to its diverse physical and chemical properties, such as ferroelectricity, piezoelectricity, and magnetoresistance, thermistor, electro-optical modulator, battery materials, and so on, playing an important role in the modern electronic industry (Li, Soh, and Wu 2004).

In an ideal cubic ABO_3 perovskite structure, the A-site is a bigger cation (such as Ca^{2+}, Sr^{2+}, Ba^{2+}), B-site is a smaller cation (such as Ti^{4+}, Zr^{4+}), in which A is located at the corner of the cubic, B occupies the center of the cubic, and oxygen (O) fills the face center position. When the O is replaced by halogen (such as F, Br, Cl, I), A and B sites should be monovalent and divalent cations in order to keep neutral. This is halide perovskite.

This section discusses the structure, optical properties, bandgap tuning, and optoelectronic properties of halide perovskites.

1.2.1 STRUCTURE OF HALIDE PEROVSKITE MATERIALS

Figure 1.3 depicts a unit crystal structure of halide perovskite. For solar cell application, the B-site is normally lead or tin (Pb^{2+} or Sn^{2+}), with lead being the majority since tin is easy to be oxidized. Germanium, in the same column where lead is in the periodic table, is even less stable than tin. Therefore, it is not very popular for solar cell applications. The cubic corner monovalent A-site cations can be anything metallic (such as Cs and Rb) or organic groups (such as methylammonium (($CH_3NH_3)^+$ or MA^+) and formamidinium ($[CH(NH_2)_2]^+$ or FA^+)) as long as the structure factor permits. The X site is, of course, a halogen: I^-, Br^-, or Cl^-.

To keep the perovskite structure stable, the ionic radius of A, B cations and X anion should meet some requirements. Goldschmidt, for the first time, proposed a "tolerance factor" (t) to study the stability of perovskite structure:

$$t = (R_A + R_X)/\left\{\sqrt{2}(R_B + R_X)\right\} \tag{1.3}$$

where the R_A, R_B, and R_X are ionic radii of the A, B, and X ions, respectively. Geometrically, for an ideal cubic perovskite structure, the ratio of bond length of A–X to B–X should be $\sqrt{2}:1$. If we assume the bond length is roughly to be the sum of two ionic radii, the tolerance factor t of an ideal perovskite should be 1.0. However, experimentally, the t value of most perovskites is between 0.8 and 1.0. The tolerance factor has guided people to design and find new perovskite materials. Even now, it has a very important role in the design and study of perovskites. When more and more perovskite materials were explored, people found that tolerance factor is a necessary but not sufficient requirement for the formation of perovskite structure. In other words, almost all known perovskites have a tolerance factor in the range of 0.8 to 1.0, but a compound with a tolerance factor in this range is not necessarily a perovskite structure. To predict more precisely the formability of perovskite structure, researchers introduced an "octahedral factor" (Li, Soh, and Wu 2004). Together with tolerance factor, this model can predict most of the known perovskite materials.

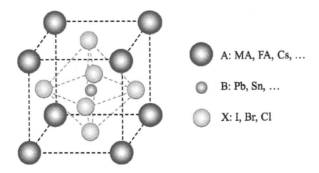

A: MA, FA, Cs, ...

B: Pb, Sn, ...

X: I, Br, Cl

FIGURE 1.3 Illustration of unit crystal structure of ABX_3 halide perovskite, in which A cation generally is MA, FA, or Cs; B cation is Pb or Sn; and X anion is a halogen including I, Br, and Cl.

As seen from Figure 1.3, BX_6 is the basic unit of perovskite. If ion B is too small, this unit may be unstable, so is the perovskite (Li, Soh, and Wu 2004). If ion B is too big, the BX_6 unit and thus the perovskite wouldn't be stable. Therefore, the ionic radii of B and X should meet the requirement of keeping this octahedral structure stable. The ratio of R_B/R_X is named "octahedral factor," and it should be another factor to determine whether the perovskite structure is stable or not. From geometric and reported data, the lowest limit of R_B/R_X is 0.414 (Rohrer 2001). By investigating 186 complex halide perovskite systems, the tolerance and octahedral factors of ABX_3 were calculated and are summarized in Figure 1.4 (Li et al. 2008). Most of perovskites fall into the area formed by the dotted line with a tolerance factor from 0.813 to 1.107 and an octahedral factor in the range of 0.442–0.895. Among all these 186 halide systems investigated, only one system ($CsF–MnF_2$) is wrongly placed inside the perovskite area, but it does not form a perovskite structure. Six systems marked in Figure 1.4, which have perovskite structure, are located outside of the perovskite region. Approximately 96% of perovskites investigated in this study are correctly included in the region formed by tolerance and octahedral factors together. If tolerance factor is the only basis, many systems that are not perovskite will be wrongly placed in the perovskite region. This further emphasizes the importance of octahedral factor in determining the formability of halide perovskite structure.

Since perovskites used in solar cells are mostly $MAPbX_3$, $FAPbX_3$, $CsPbX_3$, and their mixtures, the octahedral units are the same as those of PbX_6. Therefore, in the following section, we use only tolerance factor to identify the phase stability of these compounds without the burden of the octahedral factor.

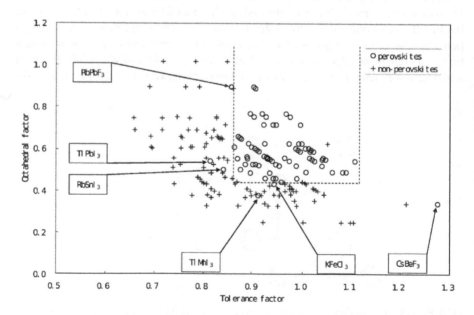

FIGURE 1.4 Classification of halide perovskite ABX_3 compounds in the tolerance-octahedral factor (μ-t) map (Li et al. 2008) (open-access).

MAPbI$_3$ is the first fully discovered halide perovskite for photovoltaic application. It is not a perfect cubic perovskite at room temperature. At higher temperatures, MAPbI$_3$ forms a cubic structure. While at a temperature of around 57°C, it changes to tetragonal phase with lattice parameters of $a = 8.800(9)$ and $c = 12.685(7)$ Å (Kawamura, Mashiyama, and Hasebe 2002, Baikie et al. 2013).

1.2.2 OPTICAL PROPERTY AND BANDGAP TUNABILITY OF HALIDE PEROVSKITE MATERIALS

The unique optical property of halide perovskite makes it a promising candidate for solar cells, lasers, light emitting diodes, and other optoelectronic applications. This section discusses the optical property along with bandgap tuning, optoelectronic properties, and defect physics.

The optical absorption coefficient for solar cell absorber materials should be as high as possible to achieve a high power conversion efficiency. Calculated optical absorption coefficients of MAPbI$_3$, CsSnI$_3$, and GaAs are shown in Figure 1.5a (Yin, Shi, and Yan 2014a). It can be seen that the optical absorption coefficient of MAPbI$_3$ is up to one order of magnitude higher than that of GaAs within the visible light range. As shown in Figure 1.2a, visible light accounts for the major usable solar spectrum for solar cell applications; high visible light absorption is critical to achieve high-efficiency cells. GaAs, as the most efficient thin film solar cell absorber currently, has an optical absorption coefficient much lower than that of perovskites. Therefore, MAPbI$_3$-based perovskite solar cells have the potential to achieve a power conversion efficiency higher than that of GaAs. Figure 1.5b shows the theoretical maximum efficiencies calculated based on the optical absorption coefficient with different film thicknesses for currently available absorber materials (Yin, Shi, and Yan 2014a). Three perovskites—MAPbI$_3$, CsSnI$_3$, and CsPbI$_3$—have a power conversion efficiency higher than that of CuInS$_2$ (CIS), CZTS, and GaAs thin film

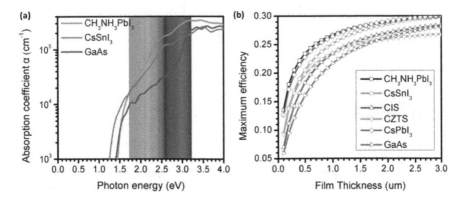

FIGURE 1.5 (a) The optical absorptions of CH$_3$NH$_3$PbI$_3$ (MAPbI$_3$), CsSnI$_3$, and GaAs and (b) calculated maximum power conversion efficiencies of halide perovskites, CuInS$_2$ (CIS), CZTS, and GaAs as a function of film thickness (Yin, Shi, and Yan 2014a). (Reprinted with permission from John Wiley and Sons.)

solar cells for any given thickness. To achieve same power conversion efficiency, perovskites normally need smaller thickness. For example, with a 0.5 μm thickness, MAPbI$_3$-based solar cells can have a maximum efficiency up to 24%, while it is only 16% for CZTS. To achieve 24% efficiency, the thickness of CZTS normally needed is 1.5 μm. These results are consistent with today's techniques for thin film solar cells, which normally need about 2 μm thick absorber layers. FAPbI$_3$, having a bandgap slightly lower than that of MAPbI$_3$ and being closer to the optimal bandgap of 1.5 eV facilitates the utilization of solar spectrum to have higher efficiency.

One of the benefits of ABX$_3$ perovskites is the tunability of their bandgap. Both composition and dimensionality will affect their bandgaps.

In MAPbI$_3$, the CBM is derived from Pb p orbitals, while the valence band maximum (VBM) is a mixture of Pb s and I p (s–p semiconductor) orbitals (Yin et al. 2015). Therefore, by changing B- and X-site elements, the bandgap is tunable in a wide range (Mitzi 2007).

Bromine (Br) substituted iodine in MAPbI$_3$ films increases the E_g. Absorption coefficients of MAPb(Br$_x$I$_{1-x}$)$_3$ ($0 \leq x \leq 1$) perovskite films are shown in Figure 1.6a, where x indicates the amount of Br in this mixed halide perovskite (Hoke et al. 2015). These curves continuously blue-shift with the increasing of Br content, indicating larger E_g values for Br-rich compositions. The bandgap shifted from 1.57 eV for MAPbI$_3$ to 2.23 eV for pure bromide MAPbBr$_3$ perovskite films. The color of perovskite films changed from dark brown/black to red then orange (inset photographs from left to right) in Figure 1.6a with increasing bromide content. This indicated that the X-site element had an important role in determining the bandgap, consistent with the theoretical analysis shown in Reference Yin et al. (2015).

However, this mixing/alloying is not limitless and sometimes causes instability issue. Thermodynamically driven halide segregation was found in a mixed-halide

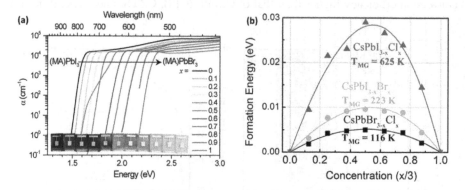

FIGURE 1.6 (a) Absorption coefficient of (MA)Pb(Br$_x$I$_{1-x}$)$_3$ measured by diffuse spectral reflection and transmission measurements on thin films and photocurrent spectroscopy of solar cells. Inset: photograph of (MA)Pb(Br$_x$I$_{1-x}$)$_3$ photovoltaic devices from $x = 0$ to $x = 1$ (left to right) (Hoke et al. 2015). (b) Formation energies of mixed halide CsPbX$_3$ (X = I, Br, Cl) for CsPb(I$_{1-x}$Br$_x$)$_3$, CsPb(I$_{1-x}$Cl$_x$)$_3$, or CsPb(Br$_{1-x}$Cl$_x$)$_3$ alloy, in which the miscibility gap temperature (T_{MG}) values are also given (Yin, Yan, and Wei 2014c). (Reprinted with permission from American Chemical Society.)

system with a formula of $MAPb(Br_xI_{1-x})_3$ under light illumination (Slotcavage, Karunadasa, and McGehee 2016). Alloying chlorine (Cl) with $MAPbI_3$ improved the stability and carrier lifetime. However, the detectable atomic ratio of Cl was <4% even though the three-fold excess Cl was used in precursor solutions (Colella et al. 2013, Ball et al. 2013). The addition of Cl almost does not change the bandgap of $MAPbI_3$. Therefore, Cl is more like a dopant, which dramatically improves the charge transport property within perovskite layers. However, more research is needed to find the mechanism behind this.

The origin of different miscibility behaviors between different types of mixed halides is not clear. To obtain more information on this, Yin et al. used first-principle calculation together with Monte Carlo simulations, taking $CsPbX_3$ as a model material, and conducted systematical studies on the structural, electronic and energetic properties of mixed-halide perovskites (Yin, Yan, and Wei 2014c).

Figure 1.6b gives the formation energy of different halides alloyed with $CsPbX_3$ (Yin, Yan, and Wei 2014c). A parabolic feature was observed for formation energy curves of all possible alloys. This indicated that compositions at both sides having lower formation energy could be more stable than those in the middle. Besides, the $CsPbBr_{3-x}Cl_x$ had the lowest formation energy and were thus easier to form (black curve in Figure 1.6b). The mixing of Cl into $CsPbI_3$ would be very difficult due to the high formation energy (blue curve in Figure 1.6b). The critical temperature at which alloys can be fully mixable, i.e. miscibility gap temperature (T_{MG}), is also shown in Figure 1.6b. The T_{MG} values for $CsPbI_{3-x}Cl_x$, $CsPbI_{3-x}Br_x$, and $CsPbBr_{3-x}Cl_x$ are 625, 223, and 116 K, respectively. Considering that halide perovskites are normally fabricated at relatively low temperature (below 600 K), both $CsPbI_{3-x}Br_x$ and $CsPbBr_{3-x}Cl_x$ can be fully mixed, while $CsPbI_{3-x}Cl_x$ is more difficult to mix, which is consistent with experimental observations.

Halide substitute is mainly tuning the VBM level of perovskites. Since the metal Pb is involved in the formation of both VBM and CBM, alloying Pb with other metals, such as Sn, should vary the VBM and CBM together, which makes the bandgap change more complex (Grånäs, Vinichenko, and Kaxiras 2016).

Figure 1.7a depicts the ultraviolet photoelectron spectroscopy (UPS) investigation of valence band energies of $MASn_{1-x}Pb_xI_3$ compounds (Hao et al. 2014). With increasing amount of Sn in the B-site (in Figure 1.7a from left to right), the VBM level slowly decreased and then raised again for pure Sn-based perovskite. The bandgap variation with fraction of Pb is shown in Figure 1.7b (Hao et al. 2014). Clearly, the bandgap change of Pb–Sn binary perovskites is not following a linear trend in between these two extremes (1.55 and 1.35 eV for $MAPbI_3$ and $MASnI_3$, respectively) but has a "bow-shape" with the lowest bandgap in between. When the Pb contents are 0.25 and 0.50, i.e. $MASn_{0.75}Pb_{0.25}I_3$ and $MASn_{0.5}Pb_{0.5}I_3$, the smallest bandgap of 1.17 eV is obtained. The lower bandgap facilitates the utilization of solar spectrum. Similar trend is found in the mixture of $FAPbI_3$ and $CsSnI_3$ compounds, as shown in Figure 1.7c (Zong et al. 2017). The smallest bandgap of 1.24 eV was obtained for composition of $(FAPbI_3)_{0.6}(CsSnI_3)_{0.4}$ although the lattice parameter a is continuously decreased with increasing amount of $CsSnI_3$.

People believe that the A-site cation is not involved in the formation of valence and conduction bands. However, there is another effect of the A-site cations. An increase

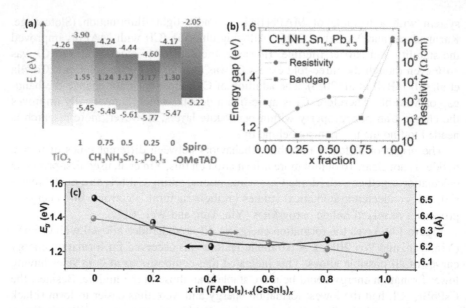

FIGURE 1.7 (a) Schematic energy level diagram of the $CH_3NH_3Sn_{1-x}Pb_xI_3$ solid solution perovskites. The VB maxima of the hybrid $CH_3NH_3Sn_{1-x}Pb_xI_3$ were extracted from UPS measurements under high vacuum (Hao et al. 2014). (b) Dependence of resistivity and optical band gap of the $CH_3NH_3Sn_{1-x}Pb_xI_3$ on the x fraction (Hao et al. 2014). (c) Change in the lattice parameter, a, of the (pseudo)cubic perovskite structures, and the bandgaps of the $(FAPbI_3)_{1-x}(CsSnI_3)_x$ perovskites, as a function of x (Zong et al. 2017). (Reprinted with permission from American Chemical Society and John Wiley and Sons.)

in the size of the A cation results in expansion of the lattice structure, decreasing the overlap between the s-orbital of B-site and p-orbital of X-site elements, thus the lower valence band level (Grånäs, Vinichenko, and Kaxiras 2016). For example, with the increasing of cation size from Cs to MA to FA, the unit cell of $APbI_3$ perovskite increases from 222 to 248 to 256 Å (Amat et al. 2014). The bandgap of lead-iodide-based perovskite decreases from 1.73 to 1.55 to 1.47 eV (Boix et al. 2015). Further first-principal calculation found that the interplay of organic cation size and hydrogen bond together affected the spin-orbit coupling and thus the optical and electronic properties of halide lead perovskites (Amat et al. 2014).

If we further increase the size of A-site cation, the 3D perovskite structure will collapse, because the tolerance factor is out of the perovskite formation range, and low-dimensional perovskite is formed. This introduces another factor, dimensionality, which could tune the bandgap of perovskites.

As shown in Figure 1.8 on the right-hand side, when the stacking of inorganic octahedral layer (n) equals to ∞, 3D perovskite is formed (Mitzi 2001). Increasing the size and amount of A-site cation converts the 3D structure into a 2D layered structure. For example, when $n=3$, three inorganic octahedral layers were separated by organic cations. When the amount of cations was increased further, the inorganic layer was formed until $n=1$ for pure 2D layered structure with mono inorganic octahedral layers in between organic cations.

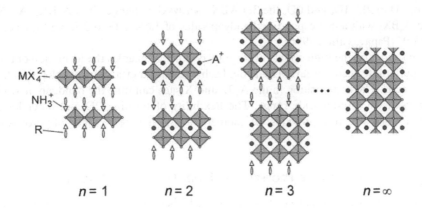

FIGURE 1.8 The stacking of inorganic octahedral layer (n) with organic cation RNH_3^+ in the ⟨100⟩-oriented 2D perovskite. When n is ∞, it is 3D structure (Mitzi 2001). (Reprinted with permission from Royal Society of Chemistry.)

When the size and amount of A-site cation were increased further, the 2D layer was further changed into a 1D chain or 0D dot with inorganic parts surrounded and isolating by organic cations, forming lower-dimensional perovskites.

As we mentioned afore, the organic cations are not effectively involved in the formation of VBM and CBM of perovskites and thus do not affect the bandgap dramatically. However, in low-dimensional perovskites, the distance between inorganic layers increases, and they are separated by organic layers. The interaction between these inorganic layers become weak, and the possibility of electron orbital overlap is less compared to the 3D structure. Therefore, bandgap values go up if the dimensionality of perovskites decreases. As shown in Figure 1.9, when the structure changes

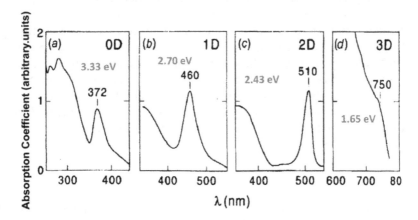

FIGURE 1.9 Room temperature absorption coefficient as a function of wavelength (λ) for (a) $(CH_3NH_3)_4PbI_3$, (b) $[(NH_2C(I)=NH)_2]_3PbI_5$, (c) $(C_9H_9NH_3)_2PbI_4$, and (d) $CH_3NH_3PbI_3$ systems consisting of 0D, 1D, 2D, and 3D extended inorganic networks of corner-sharing PbI_6 octahedra; the absorption coefficient peak positions along with their bandgaps calculated are marked in the figure (Papavassiliou 1996). (Reprinted with permission from Taylor & Francis.)

from 3D to 2D, 1D, and 0D, the 3D ABX_3 perovskite changes into A_2BX_4, A_3BX_5, and A_4BX_3 with a corresponding bandgap value of 1.65, 2.43 to 2.70 and further to 3.33 eV (Papavassiliou 1996).

In summary, the high optical absorption coefficient enables the formation of thinner perovskite absorber layers that could achieve similar efficiency as other thin film solar cells. The composition in the A, B, and X sites can tune the bandgap of ABX_3 perovskite in a quite wide range. The flexibility of forming 2D and other lower-dimensional structure provides yet another way to tune the bandgap of perovskites further.

1.2.3 Optoelectronic Property of Halide Perovskite Materials

When charges are produced and dissociated, they are to diffuse or drift to the respective electrodes and thereby power external loads. The charge transport process depends on several factors, such as the carrier mobility, lifetime, and diffusion length. Charge carrier mobility, μ, is defined as

$$\mu = \frac{e \cdot \tau_s}{m^*} \tag{1.4}$$

where τ_s is the mean scattering time, m^* is the effective mass, and e is the charge of electron. Charge carriers can be scattered by crystal defects and impurities (here ignore the phonon's scatter); hence the mobility of charge carriers can be hindered. Electrons and holes can be treated particle-like if they have effective mass. A lower mobility of charge carrier is associated with a higher effective mass. The effective mass, defined as the second derivative of the energy band, varies with the composition and crystal structure of materials. Therefore, composition, crystal structure, defect, and impurity of materials dictate the charge carrier mobility.

When the minorities diffuse or drift within materials, there are two possible recombination pathways. One is that the minorities recombine with opposite charges. A high concentration of charge carriers increases the opportunity of recombination. The other is that the minorities are captured by defects. As a result, concentrations of defects and charge carriers are two factors determining the average distance and time that minorities can travel. The average distance and time that minorities can travel prior to recombination are called effective diffusion length L and lifetime τ, respectively. They are related by (Gfroerer, Zhang, and Wanlass 2013, p. 1990, Finger 2014)

$$L = \sqrt{D\tau} \tag{1.5}$$

where D is the diffusivity. The thickness of the absorber layer should be less than the value of L to enable efficient collection by electrodes.

In view of these, charge mobility and diffusion length affect charge transport process, which are in turn influenced by defect concentration, microstructure and impurity, composition, and band structure. In this section, we discuss the influencing factors along with strategies to enhance the charge transport process.

Composition is one of the key factors that affects the charge transport property of materials. For most thin film solar cells, the effective mass of a hole is much heavier than that of an electron. Therefore, in p-type absorber, the electron, not the hole, needs to travel the whole thickness to reach the collection electrode. However, in organic–inorganic halide perovskite materials, effective masses of holes and electrons are comparable, making it a unique ambipolar conductive material. The band structure and density of states (DOS) calculated by Du are presented in Figure 1.10 (Du 2014). The edges at the top of valence bands (VBs) and bottom of conduction bands (CBs) are very dispersive, which indicate similar effective masses for electrons and holes. Further DOS image confirmed that the significant hybridization between Pb-6s and I-5p states causes the large dispersion of VBs and lowers effective mass of the hole. The CB is derived from Pb-6p states, which are spatially more extended than the usual s- or d-state-derived CB inducing smaller effective mass of electrons (Koutselas, Ducasse, and Papavassiliou 1996, Du 2014, Wang et al. 2014, Umebayashi et al. 2003). Based on Du's calculation, the average effective masses of electrons and holes are comparable with a value of 0.28 m_0 and 0.34 m_0, respectively (Du 2014). Including the spin-orbit coupling effects, m_e^* and m_h^* values are reduced to 0.23 m_0 and 0.29 m_0, respectively (Giorgi et al. 2013). Mobilities of electrons and holes in the $CH_3NH_3PbI_3$ perovskite were reported to be 5–10 and 1–5 cm^{2}V^{-1}·s, respectively (Motta et al. 2015). Due to smaller atomic size, tin (Sn) 5s has higher energy than Pb 6s state. Substituting Pb with Sn results in stronger anti-bonding mixing between Sn 5s and I 5p orbit in VB (Umari, Mosconi, and De Angelis 2014). Therefore, the VB of Sn-based perovskite is more dispersed, resulting in a smaller hole effective mass of 0.13 m_0 in $CH_3NH_3SnI_3$ compared to 0.25 m_0 in $CH_3NH_3PbI_3$.

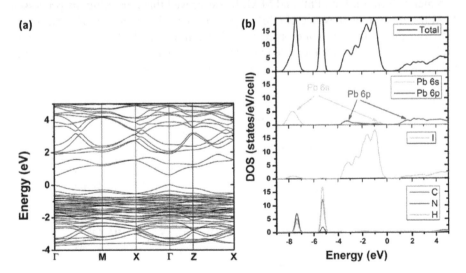

FIGURE 1.10 (a) Band structure and (b) the density of states (DOS) of β-$CH_3NH_3PbI_3$ calculated using density functional theory (DFT) with standard Perdew–Burke–Ernzerhof (PBE) functions. The energy of VB maximum is set to zero. The total DOS is projected into Pb (6s and 6p), I, C, N, and H atoms (Du 2014). (Reprinted with permission from Royal Society of Chemistry.)

Meanwhile, as the Pb $6p$ electron is more delocalized than Sn $5p$, a smaller electron effective mass in $CH_3NH_3PbI_3$ is expected, with a value of 0.19 m_0 compared to 0.28 m_0 in $CH_3NH_3SnI_3$ (Umari, Mosconi, and De Angelis 2014).

Besides composition, intrinsic defect is another factor that affects the optoelectronic properties of perovskite. Calculated energy levels of point defects of MAPbI$_3$ are given in Figure 1.11 (Yin, Shi, and Yan 2014a). Intrinsic point defects, such as iodine interstitials (I_i), CH_3NH_3(MA) molecules on Pb sites (MA$_{Pb}$), CH_3NH_3 vacancies (V_{MA}), Pb vacancies (V_{Pb}), MA interstitials (MA$_i$), anti-site defect (Pb$_{MA}$), and iodine vacancy (V_I), have low formation energy and are located near the band edge with very shallow transition energy levels. Defects with deeper energy level, such as I_{MA}, I_{Pb}, Pb$_I$, and Pb$_i$, have high formation energy and are not easy to form (Yin, Shi, and Yan 2014a,b). Therefore, point defects in MAPbI$_3$ perovskite are mainly shallow defects having less effect on the recombination and charge transport.

The morphology of materials will also affect the mobility and diffusion length/ carrier lifetime of materials. Grain boundaries are non-continuous parts inside materials, which might be accumulation sites for impurities and defects. These impurities and defects might be charged and thus scatter or capture charge carriers acting like a recombination center. Although grain boundaries inside perovskite were sometimes reported as benign (Yin, Shi, and Yan 2014a, Yun et al. 2015), longer carrier lifetime or diffusion length has always been observed for perovskites with larger grain size and less grain boundaries.

In the normal process of fabricating MAPbI$_3$ film from the reaction of PbI$_2$ with MAI layer in the traditional annealing (TA) process, the grain size is small (<300 nm), as shown in Figure 1.12b (Xiao et al. 2014). N, N-dimethylformamide (DMF) is a common solvent for both PbI$_2$ and MAI. It was proved that annealing the processed MAPbI$_3$ films in the presence of DMF vapor (named solvent anneal, SA) effectively increases the grain size to micrometers as shown in Figure 1.12a and c. With larger grains, the carrier lifetime increased to 7.2 from 1.7 μs for TA samples (Figure 1.12d). Furthermore, combining nucleation and growth into one process, a hot solution

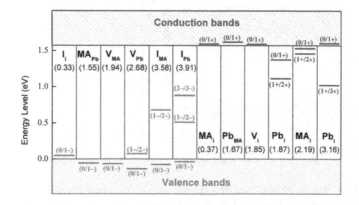

FIGURE 1.11 Calculated transition energy levels of point defects in $CH_3NH_3PbI_3$. The formation energies of neutral defects are shown in parentheses (Yin, Shi, and Yan 2014a). (Reprinted with permission from John Wiley and Sons.)

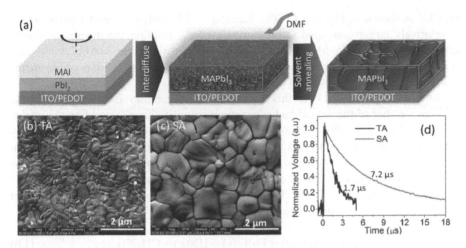

FIGURE 1.12 (a) Schematics of the approach and solvent-annealing-induced grain size increase, surface morphology of (b) traditional annealing (TA) and (c) solvent annealing (SA) of 1,015 nm-thick perovskite, and (d) photovoltage decay under 0.3 sun illumination measured by transient photovoltage technique. (Reprinted from Reference Xiao et al. (2014) with permission from John Wiley and Sons.)

spin-coating on the preheated substrate was obtained (Nie et al. 2015). Millimeter-size grains were obtained. The diffusion length further increased to 175 μm. This additionally confirmed that films with larger grains are preferred for their longer carrier lifetime and diffusion length.

1.3 STABILITY OF HALIDE PEROVSKITES

In practical applications, the temperature of solar cell devices might go up, especially in hot season or hot region. Therefore, the International Electrotechnical Commission (IEC) designed qualification testing protocol 61215 for "Crystalline Silicon Terrestrial Photovoltaic (PV) Modules" in an 85°C chamber. Although there is still a need to develop a testing protocol for perovskite solar cells, materials with suitable thermal stability are required. The next section discusses the intrinsic thermal instability of MAPbI$_3$ and efforts to design a new composition with better thermal stability. FAPbI$_3$ and all-inorganic CsPbI$_3$ developed are thermally more stable than MAPbI$_3$ albeit structurally instable. Strategies designed to stabilize structures of these perovskites will be elaborated. Here we don't cover the environmental stability issue of perovskites, and readers can refer to our recent publication for more details (Sam and Guifang 2020).

1.3.1 INTRINSIC THERMAL STABILITY OF HALIDE PEROVSKITE MATERIALS

MAPbI$_3$ is the first successfully synthesized perovskite for photovoltaic application. This is not surprising if one noticed that the tolerance factor of around 0.91 for MAPbI$_3$ is located right in the center of the perovskite formation range between 0.8

and 1.0 as shown in Figure 1.13. The bandgap of MAPbI$_3$ is around 1.58 eV, which is relatively larger compared to the optimal 1.5 eV, limiting the theoretical power conversion efficiency of MAPbI$_3$-based solar cells. Furthermore, MAPbI$_3$ undergoes a phase transformation at around 57°C, which is right in the range of possible operational temperatures of solar cell devices.

Another key drawback of this system is the instability issue of MA cation. A large number of theoretical and experimental studies have focused on the intrinsic instability of MAPbI$_3$. Although it is still under hot debate, the relatively low dissociation temperature of MA raised a serious question regarding the application of MA-contained perovskites in photovoltaics. Decomposition temperature as low as ~60°C was detectable based on Knudsen Effusion Mass Spectrometry (KEMS) technique (Brunetti et al. 2016). There are mainly two decomposition paths reported (Brunetti et al. 2016, Juarez-Perez et al. 2016):

$$CH_3NH_3PbX_3(s) = PbX_2(s) + HX(g) + CH_3NH_2(g) \tag{1.6}$$

$$CH_3NH_3PbX_3(s) = PbX_2(s) + CH_3X(g) + NH_3(g) \tag{1.7}$$

In the first path, decomposition products can form back to perovskite by acid–base neutralization reaction. Thus with suitable encapsulation to ensure no gas product escapes from the system, the lifetime of perovskite solar cell probably might not be a big problem, although the efficiency will drop if this decomposition occurred during device operational time. However, the degradation process to products of CH$_3$I and NH$_3$ is irreversible and detrimental for longtime stability of MAPbI$_3$. The observed fact that encapsulation and inert condition cannot completely avoid perovskite decomposition may well be attributed to this irreversible reaction. Therefore, total removal of MA and Br in perovskite composition seems a key toward stable solar

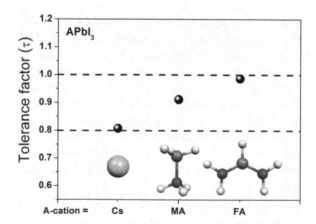

FIGURE 1.13 Calculated tolerance factors (τ) for different A-site cations of Cs, MA, and FA in APbI$_3$ perovskite system where the range between two dashed lines is the perovskite formation range. Inset: Structure and illustration of size of Cs$^+$, MA (CH$_3$NH$_3$)$^+$, and FA [CH(NH$_2$)$_2$]$^+$ cations. (Reprinted with permission from Reference Amat et al. (2014).)

cell devices. Therefore, finding a new composition with a suitable bandgap combined with higher thermal stability is crucial for practical application of perovskite solar cells.

To overcome the poor stability of MA-based perovskite, formamidinium lead iodide ($CH(NH_2)_2PbI_3$ or $FAPbI_3$) and all-inorganic halide perovskites, such as $CsPbI_3$ and $CsPbBr_3$, were developed. Table 1.1 lists the decomposition temperature along with the stable temperature of several currently used perovskite materials. Single crystal $MAPbI_3$ starts to decompose at a temperature higher than 240°C. This temperature further drops to around 85°C when it is in thin film form (Liu et al. 2015, Stoumpos, Malliakas, and Kanatzidis 2013). In contrast, there is no phase change observed for black $FAPbI_3$ films up to 150°C. The stability of all-inorganic halide perovskite has superior thermal stability with no phase change up to more than 400°C.

However, substitution of MA with larger FA cation increases the tolerance factor to 0.99 (Figure 1.13), which is very close to the upper limit of perovskite formation range. Therefore, the structure might not be stable. The experimental data did confirm that the photoactive black $FAPbI_3$ does not form as easily as $MAPbI_3$ at lower temperature. Early X-ray diffraction analysis of single crystal and polycrystalline films indicated that the $FAPbI_3$ perovskite has two polymorphs: a black perovskite-type material with trigonal symmetry (*P3m1*), lattice parameters of $a = 8.9920(9)$ Å and $c = 11.0139(7)$ Å, and a yellow hexagonal non-perovskite phase (*P6$_3$mc*) (Stoumpos, Malliakas, and Kanatzidis 2013, Koh et al. 2014). The black phase forms at higher temperature, while the yellow phase easily appears near room temperature. Later, high-resolution neutron powder diffraction confirmed the black $FAPbI_3$ as a cubic perovskite structure with $a = 6.3630(8)$ Å (Weller et al. 2015, Chen et al. 2016, Stoumpos, Malliakas, and Kanatzidis 2013, Koh et al. 2014).

Replacing MA with smaller cation Cs decreases the tolerance factor to 0.807 (Figure 1.13), which is close to the lower limit for perovskite formation range. The crystal structure of $CsPbI_3$ is more complex and has three lattice structures. The photoactive phase is normally formed at higher temperature, while yellow phase is formed at lower temperature.

TABLE 1.1
Stable Temperature of Several Perovskite Compounds

Compound	$CH_3NH_3PbI_3$	$HC(NH_2)_2PbI_3$	$CsPbI_3$	$CsPbBr_3$
Decomposition temperature (°C)	240 (single crystal) (Liu et al. 2015), 85 (thin film)	320–360(single crystal) (Han et al. 2016)	>460 (Sharma 1992)	>460 (Sharma 1992)
Perovskite phase stable temperature (°C)	55 (Stoumpos, Malliakas, and Kanatzidis 2013)	185 (single crystal) (Han et al. 2016), <150 (thin film) (Lee et al. 2014)	>300 (Sharma 1992)	>130 (Stoumpos et al. 2013, Moller 1958)

Source: Han et al. (2017). Reprinted with permission from Elsevier.

In the following sections, strategies developed to stabilize hybrid organic–inorganic FAPbI$_3$ and all-inorganic CsPbI$_3$ halide perovskite structures will be addressed.

1.3.2 PHASE STABILITY OF FAPbI$_3$ PEROVSKITE

As the black FAPbI$_3$ has suitable bandgap of 1.49 eV (thus broader absorption of solar spectrum) and desired thermal stability (much better than that of MAPbI$_3$), researchers strive to pursue the stable photoactive black α-FAPbI$_3$ perovskite phase.

Inspired by the conversion of PbI$_2$-DMSO-MAI to MAPbI$_3$, Yang et al. intercalated the dimethylsulfoxide (DMSO) molecule into PbI$_2$ framework to form a PbI$_2$-DMSO compound film (Yang et al. 2015). Toward PbI$_2$, formamidinium iodide (FAI) molecules have ionic interaction, while DMSO experiences only van der Waals interaction. Therefore, due to its higher affinity toward PbI$_2$, the external FAI further replaced DMSO molecules intercalated inside PbI$_2$. As the inorganic PbI$_2$ framework was retained during this intramolecular exchange process, uniform and dense FAPbI$_3$ films with large grains were produced. Another helpful compound for the formation of dense black FAPbI$_3$ films is N-Methyl-2-pyrrolidone (NMP) (Jo et al. 2016).

Aside from this solvent engineering approach, additives were widely investigated to stabilize the FAPbI$_3$ perovskite phase. In the fabrication of FAPbI$_3$ films with stoichiometric ratio of FAI and PbI$_2$ in DMF solvent, yellow δ-phase appeared even with very short annealing time, i.e. 150°C for 5 minutes (Figure 1.14). The amount of this δ-phase and PbI$_2$ increased with annealing time. Slowly increasing the FAI ratio consumed PbI$_2$ inside the film and compensated the loss of FAI through evaporation during annealing. When the FAI to PbI$_2$ ratio was 1.2, i.e. 20% excess FAI in precursors, pure black FAPbI$_3$ phase was formed with no δ-phase and PbI$_2$ was observed even when the annealing temperature increased from 150°C to 200°C.

When a small amount of aqueous hydroiodic acid (HI) was added into the stoichiometric FAI and PbI$_2$ precursor solution in DMF, a dense and smooth film was formed, while the control film without HI addition is porous, as shown in

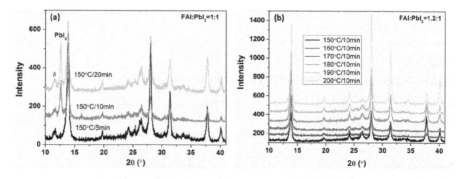

FIGURE 1.14 The effect of stoichiometric ratio of FAI to PbI$_2$ on the black phase formation of FAPbI$_3$ films. (a) Films were fabricated with equal amount of FAI and PbI$_2$ annealed at 150°C for different times, and (b) FAI to PbI$_2$ ratio is 1.2:1, and films were annealed from 150°C to 200°C for 10 minutes without any impurity phase observed.

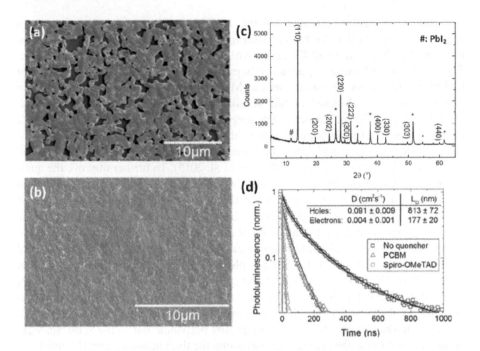

FIGURE 1.15 The surface microscopies of FAPbI$_3$ films formed from stoichiometric FAI and PbI$_2$ with (a) no additive and (b) addition of a small amount of aqueous HI to the precursor solution. (c) X-ray diffraction spectrum of FAPbI$_3$ film formed with addition of HI; no yellow phase was observed. (d) Normalized time-resolved photoluminescence intensity for FAPbI$_3$ films with and without electron and hole quenchers. Diffusion coefficients (D) and diffusion lengths (LD) are extracted from the fitting and shown in the inset (Eperon et al. 2014). (Reprinted with permission from Royal Society of Chemistry.)

Figure 1.15a and b (Eperon et al. 2014). This is a simple way to produce high-quality FAPbI$_3$ films without pinholes. The X-ray diffraction spectrum confirmed that no yellow phase appeared despite a small amount of excess PbI$_2$, as indicated in the Figure 1.15c. Time-resolved photoluminescence evaluated the quality of the FAPbI$_3$ films produced. The diffusion lengths of FAPbI$_3$ film extracted were around 813 and 177 nm for electron and hole, respectively, indicating the high quality of FAPbI$_3$ film (Figure 1.15d). Authors believed that adding aqueous HI into the perovskite precursor solution helped the solubility of inorganic components and thus slowed down the crystallization process. This slowed-down process enabled the formation of a continuous and smoother film, without having any impact on the crystal structure. Furthermore, to eliminate the effect of water inside HI solution, HPbI$_3$ was first formed by the reaction of HI with PbI$_2$ inside DMF (Wang et al. 2015). HPbI$_3$ was then used as a precursor to replace PbI$_2$. Thus the fabricated FAPbI$_3$ perovskite films exhibited a crystalline phase with strong (110) preferred orientation.

Other additives, such as rubidium, caesium, and bismuth cation, also effectively stabilized the FAPbI$_3$ black phase (Park et al. 2017, Dang et al. 2019, Zheng et al. 2016).

Actually, careful observation of the tolerance factors of $APbI_3$ shown in Figure 1.13 reveals that the highest tolerance factor is from larger FA cation and the lowest is from the small Cs cation while that of MA is located in the middle. Therefore, "mixed cation" approach was proposed to lower the tolerance factor of $FAPbI_3$ by hybrid smaller cations such as MA and Cs. It was found that the addition of a small amount of MA successfully stabilized the black $FAPbI_3$ phase (Pellet et al. 2014, Binek et al. 2015). The perovskite prepared using this "mixed cation" approach not only improved the phase purity and stability, but also boosted the conversion efficiency of devices (Jeon et al. 2015). Incorporating Cs and/or rubidium also stabilized the black $FAPbI_3$ phase (Li et al. 2016, Lee et al. 2015). To further improve the quality of perovskite films, bromine was also introduced into this double cation system and formed a mixed cation and mixed halide system with a popular composition of $MA_{0.17}FA_{0.83}Pb(I_{0.83}Br_{0.17})_3$ (Jeon et al. 2015, Jacobsson et al. 2016, Correa-Baena et al. 2015). However, the reproducibility of this composition is not very good, and the difference from lab to lab is quite significant. Successively, adding Cs to this mixed cation mixed halide system formed a triple cation compound (Saliba et al. 2016b, Service 2016, Deepa et al. 2017). Addition of only 5% of Cs (with a formula of $Cs_{0.05}$ $(MA_{0.17}FA_{0.83})_{0.95}Pb(I_{0.83}Br_{0.17})_3$) suppressed the yellow δ-phase and residual PbI_2 successfully as shown in Figure 1.16a. Compared to the control device (i.e. without Cs addition, marked as Cs_0M), larger monograins perpendicular to the substrate were obtained, facilitating the charge transport along the thickness direction (Figure 1.16b and c). The V_{oc} and J_{sc} of 5% Cs added devices improved compared to the control, indicating the higher quality of perovskite films. Along with the improved fill factor, the overall power conversion efficiency of 5% Cs added ones was much higher than that of controls (Figure 1.16d). Device stability test showed that the device efficiency with 5% Cs addition dropped from 20% to around 18% where it stayed relatively stable for at least 250 hours under continuous illumination and maximum power point tracking at room temperature. Most importantly, the reproducibility of this triple cation perovskite dramatically improved. This is a big improvement during the process and composition engineering to achieve high-quality perovskite films. This composition still holds the highest record of power conversion efficiency for perovskite-based solar cells.

Furthermore, addition of Rb to this triple cation system forms a more complex quadruple cation system (Saliba et al. 2016a). It has been noted that larger size difference between Cs, Rb, and FA could tune the effective tolerance factor of the final perovskite composition to a suitable range ($0.8 < \tau < 1$) (Li et al. 2016, Saliba et al. 2016a). X-ray photoelectron spectroscopy revealed that addition of Cs ion helped to push the composition close to stoichiometric ratio and downshift the VB position, which facilitate charge transportation (Deepa et al. 2017). Although the Rb is too small to form perovskite structure by itself, the addition of Rb into precursor not only improved the crystallinity but also suppressed defect migration. The lower trap-assisted charge carrier recombination and lower series resistance observed might be the reason for the improved device performance with the addition of Rb and Cs (Yadav et al. 2017).

Besides approaches discussed above, surface functionalization also successfully stabilized the cubic $FAPbI_3$ phase (Fu et al. 2017b). Cubic α-$FAPbI_3$ is a metastable

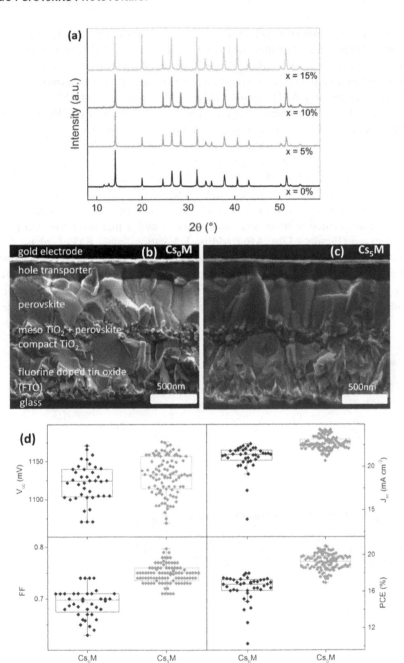

FIGURE 1.16 (a) X-ray spectra of $Cs_x(MA_{0.17}FA_{0.83})_{(100-x)}Pb(I_{0.83}Br_{0.17})_3$ (marked as CsxM) perovskite films with different amounts of Cs addition; the cross-section microstructure of (b) control ($C_{s0}M$) and (c) 5% Cs ($C_{s5}M$) added perovskite devices; and (d) statistics of 40 controls ($C_{s0}M$) and 98 Cs-based ($C_{s5}M$) devices collected over 18 different batches, which proves the reproducibility (Saliba et al. 2016a). (Reprinted with permission from Royal Society of Chemistry.)

phase where the hexagonal yellow FAPbI$_3$ is the thermodynamically stable phase at room temperature. For films with large grains, the α-FAPbI$_3$ phase tends to get converted to a more stable yellow phase. Interestingly, the surface energies of most of these metastable polymorphs are often lower than that of the thermodynamically stable phase, such as γ-Al$_2$O$_3$ versus α-Al$_2$O$_3$ (McHale et al. 1997, Navrotsky 2003). This means the surface energy played a minor role in bulk film. When grain size becomes smaller and smaller, and goes to nanoscale, the surface energy might overweigh the energy of phase transition and can thus potentially stabilize the metastable phase. Based on this, surface functionalization agents, including long-chain alkyl (n-butylammonium, n-C$_4$H$_9$NH$_3^+$ or BA$^+$) or aromatic ammonium (phenylethylammonium, 4-fluorophenylethylammonium) (together marked as LA cations), were chosen to modify the surface condition of FAPbI$_3$. As expected, hexagonal yellow phase formed without any modification, and a nanowire like morphology was observed (Figure 1.17b). After adding LA cation, cubic FAPbI$_3$ phase formed even at room temperature with a fine nanostructure shown in Figure 1.17c. However, further increase of the amount of this LA cation formed a 2D layered network with a tablet-like structure seen in Figure 1.17d. This is because the LA cation is too big to fill into the 3D PbI$_6$ cage and finally collapses the 3D cage into a layered structure.

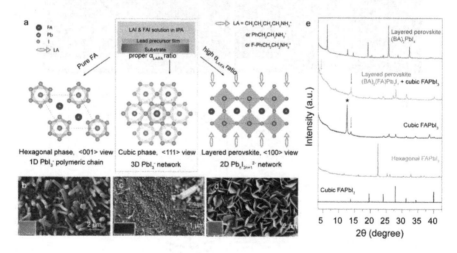

FIGURE 1.17 Schematic illustration and structural characterization of different perovskite products grown under the presence of LA with various LA/FA molar ratios. (a) Different products are grown from lead precursor films in a solution with different LA/FA ratio, and the variations of crystal structures formed, from left to right, are hexagonal, cubic FAPbI$_3$ phase, and layered (LA)$_2$(FA)Pb$_2$I$_7$ phase. (b, c, d) The surface SEM images of the hexagonal FAPbI$_3$, cubic FAPbI$_3$, and layered (BA)$_2$(FA)Pb$_2$I$_7$ perovskite phases grown from PbI$_2$ films in a pure FAI solution and a mixed BAI/FAI solution with BA/FA ratio of 0.86 and 2.57, respectively. The insets are the photographs of the as-grown samples, displaying yellow, black and red colors. (e) The corresponding XRD patterns of these three types of films obtained with the standard XRD pattern of cubic FAPbI$_3$ on the bottom, in which the peaks marked by * indicate the unreacted PbI$_2$ (Fu et al. 2017b). (Reprinted with permission from American Chemical Society.)

XRD patterns of these three types of products confirmed the phase structure presented in Figure 1.17e. Surface analysis by X-ray photoelectron spectroscopy (XPS) confirmed the presence of LA cations on the surface. With the help of theoretical calculations, authors correlated this to (1) scale effect and (2) surface functionalization, which reduced the surface energy. As the grain size decreases to nanoscale, the surface phenomena becomes dominant. At the same time, this LA cation surface functionalization further decreases the surface energy and makes this metastable cubic $FAPbI_3$ stable at room temperature. This provides a new strategy to stabilize the metastable phase by surface functionalization.

Currently, most of the high-efficiency solar cells are based on the mixed cation and mixed halogen system with a composition of $Cs_{0.05}(MA_{0.17}FA_{0.83})_{(100-x)}Pb_{(I_{0.83}Br_{0.17})_3}$. The addition of bromine in this system is, to some extent, especially crucial to suppress yellow-phase impurities. However, the bromine at the X-site replacing iodine increases the bandgap and causes blue shift, which narrows the utilization of solar spectrum. For a composition with 17% bromine, the theoretical maximum power conversion efficiency (PCE) and J_{sc} decreased from around 32% to <30% and from around 30 to $25 \, mA \, cm^{-2}$, respectively (Turren-Cruz, Hagfeldt, and Saliba 2018). In the case of $MAPbI_xBr_{3-x}$ system, halogen segregation occurred under light illumination (Slotcavage, Karunadasa, and McGehee 2016), adding in another uncertainty of perovskite solar cell under operation.

Therefore, a MA- and Br-free composition is preferred (Turren-Cruz, Hagfeldt, and Saliba 2018). Based on systematic study, $Rb_{0.05}Cs_{0.1}FA_{0.85}PbI_3$ (marked as $Rb_5Cs_{10}FAI$) performed the best. The highest efficiency achieved from this composition is 20.44% with a J_{sc} as high as $25.05 \, mA \, cm^{-2}$. Very importantly, the device from this composition shows promising stability, which sustained 1,000 hours of continuous illumination and maximum power point tracking in nitrogen atmosphere with <20% drop of their initial efficiencies. With polymer modified on surfaces of both SnO_2 and the perovskite, the stability further improved with negligible decrease in efficiency under 1,000 hours operation. This high performance came by a combination of measures: first, composition design to achieve a suitable bandgap of 1.53 eV, very close to the optimal 1.5 eV bandgap for solar spectrum; second, composition design to totally remove instable MA and Br components; third, interface modification to further decrease trap states.

In summary, the composition of perovskite undergoes evolution from pristine $MAPbI_3$ to $FAPbI_3$, mixed halide ($MAPbI_xBr_{3-x}$), and mixed cation–mixed halide system ($Cs_x(MA_{0.17}FA_{0.83})_{(100-x)}Pb_{(I_{0.83}Br_{0.17})_3}$) and finally moves back to mixed cation pure iodide system ($Rb_{0.05}Cs_{s0.1}FA_{0.85}PbI_3$), driven by stability and high performance in practical applications.

1.3.3 PHASE STABILITY OF ALL-INORGANIC CsPbI₃ PEROVSKITE

All-inorganic halide perovskite mainly refers to $CsPbI_3$ and $CsPbBr_3$ and their mixture. The relatively smaller Br makes the tolerance factor of $CsPbBr_3$ slightly higher than that of $CsPbI_3$, i.e. 0.815 vs 0.807.

These all-inorganic perovskite-based devices showed excellent thermal stability, especially when combined with the inert carbon as top electrode (Liang et al. 2016).

As shown in Figure 1.18a, TiO_2 is the electron-transporting layer, the perovskite in the middle is the all-inorganic $CsPbBr_3$, and the top electrode is carbon. The device showed good band alignment with TiO_2 and carbon electrode (Figure 1.18b). The crystal structure measured from X-ray diffraction confirmed the cubic structure as shown in Figure 1.18c. This all-inorganic solar cell, including the perovskite and electron- and hole-transporting layers, is extremely stable against moisture and temperature, even without encapsulations. Figure 1.18d shows normalized power conversion efficiency of all-inorganic devices along with the $MAPbI_3$/carbon-based and $MAPbI_3$/spiro-MeOTAD-based devices with varied storage time in a relative humidity of 90%–95% at room temperature without encapsulations. Clearly, the $CsPbBr_3$/carbon-based device had superior stability than all the rest: almost no decrease in efficiency after more than 2,500 hours storage in high-humidity environment. Besides this, the $CsPbBr_3$/carbon-based device was further heated to 100°C in this high-humidity atmosphere without encapsulation, and results are shown in Figure 1.18e. An even slightly improved power conversion efficiency was observed for $CsPbBr_3$/carbon-based device upon heating at 100°C for 840 hours. In contrast, the $MAPbI_3$/carbon-based device died after 20 hours. Although carbon is believed to protect the perovskite from degradation to some extent, the rapid degradation of $MAPbI_3$ at high temperature eventually killed the device. In real applications, solar cells have to work outdoors with high–low temperature cycles instead of constant temperature. Therefore, authors investigated the stability of $CsPbBr_3$/carbon-based devices during temperature cycles between –22°C and 100°C, and results are shown in Figure 1.18f. It turned out that the all-inorganic perovskite solar cell devices exhibited no degradation during cycles of extreme temperatures for 80 hours without encapsulations, which further proved their high stability against environmental stimuli. However, the smaller size of A-site Cs and X-site Br makes this composition having a relatively higher bandgap of 2.3 eV, which is far away from the optimal 1.5 eV for efficient utilization of solar spectrum (Figure 1.18g). Therefore, the maximum power conversion efficiency was only 6.7% with a current density of 7.4 mA cm^{-2} (c.f. Figure 1.18h).

Replacing bromine with iodine, the bandgap dramatically drops to 1.73 eV. Although this bandgap is not suitable for efficient single junction solar cells, it matches very well with current industry available low-bandgap silicon (Si) and $Cu(In_{1-x}Ga_x)Se_2$ (CIGS) solar cells for constructing tandem structure. As mentioned afore, $CsPbI_3$ has a tolerance factor relatively lower than that of $CsPbBr_3$. Therefore, $CsPbI_3$ is thermally as stable as $CsPbBr_3$ in composition, albeit structurally it has significant instability, even for iodide-rich compounds.

Hereinafter, we will thoroughly explore the phase evolution of $CsPbI_3$ with temperature and strategies developed to stabilize the black $CsPbI_3$ phase.

$CsPbI_3$ has four phases: cubic phase (α), tetragonal phase (β), and two orthorhombic phases (γ and δ). Among them, α, β, and γ phases are photoactive and collectively called black phase. Synchrotron X-ray diffraction (SXRD) of the phase transition of $CsPbI_3$ with temperature gives rise to Figure 1.19 (Marronnier et al. 2018). The room temperature stable yellow δ-$CsPbI_3$ phase got converted to the cubic black α-$CsPbI_3$ phase after heating samples above 360°C. However, during cooling, the perovskite structure got converted to the β-$CsPbI_3$ phase at temperature of 260°C and to γ-$CsPbI_3$ phase at 175°C (Figure 1.19). A few days later, thermodynamically

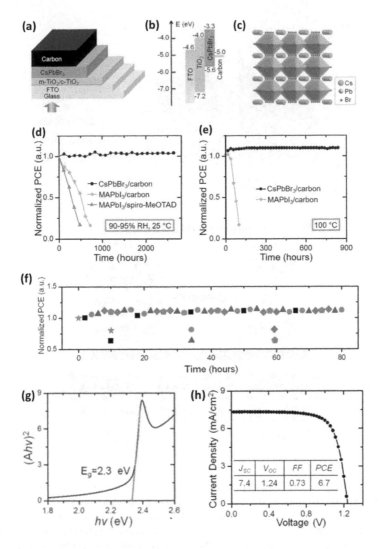

FIGURE 1.18 (a) Schematic illustration of the stacks of CsPbBr₃/carbon-based all-inorganic perovskite solar cells; (b) the energy level diagram of the all-inorganic solar cell; (c) the crystal structure of the inorganic CsPbBr₃ perovskite; (d) normalized power conversion efficiency of CsPbBr₃/carbon-based solar cells with time in a relative humidity of 90%–95% at room temperature without encapsulations; for comparison, devices of MAPbI₃/carbon-based and MAPbI₃ with spiro-MeOTAD as hole transporting layer were also included; (e) normalized PCEs of CsPbBr₃/carbon-based and MAPbI₃/carbon-based devices as a function of time heated at the temperature of 100°C in a relative humidity of 90%–95% environment without encapsulations; (f) normalized PCEs of CsPbBr₃/carbon-based all-inorganic solar cell as a function of storage time during temperature cycles between −22°C and 100°C in a relative humidity of 90%–95% at ambient condition without encapsulations; (g) Tauc plot of CsPbBr₃ film with an optical bandgap of around 2.3 eV; (h) current–voltage curve of CsPbBr₃/carbon-based all-inorganic solar cell with inset showing the corresponding photovoltaic parameters (Liang et al. 2016). (Reprinted with permission from American Chemical Society.)

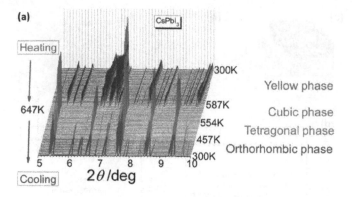

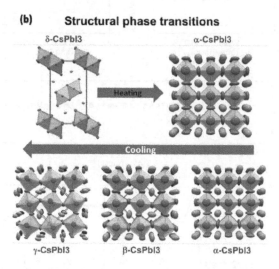

FIGURE 1.19 (a) Temperature-dependent synchrotron X-ray diffraction spectra for $CsPbI_3$ perovskite. The room temperature stable yellow δ-phase changes to cubic α-phase upon heating up, while it changes to metastable tetragonal β-phase, further orthorhombic γ-phase, and finally stable orthorhombic δ-phase during cooling. (b) Structural phase transitions in $CsPbI_3$ versus temperature, in which the initial yellow perovskite phase (δ-$CsPbI_3$) changes to black α-$CsPbI_3$ phase as temperature exceeds the transition temperature. Upon cooling, the black perovskite phase is retained till room temperature with distortions to tetragonal β-$CsPbI_3$ and further orthorhombic γ-$CsPbI_3$ phase. These phases are metastable and transform into the thermodynamic stable yellow δ-$CsPbI_3$ phase at room temperature after several days (Marronnier et al. 2018). (Reprinted with permission from American Chemical Society.)

stable yellow δ-$CsPbI_3$ phase appeared. This indicated that the black phase could appear at room temperature for some time as a metastable phase. Figure 1.19b depicts the structure transition, where the yellow phase got converted to highly symmetric cubic α-$CsPbI_3$ phase during heating. Upon cooling, the cubic structure distorted to tetragonal and further to less-ordered orthorhombic phase.

Next, we address strategies developed to stabilize black $CsPbI_3$, including α- and β/γ-phase.

Since the tolerance factor for $CsPbI_3$ is near the bottom limit of perovskite formation range, composition design was first introduced to increase the tolerance factor and thus stabilize the black phase. Process optimization and additives approaches slowed down the crystallization process and stabilized the black phase of $CsPbI_3$. Nanocrystalline films are able to stabilize the black $CsPbI_3$ due to isolation of small grains and lattice strain. Strain was then introduced intentionally to distort the lattice and stabilize the black $CsPbI_3$ phases.

As mentioned afore, Br incorporation in X-site introduced widening of bandgap and less stability under illumination, and chlorine incorporation is not easy (Beal et al. 2016, Yin, Yan, and Wei 2014c). Chlorine was introduced as a dopant to stabilize black γ-$CsPbI_3$ phase (Wang et al. 2019a. By doping 3 mol% of chlorine, the trap density was decreased with enhanced black phase stability and optoelectronic properties. The stabilized devices achieved a power conversion efficiency of 16.07%. More importantly, 94% of their initial efficiencies remained after exposure in air (relative humidity of 20%–30% at 25°C) for 60 days.

B-site doping or alloying to partially substitute Pb was also pursued to achieve stable black $CsPbI_3$ phases. Hu et al. introduced bismuth (Bi) (1.03 Å, smaller than the Pb of 1.19 Å) to $CsPbI_3$ (Hu et al. 2017). When 4 mol% of Pb was substituted with Bi, the absorption edge of films shifted from ~750 nm for the control α-$CsPbI_3$ film to ~795 nm, corresponding to a bandgap change from 1.73 to ~1.56 eV. A much stronger photoluminescence (PL) intensity was observed for $CsPb_{0.96}Bi_{0.04}I_3$ films compared to the control. These results clearly showed that the Bi substitution for Pb in the B-site not only improved the light harvesting ability with a narrower bandgap, but also improved the quality of the film itself with an enhanced PL and lower non-radiation recombination. As such, the fabricated devices demonstrated an efficiency of 13.21% and maintained 68% of their initial efficiencies after 168 hours aging under ambient conditions without encapsulations. Manganese (Mn) stabilized α-$CsPbI_3$ nanocrystals for over a month as it slightly increased the tolerance factor (Akkerman et al. 2017). The addition of Mn did not affect the optical properties of α-$CsPbI_3$ nanocrystals, as the Mn^{2+} levels fell within the CB of α-$CsPbI_3$ supported by hybrid density functional calculations. Incorporation of europium (Eu) obtained a stable $CsPbI_3$ perovskite at a lower annealing temperature of 85°C (Jena et al. 2018). The Eu-doped $CsPbI_3$ devices achieved a PCE of 13.7% under reverse scan and were stable for 30 days in ambient air.

Doping or substituting Cs at A-site with larger cations is an effective way to increase the tolerance factor and thus stabilize the α-$CsPbI_3$ phase. However, addition of large cation inside perovskite precursor often affects the crystallization process. We will come to this point later along with the process optimization approach.

In 2015, Snaith reported the first working $CsPbI_3$ device (Eperon et al. 2015). A small amount of HI was added into the precursor solution prior to spin coating, which was commonly used to enhance the solubility of perovskite precursors allowing uniform film formation. Black $CsPbI_3$ phase was obtained at only 100°C annealing temperature, which is normally more than 300°C. Absorptions of $CsPbI_3$ films fabricated with high temperature process and low temperature process with addition of HI were comparable as shown in Figure 1.20a. However, very small grain size was observed for films with HI addition, as can be seen in

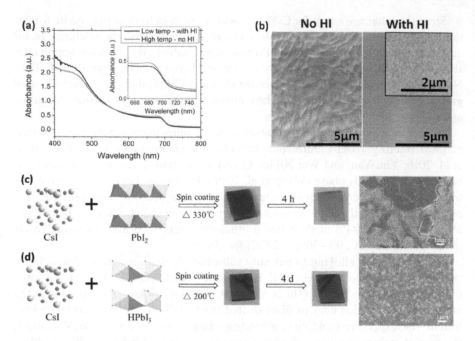

FIGURE 1.20 (a) Absorption curves of CsPbI$_3$ films processed at low temperature with addition of HI and at high temperature without HI and (b) the surface morphologies of CsPbI$_3$ films processed without and with HI (Eperon et al. 2015). Schematic illustrations of the CsPbI$_3$ films prepared from (c) PbI$_2$- and (d) HPbI$_3$-based precursor solutions along with stabilities and surface morphologies of films (Xiang et al. 2018). (Reprinted with permission from Royal Society of Chemistry and American Chemical Society.)

Figure 1.20b. Authors argued that probably the faster crystallization occurred in the solution containing HI, which might be due to the HI being driven off more rapidly than pure DMF or reduced solubility of the Cs precursor in a solution containing HI. The different crystallization processes and the smaller grain size with HI addition were reasons for the stabilization of CsPbI$_3$ black phase.

Interestingly, a small amount of H$_2$O addition manipulates the size-dependent phase formation through a so-called proton transfer reaction (Zhao et al. 2018). Thus the produced black γ-CsPbI$_3$ phase with a lower surface free energy was thermodynamically preferred over yellow δ-CsPbI$_3$ when surface area was >8,600 m^{2}mol^{-1}. The γ-CsPbI$_3$ phase was stable in ambient air over a month. This indicated that the large surface area is one of the key factors to get stable black CsPbI$_3$ phase.

Furthermore, the HI induced the formation of an intermediate phase of hydrogen lead iodide (HPbI$_{3+x}$). Therefore, HPbI$_3$ was then introduced as a lead source to fabricate CsPbI$_3$ perovskite (Xiang et al. 2018). Using PbI$_2$ as lead source, annealing at 330°C was necessary to obtain the black CsPbI$_3$ phase, and the obtained black CsPbI$_3$ phase totally degraded to yellow phase for around 4 hours of storage (Figure 1.20c). However, when HPbI$_3$ was the lead source, the annealing temperature decreased to 220°C (Figure 1.20d). Most importantly, the fabrication process was in dry air with a relative humidity of 10%–20%, while in most of reports, CsPbI$_3$ fabrication and even measurements are all inside an inert atmosphere. The resultant black $CsPbI_3$ exhibited improved stability with almost no color change after 4 days' storage in dry air condition with relative humidity of 10%–20%. The grain size of $CsPbI_3$ film fabricated from $HPbI_3$ was smaller than that from PbI_2 as lead source, which is consistent with the result with HI addition (Figure 1.20b and d).

Later, researchers revealed that the addition of aqueous HI or the consequently developed $HPbI_3$ or $HPbI_x$ approaches to improve the deposition of black phase $CsPbI_3$ perovskite will actually introduce a reaction between DMF, which is a commonly used solvent for perovskite precursors, and acids to form formic acid and dimethylamine ($(CH_3)_2NH$: DMA) (Ke et al. 2018, Noel et al. 2017). Therefore, the addition of $HI/HPbI_3$ would induce the formation of dimethylammonium iodide (DMAI). The effect of HI and $HPbI_3$ additives might be due to the dopant effect of DMAI and $DMAPbI_3$. Based on this, researchers directly added different amounts of DMAI into CsI and PbI_2 precursors and investigated the effect of DMAI on the crystallization process of $CsPbI_3$ (Wang et al. 2019a). In order to ascertain whether the addition of DMAI is a dopant or an additive effect, XPS analysis was conducted on the $CsI+DMAI+PbI_2$ precursor films with different annealing time. For comparison, the spectra of pure $DMAPbI_3$ were also included. Figure 1.21a revealed that the Pb $4f_{7/2}$ peak at 138.3 eV in the fresh precursor film could be attributed to the Pb–I binding energy in $DMAPbI_3$ perovskite. After annealing the film for 2 minutes, the peak position moved to higher binding energy. After 5 minutes annealing, the Pb $4f_{7/2}$ peak shifted to 137.7 eV, which matched well with the binding energy of Pb–I for $CsPbI_3$ perovskite. Similarly, the Pb $4f_{5/2}$ peak also followed the same trend with annealing time incensement. However, the N1s shown in Figure 1.21b had a broad peak in the precursor film. This was easy to understand as the DMAI contained NH-group. With the annealing time increased, the N1s peak diminished and finally disappeared after 5 minutes annealing. This indicated that the DMA cation is mostly like an additive, which formed an intermediate phase of $DMAPbI_3$ during the crystallization. It disappeared after annealing and had no dopant effect. Figure 1.21c schematically illustrates the mechanism where the DMAI incorporated within the PbI_6 cages forms an intermediate stage. After annealing, the volatile DMAI evaporated and left the pure $CsPbI_3$ phase behind.

The amount of DMAI affected not only the morphology of $CsPbI_3$ film obtained, but also crystal structures. In the $CsI–PbI_{2-x}DMAI$ precursor solutions, when the x was 0.5 or 0.7, the fabricated $CsPbI_3$ films had an absorption edge at wavelength of about 717 nm corresponding to a bandgap of 1.73 eV, which is similar to the reported bandgap for $CsPbI_3$ (Figure 1.21d). When x equaled 1.0 and 1.5, the absorption edge of $CsPbI_3$ films moved to 736 nm with a bandgap of approximately 1.68 eV. Different absorption edges suggested two different phases. Assisted by the XRD analysis, the 1.73 eV bandgap $CsPbI_3$ was determined to be γ-$CsPbI_3$, and the lower bandgap phase was determined to be β-$CsPbI_3$ (Figure 1.21c). This indicated that the addition of lower amount of DMAI tended to form a more distorted γ-phase while the addition of higher amount of DMAI tended to form a relatively less distorted β-$CsPbI_3$ phase. The slow evaporation of DMAI when a higher amount is added might help to form a more ordered crystal structure. Nevertheless, the relatively lower bandgap of 1.68 eV is favorable for wider solar spectrum absorption and thus higher power

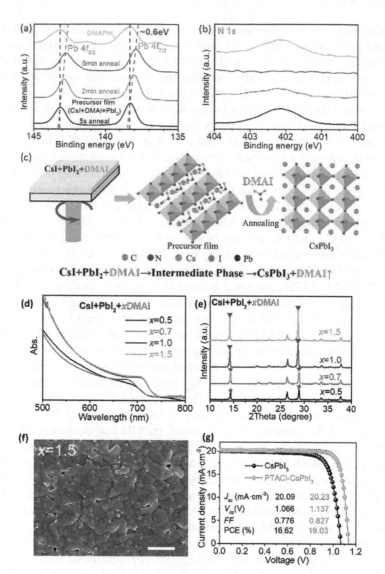

FIGURE 1.21 XPS of (a) Pb 4f and (b) N1*s* core-level spectra for Cs-1.0DMAI thin films at different annealing stages; (c) schematic mechanism for DMAI additive induced black phase CsPbI$_3$ formation; (d) absorption spectrum and (e) XRD patterns of CsPbI$_3$ films fabricated with different Cs-xDMAI ratios; (f) surface morphology of CsPbI$_3$ film fabricated with Cs-1.5DMAI, and the scale bar is 1 μm; (g) current–voltage characteristics of CsPbI$_3$-based solar cell devices with and without PTACl surface treatment under simulated AM 1.5G solar illumination of 100 mW cm^{-1} in reverse scan (Wang et al. 2019a). (Reprinted with permission from John Wiley and Sons.)

conversion efficiency. However, the surface of the synthesized β-CsPbI$_3$ film had lots of pinholes, which might act as a recombination center, lowering the performance of devices. Therefore, surface treatment by choline iodide and phenyltrimethylammonium chloride (PTACl) was introduced to refill the pinhole and passivate the interface between perovskite and hole transporting materials. The current–voltage curves of PTACl-treated and untreated β-CsPbI₃ perovskite solar cells are shown in Figure 1.21f. The V_{oc} was improved from 1.066 to 1.137 eV and the fill factor from 0.776 to 0.827 with the overall efficiency enhanced from 16.62% to 19.03%, indicating the effectiveness of this surface passivation.

Besides DMAI, larger cations, such as oleylammonium (OA) and phenylethylammonium (PEA), were also introduced during film deposition to stabilize black CsPbI₃ phase (Fu et al. 2017a). OA is a long-chain alkyl with a length of around 1.7 nm, while the PEA has an aromatic ring with a height of around 0.6 nm as shown in Figure 1.22a and b. These additives are too big to enter into the perovskite lattice but act as surface capping ligands resulting in the stabilization of the metastable

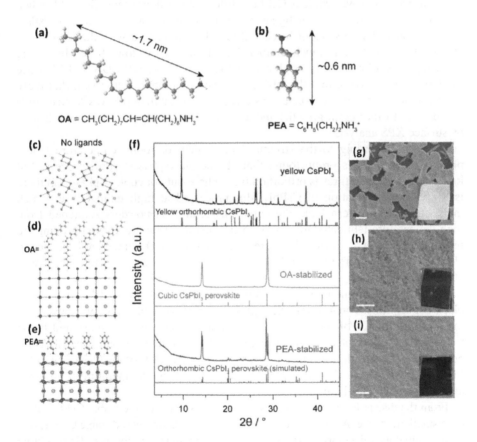

FIGURE 1.22 Chemical structures of (a) long-chain OA and (b) PEA cations; schematic crystal structure of (c) γ-CsPbI₃ without surface ligands; (d) OA-stabilized cubic and (e) PEA-stabilized orthorhombic CsPbI₃ phases; (f) powder X-ray diffraction patterns of yellow CsPbI₃ phase without additives (black curves), OA-stabilized cubic phase (red curves), and PEA-stabilized orthorhombic (blue curves) CsPbI₃ phase together with corresponding standard patterns; SEM images of the (g) γ-phase CsPbI₃, (h) OA-stabilized, and (i) PEA-stabilized CsPbI₃ thin films with a scale bar of 1 μm. Insets are the corresponding photographs of samples (Fu et al. 2017a). (Reprinted with permission from American Chemical Society.)

bulk black $CsPbI_3$ phase. One thing worthy of note is that types of ligand added also affect crystal structures of $CsPbI_3$ synthesized. When no ligands were added, yellow $CsPbI_3$ phase was formed with less symmetry as shown in Figure 1.22c and XRD pattern (black curve) as shown in Figure 1.22f. When OA and PEA were added into the precursor solution during film deposition (Figure 1.22d and e), a set of diffraction peaks that can be assigned to perovskite structures were detected without any peaks from PbI_2 and yellow phase (Figure 1.22f). Films were fabricated with the addition of OA assigned to cubic α-$CsPbI_3$. More interestingly, the peak split clearly at 2θ of ~$14°$ and $28°$ for PEA-added films, along with additional minor peaks between $14°$ and $28°$ (c.f. Figure 1.22f). These features fitted well with the orthorhombic $CsPbI_3$ perovskite structure. This means that the film deposited in the presence of OA was cubic α-$CsPbI_3$ while it was orthorhombic β-$CsPbI_3$ when PEA was used. The choice of ligand determined the crystal structure of $CsPbI_3$ thus fabricated. Grain sizes of black $CsPbI_3$ films fabricated with the addition of OA or PEA ligands were much smaller than those from yellow phase film, along with improved film stability as shown in Figure 1.22g–i. Both OA and PEA are too large to incorporate into the lattice of $CsPbI_3$; authors hypothesized that these ligands bound to the crystal surface as a capping to stabilize perovskite structures and prevent further grain growths and aggregations, which was further confirmed by surface XPS analysis.

However, according to the structural refinement of room temperature black-phase $CsPbI_3$, Snaith's group found that (1) the black-phase $CsPbI_3$ processed at low and high temperatures is orthorhombic γ-phase but not cubic α-phase; (2) there is a preferred orientation of $CsPbI_3$ film processed at high temperature, which makes the XRD pattern look like cubic phase; and (3) powders scratched from high-temperature-fabricated films show orthorhombic structure, further proving the result (2) (Sutton et al. 2018). Therefore, more careful XRD refinement is needed to clarify whether the difference in the study above (Fu et al. 2017a) is from texture or phase itself.

Not only monoammonium, but also bication, ethylenediamine (EDA) stabilized black $CsPbI_3$ films. The terminal NH_3^+ groups at both sides of EDA^{2+} were expected to cross-link the $CsPbI_3$ perovskite crystal unit, relieving it from unwanted phase transition to yellow phase (Zhang et al. 2017). Two-dimensional (2D) $EDAPbI_4$ perovskite was introduced into the precursor solution of $CsPbI_3$. Cubic α-$CsPbI_3$ films thus obtained exhibited high phase stability at room temperature for months and at $100°C$ for >150 hours.

From the discussion above, one might notice that most of the stable $CsPbI_3$ films have small grain size. Actually, the first low-temperature-processed stable $CsPbI_3$ was in quantum dots (QDs) form (Protesescu et al. 2015). In the synthesis of QDs, a large amount of organic ligands were added, which surrounded the surface of QDs preventing grain growth and agglomeration. No apparent changes were observed for $CsPbI_3$ QDs even after 60 days' storage at ambient condition. Therefore, ligands at the surface of QDs play a critical role in stabilizing the $CsPbI_3$ phase. Using methyl acetate (MeOAc) for the purifying process could effectively wash out the unreacted precursors, at same time, without full removal of the surface species on QDs (Swarnkar et al. 2016). The QDs solar cells thus fabricated exhibited a PCE of 10.77%.

Besides fabrication of QDs, organic additives were directly introduced into the precursor solution of $CsPbI_3$ for the in-situ formation of monocrystalline bulk thin films. For example, polymer poly-vinylpyrrolidone (PVP) was added into precursor to stabilize black $CsPbI_3$ phase (Li et al. 2018). The oxygen and nitrogen atoms inside PVP molecule donated electron lone pairs, which offered a large number of coordination centers. These lone electron pairs interacted with Cs ions of $CsPbI_3$ as shown in Figure 1.23a. At the initial stage (Figure 1.23b), PVP molecules attracted cations of $CsPbI_3$ precursors due to the long backbone chain and electron lone pairs. The positive and negative ions of $CsPbI_3$ tended to form a metastable cubic $CsPbI_3$ cluster. More nuclei of $CsPbI_3$ launched on PVP as time increased, and nanocrystals were formed. The long chain of PVP anchored at the surface of these nanocrystals acting as a capping to prevent aggregation (Figure 1.23c). Due to its large size, PVP cannot fit into the lattice of $CsPbI_3$. It probably accumulated at the surface or grain boundaries of $CsPbI_3$ films after annealing (Figure 1.23d). This capping effect together with the interaction between PVP and $CsPbI_3$ might isolate nanocrystals and lower surface energy, thus inhibiting phase transition. The cubic $CsPbI_3$ films exhibited a

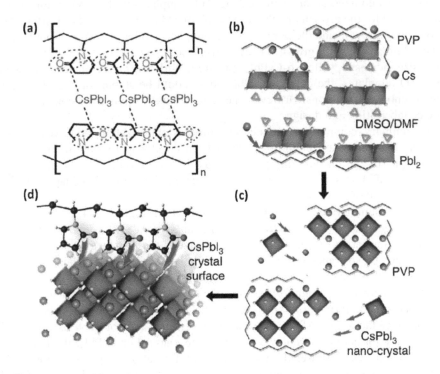

FIGURE 1.23 (a) Schematic illustration of the interaction between $CsPbI_3$ and PVP; (b) PbI_2 and Cs assembled inside DMSO/DMF solvent and interacting with PVP molecules forming a metastable state; (c) $CsPbI_3$ nanocrystal clusters attached to PVP, remaining relatively independent and stable with PVP molecules around; (d) PVP anchored on the surface of $CsPbI_3$ crystals via the interaction between N/O and Cs, thus significantly reducing the surface tension and stabilized cubic $CsPbI_3$ phase (Li et al. 2018). (Reprinted with permission from Springer Nature.)

carrier diffusion length of over 1.5 μm. The devices fabricated showed better stability against moisture and heat compared to $MAPbI_3$ devices.

In nanocrystals with capping ligands and large surface area, the strain might play another role for their outstanding stability. Strain was intentionally introduced by substrate clamping and fast quenching (Steele et al. 2019). When fast quenched from 330°C to room temperature, the $CsPbI_3$ underwent a phase transition from cubic α-phase to tetragonal β-phase with a calculated strain of 1.18%. Besides, due to the large mismatch in thermal expansion coefficients of perovskite layers (~50×10^{-6} K^{-1} for lead-iodide-based perovskites) and typical optically transparent substrates (ITO and glass of around 4×10^{-6} K^{-1} and 9×10^{-6} K^{-1}), there will be a substrate clamping strain within the perovskite film. This substrate clamping added an extra 0.47% strain for $CsPbI_3$ films. These two strains helped to stabilize and trigger the black $CsPbI_3$ phase at room temperature. However, these strained films were metastable phase. Upon mild reheating (60°C–100°C) or exposure to moisture, the metastable black phase normally turned to yellow phase.

To further improve the stability of strained $CsPbI_3$, a spatial confinement induced by anodized aluminium oxide (AAO) templates with varying pore sizes was applied for the growth of cubic $CsPbI_3$ phase (Ma et al. 2019). The $CsPbI_3$ was expected to grow inside the pores of AAO templates. When the size of pores was 30 or 41 nm, the $CsPbI_3$ grown was cubic α-phase at room temperature. When the pore size increased, yellow δ-$CsPbI_3$ appeared. With this artificial strain induced by AAO templates, the cubic α-$CsPbI_3$ phase could be stable for over 3 months under ambient conditions. However, the template used might be a constraint for device fabrication and design.

In summary, composition engineering, process optimization, additives, reduction in size, and strain engineering have been developed to stabilize the black $CsPbI_3$ phase, and encouraging progress has been achieved. The champion efficiency of $CsPbI_3$-based devices has reached around 19% with a bandgap relatively wider than that of $MAPbI_3$.

We have summarized the additive selection rule in Figure 1.24. In the middle "green range," the tolerance factor was in the range of 0.8–1.0, which is termed as perovskite formation range. When the cation size is too big, located at the upper-right corner, the tolerance factor is too big, and no perovskite forms; thus it is called upper forbidden zone to form perovskite. At the same time, when the cation size is too small, perovskite cannot form either, like the case at lower-left corner as lower forbidden zone to form perovskite. Even in the "green" perovskite formation zone, the tolerance factor for $FAPbI_3$ was near the upper limit and not stable as evidenced by the easy formation of yellow non-perovskite phase. For $CsPbI_3$, the tolerance factor approaches the lower limit for perovskite formation range. Therefore, it is not stable either as we discussed afore, and many strategies have been developed to stabilize the black $CsPbI_3$ phase. In terms of phase stabilization, complementary ionic-size cations should be considered. That is, to stabilize $FAPbI_3$, smaller cations are preferred, while for the $CsPbI_3$ case, larger cations are preferred. As we discussed afore, large cations, such as EDA, stabilized black $CsPbI_3$ phase. Addition of Cs and/or MA improved the phase stability of $FAPbI_3$. Therefore, in terms of phase stabilization, additive cations move the tolerance factor into the center range of

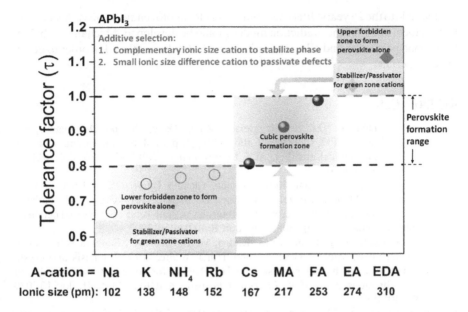

FIGURE 1.24 Calculated tolerance factors (τ) for different cations (A) in APbI$_3$ perovskite system (Han et al. 2018). (Reprinted with permission from American Chemical Society.)

"green" perovskite formation zone. For defect passivation, however, similar ionic-size cations effectively filled traps. For details, please refer to our earlier paper (Han et al. 2018).

1.4 SUMMARY

In summary, the structure and optical and optoelectronic properties of perovskite were discussed. The high absorption coefficient, long carrier lifetime, and ambipolar feature make perovskite a promising candidate for next-generation photovoltaic applications. Substitution in A-, B-, and X-site elements along with dimensionality variation enables the bandgap of perovskites tunable in a wide range, which is necessary for different photoelectronic applications, such as solar cells, light emitting diodes, lasers, and photodetectors.

The poor intrinsic stability and relatively wider bandgap of MAPbI$_3$ drove researchers pursue thermally more stable perovskites with suitable bandgap. FAPbI$_3$ and CsPbI$_3$ are thermally more stable than MAPbI$_3$ albeit structurally instable due to unsuitable tolerance factors. This chapter thoroughly explored the phase stability of FAPbI$_3$ and CsPbI$_3$, along with strategies developed to stabilize these perovskites. Process optimization, composition design, additives, surface functionalization, reduction in size, and strain engineering effectively stabilize these structures of perovskites. Tolerance factor along with octahedral factor plays an important role in determining the stability of perovskite structures.

Besides these encouraging results and vast chemistry in ABX$_3$ halide perovskites, the long-term stability of perovskite materials should be evaluated in a timescale comparable to the 25 years' lifetime requirement for photovoltaic devices. Accelerated test protocol and lifetime prediction model should be developed. Mechanisms behind these good properties and alloying limit should be further explored in order to design novel materials.

REFERENCES

Akkerman, Quinten A., Daniele Meggiolaro, Zhiya Dang, Filippo De Angelis, and Liberato Manna. 2017. Fluorescent alloy $xMn_{1-x}I_3$ perovskite nanocrystals with high structural and optical stability. *ACS Energy Letters* 2(9):2183–2186. doi: 10.1021/acsenergylett.7b00707.

Amat, Anna, Edoardo Mosconi, Enrico Ronca, Claudio Quarti, Paolo Umari, Md K. Nazeeruddin, Michael Grätzel, and Filippo De Angelis. 2014. Cation-induced bandgap tuning in organohalide perovskites: Interplay of spin–orbit coupling and octahedra tilting. *Nano Letters* 14(6):3608–3616. doi: 10.1021/nl5012992.

Baikie, Tom, Yanan Fang, Jeannette M. Kadro, Martin Schreyer, Fengxia Wei, Subodh G. Mhaisalkar, Michael Graetzel, and Tim J. White. 2013. Synthesis and crystal chemistry of the hybrid perovskite $(CH_3NH_3)PbI_3$ for solid-state sensitised solar cell applications. *Journal of Materials Chemistry A* 1(18):5628–5641. doi: 10.1039/C3TA10518K.

Ball, James M., Michael M. Lee, Andrew Hey, and Henry J. Snaith. 2013. Low-temperature processed meso-superstructured to thin-film perovskite solar cells. *Energy and Environmental Science* 6(6):1739–1743. doi: 10.1039/C3EE40810H.

Beal, Rachel E., Daniel J. Slotcavage, Tomas Leijtens, Andrea R. Bowring, Rebecca A. Belisle, William H. Nguyen, George F. Burkhard, Eric T. Hoke, and Michael D. McGehee. 2016. Cesium lead halide perovskites with improved stability for tandem solar cells. *The Journal of Physical Chemistry Letters* 7(5):746–751. doi: 10.1021/acs.jpclett.6b00002.

Binek, Andreas, Fabian C. Hanusch, Pablo Docampo, and Thomas Bein. 2015. Stabilization of the trigonal high-temperature phase of formamidinium lead iodide. *The Journal of Physical Chemistry Letters* 6(7):1249–1253. doi: 10.1021/acs.jpclett.5b00380.

Boix, Pablo P., Shweta Agarwala, Teck Ming Koh, Nripan Mathews, and Subodh G. Mhaisalkar. 2015. Perovskite solar cells: Beyond methylammonium lead iodide. *The Journal of Physical Chemistry Letters* 6(5):898–907. doi: 10.1021/jz502547f.

Brunetti, Bruno, Carmen Cavallo, Andrea Ciccioli, Guido Gigli, and Alessandro Latini. 2016. On the thermal and thermodynamic (In)stability of methylammonium lead halide perovskites. *Scientific Reports* 6:31896. doi: 10.1038/srep31896. https://www.nature.com/articles/srep31896#supplementary-information.

Chen, Tianran, Benjamin J. Foley, Changwon Park, Craig M. Brown, Leland W. Harriger, Jooseop Lee, Jacob Ruff, Mina Yoon, Joshua J. Choi, and Seung-Hun Lee. 2016. Entropy-driven structural transition and kinetic trapping in formamidinium lead iodide perovskite. *Science Advances* 2(10):e1601650. doi: 10.1126/sciadv.1601650.

Colella, Silvia, Edoardo Mosconi, Paolo Fedeli, Andrea Listorti, Francesco Gazza, Fabio Orlandi, Patrizia Ferro, Tullo Besagni, Aurora Rizzo, Gianluca Calestani, Giuseppe Gigli, Filippo De Angelis, and Roberto Mosca. 2013. $MAPbI_{3-x}Cl_x$ mixed halide perovskite for hybrid solar cells: The role of chloride as dopant on the transport and structural properties. *Chemistry of Materials* 25(22):4613–4618. doi: 10.1021/cm402919x.

Cook, Timothy R., Dilek K. Dogutan, Steven Y. Reece, Yogesh Surendranath, Thomas S. Teets, and Daniel G. Nocera. 2010. Solar energy supply and storage for the legacy and nonlegacy worlds. *Chemical Reviews* 110(11):6474–6502. doi: 10.1021/cr100246c.

Correa-Baena, Juan Pablo, Ludmilla Steier, Wolfgang Tress, Michael Saliba, Stefanie Neutzner, Taisuke Matsui, Fabrizio Giordano, T. Jesper Jacobsson, Ajay Ram Srimath Kandada, Shaik M. Zakeeruddin, Annamaria Petrozza, Antonio Abate, Mohammad Khaja Nazeeruddin, Michael Gratzel, and Anders Hagfeldt. 2015. Highly efficient planar perovskite solar cells through band alignment engineering. *Energy and Environmental Science* 8(10):2928–2934. doi: 10.1039/C5EE02608C.

Crabtree, George W., and Nathan S. Lewis. 2007. Solar energy conversion. *Physics Today* 60(3):37–42.

Dang, Hoang X., Kai Wang, Masoud Ghasemi, Mingchung Tang, Michele De Bastiani, Erkan Aydin, Emilie Dauzon, Dounya Barrit, Jun Peng, and Detlefm Smilgies. 2019. Multi-cation synergy suppresses phase segregation in mixed-halide perovskites. *Joule* 3(7):1746–1764.

Deepa, Melepurath, Manuel Salado, Laura Calio, Samrana Kazim, S. M. Shivaprasad, and Shahzada Ahmad. 2017. Cesium power: Low Cs$^+$ levels impart stability to perovskite solar cells. *Physical Chemistry Chemical Physics* 19(5):4069–4077. doi: 10.1039/C6CP08022G.

Du, M. H. 2014. Efficient carrier transport in halide perovskites: Theoretical perspectives. *Journal of Materials Chemistry A* 2(24):9091–9098. doi: 10.1039/C4TA01198H.

Eperon, Giles E., Giuseppe M. Paternò, Rebecca J. Sutton, Andrea Zampetti, Amir Abbas Haghighirad, Franco Cacialli, and Henry J. Snaith. 2015. Inorganic caesium lead iodide perovskite solar cells. *Journal of Materials Chemistry A* 3(39):19688–19695. doi: 10.1039/C5TA06398A.

Eperon, Giles E., Samuel D. Stranks, Christopher Menelaou, Michael B. Johnston, Laura M. Herz, and Henry J. Snaith. 2014. Formamidinium lead trihalide: A broadly tunable perovskite for efficient planar heterojunction solar cells. *Energy and Environmental Science* 7(3):982–988. doi: 10.1039/C3EE43822H.

Finger, Ashley N. 2014. *Using a Semiconductor Defect to Connect Diffusion Lengths and Lifetimes.* Davidson, NC: Davidson College.

French, A. P., and E. F. Taylor. 1978. *An Introduction to Quantum Physics.* London: Van Nostrand Reinhold.

Fu, Yongping, Morgan T. Rea, Jie Chen, Darien J. Morrow, Matthew P. Hautzinger, Yuzhou Zhao, Dongxu Pan, Lydia H. Manger, John C. Wright, Randall H. Goldsmith, and Song Jin. 2017a. Selective stabilization and photophysical properties of metastable perovskite polymorphs of CsPbI$_3$ in thin films. *Chemistry of Materials* 29(19):8385–8394. doi: 10.1021/acs.chemmater.7b02948.

Fu, Yongping, Tao Wu, Jue Wang, Jianyuan Zhai, Melinda J. Shearer, Yuzhou Zhao, Robert J. Hamers, Erjun Kan, Kaiming Deng, X. Y. Zhu, and Song Jin. 2017b. Stabilization of the metastable lead iodide perovskite phase via surface functionalization. *Nano Letters* 17(7):4405–4414. doi: 10.1021/acs.nanolett.7b01500.

Gfroerer, T. H., Yong Zhang, and Mark W. Wanlass. 2013. An extended defect as a sensor for free carrier diffusion in a semiconductor. *Applied Physics Letters* 102(1):012114. doi: 10.1063/1.4775369.

Giorgi, Giacomo, Jun-Ichi Fujisawa, Hiroshi Segawa, and Koichi Yamashita. 2013. Small photocarrier effective masses featuring ambipolar transport in methylammonium lead iodide perovskite: A density functional analysis. *The Journal of Physical Chemistry Letters* 4(24):4213–4216. doi: 10.1021/jz4023865.

Grånäs, Oscar, Dmitry Vinichenko, and Efthimios Kaxiras. 2016. Establishing the limits of efficiency of perovskite solar cells from first principles modeling. *Scientific Reports* 6(1):36108. doi: 10.1038/srep36108.

Green, M.A. 1982. *Solar Cells: Operating Principles, Technology, and System Applications.* Upper Saddle River, NJ: Prentice-Hall.

Han, G. F., S. Zhang, P. P. Boix, L. H. Wong, L. D. Sun, and S. Y. Lien. 2017. Towards high efficiency thin film solar cells. *Progress in Materials Science* 87:246–291. doi: 10.1016/j.pmatsci.2017.02.003.

Han, Guifang, Harri Dharma Hadi, Annalisa Bruno, Sneha Avinash Kulkarni, Teck Ming Koh, Lydia Helena Wong, Cesare Soci, Nripan Mathews, Sam Zhang, and Subodh G. Mhaisalkar. 2018. Additive selection strategy for high performance perovskite photovoltaics. *The Journal of Physical Chemistry C* 122(25):13884–13893. doi: 10.1021/acs.jpcc.8b00980.

Han, Qifeng, Sang-Hoon Bae, Pengyu Sun, Yao-Tsung Hsieh, Yang Yang, You Seung Rim, Hongxiang Zhao, Qi Chen, Wangzhou Shi, Gang Li, and Yang Yang. 2016. Single Crystal Formamidinium Lead Iodide (FAPbI$_3$): Insight into the structural, optical, and electrical properties. *Advanced Materials* 28(11):2253–2258. doi: 10.1002/adma.201505002.

Hao, Feng, Constantinos C. Stoumpos, Robert P. H. Chang, and Mercouri G. Kanatzidis. 2014. Anomalous band gap behavior in mixed Sn and Pb perovskites enables broadening of absorption spectrum in solar cells. *Journal of the American Chemical Society* 136(22):8094–8099. doi: 10.1021/ja5033259.

Hoke, Eric T., Daniel J. Slotcavage, Emma R. Dohner, Andrea R. Bowring, Hemamala I. Karunadasa, and Michael D. McGehee. 2015. Reversible photo-induced trap formation in mixed-halide hybrid perovskites for photovoltaics. *Chemical Science* 6(1):613–617. doi: 10.1039/C4SC03141E.

Hu, Yanqiang, Fan Bai, Xinbang Liu, Qingmin Ji, Xiaoliang Miao, Ting Qiu, and Shufang Zhang. 2017. Bismuth Incorporation Stabilized α-CsPbI3 for fully inorganic perovskite solar cells. *ACS Energy Letters* 2(10):2219–2227. doi: 10.1021/acsenergylett.7b00508.

Jacobsson, T. Jesper, Juan Pablo Correa-Baena, Meysam Pazoki, Michael Saliba, Kurt Schenk, Michael Grätzel, and Anders Hagfeldt. 2016. Exploration of the compositional space for mixed lead halogen perovskites for high efficiency solar cells. *Energy and Environmental Science* 9(5):1706–1724.

Jena, Ajay Kumar, Ashish Kulkarni, Yoshitaka Sanehira, Masashi Ikegami, and Tsutomu Miyasaka. 2018. Stabilization of α-CsPbI$_3$ in ambient room temperature conditions by incorporating Eu into CsPbI$_3$. *Chemistry of Materials* 30(19):6668–6674. doi: 10.1021/acs.chemmater.8b01808.

Jeon, Nam Joong, Jun Hong Noh, Woon Seok Yang, Young Chan Kim, Seungchan Ryu, Jangwon Seo, and Sang Il Seok. 2015. Compositional engineering of perovskite materials for high-performance solar cells. *Nature* 517(7535):476–480. doi: 10.1038/nature14133.

Jo, Yimhyun, Kyoung Suk Oh, Minjin Kim, Ka-Hyun Kim, Heon Lee, Chan-Woo Lee, and Dong Suk-Kim. 2016. High performance of planar perovskite solar cells produced from PbI$_2$(DMSO) and PbI$_2$(NMP) complexes by intramolecular exchange. *Advanced Materials Interfaces* 3(10):1500768. doi: 10.1002/admi.201500768.

Juarez-Perez, Emilio J., Zafer Hawash, Sonia R. Raga, Luis K. Ono, and Yabing Qi. 2016. Thermal degradation of CH$_3$ NH$_3$ PbI$_3$ perovskite into NH$_3$ and CH$_3$ I gases observed by coupled thermogravimetry–mass spectrometry analysis. *Energy and Environmental Science* 9(11):3406–3410.

Kawamura, Yukihiko, Hiroyuki Mashiyama, and Katsuhiko Hasebe. 2002. Structural study on cubic–tetragonal transition of CH$_3$NH$_3$PbI$_3$. *Journal of the Physical Society of Japan* 71(7):1694–1697. doi: 10.1143/JPSJ.71.1694.

Ke, Weijun, Ioannis Spanopoulos, Constantinos C. Stoumpos, and Mercouri G. Kanatzidis. 2018. Myths and reality of HPbI$_3$ in halide perovskite solar cells. *Nature Communications* 9(1):4785. doi: 10.1038/s41467-018-07204-y.

Koh, Teck Ming, Kunwu Fu, Yanan Fang, Shi Chen, Tze Chien Sum, Nripan Mathews, S.G. Mhaisalkar, Pablo P. Boix, and Tom Baikie. 2014. Formamidinium-containing metal-halide: An alternative material for near-IR absorption perovskite solar cells. *Journal of Physical Chemistry C* 118(30):16458–16462.

Koutselas, I. B., L. Ducasse, and G. C. Papavassiliou. 1996. Electronic properties of three-and low-dimensional semiconducting materials with Pb halide and Sn halide units. *Journal of Physics: Condensed Matter* 8(9):1217.

Lee, Jin-Wook, Deok-Hwan Kim, Hui-Seon Kim, Seung-Woo Seo, Sung Min Cho, and Nam-Gyu Park. 2015. Formamidinium and cesium hybridization for photo- and moisture-stable perovskite solar cell. *Advanced Energy Materials* 5(20):1501310. doi: 10.1002/aenm.201501310.

Lee, Jin-Wook, Dong-Jin Seol, An-Na Cho, and Nam-Gyu Park. 2014. High-efficiency perovskite solar cells based on the black polymorph of HC(NH$_2$)$_2$PbI$_3$. *Advanced Materials* 26(29):4991–4998. doi: 10.1002/adma.201401137.

Lewis, Nathan S, and George Crabtree. 2005. Basic research needs for solar energy utilization: report of the basic energy sciences workshop on solar energy utilization, April 18–21, 2005. US Department of Energy, Office of Basic Energy Science.

Lewis, Nathan S., and Daniel G. Nocera. 2006. Powering the planet: Chemical challenges in solar energy utilization. *Proceedings of the National Academy of Sciences* 103(43):15729–15735. doi: 10.1073/pnas.0603395103.

Li, Bo, Yanan Zhang, Lin Fu, Tong Yu, Shujie Zhou, Luyuan Zhang, and Longwei Yin. 2018. Surface passivation engineering strategy to fully-inorganic cubic CsPbI$_3$ perovskites for high-performance solar cells. *Nature Communications* 9(1):1076. doi: 10.1038/s41467-018-03169-0.

Li, Chonghe, Kitty Chi Kwan Soh, and Ping Wu. 2004. Formability of ABO$_3$ perovskites. *Journal of Alloys and Compounds* 372(1):40–48. doi: https://doi.org/10.1016/j.jallcom.2003.10.017.

Li, Chonghea, Xionggang Lu, Weizhong Ding, Liming Feng, Yonghui Gao, and Ziming Guo. 2008. Formability of ABX$_3$ (X=F, Cl, Br, I) halide perovskites. *Acta Crystallographica Section B: Structural Science* 64(6):702–707.

Li, Zhen, Mengjin Yang, Ji-Sang Park, Su-Huai Wei, Joseph J. Berry, and Kai Zhu. 2016. Stabilizing perovskite structures by tuning tolerance factor: Formation of formamidinium and cesium lead iodide solid-state alloys. *Chemistry of Materials* 28(1):284–292. doi: 10.1021/acs.chemmater.5b04107.

Liang, Jia, Caixing Wang, Yanrong Wang, Zhaoran Xu, Zhipeng Lu, Yue Ma, Hongfei Zhu, Yi Hu, Chengcan Xiao, Xu Yi, Guoyin Zhu, Hongling Lv, Lianbo Ma, Tao Chen, Zuoxiu Tie, Zhong Jin, and Jie Liu. 2016. All-inorganic perovskite solar cells. *Journal of the American Chemical Society* 138(49):15829–15832. doi: 10.1021/jacs.6b10227.

Liu, Yucheng, Zhou Yang, Dong Cui, Xiaodong Ren, Jiankun Sun, Xiaojing Liu, Jingru Zhang, Qingbo Wei, Haibo Fan, Fengyang Yu, Xu Zhang, Changming Zhao, and Shengzhong Liu. 2015. Two-inch-sized perovskite CH$_3$NH$_3$PbX$_3$ (X=Cl, Br, I) crystals: Growth and characterization. *Advanced Materials* 27(35):5176–5183. doi: 10.1002/adma.201502597.

Ma, Sunihl, Seong Hun Kim, Beomjin Jeong, Hyeok-Chan Kwon, Seong-Cheol Yun, Gyumin Jang, Hyunha Yang, Cheolmin Park, Donghwa Lee, and Jooho Moon. 2019. Strain-mediated phase stabilization: A new strategy for ultrastable α-CsPbI3 perovskite by nanoconfined growth. *Small* 15(21):1900219. doi: 10.1002/smll.201900219.

Marronnier, Arthur, Guido Roma, Soline Boyer-Richard, Laurent Pedesseau, Jean-Marc Jancu, Yvan Bonnassieux, Claudine Katan, Constantinos C. Stoumpos, Mercouri G. Kanatzidis, and Jacky Even. 2018. Anharmonicity and disorder in the black phases of cesium lead iodide used for stable inorganic perovskite solar cells. *ACS Nano* 12(4):3477–3486. doi: 10.1021/acsnano.8b00267.

McHale, J. M., A. Auroux, A. J. Perrotta, and A. Navrotsky. 1997. Surface energies and thermodynamic phase stability in nanocrystalline aluminas. *Science* 277(5327):788–791. doi: 10.1126/science.277.5327.788.

Mitzi, David B. 2001. Templating and structural engineering in organic-inorganic perovskites. *Journal of the Chemical Society, Dalton Transactions* (1):1–12. doi: 10.1039/B007070J.

Mitzi, David B. 2007. Synthesis, structure, and properties of organic-inorganic perovskites and related materials. In Progress in Inorganic Chemistry, 1–121. Hoboken, NJ: John Wiley & Sons, Inc.

Moller, Chr Kn. 1958. Crystal structure and photoconductivity of caesium plumbohalides. *Nature* 182(4647):1436.

Motta, Carlo, Fedwa El-Mellouhi, and Stefano Sanvito. 2015. Charge carrier mobility in hybrid halide perovskites. *Scientific reports* 5:12746-12746. doi: 10.1038/srep12746.

Navrotsky, Alexandra. 2003. Energetics of nanoparticle oxides: interplay between surface energy and polymorphism. *Geochemical Transactions* 4(1):34. doi: 10.1186/1467–4866–4–34.

Nie, Wanyi, Hsinhan Tsai, Reza Asadpour, Jean-Christophe Blancon, Amanda J. Neukirch, Gautam Gupta, Jared J. Crochet, Manish Chhowalla, Sergei Tretiak, Muhammad A. Alam, Hsing-Lin Wang, and Aditya D. Mohite. 2015. High-efficiency solution-processed perovskite solar cells with millimeter-scale grains. *Science* 347(6221):522–525. doi: 10.1126/science.aaa0472.

Noel, Nakita K., Martina Congiu, Alexandra J. Ramadan, Sarah Fearn, David P. McMeekin, Jay B. Patel, Michael B. Johnston, Bernard Wenger, and Henry J. Snaith. 2017. Unveiling the influence of pH on the crystallization of hybrid perovskites, delivering low voltage loss photovoltaics. *Joule* 1(2):328–343. doi: 10.1016/j.joule.2017.09.009.

Papavassiliou, G. C. 1996. Synthetic three-and lower-dimensional semiconductors based on inorganic units. *Molecular Crystals and Liquid Crystals Science and Technology, Section A* 286(1):231–238. doi: 10.1080/10587259608042291.

Park, Yun Hee, Inyoung Jeong, Seunghwan Bae, Hae Jung Son, Phillip Lee, Jinwoo Lee, Chul-Ho Lee, and Min Jae Ko. 2017. Inorganic rubidium cation as an enhancer for photovoltaic performance and moisture stability of $HC(NH_2)_2PbI_3$ perovskite solar cells. *Advanced Functional Materials* 27(16):1605988. doi: 10.1002/adfm.2016 05988.

Pellet, N., P. Gao, G. Gregori, T. Y. Yang, M. K. Nazeeruddin, J. Maier, and M. Gratzel. 2014. Mixed-organic-cation perovskite photovoltaics for enhanced solar-light harvesting. *Angewandte Chemie International Edition in English* 53(12):3151–3157. doi: 10.1002/ anie.201309361.

Protesescu, Loredana, Sergii Yakunin, Maryna I. Bodnarchuk, Franziska Krieg, Riccarda Caputo, Christopher H. Hendon, Ruo Xi Yang, Aron Walsh, and Maksym V. Kovalenko. 2015. Nanocrystals of cesium lead halide perovskites ($CsPbX_3$, X=Cl, Br, and I): Novel optoelectronic materials showing bright emission with wide color gamut. *Nano Letters* 15(6):3692–3696. doi: 10.1021/nl5048779.

Rohrer, Gregory S. 2001. *Structure and Bonding in Crystalline Materials*. Cambridge: Cambridge University Press.

Roy, R., and O. Muller. 1974. *The Major Ternary Structural Families*. Berlin: Springer-Verlag.

Sharma, S., N. Weiden, and A. Weiss. 1992. Phase diagrams of Quasibinary systems of the type: ABX_3 — $A'BX_3$; ABX_3 — $AB'X_3$, and ABX_3 — ABX'_3; X=Halogen. *Zeitschrift für Physikalische Chemie* 175(1):63–80.

Saliba, Michael, Taisuke Matsui, Konrad Domanski, Ji-Youn Seo, Amita Ummadisingu, Shaik M. Zakeeruddin, Juan-Pablo Correa-Baena, Wolfgang R. Tress, Antonio Abate, Anders Hagfeldt, and Michael Grätzel. 2016a. Incorporation of rubidium cations into perovskite solar cells improves photovoltaic performance. *Science* 354(6309):206–209. doi: 10.1126/science.aah5557.

Saliba, Michael, Taisuke Matsui, Ji-Youn Seo, Konrad Domanski, Juan-Pablo Correa-Baena, Mohammad Khaja Nazeeruddin, Shaik M. Zakeeruddin, Wolfgang Tress, Antonio Abate, Anders Hagfeldt, and Michael Grätzel. 2016b. Cesium-containing triple cation perovskite solar cells: improved stability, reproducibility and high efficiency. *Energy and Environmental Science* 9(6):1989–1997. doi: 10.1039/C5EE03874J.

Sam, Zhang, and Han Guifang. 2020. Intrinsic and environmental stability issues of perovskite photovoltaics. *Progress in Energy* 2 (2):022002. doi: 10.1088/2516-1083/ab70d9.

Service, Robert F. 2016. Cesium fortifies next-generation solar cells. *Science* 351(6269):113–114. doi: 10.1126/science.351.6269.113.

Shah, Arvind. 2010. *Thin-Film Silicon Solar Cells*. New York: EPFL Press, https://doi.org/10.1201/b16327

Shockley, William, and Hans J. Queisser. 1961. Detailed balance limit of efficiency of p-n junction solar cells. *Journal of Applied Physics* 32(3):510–519. doi: 10.1063/1.1736034.

Slotcavage, Daniel J., Hemamala I. Karunadasa, and Michael D. McGehee. 2016. Light-induced phase segregation in halide-perovskite absorbers. *ACS Energy Letters* 1(6):1199–1205. doi: 10.1021/acsenergylett.6b00495.

Steele, Julian A., Handong Jin, Iurii Dovgaliuk, Robert F. Berger, Tom Braeckevelt, Haifeng Yuan, Cristina Martin, Eduardo Solano, Kurt Lejaeghere, Sven M. J. Rogge, Charlotte Notebaert, Wouter Vandezande, Kris P. F. Janssen, Bart Goderis, Elke Debroye, Ya-Kun Wang, Yitong Dong, Dongxin Ma, Makhsud Saidaminov, Hairen Tan, Zhenghong Lu, Vadim Dyadkin, Dmitry Chernyshov, Veronique Van Speybroeck, Edward H. Sargent, Johan Hofkens, and Maarten B. J. Roeffaers. 2019. Thermal unequilibrium of strained black CsPbI$_3$ thin films. *Science* 365(6454):679–684. doi: 10.1126/science.aax3878.

Stoumpos, Constantinos C., Christos D. Malliakas, and Mercouri G. Kanatzidis. 2013. Semiconducting tin and lead iodide perovskites with organic cations: Phase transitions, high mobilities, and near-infrared photoluminescent properties. *Inorganic Chemistry* 52(15):9019–9038.

Stoumpos, Constantinos C., Christos D. Malliakas, John A. Peters, Zhifu Liu, Maria Sebastian, Jino Im, Thomas C. Chasapis, Arief C. Wibowo, Duck Young Chung, Arthur J. Freeman, Bruce W. Wessels, and Mercouri G. Kanatzidis. 2013. Crystal growth of the perovskite semiconductor CsPbBr3: A new material for high-energy radiation detection. *Crystal Growth and Design* 13(7):2722–2727. doi: 10.1021/cg400645t.

Sutton, Rebecca J., Marina R. Filip, Amir A. Haghighirad, Nobuya Sakai, Bernard Wenger, Feliciano Giustino, and Henry J. Snaith. 2018. Cubic or orthorhombic? Revealing the crystal structure of metastable black-phase CsPbI$_3$ by theory and experiment. *ACS Energy Letters* 3(8):1787–1794. doi: 10.1021/acsenergylett.8b00672.

Swarnkar, Abhishek, Ashley R. Marshall, Erin M. Sanehira, Boris D. Chernomordik, David T. Moore, Jeffrey A. Christians, Tamoghna Chakrabarti, and Joseph M. Luther. 2016. Quantum dot–induced phase stabilization of α-CsPbI$_3$ perovskite for high-efficiency photovoltaics. *Science* 354(6308):92–95. doi: 10.1126/science.aag2700.

Tanaka, Hirohisa, and Makoto Misono. 2001. Advances in designing perovskite catalysts. *Current Opinion in Solid State and Materials Science* 5(5):381–387.

Turren-Cruz, Silver-Hamill, Anders Hagfeldt, and Michael Saliba. 2018. Methylammonium-free, high-performance, and stable perovskite solar cells on a planar architecture. *Science* 362(6413):449–453. doi: 10.1126/science.aat3583.

Umari, P., E. Mosconi, and F. De Angelis. 2014. Relativistic GW calculations on CH$_3$NH$_3$PbI$_3$ and CH$_3$NH$_3$SnI$_3$ perovskites for solar cell applications. *Scientific Repotrs* 4:4467. doi: 10.1038/srep04467.

Umebayashi, T., K. Asai, T. Kondo, and A. Nakao. 2003. Electronic structures of lead iodide based low-dimensional crystals. *Physical Review B* 67(15):155405.

Würfel, Peter. 2007. Limitations on energy conversion in solar cells. In *Physics of Solar Cells*, 137–153. Weinheim, Germany: Wiley-VCH Verlag GmbH.

Wang, Feng, Hui Yu, Haihua Xu, and Ni Zhao. 2015. HPbI$_3$: A new precursor compound for highly efficient solution-processed perovskite solar cells. *Advanced Functional Materials* 25(7):1120–1126. doi: 10.1002/adfm.201404007.

Wang, Kang, Zhiwen Jin, Lei Liang, Hui Bian, Haoran Wang, Jiangshan Feng, Qian Wang, and Shengzhong Liu. 2019a. Chlorine doping for black γ-CsPbI$_3$ solar cells with stabilized efficiency beyond 16%. *Nano Energy* 58:175–182. doi: 10.1016/j.nanoen.2019.01.034.

Wang, Yong, Xiaomin Liu, Taiyang Zhang, Xingtao Wang, Miao Kan, Jielin Shi, and Yixin Zhao. 2019b. The role of dimethylammonium iodide in CsPbI$_3$ perovskite fabrication: Additive or dopant? *Angewandte Chemie International Edition* 58(46). doi: 10.1002/anie.201910800.

Wang, Yun, Tim Gould, John F. Dobson, Haimin Zhang, Huagui Yang, Xiangdong Yao, and Huijun Zhao. 2014. Density functional theory analysis of structural and electronic properties of orthorhombic perovskite CH$_3$NH$_3$PbI$_3$. *Physical Chemistry Chemical Physics* 16(4):1424–1429. doi: 10.1039/C3CP54479F.

Weller, Mark T., Oliver J. Weber, Jarvist M. Frost, and Aron Walsh. 2015. Cubic perovskite structure of black formamidinium lead iodide, α-[HC(NH$_2$)$_2$]PbI$_3$, at 298 K. *Journal of Physical Chemistry Letters* 6(16):3209–3212.

Xiang, Sisi, Zhongheng Fu, Weiping Li, Ya Wei, Jiaming Liu, Huicong Liu, Liqun Zhu, Ruifeng Zhang, and Haining Chen. 2018. Highly air-stable carbon-based α-CsPbI$_3$ perovskite solar cells with a broadened optical spectrum. *ACS Energy Letters* 3(8):1824–1831. doi: 10.1021/acsenergylett.8b00820.

Xiao, Z., Q. Dong, C. Bi, Y. Shao, Y. Yuan, and J. Huang. 2014. Solvent annealing of perovskite-induced crystal growth for photovoltaic-device efficiency enhancement. *A dvanced Materials* 26(37):6503–6509. doi: 10.1002/adma.201401685.

Yadav, Pankaj, M. Ibrahim Dar, Neha Arora, Essa A. Alharbi, Fabrizio Giordano, Shaik Mohammed Zakeeruddin, and Michael Grätzel. 2017. The role of rubidium in multiple-cation-based high-efficiency perovskite solar cells. *Advanced Materials* 29(40):1701077-n/a. doi: 10.1002/adma.201701077.

Yang, Woon Seok, Jun Hong Noh, Nam Joong Jeon, Young Chan Kim, Seungchan Ryu, Jangwon Seo, and Sang Il Seok. 2015. High-performance photovoltaic perovskite layers fabricated through intramolecular exchange. *Science* 348(6240):1234–1237. doi: 10.1126/science.aaa9272.

Yin, Wan-Jian, Tingting Shi, and Yanfa Yan. 2014a. Unique properties of halide perovskites as possible origins of the superior solar cell performance. *Advanced Materials* 26(27):4653–4658. doi: 10.1002/adma.201306281.

Yin, Wan-Jian, Tingting Shi, and Yanfa Yan. 2014b. Unusual defect physics in CH$_3$NH$_3$PbI$_3$ perovskite solar cell absorber. *Applied Physics Letters* 104(6):063903. doi: 10.1063/1.4864778.

Yin, Wan-Jian, Yanfa Yan, and Su-Huai Wei. 2014c. Anomalous alloy properties in mixed halide perovskites. *The Journal of Physical Chemistry Letters* 5(21):3625–3631. doi: 10.1021/jz501896w.

Yin, Wan-Jian, Ji-Hui Yang, Joongoo Kang, Yanfa Yan, and Su-Huai Wei. 2015. Halide perovskite materials for solar cells: A theoretical review. *Journal of Materials Chemistry A* 3(17):8926–8942. doi: 10.1039/C4TA05033A.

Yun, Jae S., Anita Ho-Baillie, Shujuan Huang, Sang H. Woo, Yooun Heo, Jan Seidel, Fuzhi Huang, Yi-Bing Cheng, and Martin A. Green. 2015. Benefit of grain boundaries in organic–inorganic halide planar perovskite solar cells. *The Journal of Physical Chemistry Letters* 6(5):875–880. doi: 10.1021/acs.jpclett.5b00182.

Zhang, Taiyang, M. Ibrahim Dar, Ge Li, Feng Xu, Nanjie Guo, Michael Grätzel, and
   Yixin Zhao. 2017. Bication lead iodide 2D perovskite component to stabilize inorganic α-CsPbI$_3$ perovskite phase for high-efficiency solar cells. *Science Advances*
   3(9):e1700841. doi: 10.1126/sciadv.1700841.
Zhao, Boya, Shi-Feng Jin, Sheng Huang, Ning Liu, Jing-Yuan Ma, Ding-Jiang Xue, Qiwei Han,
   Jie Ding, Qian-Qing Ge, Yaqing Feng, and Jin-Song Hu. 2018. Thermodynamically
   stable orthorhombic γ-CsPbI$_3$ thin films for high-performance photovoltaics. *Journal
   of the American Chemical Society* 140(37):11716–11725. doi: 10.1021/jacs.8b06050.
Zheng, Xiaojia, Congcong Wu, Shikhar K. Jha, Zhen Li, Kai Zhu, and Shashank Priya. 2016.
   Improved phase stability of formamidinium lead triiodide perovskite by strain relaxation. *ACS Energy Letters* 1(5):1014–1020. doi: 10.1021/acsenergylett.6b00457.
Zong, Yingxia, Ning Wang, Lin Zhang, Ming-Gang Ju, Xiao Cheng Zeng, Xiao Wei Sun,
   Yuanyuan Zhou, and Nitin P. Padture. 2017. Homogenous alloys of formamidinium
   lead triiodide and cesium tin triiodide for efficient ideal-bandgap perovskite solar
   cells. *Angewandte Chemie International Edition* 56(41):12658–12662. doi: 10.1002/
   anie.201705965.

2 Carbon Nanomaterials for Flexible Energy Storage Devices

Huijuan Lin, Hui Li, and Chenjun Zhang
Nanjing Tech University

CONTENTS

2.1 Overview of Flexible Energy Storage Devices 46
2.2 Flexible Supercapacitors 46
    2.2.1 Mechanism and Advancement of Supercapacitors 46
    2.2.2 Electrodes 48
        2.2.2.1 Carbon Materials 48
        2.2.2.2 Conducting Polymers 49
        2.2.2.3 Transition Metal Oxides 49
    2.2.3 Flexible Wire-Shaped Supercapacitors 50
        2.2.3.1 Wire-Shaped Supercapacitors in a Parallel Structure 50
        2.2.3.2 Wire-Shaped Supercapacitors in a Twisted Structure 51
        2.2.3.3 Wire-Shaped Supercapacitors in a Coaxial Structure 52
    2.2.4 Applications 52
        2.2.4.1 Supercapacitor Textiles 52
        2.2.4.2 Flexible Microsupercapacitors 55
2.3 Flexible Batteries 56
    2.3.1 Mechanism and Development of Batteries 57
    2.3.2 Electrodes 57
        2.3.2.1 Carbon Nanotube 57
        2.3.2.2 Graphene 58
        2.3.2.3 Carbon Paper/Carbon Cloth 61
    2.3.3 Configuration 62
        2.3.3.1 Flexible Planar Batteries 62
        2.3.3.2 Flexible Wire-Shaped Batteries 62
    2.3.4 Application 64
        2.3.4.1 Textile Batteries 64
        2.3.4.2 Multifunctional Batteries 66
2.4 Integrated Energy Device 68
2.5 Summary and Outlook 70
References 71

This chapter discusses the necessity to develop flexible energy storage devices. The main efforts are first paid to describe the history and recent advancements in flexible supercapacitors/batteries. Detailed discussion is focused on electrode materials, device configuration/structure (1-dimensional (1D) fiber and 2D planar shapes), and their mechanical/electrochemical performances (flexibility, stability, and capacity). Special attention is paid to carbon-based materials (e.g., carbon nanotubes (CNTs), graphene, and carbon composites) for supercapacitors, conductive carbon cloth-, paper-, and insulating textile-based electrodes for batteries, etc. Some recent applications of flexible supercapacitors/batteries are further described. Finally, emphasis is given to integrated energy storage devices, and the remaining difficulties are summarized.

2.1 OVERVIEW OF FLEXIBLE ENERGY STORAGE DEVICES

Nowadays, the advent of flexible electronic devices such as smart watches, bendable screens, and wearable sensors has attracted great interest. Compared with conventional electronic facilities, flexible electronics possess many advantages of being flexible, lightweight, wearable, and even implantable. To satisfy the requirements of high-performance flexible electronics, considerable efforts have been devoted to seeking compatible power systems such as flexible supercapacitors and lithium-ion batteries (LIBs).

2.2 FLEXIBLE SUPERCAPACITORS

Traditional supercapacitors in rigid planar structures cannot meet the rapid development of portable and wearable electronic devices which demand the power sources of small size, light weight, and high flexibility. Hence, it is urgent to develop flexible supercapacitors.

2.2.1 Mechanism and Advancement of Supercapacitors

According to the energy storage mechanism, supercapacitors can be divided into two categories, namely electrochemical double-layer capacitors (EDLCs) and pseudocapacitors [1,2]. As illustrated in Figure 2.1a, the capacitance in EDLCs mainly relies on the accumulation/desorption of ions at the electrode/electrolyte interface. Differentially, the capacitance in pseudocapacitors stems from fast and reversible redox reactions occurring on the surface of electroactive materials (such as transition metal oxides and conducting polymers) (Figure 2.1b), ensuring good rate capability and high power density.

The pseudocapacitors can be further classified into three faradaic mechanisms based on the various charge transfer processes, including underpotential deposition, redox pseudocapacitance, and intercalation pseudocapacitance as illustrated in Figure 2.2. The underpotential deposition happens when the surface of metallic materials is covered with a monolayer of metal ions or protons whose potential is beyond the redox potential of the materials. Redox pseudocapacitance indicates that active materials (transition metal oxides or conducting polymers) are electrochemically

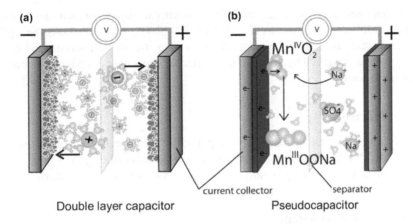

FIGURE 2.1 Schematics of (a) carbon-based EDLC and (b) MnO₂-based pseudocapacitor. (Reprinted with permission from Ref. [3]. Copyright 2014, Royal Society of Chemistry.)

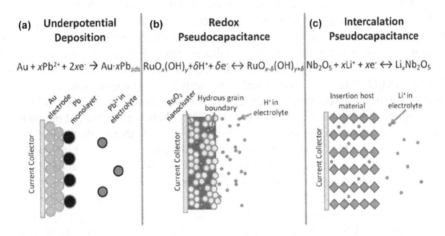

FIGURE 2.2 Schematic of three types of charge storage mechanisms for pseudocapacitors: (a) underpotential deposition, (b) redox pseudocapacitance, and (c) intercalation pseudocapacitance. (Reprinted with permission from Ref. [4]. Copyright 2014, Royal Society of Chemistry.)

absorbed by ions on or near the surface, accompanied by faradaic charge transfer process. In the case of intercalation pseudocapacitance, cations are intercalated into the lattice framework of active materials involving a faradaic charge transfer without structural distortion.

To assess the electrochemical performance of supercapacitors, cyclic voltammetry (CV) and galvanostatic charge and discharge tests are usually conducted. The specific capacitance of the supercapacitor can be obtained based on the charge–discharge curve: $C_{sp} = 2i/(m\Delta V/\Delta t)$, where m is the mass of the active material in a single electrode, i is the discharge current, and $\Delta V/\Delta t$ corresponds to the slope of the discharge curve. Specific volumetric or areal capacitance of the supercapacitor can also be obtained. Energy density and power density of supercapacitor can be calculated according to the equations $E = 1/2CV^2$, $P = V^2/4R_s$, where R_s (Ω) is the equivalent series resistance of the system. According to the above analysis, the energy density can be enhanced by increasing the specific capacitances and/or widening the potential window.

2.2.2 ELECTRODES

Typically, supercapacitors consist of four parts: two electrodes, electrolyte, and separator. Particularly, electrode materials are crucial in determining the supercapacitor performance. They can be generally categorized into carbon-based materials, transition metal oxides, and conducting polymers.

2.2.2.1 Carbon Materials

Carbon materials (e.g., CNTs and graphene), possessing attractive advantages of lightweight, high mechanical properties, and high electrical conductivity, are promising electrode materials for high-performance flexible supercapacitors.

CNTs have been widely studied for various fields due to their unique properties (e.g., excellent electrical properties and superb mechanical strength) and open tubular network structure since they were discovered in 1991 [5]. A popular method to prepare CNTs is chemical vapor deposition, using ethylene as carbon source and Fe (1 nm)/Al_2O_3 (10 nm) on silicon wafer as catalyst [6]. The spinnable CNT arrays obtained in the above-mentioned approach can be easily transformed into CNT sheets with aligned direction or further twisted into CNT fibers with enhanced mechanical property. The more twisted the angle of the formed CNT fiber is, the higher tensile strain can be. As to CNTs either in sheet or fiber shape, the merits of high mechanical flexibility and electronic conductivity enable the CNT-based electrodes to operate normally under various mechanical deformations as well as achieve excellent device performance by introduction of active materials into CNT matrix through physical electron-beam evaporation (e.g., Si/CNT [7]), co-spinning process (e.g., $LiMn_2O_4$/CNT (LMO/CNT), $Li_4Ti_5O_{12}$/CNT (LTO/CNT) [8]), and electrochemical deposition (e.g., polyaniline/CNT (PANI/CNT) [9]). CNTs can also be incorporated with graphene to form a hybrid electrode for construction of free-standing and metal-current-collector-free supercapacitors. Compared with electrodes with only a single component, the hybrid electrode displays higher electrical and mechanical properties, which are conducive to building high-performance supercapacitors.

Graphene is a monolayer 2D carbon sheet with a sp^2 conjugated configuration and has been studied for supercapacitors for many years. It features good flexibility, chemical and thermal stability, outstanding conductivity, wide potential window, and specific surface area of up to 2,630 m^2g^{-1} [10,11]. With the 2D structure and excellent conductivity, graphene can be also developed into a building block for flexible and self-standing supercapacitor electrodes without polymeric binder and current collector. However, it is easy to restack and aggregate due to the van der Waals interaction between graphene sheets, leading to the electrochemical performance fading. A facile way to avoid the restacking of the sheets is to add surfactants to disperse graphene more easily in aqueous solution and stabilize the morphology of single layer graphene sheet with surfactant intercalation [12]. Another effective strategy is to engineer the graphene sheets into a 3D material by incorporating other carbonaceous materials such as carbon spheres into the graphene sheets [13]. In addition, a variety of approaches, such as thermal reduction [14], chemical activation [15], and direct laser treatment [16], have also been used to achieve superior energy storage capability of graphene-based supercapacitors.

2.2.2.2 Conducting Polymers

Conducting polymers feature high electrical conductivity with conjugated bonds along the polymer chain. Nowadays, conducting polymers applied for flexible supercapacitors mainly include polyaniline (PANI), polypyrrole (PPy), polythiophene (PTh), and their derivatives. They can store energy via reversible doping and undoping reactions. In addition, they usually cost less than metal oxides and are easy to synthesize. Typically, they can be synthesized via electrochemical or chemical polymerization of their monomers. Polymers synthesized via chemical polymerization can be easily produced on a large scale. The crystal structures, surface properties, and the doped state of the polymer during the polymerization, will affect their electrochemical performance [12].

PANI is one of the ideal conducting polymers for electrode materials for its high theoretical specific capacitance (over 2,000 F g^{-1}) and low cost [17]. PANI has viable oxidation states and protonic acid doping/undoping properties which contribute to the excellent pseudocapacitance. The redox reaction is associated with its interconvertible structures between benzenoid and quinoid rings. In addition, PANI-based materials exhibit high surface area and porous structure, which favor the enhanced electrochemical performance of supercapacitors. PPy has received much attention for pseudocapacitors due to its fast charge/discharge rate and low cost. The doping process of PPy usually occurs with the existence of single charged anions. When multiple charged anions such as SO_4^{2-} are used, the polymer will be crosslinked, which will reduce the active sites of the polymer backbone, leading to a capacitance reduction [18]. Compared with PANI and PPy, PTh shows lower conductivity and theoretical specific capacitance [19]. However, the stable chemical properties of PTh against the air and moisture and a wide potential window of around 1.2 V indicate PTh as a promising polymer for supercapacitors.

2.2.2.3 Transition Metal Oxides

The pseudocapacitance of metal oxides comes from the faradic redox reaction and the charge separation at the electrode/electrolyte interface. Recently, many researchers have reported supercapacitors made from various metal oxides (such as RuO_2, MnO_2, V_2O_5, SnO_2, and Fe_2O_3) and metal hydroxides (such as $Co(OH)_2$, $Ni(OH)_2$, or their complexes) [20,21]. Amongst them, RuO_2 is the most studied material, due to its high pseudocapacitance, wide potential window, high proton conductivity, highly reversible redox reaction, good thermal stability, long cycle life, high metallic conductivity, and high rate capability [22]. The RuO_2-based supercapacitor's performance depends on many aspects, such as the crystallinity degree and specific surface area of RuO_2. RuO_2 with high crystallinity is not conducive to the transport and diffusion of protons, resulting in large diffusion resistance. Since the capacitance of RuO_2 with good crystallinity is mainly contributed by surface reactions, RuO_2 prepared in films on various substrates or in nanosized particles [23,24] can provide high specific surface area with more accessible active sites that can participate in redox reactions and thus improve the specific capacitance. For example, amorphous RuO_2 films deposited on a stainless steel substrate by anodic deposition showed stable electrochemical performance, achieving a maximum specific capacitance of 1,190 F g^{-1} in H_2SO_4 electrolyte. Compared to RuO_2, manganese oxide has a relatively low cost, less toxicity, and a higher theoretical capacitance, up to 1,100–1,300 F g^{-1}, which has been widely considered as a promising electrode material [25]. Manganese oxides with other valence states have also shown great potential for applications in supercapacitors. For example, a flexible solid-state supercapacitor was assembled with two nanosheet-like Mn_3O_4 electrodes in PVA/ H_2SO_4 gel electrolyte [26], showing a high specific capacitance of 127 F g^{-1}. A flexible supercapacitor [27] based on a carbon nanoparticle and MnO_2 nanorod hybrid structure using PVA/H_3PO_4 electrolyte showed good electrochemical performance. The specific surface area and electrical conductivity of MnO_2 can also be improved by doping other metal elements (Co, Ni, Al, Mo, V, Fe, etc.) on MnO_2 to form mixed oxides with more defects and charge carriers.

Pseudocapacitive materials are usually afflicted by poor cycling stability and low power density due to the limited conductivity and tremendous volumetric expansion during charge and discharge process. An efficient strategy has been proposed to construct composite materials. A great deal of research efforts have been made to construct conducting polymer/carbon materials (e.g., PPy/carbon aerogel composite, PANI/CNT sheets) and metal oxides/carbon materials (e.g., MnO_2/CNTs hybrid fiber, hierarchical Mn_3O_4/3D graphene aerogels), which combine synergistically the high pseudocapacitance of metal oxides or conducting polymers plus the outstanding characters of carbon nanomaterials (excellent electric conductivity, mechanical properties, stable cycle life, etc.), thus creating fast ion and electron transport pathways and reaching a significant improvement in electrochemical performance.

2.2.3 Flexible Wire-Shaped Supercapacitors

To meet the requirement of wearable electronics, wire-shaped supercapacitors are actively exploited as they can stably operate even under various deformations and are easily woven into textiles. Commonly, the wire-shaped supercapacitors can be classified into parallel, twisted, and coaxial structures.

2.2.3.1 Wire-Shaped Supercapacitors in a Parallel Structure

Among the three kinds of wire-shaped structures, supercapacitors in parallel are thought of as a simple design in which they are just assembled by grouping two wire electrodes together leaving a space between them and then filling the gel electrolyte into the two electrodes without encapsulation. Besides, multiple supercapacitors in parallel structures can be readily integrated in series or parallel to further improve the output potential or current of the devices for wider applications. A wire-shaped supercapacitor in a parallel pattern was developed by the paired symmetric MnO_2/ oxidized CNT hybrid electrodes (Figure 2.3a) [28]. These two wire-shaped electrodes

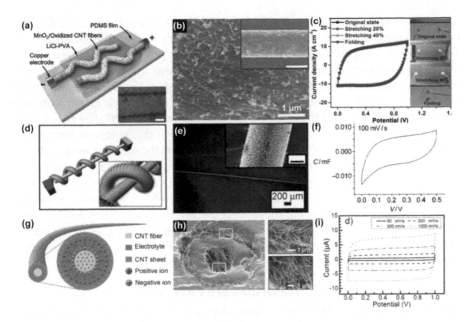

FIGURE 2.3 Architectural design of flexible wire-shaped supercapacitors. (a) Structure of the stretchable wire-shaped supercapacitor in parallel pattern. (b) SEM images of MnO_2/oxidized CNT fiber. (c) CV curves of the parallel device at $50\,mV\ s^{-1}$ under stretching and folding, respectively. (Reprinted with permission from Ref. [28]. Copyright 2017, Wiley.) (d) Structure of the wire-shaped supercapacitor in twisted pattern. (e) SEM images of a Kevlar wire grown with ZnO nanowire arrays. (f) CV curves of the twisted device using PVA/H_3PO_4 gel electrolyte at a scan rate of $100\,mV\ s^{-1}$. (Reprinted with permission from Ref. [29]. Copyright 2011, Wiley.) (g) Schematic of the cross-sectional structure and mechanism of a coaxial wire-shaped supercapacitor. (h) Cross-sectional SEM images of a coaxial structure. (i) CV curves at scan rates of 50, 200, 500, and $1,000\,mV\ s^{-1}$. (Reprinted with permission from Ref. [32]. Copyright 2013, Wiley.) Scale bars of insets: (a) $200\,\mu m$, (b) $10\,\mu m$, and (e)$10\,\mu m$.

were fabricated by simple solution method and assembled in parallel with a pre-strain process, forming a buckled structure. As shown in Figure 2.3b, nanostructured MnO_2 particles were uniformly anchored to the oxidized CNT fiber surface. The resultant supercapacitors exhibited an excellent specific volumetric capacitance of $409.4\ F\ cm^{-3}$ at $0.75\ A\ cm^{-3}$ in LiCl-PVA gel electrolyte. Particularly, CV curves of the parallel-structure supercapacitors (Figure 2.3c) tested under different degrees of stretching and folding were observed with nearly no curve change compared with the original CV plots, indicating high mechanical properties and stable capacitance retention especially under a tensile strain of up to 40%.

2.2.3.2 Wire-Shaped Supercapacitors in a Twisted Structure

Twisting two wire-shaped electrodes into a twisted structure device is an effective approach to construct a flexible supercapacitor. Compared with the supercapacitor assembled in a parallel structure which needs planar supporting substrate, the twisted structure device is able to self-support and easily meet the practical requirement of being woven into a fabric or textile. A twisted supercapacitor was fabricated using a flexible Au-coated plastic wire and a Kevlar fiber substrate with ZnO nanowires (Figure 2.3d) [29]. During the twisting process, the Kevlar fiber was entangled around the straight plastic carefully which was fixed on a stage. To avoid the direct contact between two electrodes causing a short circuit, a gel electrolyte was coated to separate them. The surface morphology of electrode grown with ZnO nanowires can be seen in Figure 2.3e. A low magnification of SEM showed a flexible plastic wire with a diameter of around 200 um. A high magnification observation showed hexagonal-shaped ZnO nanowires had grown on the substrate uniformly and densely. In addition, the wire-shaped supercapacitor was assembled with a gel PVA/H_3PO_4 electrolyte, delivering a specific capacitance of around 2.4 mF cm^{-2} according to the CV curve (Figure 2.3f) at a scan rate of 100 mV s^{-1}. However, the twisted wire electrodes often cannot ensure enough contact with each other, resulting in a high internal resistance. When they suffer from bending or folding deformations, the twisted electrodes may physically detach from each other, giving rise to a decrease in device performance [30].

2.2.3.3 Wire-Shaped Supercapacitors in a Coaxial Structure

Instead of paralleling or twisting two wire electrodes into a device, the supercapacitor in a coaxial structure can be assembled by paving one layer onto another inner layer with a separator between them. In addition, the coaxial structure provides wider contact area between the wire-shaped electrodes with reduced internal resistance [31], as well as more stable performance under bending deformation than the device in twisted structure. A coaxial wire-shaped supercapacitor was developed, comprising aligned CNT fiber as an inner layer and CNT sheet as an outer layer with a gel electrolyte sandwiched between the two electrodes [32]. The schematic of the cross-sectional view (Figure 2.3g) shows a coaxial structure. The inner CNT fiber was fabricated by spinning process with a diameter modified from 6 to 40 μm depending on the width of spinnable CNT arrays. The cross-sectional SEM image (Figure 2.3h) evidences the designed coaxial structure, clearly showing the morphology of the inner CNT and the outer CNT sheet which were distinguished by two concentric cycles. The device maintained a rectangular shape of CV with increasing scan rates from 50 to 1,000 mV s^{-1} (Figure 2.3i), indicating a double-layer capacitor with a low internal resistance and stable performance under rapid charge–discharge operation. Moreover, the coaxial-structure supercapacitor yielded a maximum capacitance of 59 F g^{-1} with negligible capacitance drop after 11,000 long-term cycles and operated well under bent and stretched states. Technically, it is still a challenge to fabricate the coaxial wire-shaped supercapacitors on a large scale.

2.2.4 Applications

2.2.4.1 Supercapacitor Textiles

The emerging portable and wearable electronics, such as health monitoring devices and smart garments, demand energy supply units with wearable and lightweight properties; therefore a supercapacitor in textiles is an ideal candidate for promising applications in these electronics. Supercapacitor textiles commonly contain three aspects: textile materials, structure, and processing technologies. The textile materials used for textile supercapacitors can be cotton and polyester, which are cheap and possess excellent mechanical properties [33,34]. The main strategies used to construct flexible supercapacitor textiles include weaving the wire-shaped supercapacitors into flexible textile substrate [35] or fabricating the supercapacitor textile directly using active material-coated textile materials [36].

Wire-shaped supercapacitors with small fiber electrodes in diameters ranging from micrometers to millimeters and high flexibility can be deformed (e.g., by bending, twisting, or stretching) in order to be sewed into wearable textile or other architectures. Even though the wire-shaped supercapacitors often demonstrate excellent gravimetric capacitances and energy densities, it is still hard to satisfy the requirement of flexible electronic devices for practical applications. By means of weaving several wire-shaped supercapacitors into a textile, the energy output can be greatly enhanced. For example, a wire-shaped supercapacitor was fabricated and composited by hollow reduced graphene oxide (rGO)/conducting polymer electrodes [37]. The electrode was flexible enough to be tied into knots and even rolled up. As shown in Figure 2.4a, the hollow graphene electrode was twisted into belts, and no obvious cracks appeared. The fiber supercapacitor in PVA/H_3PO_4 gel electrolyte illustrated a high areal capacitance of 304.5 mF cm^{-2} at a current density of 0.08 mA cm^{-2}. It can retain 96% capacitance after 10,000 cycles, and almost no attenuation happens under 500 bending cycles. Furthermore, several fiber supercapacitors were woven into flexible textiles (Figure 2.4b). The energy and power of the textile device could be

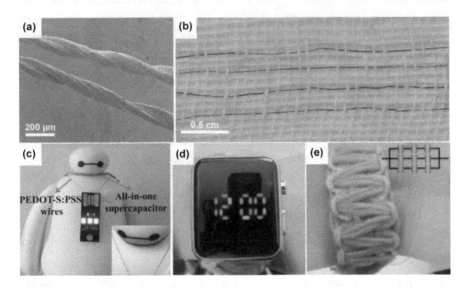

FIGURE 2.4 (a) SEM image of two twisted hollow composite fibers. (b) Photograph of several fiber supercapacitors woven into a cotton cloth. (Reprinted with permission from Ref. [37]. Copyright 2016, Wiley.) (c) Photograph of the all-in-one fiber supercapacitors to light up a light source. (d and e) Photographs of a commercial digital watch powered by the fabric strap woven with some connected textile devices. (Reprinted with permission from Ref. [38]. Copyright 2018, Elsevier.)

increased many times compared to a single device. Another all-in-one fiber-shaped supercapacitor was fabricated using an elastic polymer substrate with outstanding stretchability [38]. The fiber electrode was composed of poly(3,4-ethylenedioxy-thiophene): poly(styrene sulfonate) (PEDOT:PSS) conductive polymer prepared by a wet-spinning method, displaying high tensile strength and electrical conductivity. The assembled flexible supercapacitor showed a wide potential window of up to 1.6 V and areal capacitance of up to 93.1 mF cm^{-2}. Even when it was stretched to 400%, no obvious capacitance degradation was observed. As presented in Figure 2.4c, the supercapacitors connected in series can light up a lamp with a minimum voltage of 2.5 V. Furthermore, three supercapacitors connected in parallel were easily woven into a textile fabric to power a commercial watch with a minimum voltage of 3.0 V (Figure 2.4d and e).

In addition, several attempts have been explored to fabricate the textile super-capacitor directly from fabric electrodes, which seems to be more convenient for scaling up production. A facile method has been reported to prepare textile super-capacitors via dip-coating process (Figure 2.5a). A conductive electrode was firstly obtained by dipping an insulating textile into the SWNT ink [39]. From the SEM image (Figure 2.5b), the textile demonstrated 3D porous structure, which was benefi-cial for the SWNTs to be conformally coated onto the substrate. The SWNTs-coated textile showed a superior conductivity of 125 S cm^{-1}. Then, a flexible and stretch-able supercapacitor can also be fabricated by putting a fabric separator in between two conductive and stretchable fabrics (Figure 2.5c), showing a stable capacitance retention under 120% strain for 100 cycles. The areal capacitance of the textile

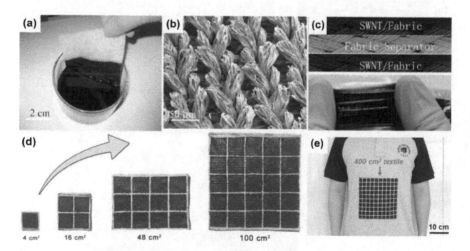

FIGURE 2.5   (a) Conductive textile prepared by dipping textile into SWCNT ink. (b) SEM image of the conductive textile covered with SWCNT evenly. (c) Schematic illustration of the structure of the supercapacitor textile (top) and photograph of the supercapacitor under stretching to 120% (bottom). (Reprinted with permission from Ref. [39]. Copyright 2010, American Chemical Society). (d) Photographs of supercapacitor textiles in an increased area. (e) Photograph of a supercapacitor textile woven into cloth. (Reprinted with permission from Ref. [40]. Copyright 2016, Wiley.)

supercapacitor was up to 0.48 F cm^{-2}, and it could be further improved by loading additional pseudocapacitive materials. Another textile supercapacitor based on the PANI/rGO/polyester textile electrode in a hierarchical structure was also created by dipping of the polyester textile into GO dispersion and subsequent reduction to rGO in HI, followed by the incorporation of PANI via in situ polymerization [40]. The supercapacitor textile can be easily assembled, from 4 to 100 cm^2 (Figure 2.5d), indicating a potential for large-scale production, and reaching superior performances (capacitance: 69.3 F, power: 80.7 mW and energy: 5.4 mWh). Furthermore, the intrinsic textile characters of the supercapacitors offer the possibility of combining them with daily clothes. As shown in Figure 2.5e, an area of 400 cm^2 supercapacitor textile can be woven into the clothes, presenting a potential for practical application in wearable electronic devices.

2.2.4.2 Flexible Microsupercapacitors

A microsupercapacitor (MSC) is another promising energy storage device for the emerging miniaturized electronics because of the following advantages: (1) MSCs can be integrated in series or parallel onto various flexible or stretchable chips to increase the potential output. (2) The miniaturized device shortens ion diffusion path and enables accessibility of the electrolyte to the electrodes, thereby making more efficient use of the electrochemical surface area. (3) It also demonstrates the feasibility of reducing the complexity of supercapacitor devices by eliminating complex interconnections to bulky energy devices.

Recently, planar MSCs have gained widespread interest, which can provide power densities that are orders of magnitude larger than conventional batteries and supercapacitors. Generally, the methods for making planar MSCs include spraying [41] and electrophoretic deposition [42]. Researchers have applied carbon materials (e.g., CNTs, graphene, and mesoporous carbon) to construct planar MSCs. Planar MSCs based on stretchable single-walled carbon nanotube (SWCNT) electrodes have a high capacitance of 100 μF and good cycle stability even under bending [43]. Especially, graphene has been widely used to make microelectrodes due to its excellent electron mobility (15,000 cm^{2}V^{-1}s^{-1}), high surface area (~2,630 m^{2}g^{-1}), and high theoretical capacitance (~550 F g^{-1}) [44]. Chen et al. reported a flexible and ultrathin MSC from rGO interdigital electrodes (Figure 2.6), which were prepared via the combination of photolithography and selective electrophoresis [45]. The shorter ion diffusion path in MSC results in a higher power density compared with the traditional supercapacitor. In this design, the micro patterned rGO electrodes with interdigitated structure not only shortens the ion diffusion path to result in high power density, but also greatly accommodates the stress in flexible devices during deformation.

Ogale et al. constructed graphene-based MSC using metal organic framework (MOF) as a precursor via laser writing approach to produce few-layer graphene electrodes with metal anchoring, heteroatom doping, and porous structure [46]. The MSC device exhibited quite stable capacitance retention (~100%) even after 200,000 cycles. Kaner et al. also produced planar MSCs by laser reduction and patterning of the GO film [47]. Firstly, GO solution was casted on polyethylene terephthalate (PET) sheet glued on DVD disc and dried overnight. After inserting GO-coated DVD disc into DVD driver upon laser patterning, the GO film was converted into

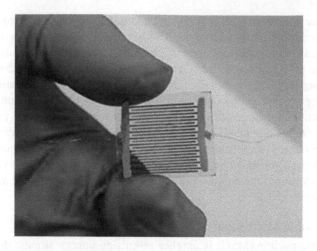

FIGURE 2.6 Graphene-based MSCs on PET. (Reprinted with permission from Ref. [45]. Copyright 2013, Wiley.)

graphene with an interdigital structure guided by computer software. Planar MSCs were produced finally after coating the ionic gel electrolyte. By tuning the number of the interdigital microelectrodes per unit area, the energy and power density of the microdevice can be controlled. The prepared planar MSC generated ultra-high power of 200 W cm^{-3} and excellent frequency response with resistor-capacitor (RC) time constant of only 19 ms.

However, graphene-based electrode materials showed a limited electrochemical performance, caused by the easy aggregation of graphene sheets during processing, and a reduced accessible surface area of the electrodes. An effective strategy to avoid this problem is to add spacers between graphene sheets to prevent them from restacking. Using CNTs as spacers to form nano-channels between graphene sheets can greatly reduce the accumulation of graphene sheets [48]. Nanochannels can expand the surface area of the electrodes and provide a pathway for efficient parallel ion transportation to the graphene plane, thus improving the electrochemical performance of MSC. Besides, electrode materials should be carefully designed with suitable nanostructures, which may affect the cyclic stability and power density of materials. The MSC is still in its initial stage with unsatisfactory energy density and far away from manufacturing maturity to cater to the needs of on-chip electronics.

2.3 FLEXIBLE BATTERIES

Batteries play a vital role in modern life, for instance, as power systems of electronic devices ranging from mobile phones, computers, digital cameras to electric vehicles and even large-scale energy storage networks. Traditional rigid batteries (e.g., lead-acid batteries) are easily damaged under conditions of bending, twisting, or stretching. The nonflexibility of traditional batteries is closely correlated to rigid active materials, the easy leakage of liquid electrolyte, and poor interfaces between battery materials. Besides, an irreversible change in internal structure of batteries upon deformation will occur, leading to performance decay or even battery failure. Therefore, to satisfy the requirements of portable and wearable electronics, batteries with flexibility are needed.

2.3.1 MECHANISM AND DEVELOPMENT OF BATTERIES

With the continuous development of flexible and wearable electronics, considerable efforts have been made to develop flexible batteries especially LIBs to meet practical needs. Typically, flexible batteries contain two electrodes (cathode and anode), separator, electrolyte, and flexible packaging. LIBs work mainly by moving Li^+ between the cathode and anode. During charging, Li^+ is deintercalated from the cathode through the electrolyte to be inserted into the anode to form a lithium-rich state. During discharging, a reverse reaction occurs. Through reversible Li^+ intercalation and deintercalation processes, flexible LIBs realize energy storage and conversion.

2.3.2 ELECTRODES

In a conventional LIB, the electrode material is usually a lithium-containing transition metal oxide, which is uniformly ground with a conductive agent and a binder and coated on the metal current collector. The main disadvantages of the electrode prepared by this process are as follows: (1) the limited energy density of the battery due to the heavy metal foil, (2) the poor contact between the metal foil and the electrode material, and (3) non-flexibility of materials and structure. Hence, in flexible batteries, the design of the electrode material needs to be self-sustaining without using the current collector to accommodate deformations such as bending and folding. On this basis, two main solutions to construct flexible electrodes include the use of electrochemically active substances with intrinsic flexibility and deposition of active materials on flexible substrates.

2.3.2.1 Carbon Nanotube

It has been demonstrated that CNTs are very promising materials to improve the performance of flexible electrodes. Compared with metallic current collectors, CNTs possess distinct properties of lightweight, flexibility, and large surface area as mentioned before. They can form a network structure with high mechanical properties and sufficient spaces for ion and electron transport. Many researches have attempted to prepare CNT composite electrodes via either mechanical mixing [49] or chemical bonding with active materials. The introduction of CNTs not only facilitates the electron transport, but also provides enough mechanical connections and electric contacts between active materials, thus realizing binder-free electrodes. Besides, the CNT network is quite flexible that can buffer volume changes of the electrodes during charge and discharge. Binder-free CNT composite electrodes can be readily obtained by vacuum filtration of the CNT-active material dispersion [50,51] or spraying of the active materials on the aligned CNT sheets [52]. For instance, a self-supporting single-walled CNT/(34 wt%)SnO_2 paper was prepared through vacuum filtration of hybrid materials of single-walled CNT and SnO_2 in a two-step process, exhibiting a high specific capacity of 454 mAh g^{-1} at a current density of 25 mA g^{-1}

after 100 cycles, much higher than the corresponding pure CNT paper [50]. To simplify the fabrication process, Luo et al. [51] developed a CNT–LiCoO$_2$ (LCO) electrode comprising 95 wt% LCO active material and only 5 wt% super-aligned CNT (SACNT) as conductive additive and structural matrix via the co-deposition method (Figure 2.7a). Compared to the classical LCO-Super P cathode with heavy agglomeration of Super P powders (Figure 2.7b), LCO particles were uniformly distributed in the CNT network (Figure 2.7c) to speed up the ion and electron transport. As expected, this LCO/CNT used as a cathode in LIBs delivered a specific capacity of 151.4 mAh g^{-1} at 0.1°C, with a capacitance retention of 98.4% over 50 cycles (Figure 2.7d). The outstanding electrochemical performance can be explained by the rational design of the electrodes, in which a 3D porous conductive CNT network provides the composite electrodes with high mechanical strength and flexibility and sufficient free space to adapt to volume changes of the electrode, promotes Li-ion transport and electrolyte infiltration, and thus guarantees better battery capability.

2.3.2.2 Graphene

Graphene is a 2D material with good electrical and thermal conductivity, super chemical stability, high temperature resistance, and a large theoretical surface area, making it ideal for flexible supercapacitors and batteries. By surface modification [53], doping [54], and changing the morphology of graphene [55], the restacking of graphene sheets can be suppressed. Typically, 2D graphene materials are usually prepared by vacuum infiltration of rGO suspension as a raw material with the addition

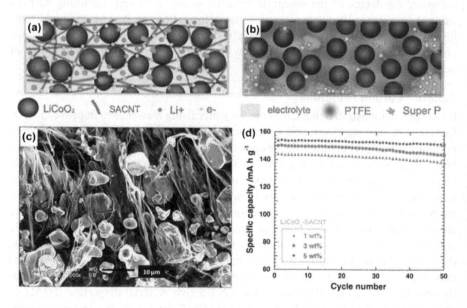

FIGURE 2.7 The binder-free SACNT-LCO composite cathode: (a) schematic of the structure, (b) the classical LCO cathode with Super P conductor and PTFE binder, (c) cross-sectional SEM image, and (d) cycling test at 0.1°C. (Reprinted with permission from Ref. [51]. Copyright 2012, Wiley.)

of surfactants to disperse rGO [56]. When applied for battery test, the residual functional groups from the added surfactants result in irreversible cathodic reaction at 1.47 V, reducing battery capacity and efficiency. Due to the aforementioned merits of graphene, it can be easily processed into films. To improve the graphene-based electrode, graphene can also be composited with an electrochemically active phase/other carbon material/polymer to form a flexible electrode through various methods such as coating, solvothermal method, deposition, and mechanical mixing [57]. For example, Shi et al. [58] prepared flexible and free-standing graphene-based electrodes via a two-step vacuum filtration method (Figure 2.8). A graphene membrane was firstly fabricated by filtration of aqueous graphene dispersion in organic solvent. Then, an anode or cathode slurry comprising LTO or LiFePO$_4$ (LFP), carbon black Super P, and polyvinylidene fluoride (PVDF) (mass ratio of 8:1:1) in N-methyl-2-pyrrolidinone (NMP) solvent was deposited on the graphene membrane by subsequent filtration. Notably, the PVDF binder was removed by NMP during continuous filtration. Finally, the free-standing G-electrode of G-LTO or G-LFP was obtained after drying and peeling off from the filter membrane. The full battery was assembled with these two electrodes, and it showed high energy density and rate performance and long-term stability.

However, this system still needs the addition of a conductive agent and a binder. To further improve the overall specific capacity of battery, graphene-based batteries without conductive agents were developed with processes including the filtration of nanosized SnO$_2$ particles–GO dispersion and subsequent reduction at elevated temperature to form flexible, free-standing rGO/SnO$_2$ electrode membrane [59] and the filtration of hollow Fe$_3$O$_4$ nanorods via hydrothermal synthesis of GO and subsequent reduction to obtain a rGO/hollow Fe$_3$O$_4$ composite electrode membrane [60]. Through the physical mixing of active materials and graphene plus subsequent filtration, free-standing, flexible electrodes can be readily obtained. This method is simple and easy to implement. However, physical combination of the active material with graphene cannot ensure the structural integrity during repeated deformations, which will lead to the degradation of battery performance.

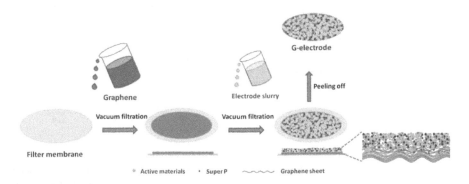

FIGURE 2.8 Schematic of the preparation of a graphene-based flexible electrode. (Reprinted with permission from Ref. [58]. Copyright 2015, IOP Publishing.)

In addition, graphene can also be made into a 3D structure to provide a channel for fast transport of electrons and ions. Cheng et al. [61] used a 3D porous conductive graphene foam (GF) to develop flexible LFP/GF and LTO/GF electrodes (Figure 2.9). Firstly, a layer of 3D graphene was grown on the surface of a nickel foam by chemical vapor deposition method, obtaining a highly porous GF after the removal of the nickel foam treated by acidic solution. Then, the hybrid electrodes were prepared by deposition of LFP or LTO on GF via in situ hydrothermal process. Due to the advantage of GF, the prepared hybrid materials were flexible and retained the 3D structure of GF well under bending. A full battery was assembled with a high energy density of ~110 Wh kg^{-1} on the basis of the total mass of the electrodes, ascribed to the use of lightweight GF to replace the conventional metallic current collector, conducting agent, and polymeric binder. Moreover, the flexible battery provided good cycle stability under repeated deformation, maintaining a capacitance retention of ~95% after 15 cycles under bending with a radius of 5 mm.

Another strategy to prepare porous 3D graphene is through freeze-drying of GO dispersion and subsequent thermal treatment. Botas et al. [55] reported flexible SnO$_2$–GFs as binder-free anodes for LIBs. The aerogels were obtained by freeze-drying a suspension of tin precursor and GO, followed by calcination at high temperature under an argon atmosphere to form SnO$_2$–rGO composite foam. When applied as anode for LIBs, it displayed a high specific capacity of 1,010 mAh g^{-1} at a charge–discharge current density of 0.05 A g^{-1}.

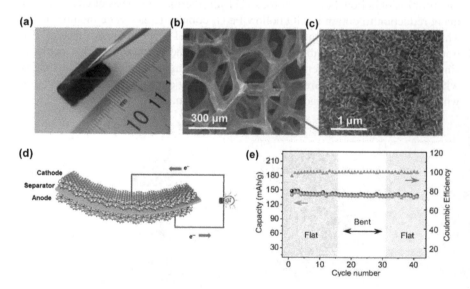

FIGURE 2.9   (a) Photograph of a flexible LTO/GF upon bending. (b and c) SEM images of the LTO/GF. (d) Schematic structure of a flexible battery based on a cathode and an anode made from 3D GF. (e) Long-term performance of the LTO/GF//LFP/GF full battery under flat and bent states. (Reprinted with permission from Ref. [61]. Copyright 2012, National Academy of Sciences.)

2.3.2.3　Carbon Paper/Carbon Cloth

Carbon paper or carbon cloth is made of intertwined carbon fibers, possessing high electrical conductivity, mechanical strength, and flexibility. Flexible binder-free electrodes can be easily obtained using carbon paper as the matrix to load a variety of electrochemically active materials. Liu et al. [62] designed the electrodes of 3D needle-like $ZnCo_2O_4$ grown on the surface of carbon cloth under hydrothermal condition for high-performance binder-free lithium battery anodes. It had a reversible capacity of 1,300–1,400 mAh g^{-1} and excellent cycle ability. Full batteries were also manufactured, exhibiting high flexibility, excellent electrical stability, and superior electrochemical performance.

Cellulose paper/fabric materials can also be used as the structural framework for flexible batteries because of their distinctive properties of being foldable, recyclable, environmentally friendly and easy to handle, lightweight, and low in cost and having abundant surface functional groups and large specific surface area. Prior to using papers/fabrics as matrices, they need to be made conductive. Popular strategies for this are as follows: (1) deposition of conductive materials such as metals on the paper/fabric surface by chemical vapor deposition [63]. However, in this method, it is easy to get the nanopores blocked; (2) compounding a conductive material such as carbon on the paper/fabric surface by coating and printing [64,65] or absorbing the conductive phase into a non-conductive substrate by impregnation method, such as immersion of papers/fabrics into CNT or graphene inks [66]; and (3) imitating the papermaking process to directly obtain the conductive composite material in the paper preparation stage inspired by the method of filtering cellulose–graphene/CNT slurry, which can combine the mechanical properties of cellulose and the electrical conductivity of graphene/CNT. In addition to being a substrate, the porous nature of paper and fabric allows them to be applied as separators having lower impedance than commercial separators. Cui et al. [67] used CNT/LCO and LTO/CNT paper-like material electrodes and obtained ultra-thin flat LIBs with low cost, low impedance, and high flexibility, as shown in Figure 2.10. Furthermore, the idea of integrating all the components required for a battery/capacitor on a piece of paper by a printing process has been proposed [39] to meet the expectation of an energy device with high energy density and low cost.

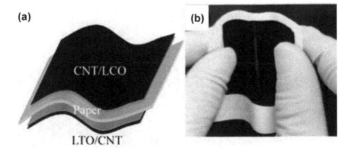

FIGURE 2.10 Schematic of a Li-ion paper battery with the paper substrate laminated between CNT/LiCoO$_2$ (LCO) and Li$_4$Ti$_5$O$_{12}$ (LTO)/CNT layers. (Reprinted with permission from Ref. [67]. Copyright 2010, American Chemical Society.)

2.3.3 CONFIGURATION

2.3.3.1 Flexible Planar Batteries

Constructing LIBs in 2D planar structure is a facile way to achieve flexibility. In a typical planar battery, liquid or gel electrolyte is sandwiched between the two battery poles of electrode materials coated or self-grown on flexible current collectors, forming a flat battery device with a sandwich or laminated structure. 2D graphene sheets have been demonstrated as effective Li^+ intercalation materials with high capacity and rate performance. Wei et al. [68] prepared ultra-thin flexible solid-state batteries in planar structure as depicted in Figure 2.11. A monolayer of graphene grown on a metallic current collector (Cu foil) by chemical vapor deposition can be directly paired with lithium foil anode to make a mechanically flexible all-solid-state LIB with a total thickness of about 50 um. Such ultra-thin batteries produce energy and power densities of up to 10 Wh L^{-1} and 300 W L^{-1} and superb cycling stability at 100 $\mu A\ cm^{-2}$ for more than 100 cycles. This design offers a strategy to fabricate a binder-free and flexible battery via direct growth of a graphene electrode on a current collector substrate. Despite these progresses, flexible batteries configured in a traditional planar structure still cannot satisfy the ongoing needs of being lightweight and weaveable in novel electronics.

2.3.3.2 Flexible Wire-Shaped Batteries

A flexible battery in wire format is often known as power fiber, fiber battery, or cable-type battery and has a very large aspect ratio and a very small diameter that can serve the needs of electronic devices of different shapes. Wire-shaped battery can be straight, curved, and wrapped around and can even be knotted when needed. A single wire-shaped battery is generally fabricated by coating a thin and uniform active material layer on the surface of a single fiber or fiber bundle, with the diameter of a single fiber electrode not exceeding 100 µm [7,69]. In recent years, wire-shaped flexible LIBs have achieved preliminary exploratory results.

Peng et al. developed a wire-shaped battery in parallel comprising Si/CNTs hybrid fiber and Li wire [7]. The Si/CNTs fiber was prepared by atomic layer deposition of Si on aligned CNT sheets and subsequent twisting of the composite sheets to

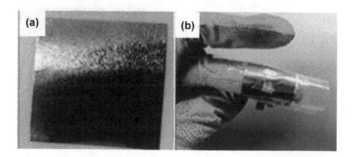

FIGURE 2.11 Photographs of (a) monolayer graphene grown on Cu and (b) flexible graphene battery under the deformation. (Reprinted with permission from Ref. [68]. Copyright 2013, Royal Society of Chemistry.)

form the hybrid fiber. The assembled flexible LIB can not only have advantages of high specific capacity of Si and excellent conductivity of CNTs, but also efficiently alleviate the volumetric expansion of Si during the charge and discharge processes. A wire battery with such a design showed satisfactory flexibility and no significant capacitance loss after hundred cycles of bending. Furthermore, a full battery with parallelly coupled LMO/CNT hybrid fiber (cathode) and LTO/CNT hybrid fiber (anode) [8] (Figure 2.12a) had a capacity of 78.8 mAh g⁻¹ (length: 1 cm) at a current of 0.05 mA and maintained a capacity retention of 85% after 100 cycles. Again, this wire-shaped battery was flexible enough to endure various deformations such as bending and able to power LED lights with a length of 10 cm.

When fiber battery is assembled with two twisted electrodes, it can achieve high mechanical performance. Ren et al. [70] introduced a twisted battery on the basis of the flower-like MnO_2 nanoparticles homogeneously grown on the CNT fiber (CNT/MnO_2 hybrid fiber) and Li wire (Figure 2.12b) having a capacity of 109.6 mAh cm⁻³ at 5×10^{-4} mA. To further improve the mechanical property of electrodes, a full battery made by twisting LMO/CNT and LTO/CNT possessed excellent mechanical strength and outstanding electrochemical performance at a stretched state and achieved a capacity of 91.3 mAh g⁻¹ and a capacity retention of up to 88% after stretching by 600%. The stretching strain of the twisted batteries can be tuned by adjusting the helix angle between the fiber substrate and the wrapped twisted electrode.

Batteries in cable shape is another unique architecture with high flexibility. Kim et al. designed a mechanically flexible cable-type battery (Figure 2.12c) [71]. The anode in a hollow-spiral structure was prepared by winding the twisted bundle of Ni–Sn-coated Cu wires around a circular rod. A modified PET nonwoven support with high thermal stability was used as the separator. Then, Al wire was wound on the hollow-spiral anode, followed by the coating of a LCO cathode slurry. Finally, a cable-type battery was produced by injecting the liquid $LiPF_6$-based organic electrolyte into the hollow space. The cross-sectional optical images showed the cable battery was well-constructed with a hollow-helix anode, modified PET layer, Al wire, tubular cathode, and shrunken packaging tube, with a total outer diameter of up to several millimeters. Compared with non-hollow anode, the cable battery with

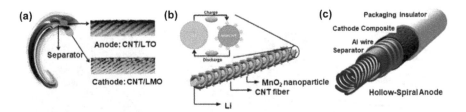

FIGURE 2.12 (a) A fiber-shaped LIB with aligned CNT/LTO and CNT/LMO hybrid fibers in a parallel pattern. (Reprinted with permission from Ref. [8]. Copyright 2014, Wiley.) (b) A fiber-shaped LIB with CNT/MnO_2 hybrid fiber and Li wire in a twisted pattern. (Reprinted with permission from Ref. [70]. Copyright 2013, Wiley.) (c) A cable battery with a hollow spiral anode. (Reprinted with permission from Ref. [71]. Copyright 2012, Wiley.)

the hollow anode exhibited a higher capacity and more stable capacity retention, originating from the improved permeability/accessibility of the electrolyte to active materials. Besides, the cable battery can still work stably under large strain such as bending or twisting deformation, demonstrating excellent mechanical flexibility.

The three types of configurations discussed above have their merits and shortcomings. Wire-shaped batteries have high flexibility and can be deformed into a bent or twisted form. In addition, they can easily be woven into textiles for applications in wearable electronics. Twisted batteries, in addition to bending and twisting, can also withstand a certain degree of stretching without damaging the overall structure. In comparison, cable-shaped batteries are usually made in a helically coaxial structure with hollow space, which allows all active materials to be well penetrated by the electrolyte. Besides, the capacity match between the anode and the cathode can be readily adjusted by tuning the strand number of anode and the thickness of cathode. Although they can demonstrate stable battery behavior under repeated deformation due to bending, they are too bulky to be woven and inconsistent with the development of miniaturized and wearable electronic devices.

2.3.4 APPLICATION

2.3.4.1 Textile Batteries

To increase the volumetric energy density, the above-mentioned flexible wire-shaped batteries could be further assembled into a high output device in novel 2D and 3D textile structures. Wire-shaped batteries can be woven into textiles to form energy storage textiles and can be easily sewn into commercial clothes for wearable electronic devices. For example, Wang et al. [72] designed a highly flexible lithium-air battery as shown in Figure 2.13, which was successfully prepared by coating low-density polyethylene film (LDPE) on the CNTs' surfaces to block water vapor and combining with LiI solid electrolyte. In this design, the introduction of LDPE effectively inhibits the reaction of the discharge product Li_2O_2 with water molecules and CO_2 in air to form Li_2CO_3. Besides, the introduction of LiI is conducive to the decomposition of Li_2O_2, which greatly improves the cycle life of Li-air batteries. Through further assembly into a linear flexible lithium-air battery, it was found that the existence of LDPE had excellent waterproof properties, allowing the battery

FIGURE 2.13 (a) The preparation of flexible fiber-shaped Li–air battery. (b and c) Digital images of flexible fiber-shaped Li–air batteries being woven into commercial clothes to charge a smartphone. (Reprinted with permission from Ref. [72]. Copyright 2018, Wiley.)

to stably work in ambient air up to 610 cycles. The performance did not degrade even after conducting the bending test 1,000 times. Due to the high flexibility of these batteries, they can be woven into cloth and used as a power source to charge a smartphone.

Apart from the weaving of wire-shaped batteries into textiles, a fabric battery can be directly fabricated from fabric electrodes, e.g., from insulating polymer fabrics coated with conductive material. Commonly, a fabric is composed of polymer fibers with flexible porous structure. The porous structure and surface chemistry enable polymer fibers in fabric to tightly absorb a great amount of active materials in solution and provide a fast ion transport pathway. As a typical example of flexible fabric LIBs, a 3D porous fabric was coated with CNT ink to transform it into a conductive fabric so that it acts as a current collector [73]. The porous structure of the fabric allowed the active materials (LFP and LTO) to get loaded into the conductive fabric by a simple dipping and drying process. For a porous 3D fabric, the mass loading of active materials can reach up to 168 mg cm^{-2}, which shows a far higher weight percentage of active materials in the whole battery than that obtained for a traditional metallic current collector (Figure 2.14a). Figure 2.14b shows clearly the detailed interface among the fabric electrode (LFP, polyester fibers with conductive additive Super-P and coated CNTs), where Super-P with a percolative network connect battery particles to allow for charge transport, and the inset is a prepared fabric filled with LFP. The resulting fabrics coated with LTO and LFP as anode and cathode, respectively, are highly flexible and can be assembled into a full battery. Owing to the combined advantages of porous structure and high mass loading, the full battery exhibited a high capacity of 160 mAh g^{-1} and 88.5% retention of capacity after 30 cycles (Figure 2.14c).

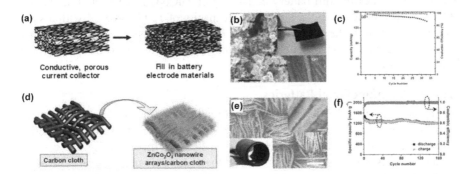

FIGURE 2.14 Fabric batteries from fabric electrodes. (a) Design and fabrication of 3D porous current collectors filled with battery electrode materials. (b) A SEM image and photograph of the fabric filled with LFO. (c) Long-term cycling of the discharge capacity and coulombic efficiency for full batteries. (Reprinted with permission from Ref. [73]. Copyright 2011, Wiley.) (d) Schematic illustration of the synthesis of flexible 3D ZnCo$_2$O$_4$ nanowire arrays/carbon cloth. (e) SEM images of the ZnCo$_2$O$_4$ nanowire arrays growing on carbon cloth at different magnifications, scale bars: 200, 20 µm (left insert). Right inset in (e) is a photograph of the resulting carbon-cloth-based materials with good flexibility. (f) Long-term performance of the ZnCo$_2$O$_4$ nanowire arrays/carbon cloth electrode. (Reprinted with permission from Ref. [62]. Copyright 2012, American Chemical Society.)

To conclude, the main strategies for building textile batteries can be divided into bottom-up and top-down approaches. The bottom-up approach is used for the native fibers to transform them into conductive fiber electrodes. These fiber electrodes can be paired to fabricate a fiber-shaped full battery, which can be woven into textiles to form wearable electronics. This approach demands that the electrical conductivity and tensile strength of fibers be high enough to bear further knitting and weaving processes. Unlike the bottom-up approach, the top-down approach starts from a fabric product transformed into the battery textiles. Commonly, a non-conductive fabric is coated with a conductive layer, followed by the incorporation of active materials into the conductive fabric to improve the battery capacity. An obvious advantage of this approach is to easily achieve an energy storage textile with a large area and high mass loading.

Besides of the fabric, carbon cloth is also widely applied for constructing fabric batteries. Carbon cloth is a conducting fabric with excellent mechanical flexibility. In addition, a large number of pores on the cloth fibers greatly facilitate the diffusion of electrolyte into the electrode material, demonstrating its potential as a conductive platform for loading active materials for high-performance energy storage devices. For instance, a flexible carbon-cloth-based anode material was synthesized by growing the hierarchical 3D $ZnCo_2O_4$ nanowire arrays on carbon cloth through a facile hydrothermal route (Figure 2.14d and e) [62]. It produced a capacity of about 1,200 mAh^{-1} and a stable capacity retention of 99% after 160 cycles at 200 mA g^{-1} (Figure 2.14f). The tight adhesion of $ZnCo_2O_4$ nanowire arrays on carbon cloth with fast charge transfer, the loose textures and open spaces between nanowire arrays with facile diffusion of electrolyte, and the shorter Li$^+$ ion diffusion paths in the nanowires greatly favor and improve the electrochemical performance. Furthermore, a flexible full battery consisting of $ZnCo_2O_4$/carbon cloth anode and LCO/Al foil cathode exhibited a discharge capacity of 1,314 mAh g^{-1} with a stable capacity retention in the following cycles. Besides, it shows excellent mechanical flexibility and stable electrical stability and can still light up LED and mobile phone screens under bending.

2.3.4.2 Multifunctional Batteries

During practical use, flexile batteries may break under various deformations (e.g., bending, twisting, cutting), leading to the battery failure or even serious safety issues. Therefore, flexible batteries with self-healing functionality are urgently needed to withstand these deformations. Peng et al. developed a novel type of all-solid-state and flexible aqueous LIBs with self-healing property [74]. The battery was designed by coupling aligned CNTs sheets loaded with LMO and LiTi$_2$(PO$_4$)$_3$ (LTP) nanoparticles on a self-healing polymer substrate as the cathode and anode and aqueous lithium sulfate/sodium carboxymethylcellulose (Li$_2$SO$_4$/CMC) as gel electrolyte/separator. The battery can regain its electrochemical performance (specific capacity, rate capability, and cycle performance) after several cutting–contacting process. This superb self-healing performance can be attributed to the reconstruction of hydrogen bonds at fractures in polymer substrate as well as the gel electrolyte to realize high flexibility and good recovery of electrochemical properties.

To observe the reconnection more easily, Gao et al. [75] designed a flexible and self-healing all-fiber quasi-solid-state LIBs in diameters of hundred micrometers, as displayed in Figure 2.15a. The novel battery was assembled by the anode composed of macroporous rGO fibers containing SnO_2 quantum dots, the cathode made up of spring-like rGO fibers containing LCO nanoparticles, gel electrolyte containing poly (vinylidenefluoride-*co*-hexafluoropropylene, PVDF-*co*-HFP) and $LiClO_4$ in solvents of diethyl carbonate (DEC)/ethylene carbonate (EC) and a self-healing package layer of carboxylated polyurethane (PU). The length of the spring-shaped cathode can be readily adjusted to easily match the capacity of the anode. Besides, the anode and cathode fibers were designed with diameters of 750 and 250 μm, thick enough to reconnect the broken electrodes. Similarly, the self-healing mechanism in this design can be explained by rich hydrogen bonds existing in the supramolecular network which can reconnect at the broken surface. The as-obtained LIB displayed a capacity of 82.6 mAh g^{-1} with a capacity retention of 82.2% over 50 cycles under deformations of bending and twisting (Figure 2.15b) and maintained a capacity of 50.1 mAh g^{-1} with a capacity retention of 50.3% after five healing cycles at a current density of 0.1 A g^{-1} (Figure 2.15c and d). These above results demonstrate outstanding flexibility in complex deformations and self-healing ability, opening new opportunity to design flexible power sources for the next-generation portable and wearable devices.

At present, most of the reported flexible batteries use strong acid/base or flammable organic solvents as electrolytes. When applied to wearable or implantable electronic devices, electrolyte leakage will cause great security risks. On this basis, Peng and Wang et al. [76] proposed high-safety flexible aqueous sodium-ion batteries in belt and fiber shapes using $Na_{0.44}MnO_2$ as cathode, carbon-coated nano $NaTi_2(PO_4)_3$ (denoted as $NaTi_2(PO_4)_3@C$) as anode, and aqueous Na^+ solution as the electrolyte. Both types of sodium-ion batteries exhibit high energy and power density, remarkable long-term stability, as well as excellent flexibility and are expected to be used in wearable electronics. When using a biocompatible solution containing Na^+, such as normal saline, cell-culture fluid, as the electrolyte, these batteries can still work

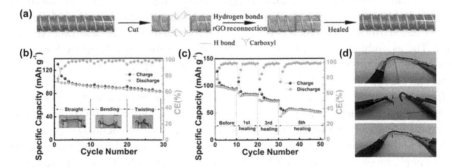

FIGURE 2.15 Fiber-shaped LIB with flexibility and self-healing property. (a) Schematic diagrams for the self-healable mechanism of the fiber-shaped LIB with the PU packaging layer. (b) Long-term performance under various deformations (straight, bending, and twisting). (c) Long-term performance with different healing times. (d) The detailed self-healing procedure upon cutting. (Reprinted with permission from Ref. [75]. Copyright 2018, Elsevier.)

well, demonstrating their potential in implantable electronic devices. Additionally, the as-obtained fiber electrode exhibited an oxygen-eliminating function in an aqueous Na$^+$ solution, which can consume dissolved oxygen by electrocatalytic oxygen reduction and adjust the local pH of electrolyte, implying that the flexible battery with the electrochemical deoxygenation function is promising for applications in biological or medical field.

2.4 INTEGRATED ENERGY DEVICE

Recently, flexible energy supply devices integrated with other functional components in portable/wearable electronics have received great interest due to the realization of their multifunctional effect within small volume for manufacturing self-powered systems without external power supply. Different devices can be connected through the same or compatible current collectors, electrode materials, etc. or can be grouped into similar components to achieve the integration of multiple functions, thereby achieving miniaturization and multifunctionality of the entire system. According to the application, the flexible multifunctional system can be a self-powered system containing an energy conversion device plus an energy storage device, or it can be a passive device containing an energy storage device plus a sensor, controller, or detector [77].

Flexible energy storage devices can be integrated into a self-powered system, allowing the combination of energy conversion and storage to be converted into electrical energy through the use of external mechanical movement, sunlight, thermal energy, etc., and then this energy can be stored in these devices. Sun et al. [78] reported an integrated self-charging power unit having the combination of a hybrid Si nanowire/polymer heterojunction solar cell and a PPy-based supercapacitor, with Ti film acting as conjunct electrode in between. The integrated system is efficient in solar energy conversion and energy storage, achieving a total photoelectric conversion to storage efficiency of 10.5%. This system can not only buffer the unstable sunlight and reduce the fluctuation of light intensity to solar cells, but also pave the way to efficient self-charging units. Furthermore, Wang et al. [79] developed an ultralight, rhombic-shaped, cutting-paper-based self-charging power unit that can simultaneously harvest and store mechanical energy from body movements by coupling paper-based triboelectric nanogenerators (TENGs) and a graphite-based supercapacitor. The assembled nanogenerator using paper as a substrate efficiently lowers the weight of the entire power unit, achieves a high charge output (82 nC g^{-1}/75 nC cm^{-3}), and can charge the supercapacitor having a capacity of $\sim$1 m F to 1 V in minutes. This self-charging power unit has subsequently proven to be a sustainable power source to drive wearable and portable electronic devices. Peng et al. [80] proposed a solid-state, coaxial, and self-powered fiber which could simultaneously capture solar energy and store the electric energy. This "energy fiber" mainly comprises the two parts of photovoltaic conversion (PC) and energy storage (ES). It was realized using vertically aligned TiO$_2$ nanotube-modified Ti wire and aligned multi-walled CNT sheet as two electrodes (Figure 2.16a). In this rational design, the coaxial structure greatly contributes to the stable performance of flexible devices. The device structure can be well preserved under the deformation, with <10% of the entire photoelectric conversion and storage efficiency after 1,000 bending cycles. This performance

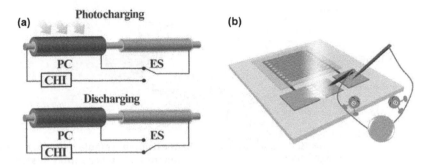

FIGURE 2.16 (a) A self-powered fiber can simultaneously capture solar energy and store the electrical energy. (Reprinted with permission from Ref. [80]. Copyright 2014, Wiley.) (b) The schematic illustration of the self-powered photodetector driven by the MSC. (Reprinted with permission from Ref. [81]. Copyright 2015, Elsevier.)

was much superior over the twisted wire-shaped device which failed to work after bending for only five cycles even with sealing and the flexible membrane device with a photoelectric conversion and storage efficiency of 0.79% decreasing to 0.05% after bending for 100 cycles. Additionally, due to the development of textile technology, these "energy fibers" can be easily scaled up into flexible textiles for practical usage by intertwining the fibers with each other.

To further achieve flexible devices with high integrity and miniaturization, Shen and Xu [81] designed a flexible multifunctional microsystem on a chip with integrated MSCs and photodetectors (Figure 2.16b). The MSC with planar interdigital structure of rGO electrodes, which were prepared via UV lithography technology combined with oxygen plasma treatment, can shorten the ion diffusion path and accommodate the mechanical stress to obtain high performance. When this MSC (areal capacitance: 896.77 $\mu F\ cm^{-2}$) was applied as a micropower source for CdS nanowire-based photodetector, the integrated device revealed a stable response to light illumination with a current on/off ratio of 34.50, a comparable result when using the conventional external energy storage unit. A higher current on/off ratio of 79.81 can be obtained for a tandem microdevice. These results demonstrate the feasibility of integrating MSCs and photodetector systems on a chip.

Nowadays, the integrated system has emerged to stably perform complex tasks including not only power supply, but also its monitoring. Apart from the light detectors, the multifunctional system has also been designed with sensing properties. Ha et al. [82] reported an integrated system of multifunctional sensors and a supercapacitor, constructing from polydimethylsiloxane-coated microporous PPy/GF composites (PDMS/PPy/GF) integrated on a skin-attachable flexible substrate to detect the motions of hand (Figure 2.17). The dual-mode sensor based on PDMS/PPy/GF can measure pressure and temperature through the variations of current and voltage without interfering with each other, showing high sensitivity, fast response/recovery, and high durability (up to 10,000 pressure-loading cycles). In addition, strain sensors were made using the same PDMS/PPy/GF that can detect strains up to 50%. Flexible supercapacitors were prepared using PPy/GF electrodes as the power source

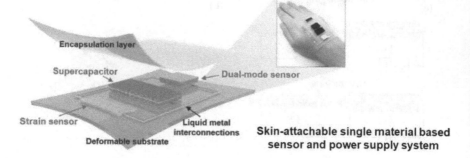

FIGURE 2.17 A PPy/GF-based integrated system of multifunctional sensors and flexible supercapacitors. (Reprinted with permission from Ref. [82]. Copyright 2018, Wiley.)

to exhibit high performance. This successful integration can be ascribed to the porous 3D GF with high surface area and excellent electrical/mechanical properties as well as high pseudocapacitance of PPy, demonstrating that the proper selection of a functional material can realize multifunctional sensors for detecting pressure/temperature/strain as well as supercapacitors for wireless wearable devices.

2.5   SUMMARY AND OUTLOOK

The boosting proliferation of portable electronic devices urgently requires flexible supercapacitors and batteries to offer power supply. The performances of energy devices are strongly dependent on electrode materials, the electrolyte, and device configuration. Carbon materials are ideal electrode materials with very prominent advantages of established manufacturing process, excellent electrical conductivity, stable physical and chemical properties, etc. To obtain high electrochemical performance, hybrid electrodes of carbon materials and conductive polymers or transition metal oxides have been explored in combination of the former's long operation life and the latter's high capacitance. Special care is also needed to design electrode materials with enough mechanical properties that can adapt to a wide range of strains to exhibit good performance retention and in innovative nanostructures with outstanding electronic conductivity and accessible surface areas to accelerate ion/charge transport and penetration of electrolyte ions to active materials. Also, a considerable amount of active material should be deposited onto the current collector to ensure a high-quality loading for high-energy-density devices. Besides, the device configuration is critical to determining the energy density within the storage material of given mass or volume. Typically, a carbon-based flexible device can be designed in a planar or fiber shape. Energy devices in fiber shape can be further woven into textiles to enhance the potential/current output. Further attempts need to be exerted to develop integrated systems with high flexibility and multiple functions to satisfy the needs of portable and wearable electronic devices, intelligent sensing, automatic monitoring, etc. Additionally, the type of electrolytes and packaging materials are also important issues to be tackled to realize flexible energy devices with superior performance.

REFERENCES

1. H. Wang, A. C. Forse, J. M. Griffin, N. M. Trease, L. Trognko, P.-L. Taberna, P. Simon, C. P. Grey, In situ NMR spectroscopy of supercapacitors: Insight into the charge storage mechanism, *J. Am. Chem. Soc.* 135 (2013) pp. 18968–18980.
2. J. R. Miller, P. Simon, Electrochemical capacitors for energy management, *Science* 321 (2008) p. 651.
3. K. Jost, G. Dion, Y. Gogotsi, Textile energy storage in perspective, *J. Mater. Chem. A* 2 (2014) pp. 10776–10787.
4. V. Augustyn, P. Simon, B. Dunn, Pseudocapacitive oxide materials for high-rate electrochemical energy storage, *Energy Environ. Sci.* 7 (2014) pp. 1597–1614.
5. S. Iijima, Helical microtubules of graphitic carbon, *Nature* 354 (1991) pp. 56–58.
6. T. Chen, S. Wang, Z. Yang, Q. Feng, X. Sun, L. Li, Z.-S. Wang, H. Peng, Flexible, lightweight, ultrastrong, and semiconductive carbon nanotube fibers for a highly efficient solar cell, *Angew. Chem. Int. Ed.* 50 (2011) pp. 1815–1819.
7. H. Lin, W. Weng, J. Ren, L. Qiu, Z. Zhang, P. Chen, X. Chen, J. Deng, Y. Wang, H. Peng, Twisted aligned carbon nanotube/silicon composite fiber anode for flexible wire-shaped lithium-ion battery, *Adv. Mater.* 26 (2014) pp. 1217–1222.
8. J. Ren, Y. Zhang, W. Bai, X. Chen, Z. Zhang, X. Fang, W. Weng, Y. Wang, H. Peng, Elastic and wearable wire-shaped lithium-ion battery with high electrochemical performance, *Angew. Chem. Int. Ed.* 53 (2014) pp. 7864–7869.
9. H. Lin, L. Li, J. Ren, Z. Cai, L. Qiu, Z. Yang, H. Peng, Conducting polymer composite film incorporated with aligned carbon nanotubes for transparent, flexible and efficient supercapacitor, *Sci. Rep.* 3 (2013) p. 1353.
10. X. Wang, J. Li, Y. Luo, M. Huang, A novel ammonium perchlorate/graphene aerogel nanostructured energetic composite: Preparation and thermal decomposition, *Sci. Adv. Mater.* 6 (2014) pp. 530–537.
11. M. J. Allen, V. C. Tung, R. B. Kaner, Honeycomb carbon: A review of graphene, *Chem. Rev.* 110 (2010) pp. 132–145.
12. Z. Wu, L. Li, J. M. Yan, X. B. Zhang, Materials design and system construction for conventional and new-concept supercapacitors, *Adv. Sci.* 4 (2017) p. 1600382.
13. K. Zhang, L. Mao, L. L. Zhang, H. S. O. Chan, X. S. Zhao, J. Wu, Surfactant-intercalated, chemically reduced graphene oxide for high performance supercapacitor electrodes, *J. Mater. Chem.* 21 (2011) pp. 7302–7307.
14. Z.-L. Wang, D. Xu, Y. Huang, Z. Wu, L.-M. Wang, X.-B. Zhang, Facile, mild and fast thermal-decomposition reduction of graphene oxide in air and its application in high-performance lithium batteries, *Chem. Commun.* 48 (2012) pp. 976–978.
15. E. Raymundo-Pinero, P. Azais, T. Cacciaguerra, D. Cazorla-Amorós, A. Linares-Solano, F. Béguin, KOH and NaOH activation mechanisms of multiwalled carbon nanotubes with different structural organisation, *Carbon* 43 (2005) pp. 786–795.
16. M. F. El-Kady, V. Strong, S. Dubin, R. B. Kaner, Laser scribing of high-performance and flexible graphene-based electrochemical capacitors, *Science* 335 (2012) pp. 1326–1330.
17. J.-W. Jeon, Y. Ma, J. F. Mike, L. Shao, P. B. Balbuena, J. L. Lutkenhaus, Oxidatively stable polyaniline: Polyacid electrodes for electrochemical energy storage, *Phys. Chem. Chem. Phys.* 15 (2013) pp. 9654–9662.
18. H.-H. Chang, C.-K. Chang, Y.-C. Tsai, C.-S. Liao, Electrochemically synthesized graphene/polypyrrole composites and their use in supercapacitor, *Carbon* 50 (2012) pp. 2331–2336.
19. G. A. Snook, G. Z. Chen, The measurement of specific capacitances of conducting polymers using the quartz crystal microbalance, *J. Electroanal. Chem.* 612 (2008) pp. 140–146.

20. H. Choi, H. Yoon, Nanostructured electrode materials for electrochemical capacitor applications, *Nanomaterials* 5 (2015) pp. 906–936.
21. S. Faraji, F. N. Ani, Microwave-assisted synthesis of metal oxide/hydroxide composite electrodes for high power supercapacitors–a review, *J. Power Sources* 263 (2014) pp. 338–360.
22. H. Lee, M. S. Cho, I. H. Kim, J. Do Nam, Y. Lee, RuO_x/polypyrrole nanocomposite electrode for electrochemical capacitors, *Synth. Met.* 160 (2010) pp. 1055–1059.
23. C.-C. Hu, K.-H. Chang, M.-C. Lin, Y.-T. Wu, Design and tailoring of the nanotubular arrayed architecture of hydrous RuO_2 for next generation supercapacitors, *Nano Lett.* 6 (2006) pp. 2690–2695.
24. W. Sugimoto, S. Makino, R. Mukai, Y. Tatsumi, K. Fukuda, Y. Takasu, Y. Yamauchi, Synthesis of ordered mesoporous ruthenium by lyotropic liquid crystals and its electrochemical conversion to mesoporous ruthenium oxide with high surface area, *J. Power Sources* 204 (2012) pp. 244–248.
25. A. González, E. Goikolea, J. A. Barrena, R. Mysyk, Review on supercapacitors: Technologies and materials, *Renewable Sustainable Energy Rev.* 58 (2016) pp. 1189–1206.
26. D. P. Dubal, R. Holze, All-solid-state flexible thin film supercapacitor based on Mn_3O_4 stacked nanosheets with gel electrolyte, *Energy* 51 (2013) pp. 407–412.
27. L. Yuan, X.-H. Lu, X. Xiao, T. Zhai, J. Dai, F. Zhang, B. Hu, X. Wang, L. Gong, J. Chen, Flexible solid-state supercapacitors based on carbon nanoparticles/MnO_2 nanorods hybrid structure, *ACS Nano* 6 (2011) pp. 656–661.
28. M. Li, M. Zu, J. Yu, H. Cheng, Q. Li, Stretchable fiber supercapacitors with high volumetric performance based on buckled MnO_2/oxidized carbon nanotube fiber electrodes, *Small* 13 (2017) p. 1602994.
29. J. Bae, M. K. Song, Y. J. Park, J. M. Kim, M. Liu, Z. L. Wang, Fiber supercapacitors made of manowire-fiber hybrid structures for wearable/flexible energy storage, *Angew. Chem. Int. Ed.* 50 (2011) pp. 1683–1687.
30. M. Acerce, D. Voiry, M. Chhowalla, Metallic 1T phase MoS_2 nanosheets as supercapacitor electrode materials, *Nat. Nanotechnol.* 10 (2015) p. 313.
31. V. T. Le, H. Kim, A. Ghosh, J. Kim, J. Chang, Q. A. Vu, D. T. Pham, J.-H. Lee, S.-W. Kim, Y. H. Lee, Coaxial fiber supercapacitor using all-carbon material electrodes, *ACS Nano* 7 (2013) pp. 5940–5947.
32. X. Chen, L. Qiu, J. Ren, G. Guan, H. Lin, Z. Zhang, P. Chen, Y. Wang, H. Peng, Novel electric double-layer capacitor with a coaxial fiber structure, *Adv. Mater.* 25 (2013) pp. 6436–6441.
33. L. Dong, C. Xu, Y. Li, C. Wu, B. Jiang, Q. Yang, E. Zhou, F. Kang, Q. H. Yang, Simultaneous production of high-performance flexible textile electrodes and fiber electrodes for wearable energy storage, *Adv. Mater.* 28 (2016) pp. 1675–1681.
34. L. Dong, C. Xu, Q. Yang, J. Fang, Y. Li, F. Kang, High-performance compressible supercapacitors based on functionally synergic multiscale carbon composite textiles, *J. Mater. Chem. A* 3 (2015) pp. 4729–4737.
35. W. Weng, P. Chen, S. He, X. Sun, H. Peng, Smart electronic textiles, *Angew. Chem. Int. Ed.* 55 (2016) pp. 6140–6169.
36. Q. Huang, D. Wang, Z. Zheng, Textile-based electrochemical energy storage devices, *Adv. Energy Mater.* 6 (2016) p. 1600783.
37. G. Qu, J. Cheng, X. Li, D. Yuan, P. Chen, X. Chen, B. Wang, H. Peng, A fiber supercapacitor with high energy density based on hollow graphene/conducting polymer fiber electrode, *Adv. Mater.* 28 (2016) pp. 3646–3652.
38. Z. Wang, J. Cheng, Q. Guan, H. Huang, Y. Li, J. Zhou, W. Ni, B. Wang, S. He, H. Peng, All-in-one fiber for stretchable fiber-shaped tandem supercapacitors, *Nano Energy* 45 (2018) pp. 210–219.

39. L. Hu, M. Pasta, F. La Mantia, L. Cui, S. Jeong, H. D. Deshazer, J. W. Choi, S. M. Han, Y. Cui, Stretchable, porous, and conductive energy textiles, *Nano Lett.* 10 (2010) pp. 708–714.

40. H. Sun, S. Xie, Y. Li, Y. Jiang, X. Sun, B. Wang, H. Peng, Large-area supercapacitor textiles with novel hierarchical conducting structures, *Adv. Mater.* 28 (2016) pp. 8431–8438.

41. Z. Liu, Z. S. Wu, S. Yang, R. Dong, X. Feng, K. Müllen, Ultraflexible in-plane micro-supercapacitors by direct printing of solution-processable electrochemically exfoliated graphene, *Adv. Mater.* 28 (2016) pp. 2217–2222.

42. D. Pech, M. Brunet, H. Durou, P. Huang, V. Mochalin, Y. Gogotsi, P.-L. Taberna, P. Simon, Ultrahigh-power micrometre-sized supercapacitors based on onion-like carbon, *Nat. Nanotechnol.* 5 (2010) p. 651.

43. D. Kim, G. Shin, Y. J. Kang, W. Kim, J. S. Ha, Fabrication of a stretchable solid-state micro-supercapacitor array, *ACS Nano* 7 (2013) pp. 7975–7982.

44. R. Raccichini, A. Varzi, S. Passerini, B. Scrosati, The role of graphene for electrochemical energy storage, *Nat. Mater.* 14 (2015) p. 271–279.

45. Z. Niu, L. Zhang, L. Liu, B. Zhu, H. Dong, X. Chen, All-solid-state flexible ultrathin micro-supercapacitors based on graphene, *Adv. Mater.* 25 (2013) pp. 4035–4042.

46. A. Basu, K. Roy, N. Sharma, S. Nandi, R. Vaidhyanathan, S. Rane, C. Rode, S. Ogale, CO_2 laser direct written mof-based metal-decorated and heteroatom-doped porous graphene for flexible all-solid-state microsupercapacitor with extremely high cycling stability, *ACS Appl. Mater. Interfaces* 8 (2016) pp. 31841–31848.

47. M. F. El-Kady, R. B. Kaner, Scalable fabrication of high-power graphene micro-supercapacitors for flexible and on-chip energy storage, *Nat. Commun.* 4 (2013) p. 1475.

48. J. Chang, S. Adhikari, T. H. Lee, B. Li, F. Yao, D. T. Pham, V. T. Le, Y. H. Lee, Leaf vein-inspired nanochanneled graphene film for highly efficient micro-supercapacitors, *Adv. Energy Mater.* 5 (2015) p. 1500003.

49. S. R. Sivakkumar, D.-W. Kim, Polyaniline/carbon nanotube composite cathode for rechargeable lithium polymer batteries assembled with gel polymer electrolyte, *J. Electrochem. Soc.* 154 (2007) p. A134.

50. L. Noerochim, J.-Z. Wang, S.-L. Chou, D. Wexler, H.-K. Liu, Free-standing single-walled carbon nanotube/SnO_2 anode paper for flexible lithium-ion batteries, *Carbon* 50 (2012) pp. 1289–1297.

51. S. Luo, K. Wang, J. Wang, K. Jiang, Q. Li, S. Fan, Binder-free $LiCoO_2$/carbon nanotube cathodes for high-performance lithium ion batteries, *Adv. Mater.* 24 (2012) pp. 2294–2298.

52. H. Wu, L. Hu, M. W. Rowell, D. Kong, J. J. Cha, J. R. McDonough, J. Zhu, Y. Yang, M. D. McGehee, Y. Cui, Electrospun metal nanofiber webs as high-performance transparent electrode, *Nano Lett.* 10 (2010) pp. 4242–4248.

53. N. T. Wu, W. Z. Du, X. Gao, L. Zhao, G. L. Liu, X. M. Liu, H. Wu, Y. B. He, Hollow SnO_2 nanospheres with oxygen vacancies entrapped by a N-doped graphene network as robust anode materials for lithium-ion batteries, *Nanoscale* 10 (2018) pp. 11460–11466.

54. J. W. Lee, S. Y. Lim, H. M. Jeong, T. H. Hwang, J. K. Kang, J. W. Choi, Extremely stable cycling of ultra-thin V_2O_5 nanowire–graphene electrodes for lithium rechargeable battery cathodes, *Energy Environ. Sci.* 5 (2012) p. 9889.

55. C. Botas, D. Carriazo, G. Singh, T. Rojo, Sn– and SnO_2–graphene flexible foams suitable as binder-free anodes for lithium ion batteries, *J. Mater. Chem. A* 3 (2015) pp. 13402–13410.

56. M. Liang, L. Zhi, Graphene-based electrode materials for rechargeable lithium batteries, *J. Mater. Chem.* 19 (2009) p. 5871.

57. Z.-S. Wu, G. Zhou, L.-C. Yin, W. Ren, F. Li, H.-M. Cheng, Graphene/metal oxide composite electrode materials for energy storage, *Nano Energy* 1 (2012) pp. 107–131.

58. Y. Shi, L. Wen, G. Zhou, J. Chen, S. Pei, K. Huang, H.-M. Cheng, F. Li, Graphene-based integrated electrodes for flexible lithium ion batteries, *2D Materials* 2 (2015) p. 024004.
59. J. Liang, Y. Zhao, L. Guo, L. Li, Flexible free-standing graphene/SnO(2) nanocomposites paper for Li-ion battery, *ACS Appl. Mater. Interfaces* 4 (2012) pp. 5742–5748.
60. R. Wang, C. Xu, J. Sun, L. Gao, C. Lin, Flexible free-standing hollow Fe_3O_4/graphene hybrid films for lithium-ion batteries, *J. Mater. Chem. A* 1 (2013) pp. 1794–1800.
61. N. Li, Z. Chen, W. Ren, F. Li, H.-M. Cheng, Flexible graphene-based lithium ion batteries with ultrafast charge and discharge rates, *Proc. Natl. Acad. Sci U. S. A.* 109 (2012) pp. 17360–17365.
62. B. Liu, J. Zhang, X. Wang, G. Chen, D. Chen, C. Zhou, G. Shen, Hierarchical three-dimensional $ZnCo_2O_4$ nanowire arrays/carbon cloth anodes for a novel class of high-performance flexible lithium-ion batteries, *Nano Lett.* 12 (2012) pp. 3005–3011.
63. L. Hu, Y. Cui, Energy and environmental nanotechnology in conductive paper and textiles, *Energy Environ. Sci.* 5 (2012) p. 6423.
64. G. Yu, L. Hu, M. Vosgueritchian, H. Wang, X. Xie, J. R. McDonough, X. Cui, Y. Cui, Z. Bao, Solution-processed graphene/MnO_2 nanostructured textiles for high-performance electrochemical capacitors, *Nano Lett.* 11 (2011) pp. 2905–2911.
65. X. Wang, B. Liu, X. Hou, Q. Wang, W. Li, D. Chen, G. Shen, Ultralong-life and high-rate web-like $Li_4Ti_5O_{12}$ anode for high-performance flexible lithium-ion batteries, *Nano Res.* 7 (2014) pp. 1073–1082.
66. L. Hu, G. Zheng, J. Yao, N. Liu, B. Weil, M. Eskilsson, E. Karabulut, Z. Ruan, S. Fan, J. T. Bloking, M. D. McGehee, L. Wågberg, Y. Cui, Transparent and conductive paper from nanocellulose fibers, *Energy Environ. Sci.* 6 (2013) pp. 513–518.
67. L. Hu, H. Wu, F. La Mantia, Y. Yang, Y. Cui, Thin, flexible secondary Li-ion paper batteries, *ACS Nano* 4 (2010) pp. 5843–5848.
68. D. Wei, S. Haque, P. Andrew, J. Kivioja, T. Ryhänen, A. Pesquera, A. Centeno, B. Alonso, A. Chuvilin, A. Zurutuza, Ultrathin rechargeable all-solid-state batteries based on monolayer graphene, *J. Mater. Chem. A* 1 (2013) pp. 3177–3181.
69. X. Lin, Q. Kang, Z. Zhang, R. Liu, Y. Li, Z. Huang, X. Feng, Y. Ma, W. Huang, Industrially weavable metal/cotton yarn air electrodes for highly flexible and stable wire-shaped $Li–O_2$ batteries, *J. Mater. Chem. A* 5 (2017) pp. 3638–3644.
70. J. Ren, L. Li, C. Chen, X. Chen, Z. Cai, L. Qiu, Y. Wang, X. Zhu, H. Peng, Twisting carbon nanotube fibers for both wire-shaped micro-supercapacitor and micro-battery, *Adv. Mater.* 25 (2013) pp. 1155–1159.
71. Y. H. Kwon, S.-W. Woo, H.-R. Jung, H. K. Yu, K. Kim, B. H. Oh, S. Ahn, S.-Y. Lee, S.-W. Song, J. Cho, H.-C. Shin, J. Y. Kim, Cable-type flexible lithium ion battery based on hollow multi-helix electrodes, *Adv. Mater.* 24 (2012) pp. 5192–5197.
72. L. Wang, J. Pan, Y. Zhang, X. Cheng, L. Liu, H. Peng, A Li–air battery with ultralong cycle life in ambient air, *Adv. Mater.* 30 (2018) p. 1704378.
73. L. Hu, F. La Mantia, H. Wu, X. Xie, J. McDonough, M. Pasta, Y. Cui, Lithium-ion textile batteries with large areal mass loading, *Adv. Energy Mater.* 1 (2011) pp. 1012–1017.
74. Y. Zhao, Y. Zhang, H. Sun, X. Dong, J. Cao, L. Wang, Y. Xu, J. Ren, Y. Hwang, I. H. Son, X. Huang, Y. Wang, H. Peng, A self-healing aqueous lithium-ion battery, *Angew. Chem. Int. Ed.* 55 (2016) pp. 14384–14388.
75. J. Rao, N. Liu, Z. Zhang, J. Su, L. Li, L. Xiong, Y. Gao, All-fiber-based quasi-solid-state lithium-ion battery towards wearable electronic devices with outstanding flexibility and self-healing ability, *Nano Energy* 51 (2018) pp. 425–433.
76. Z. Guo, Y. Zhao, Y. Ding, X. Dong, L. Chen, J. Cao, C. Wang, Y. Xia, H. Peng, Y. Wang, Multi-functional flexible aqueous sodium-ion batteries with high safety, *Chem* 3 (2017) pp. 348–362.

77. M. Schreiter, R. Gabl, J. Lerchner, C. Hohlfeld, A. Delan, G. Wolf, A. Blüher, B. Katzschner, M. Mertig, W. Pompe, Functionalized pyroelectric sensors for gas detection, sensors actuat. *B-Chem.* 119 (2006) pp. 255–261.
78. R. Liu, J. Wang, T. Sun, M. Wang, C. Wu, H. Zou, T. Song, X. Zhang, S. T. Lee, Z. L. Wang, B. Sun, Silicon nanowire/polymer hybrid solar cell-supercapacitor: A self-charging power unit with a total efficiency of 10.5, *Nano Lett.* 17 (2017) pp. 4240–4247.
79. H. Guo, M. H. Yeh, Y. Zi, Z. Wen, J. Chen, G. Liu, C. Hu, Z. L. Wang, Ultralight cut-paper-based self-charging power unit for self-powered portable electronic and medical systems, *ACS Nano* 11 (2017) pp. 4475–4482.
80. Z. Zhang, X. Chen, P. Chen, G. Guan, L. Qiu, H. Lin, Z. Yang, W. Bai, Y. Luo, H. Peng, Integrated polymer solar cell and electrochemical supercapacitor in a flexible and stable fiber format, *Adv. Mater.* 26 (2014) pp. 466–470.
81. J. Xu, G. Shen, A flexible integrated photodetector system driven by on-chip microsupercapacitors, *Nano Energy* 13 (2015) pp. 131–139.
82. H. Park, J. W. Kim, S. Y. Hong, G. Lee, D. S. Kim, J. h. Oh, S. W. Jin, Y. R. Jeong, S. Y. Oh, J. Y. Yun, J. S. Ha, Microporous polypyrrole-coated graphene foam for high-performance multifunctional sensors and flexible supercapacitors, *Adv. Funct. Mater.* 28 (2018) p. 1707013.

3 Triboelectric Materials for Nanoenergy

Xiude Yang
Southwest University
Zunyi Normal College

Jun Dong
Southwest University
Yangtze Normal University

Juanjuan Han and Qunliang Song
Southwest University

CONTENTS

3.1 Introduction 78
    3.1.1 Nanoenergy 78
    3.1.2 Triboelectric Nanogenerator 79
        3.1.2.1 Basic Working Models 80
        3.1.2.2 Theory Basis 82
        3.1.2.3 Material Sources 85
3.2 The Development of Materials for TENGs 86
    3.2.1 Dielectric-to-Dielectric Device Structure 86
        3.2.1.1 The Working Principle of Dielectric-to-Dielectric Device 87
        3.2.1.2 The Advantages of Dielectric-to-Dielectric Device 87
        3.2.1.3 Development of Dielectric-to-Dielectric Paired
                 Materials for TENGs 88
    3.2.2 Dielectric-to-Conductor Device Structure 93
        3.2.2.1 The Working Principle of Dielectric-to-Conductor
                 Device 93
        3.2.2.2 The Advantages of Dielectric-to-Conductor Device 94
        3.2.2.3 Development of Dielectric-to-Conductor Paired
                 Materials for TENGs 94
    3.2.3 Semiconductor Device Structure 97
        3.2.3.1 The Role of Semiconductor for TENGs 97
        3.2.3.2 Semiconductor-to-Dielectric Device 100
        3.2.3.3 Semiconductor-to-Conductor Device 101
3.3 The Output Enhancement Mechanism in TENGs 104
    3.3.1 Selection of Paired Materials 104

        3.3.2   Enhancement in Effective Contact 104
        3.3.3   Modification of Material Composition 105
        3.3.4   Control of Environmental Conditions 106
        3.3.5   Designs of Hybrid Cells.. 107
3.4   Recent Advancement of TENGs for Nanoenergy 108
        3.4.1   Micro/Nano Power Source 109
        3.4.2   Self-Powered Sensors 109
        3.4.3   Blue Energy 113
3.5   Challenge and Perspective... 117
        3.5.1   Research on High-Performance Semiconductor Materials
                   for TENGs 118
        3.5.2   In-Depth Study on Working Principle and Simulation Models........ 118
        3.5.3   Research on Hybrid Energy Units 118
        3.5.4   Exploration of Energy Collection Circuits 118
References.. 118

3.1   INTRODUCTION

3.1.1   NANOENERGY

It is well known that energy is an essential basis for human survival and development. As shown in Figure 3.1, the macroscale energy in the magnitude of kilowatt, megawatt, gigawatt, etc. plays an irreplaceable role in running a household appliance, a factory, a city, or a country. With the rapid development of microelectronics technique, radio technology, Internet of Things (IoTs), and artificial intelligence, the microelectronic

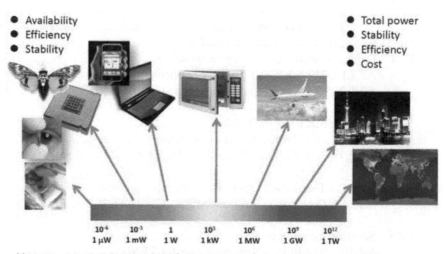

Nanoenergy: energy required for the sustainable, maintains-free and self-powered operation of micro/nanosystems

FIGURE 3.1   Magnitude of power energy and its corresponding applications. (Reproduced with permission from Wiley Ref. [1].)

devices are tending toward the trend of more and more miniaturization, mobility, and multifunctional capability. Obviously, traditional batteries such as nickel metal hydride battery and lithium battery are no longer suitable to drive the worldwide sensor network that will emerge in the near future. This is because one has to trace the location, replace the batteries, and inspect the real-time working status. Meanwhile, the recycling of batteries will lead to environmental pollution and health hazards to humans. Therefore, producing and widely distributing microscale energy in the magnitude of watt, milliwatt, and even microwatt to power small low-energy electronics becomes critical and is referred to as "nanoenergy" [1]. This is a brand new field, which provides power for sustainable, maintenance-free, and self-powered operation of micro/nanosystems. Nanoenergy can come from thermal energy, wind energy, or mechanical trigger/vibration in a device's working environment. Even though it is not enough to provide consistent power to the main grid, it is very suitable for the mobile sensors and self-powered systems. For these reasons, it can be said that the development of distributed mobile nanoenergy is the most appropriate solution to power supply for future microelectronic products.

3.1.2 TRIBOELECTRIC NANOGENERATOR

Triboelectric nanogenerator (TENG) is a new type of mechanical energy harvester invented by Wang's group in 2012 [2] and is shown in Figure 3.2. It uses the ubiquitous triboelectrification-induced polarization changes of surface charges as materials' contact–separation varies, thus producing the displacement current that drives electrons to flow between electrodes. In such a working process based on the coupling of triboelectrification and electrostatic induction effects, the mechanical energy existing in the environment or organisms can be converted into electricity directly. In our environment, there exists a large amount of renewable and clean energy as shown in Figure 3.3, such as wind energy, tidal energy, the movement of cars and trains, and human walking and typing, which can be used as direct sources of mechanical energy collected by TENG. And other forms of energy, such as solar

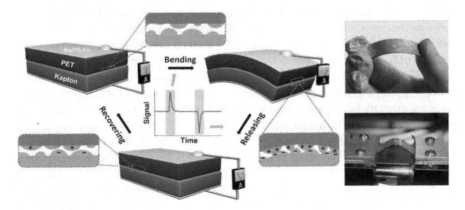

FIGURE 3.2 The first TENG invented by Wang's group. (Reproduced with permission from Elsevier Ref. [2].)

FIGURE 3.3 Renewable and clean energy existing in our environment. (Reproduced with permission from Royal Society of Chemistry Ref. [3].)

energy, geothermal energy, chemical energy, and biomass energy can also be indirectly harvested into electricity. It is evident that this is the new era energy to adapt to the development of IOT. Therefore, TENG as a new type of energy collecting device is considered to be one of the most effective ways to realize distributed nanoenergy. For these reasons, we can say that the invention of TENG is an important milestone in the field of the mechanical energy to generate electricity and produce self-powered applications, showing a great commercial prospect in wearable electronic products, IOTs, environmental monitoring, infrastructure monitoring, medical monitoring, information security monitoring, etc.

3.1.2.1 Basic Working Models

Generally, TENG is composed of two different dielectric films with a back electrode. Based on the coupling of triboelectrification and electrostatic induction effects, it converts mechanical energy into electrical energy. As shown in Figure 3.4, TENG can operate in four fundamental modes [4]: vertical contact–separation (a), lateral sliding (b), as single electrode (c), and as freestanding triboelectric layer(d). The two different dielectric films are the basic components that can provide micro/nanoenergy for electronic devices. If these basic components with different advantages are properly integrated, a greater increase in device output and even large-scale power generation can be achieved.

Here, we take a vertical contact–separation mode TENG as an example to describe the formation principle of the open-circuit voltage (V_{oc}) and the short-circuit current (I_{sc}) in a cycle, as shown in Figure 3.5. In such a structure, the device converts mechanical energy into electrical energy through vertical polarization. At the initial state, two dielectric films separate; therefore, no triboelectric charge is generated on respective surfaces. When they are brought into physical contact under external force, due to the significant difference in electron affinity, their surfaces will acquire

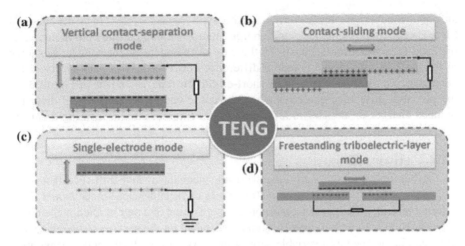

FIGURE 3.4 The four fundamental modes of TENGs. (a) Vertical contact–separation mode, (b) lateral sliding mode, (c) single electrode mode (c), and (d) freestanding triboelectric layer mode. (Reproduced with permission from Elsevier Ref. [4].)

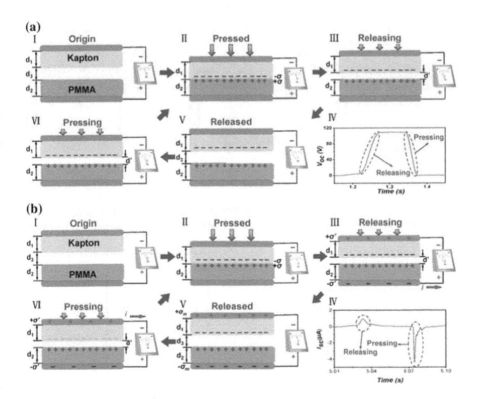

FIGURE 3.5 The operating principles of the open-circuit voltage (a) and the short-circuit current (b) for a TENG. (Reproduced with permission from American Chemical Society Ref. [5].)

an equal amount of charge with opposite signs, respectively. Once the external force is cancelled, the triboelectric charges will be separated, thus producing electric field in space simultaneously. In the case of the open-circuit voltage, as shown in Figure 3.5a, an induced potential difference between the two back electrodes is generated (i.e., V_{oc}). In the case of short-circuit current, as shown in Figure 3.5b, the potential difference generated by triboelectric charge separation will drive the induced charges on the electrode to flow in the external circuit, forming a current (i.e., I_{sc}).

3.1.2.2 Theory Basis

At the beginning of the study, the working mechanism and influence factors of TENG were mainly studied based on the theory of variable capacitance model in the external circuit. This is because the contact–separation of a pair of frictional charges with an equal amount but opposite signs distributed on the surface of two dielectric films can be deemed as a variable capacitance. As for a vertical contact–separation TENG in Figure 3.6, according to the intuitive variable model shown in Figure 3.7, the electric current I in the external circuit can be expressed as

$$I = \frac{dQ}{dt} = A\frac{d\sigma_I}{dt} \tag{3.1}$$

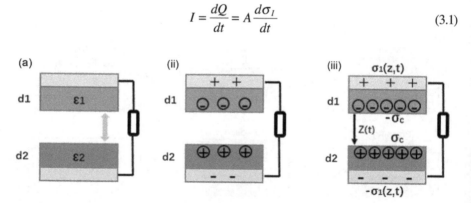

FIGURE 3.6 The principle of TENG: according to the contact electrification effect, the output current of TENG increases as the contact period increases.

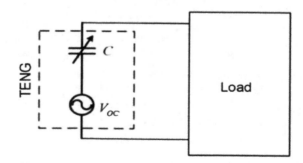

FIGURE 3.7 Equivalent open-circuit variable-voltage capacitance model. (John Wiley and Sons Ref. [6])

where Q, A, σ_I are the transferred charge amount between two electrodes, surface area of dielectric film, and surface charge density of electrode, respectively. Considering a load resistance R, according to the Ohm's law, the output voltage equation can be described as

$$RA\frac{d\sigma_I(z,t)}{dt} = V_{oc} - A\sigma_I(z,t)/C \qquad (3.2)$$

where V_{oc} and C are the open-circuit voltage and the capacitance, respectively. According to the above formula, the output characteristics of the various TENGs can be systematically obtained.

Through continuous research, the basic theory of TENG has been made clearer. In 2017, for the first time, the Maxwell displacement current was proposed to be the theoretical basis of TENG [7].That is, the surface polarized charges generated by friction due to the contact–separation of functional materials would generate displacement current and drive the charges of external circuits to flow between electrodes.

The formula of displacement current is

$$J_D = \frac{\partial D}{\partial t} = \varepsilon_0 \frac{\partial E}{\partial t} + \frac{\partial P_s}{\partial t} \qquad (3.3)$$

where E, D, P_s, ε_0 are electric field, displacement field, polarization field, and the permittivity of vacuum. From previous research, we have known for a long time that the first term $\varepsilon_0 \frac{\partial E}{\partial t}$ is the induction current generated by the change of electric field, which is the theoretical basis of electromagnetic wave, antenna broadcasting, television telegraphy, radar microwave, wireless communication, and space technology (see the left part in Figure 3.8); while the second term $\frac{\partial P_s}{\partial t}$ is the current caused by the polarization field of static charges on the surface. Now, it has been clarified that it is the fundamental theory and source of nanogenerator (including triboelectric, piezoelectric, and pyroelectric nanogenerators), which leads to the significant application of displacement current in energy and sensing technology fields (see the right part in Figure 3.8).

For the contact–separation model TENG in Figure 3.6, according to the Gauss theorem of electromagnetic fields, the electric field in dielectric 1, dielectric 2, and air gap are $E_{1z} = \sigma_I(z,t)/\varepsilon_1$, $E_{2z} = \sigma_I(z,t)/\varepsilon_2$, and $E_{gz} = [\sigma_I(z,t) - \sigma_c]/\varepsilon_0$, respectively. Thus, the relative voltage difference between the two back electrodes can be written as

$$V = \sigma_I(z,t)\left[\varepsilon_1/d_1 + \varepsilon_2/d_2\right] + z\left[\sigma_I(z,t) - \sigma_c\right]/\varepsilon_0 \qquad (3.4)$$

where ε_1 and ε_2 are the dielectric constants of dielectric 1 and dielectric 2, respectively, and d_1 and d_2 are their corresponding thicknesses, respectively.

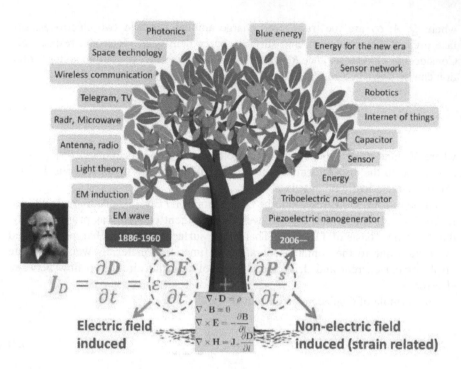

FIGURE 3.8 A tree diagram to illustrate the Maxwell's displacement current: The first component of displacement current $\varepsilon\partial E/\partial t$ is the theoretical basis of electromagnetic waves, from which wireless communication and optical technology are derived; the second component $\partial Ps/\partial t$ is the theoretical basis of nanogenerators, from which micro-nano energy and sensing technology are derived. (Reproduced with permission from Elsevier Ref. [8].)

In the case of short-circuit current, due to the potential difference between the two back electrodes $V = 0$, the transferred charge amount in the external circuit can be obtained as

$$\sigma_I(z,t) = \frac{z\sigma_c}{d_1\varepsilon_0/\varepsilon_1 + d_2\varepsilon_0/\varepsilon_2 + z} \tag{3.5}$$

Surface charge density σ_c on dielectric film may be maintained for a long time because of its good insulation properties and therefore can be regarded as a constant. Thus, the displacement current in device can be expressed as

$$J_D = \frac{\partial D_z}{\partial t} = \frac{\partial \sigma_I(z,t)}{\partial t} = \sigma_c \frac{dz}{dt} \frac{d_1\varepsilon_0/\varepsilon_1 + d_2\varepsilon_0/\varepsilon_2}{\left[d_1\varepsilon_0/\varepsilon_1 + d_2\varepsilon_0/\varepsilon_2 + z\right]^2} \tag{3.6}$$

Then, the output voltage equation can be described as

$$RA\frac{d\sigma_I(z,t)}{dt} = z\sigma_c/\varepsilon_0 - \sigma_I(z,t)\left[d_1/\varepsilon_1 + d_2/\varepsilon_2 + z/\varepsilon_0\right] \tag{3.7}$$

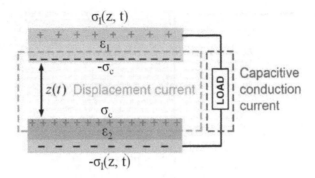

FIGURE 3.9 The displacement current model of TENG. (Reproduced with permission from Elsevier Ref. [7].)

By comparing the Eqs. (3.1) and (3.6) to (3.2) and (3.7), it can be seen that the results of the two theoretical models are exactly the same.

Therefore, it can be said that the fundamental theory of the capacitive model is displacement current, revealing the output characteristics of TENG from the device's internal materials, while capacitance model reflects the dynamic transport process from the external circuit. As shown in Figure 3.9, it can be seen that the internal circuit in TENG is dominated by the displacement current, and the measured current in external circuit is the capacitive conduction current. The internal circuit and external circuit will form a complete loop when meet at the two electrodes. Therefore, the displacement current inside TENG is the physical essence that generates the internal driving force; while the capacitance conduction current in the external circuit is the external manifestation of the displacement current [7].

3.1.2.3 Material Sources

Based on our experience, we know that almost all materials in nature such as metal, polymer, silk, and wood have triboelectrification effect. Theoretically, all of them can be used to fabricate TENG, which indicates that a wide range of materials can be selected for TENG. In 1957, John Carl Wilcke proposed the first triboelectric series on static charge based on the difference in capacity of materials to gain electrons or lose electrons [9,10]. The triboelectric series of some common materials are given in the Figure 3.10. During triboelectrification process, the materials close to the top of figure tend to gain electrons and accumulate negative charges due to their large electron affinity, while the materials close to the bottom of figure tend to lose electrons and accumulate positive charges. Previous researches have shown that the farther apart two materials are in this table, the more transferred charges they generate in the friction process. Consequently, in order to obtain more triboelectric charges and thus larger output, we usually choose a pair of tribomaterials with a large difference in electronegativity (or electron affinity) as the functional layer of TENG (such as a metal and a polymer). Especially, polytetrafluoroethylene (PTFE), polydimethylsiloxane (PDMS), polyvinyl fluoride (PVC), and polyimide (Kapton), which are located at the bottom of the table, are often studied intensively as negative polarity materials owing to their good electronegativity.

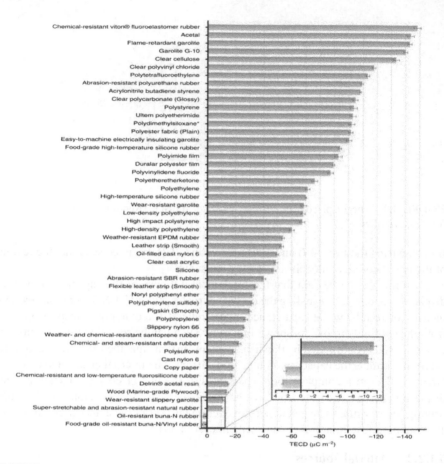

FIGURE 3.10 Triboelectric series for some common materials having a tendency to easily lose electrons (and become positive) and gain electrons (and become negative). (Reproduced with permission from Springer Nature Ref. [11].)

3.2 THE DEVELOPMENT OF MATERIALS FOR TENGs

For its working, a TENG always requires a pair of different materials to achieve mechanical energy conversion through friction. According to the differences in paired materials, it usually has three operating device structures: dielectric-to-dielectric structure, dielectric-to-conductor structure, and semiconductor-to-dielectric (or conductor) structure.

3.2.1 DIELECTRIC-TO-DIELECTRIC DEVICE STRUCTURE

The first TENG was reported by Wang's research group [2] in *Nano Energy* magazine and is shown in Figure 3.27. It is just a dielectric-to-dielectric structure TENG, which consists of a polyester (PET) film and a Kapton film with a metal electrode on their back. By coupling triboelectrification and electrostatic induction effects, the TENG has successfully converted the mechanical energy often wasted in our life in pressing, bending, etc. into electrical energy.

3.2.1.1 The Working Principle of Dielectric-to-Dielectric Device

Here, we take the device structure in Figure 3.5 as an example to describe the working principle of a dielectric-to-dielectric TENG. The two polymeric materials for the dielectric films of this structure of TENG are Kapton and polymethyl methacrylate (PMMA) films. At the initial state, no charge is generated; thus no potential difference occurs between two electrodes, as shown in Figure 3.5a-I. When the two polymers come into physical contact under an external trigger, electrons on PMMA surface would be transferred to the surface of Kapton due to the triboelectrification phenomenon. Therefore, Kapton surface gains negative charges, and PMMA surface accumulates the same amount of positive charges. It is worth noting that the triboelectric charges on these polymer surfaces can be maintained for hours or even days because of their good insulation. But, these charges are limited to polymer surfaces. Thus, an equal amount of charges with opposite signs in the same plane still does not create potential difference between the two electrodes, as shown in Figure 3.5a-II. When the external force is removed, Kapton film will go back to its original position.

In the case of open-circuit voltage, the separation of two charged surfaces creates a potential difference, as shown in Figure 3.5a-III. As they separate continuously, the V_{oc} increases continuously. When they reach the original position, the V_{oc} reaches saturation, as shown in Figure 3.5a-IV and a-V. As long as the internal resistance of the voltmeter is infinite, the measured voltage is maintained at the maximum position. Then, if the external force is loaded onto the device, the two polymer surfaces get closer to each other, and the generated potential difference begins to disappear gradually. When they are fully in contact again, the V_{oc} will be zero.

In the case of short-circuit current, the potential difference between two electrodes caused by the separation of two charged polymers will drive electrons to flow from the top electrode to the bottom electrode, as shown in Figure 3.5b-III. This means that a positive current is generated when the external force is withdrawn, as shown in Figure 3.5b-IV. Once the pressure is applied to them again, the distance between the two polymer films gradually decreases, making the potential of upper electrode higher than that of the lower electrode. Thus, electrons will flow toward the reverse direction, and the induction charges will be reduced (Figure 3.5b-V). This indicates that a negative current is generated as the two polymers get closer under external force (Figure 3.5b-VI), and ultimately all induced charges will get neutralized if they return to the full contact state.

3.2.1.2 The Advantages of Dielectric-to-Dielectric Device

In dielectric-to-dielectric device designs, the functional materials of device are mainly polymers. So, we used to call dielectric-to-dielectric TENG as polymer-to-polymer TENG. Because of this, dielectric-to-dielectric TENGs have many advantages over the other structures of devices, which are as follows:

i. **Low cost**: It is well known that most of the polymer materials are considerably cheap, and it is very easy to get them from a merchant; thus devices can be fabricated with low cost.

ii. **Easy fabrication**: The preparation of polymer films doesn't require high vacuum and high temperature sintering. Likewise, the assembly of device is very simple.

iii. **Good stability**: The physical and chemical properties of polymers are very stable in the range from −200°C to 200°C. In addition, it is very important that the polymer has good anti-abrasion property, which can ensure a continuous and stable output for a TENG.

iv. **Transparency and flexibility**: Most of the polymer films are transparent and flexible. Therefore, it can be speculated that polymer-to-polymer TENGs can be applied widely in flexible electron products.

Based on the above advantages, many dielectric-to-dielectric TENGs have been investigated and reported by many research groups in recent years.

3.2.1.3 Development of Dielectric-to-Dielectric Paired Materials for TENGs

Owing to polymer films' good flexibility, nontoxicity, biocompatibility, etc., dielectric-to-dielectric TENG has shown huge potential for powering flexible electronic devices, wearable products, and implantable medical devices.

PET–Kapton device: In 2012, Wang's group [2] reported a flexible dielectric-to-dielectric TENG, which is the first TENG as well, as shown in Figure 3.11. The device was fabricated by stacking PET and Kapton sheets covered with a layer of Au on the top and bottom surfaces of the device structure. Since the two polymers have distinctly different triboelectric characteristics, equal amounts of electrostatic

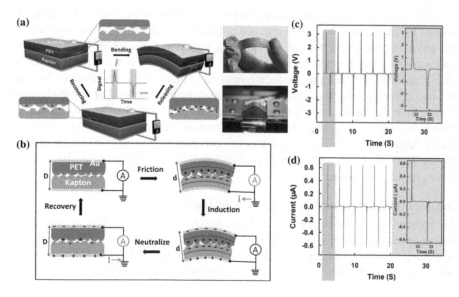

FIGURE 3.11 Schematic illustration of the structure (a) and working principle (b) of the TENG; output voltage (c) and output current (d) of such a TENG. (Reproduced with permission from Elsevier Ref. [2].)

charges with opposite signs would be generated at their surfaces as the external force is applied to the device during the deformation process. Subsequently, the periodic contacting–separation of surface charges will create a potential difference in space and thus drive the flow of induced charges from one electrode to another. Such a flexible polymer device gives a peak voltage of 3.3 V and current of 0.6 mA with a peak power density of 10.4 mW cm⁻³. Thus, it has the potential of harvesting energy from human activities, rotating tires, ocean waves, mechanical vibration, and more and is promising for applications in self-powered systems for personal electronics, environmental monitoring, and medical science.

Kapton–PMMA device: G. Zhu et al. [5] designed a Kapton–PMMA TENG of contact–separation model, as shown in Figure 3.12. Firstly, a thin Al was deposited on a glass substrate as bottom electrode by using an electron beam evaporator. Then, a layer of PMMA was spin-coated. Afterward, an insulative spacer was added at the edges, forming a square cavity at the center. A Kapton deposited with a layer of Al as top electrode was anchored on the spacer with the bottom electrode face to face. When triggered by a vibration source, the peak values of V_{oc} and I_{sc} were obtained as 110 V and 6 µA, respectively. With the external matching load resistance on the order of MΩ, the electrical power reached an instantaneous peak value of 110 µW corresponding to a power density of 31.2 mW cm⁻³. This demonstrated TENG features of simple fabrication/implementation, low cost, and strong performance.

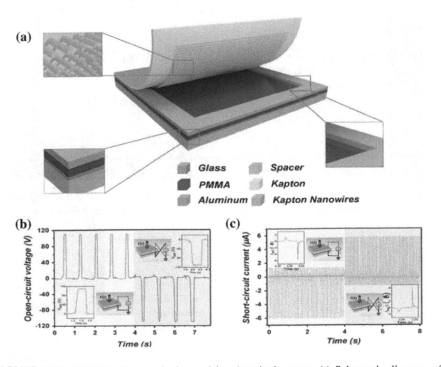

FIGURE 3.12 PMMA–Kapton device and its electrical output. (a) Schematic diagram of device, (b) open-circuit voltage V_{oc}, and (c) short-circuit current I_{sc}. (Reproduced with permission from American Chemical Society Ref. [5].)

Nylon–PTFE device: S.H. Wang et al. [12] demonstrated a sliding-triboelectric nanogenerator composed of polyamide 6,6 (Nylon) and PTFE films, the two polymers as triboelectric layers adhered on the surfaces of glass slides, as shown in Figure 3.13. Prior to the growth of polymer films, Au films were deposited on nylon as the top electrode and PTFE as the bottom electrode on the side next to glass slide, separately. Initially, two plates are fully in contact. During the experimental measurement, the PTFE-covered plate is bonded on a stationary stage, and the Nylon-covered plate is fixed to a flat rail guide connected to a line motor for inducing an in-plane motion. As the contact area changes periodically, the device can reach the peak V_{oc} of ~1,300 V with the acceleration rate of ±20 m s^{-2} and the peak J_{sc} of 4.1 mA m^{-2} at a maximum sliding velocity of 1.2 m s^{-1}. With such a power output from the sliding motion, hundreds of light-emitting diodes (LEDs) can be lit instantaneously by a single TENG. In addition, the instantaneous power density on the load reaches the maximum value of 0.42 W m^{-2} at a resistance of ~50 MΩ. This device has demonstrated shown promise in applications in wearable clothing for harvesting biomechanical energy from human body.

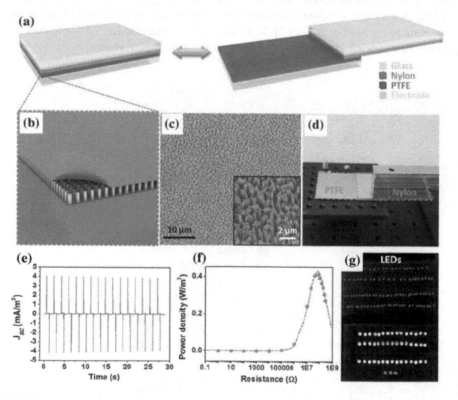

FIGURE 3.13 Nylon–PTFE device and corresponding output performance. (a–d) The sketch map of device structure and the photograph of a TENG; (e–f) The output performance: (e) the short-circuit current density, (f) the output power density; (g) the test photo of TENG's supply capability to LEDs. (Reproduced with permission from American Chemical Society [12].)

Paper–PTFE device: By applying the origami approach, Yang et al. have developed a new type of paper-based TENG with slinky shape [13]. The structure of slinky TENG is schematically illustrated in Figure 3.14. It is clearly shown that it consists of three parts: a paper substrate, a PTFE thin film, and an Al film. In such a device structure, the origami technique provides an innovative path to fabricate a stacked TENG without expanding the area or complicating the fabrication process. The I_{sc} and V_{oc} of a slinky TENG with seven units can reach 2 µA and 20 V, respectively. And the corresponding maximum output power density of 0.14 W m^{-2} is obtained at a load resistance of 400 MΩ. Furthermore, the slinky TENG can convert energy in various kinds of human motion, such as stretching, lifting, and twisting, and can serve as a self-powered pressure sensor to estimate the weight of coins. This study opens up the possibility to utilize origami configuration into TENG design, extending its application scope, especially for paper-based portable power sources and pressure sensors.

Skin–PDMS device: In the light of triboelectric series, human skin can be used as triboelectric material for fabricating TENG to harvest biomechanical energy. Based on this, Y. Yang et al. [14] demonstrated a first human-skin-based TENG, as

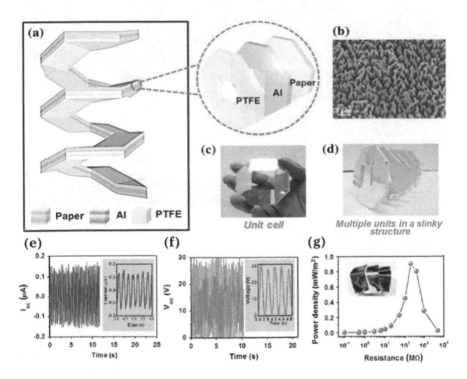

FIGURE 3.14 Structure and photographs of a slinky TENG based on paper–PTFE: (a) schematic diagram of a TENG; (b) SEM image of the etched PTFE surface; (c, d) photograph of a TENG with different units; (e–g) electrical output capability of a TENG with seven units: (e) I_{sc}, (f) V_{oc}, (g) power density. (Reproduced with permission from American Chemical Society [13].)

shown in Figure 3.15. The device is made up of a human skin patch and a PDMS film with a micropyramid structure on an ITO (indium tin oxides) transparent flexible electrode. To solve the problem that metal electrode can't be grown on human skin, a single-electrode-based TENG was designed to collect biomechanical energy based on the periodic contact–separation between PDMS and human skin patch. Since PDMS is more triboelectrically negative than skin, electrons are injected from skin to PTFE during the contact electrification process. When they are in full contact state, the generated triboelectric charges with opposite polarities on their surface are just neutralized, causing no electron flow in the external circuit. Once they are separated under an external force, the electric field produced in space will induce

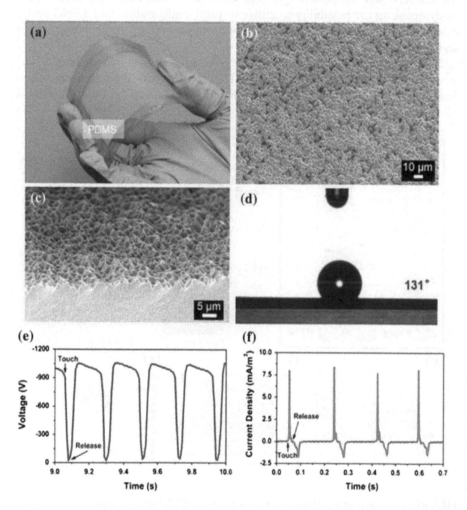

FIGURE 3.15   (a) Photograph of the TENG based on skin–PDMS, (b) SEM image of PDMS surface with microstructures, (c) SEM image of PDMS surface, (d) the contact angle between PDMS and water, (e) the V_{oc}, and (f) the J_{sc}. (Reproduced with permission from American Chemical Society Ref. [14].)

positive charges on the ITO electrode and drive free electrons flowing from ITO to the ground, which gives an output electrical signal. When skin is reverted to reach PDMS, the electrons flow from the ground to ITO electrode until they are fully in contact with each other again, leading to a reverse electrical output signal. By converting the biomechanical energy into electricity, the TENG can produce a V_{oc} up to ~1,000 V, a J_{sc} of 8 mA m^{-2}, and a power density of 500 mW m^{-2} on a load resistance of 100 MΩ. The experimental results have shown that the improvement of contact angle plays an important role in increasing device output performance. The obtained output power can be used to directly light up tens of green LEDs. This work moves a significant step forward toward the practical applications of skin-based biomechanical energy harvesting techniques and self-powered touch pad technology.

3.2.2 DIELECTRIC-TO-CONDUCTOR DEVICE STRUCTURE

The conductor and dielectric have a huge difference in electron affinity, and this is very beneficial for preparing TENG. In many TENG researches, they are chosen as a pair of triboelectric materials, hence the name dielectric-to-conductor TENG.

3.2.2.1 The Working Principle of Dielectric-to-Conductor Device

Unlike the dielectric-to-dielectric TENG, there is only one dielectric for this type of device. The conductor performs two roles; it acts both as an electrode and as a triboelectric layer. If the conductor layer and dielectric layer come into physical contact, the triboelectrification process occurs. In the meantime, the conductor layer tends to lose electrons and becomes positively charged. On the contrary, the dielectric layer captures electrons and is negatively charged, as shown in Figure 3.16. Once the two triboelectric layers are separated by external force, the tribocharges on the dielectric surface will remain for a long time with a negligible decay because of its insulating nature, and a potential difference is induced, driving electron flow from one electrode to another electrode to generate current. Hence, the total charge on the triboelectric metal layer includes two parts: one is tribocharges, and the other is charges transferred between the two electrodes. As the distance between two triboelectric

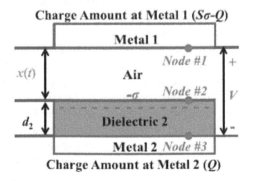

FIGURE 3.16 Dielectric-to-conductor attached-electrode parallel-plate contact-mode TENG. (Reproduced with permission from Royal Society of Chemistry Ref. [15].)

layers can be varied by the agitation of mechanical force, the mechanical energy is transformed into electricity continuously.

3.2.2.2 The Advantages of Dielectric-to-Conductor Device

Compared to dielectric-to-dielectric type of device, dielectric-to-conductor TENG has the following advantages:

  i. In a conductor, the outer electrons of metal atoms can move freely inside the metal. During the triboelectrification process with dielectric, electrons can be injected readily from conductor to dielectric, which is very beneficial to increase the density of tribocharges on contact surfaces and further the device's output capability.
  ii. Conductor as one tribolayer avoids the difficulty of growing the electrode on many dielectrics.
  iii. In addition, because of only one dielectric in the device, the manufacturing cost reduces to some extent.

3.2.2.3 Development of Dielectric-to-Conductor Paired Materials for TENGs

PDMS–Al device: S.H. Wang et al. [16] reported a dielectric-to-conductor TENG based on micromachining PDMS and Al film as a pair of triboelectric layers, as shown in Figure 3.17. Because they are distributed at both ends of the triboelectric series, there is a strong charge transfer generated in the triboelectrification process. For such a device with an active area of 3 cm×2.8 cm, a V_{oc} of 230 V and an I_{sc} of 94 μA are obtained at the frequency of 6 Hz. On increasing the frequency to 10 Hz, the I_{sc} can reach 130 μA; meanwhile the instantaneous power can reach 128 mW cm^{-3}. The electrical output is applied directly to drive small portable electronic devices, such as LEDs.

PDMS–Au device: J. Chen et al. [17] fabricated a microcavity–nanoparticles (NPs) assembled structure of PDMS–Au paired TENG, as shown in Figure 3.18. A PDMS mesoporous film impregnated with gold NPs was adopted as an effective dielectric for enhancing the device's output performance. In this structure, the periodic contact and separation between PDMS and Au nanoparticles can be realized via the microcavity without the need of the presence of an air gap. Before they contact inside pores, there is no charge transfer, as well as no potential. Under compressive force, friction occurs, and electrons are transferred from Au NPs to PDMS film, inducing positive and negative charges on their surfaces, respectively. Once the force is withdrawn, the separation of surface charges produces a distributed electric field in space, leading to the induced electrons flowing from top electrode to bottom electrode and thus generating a negative current signal. If the force is loaded again, the Au NPs will contact PDMS again, and the electrons are driven from bottom electrode to top electrode, corresponding to a positive current signal. For such a TENG, the output instantaneous power reaches a peak value of 6 mW (78 mW cm^{-2}) at a resistance of 10 MΩ. Experimental results have shown that the pores in the mesoporous film have a very important role in determining the device's electrical output performance.

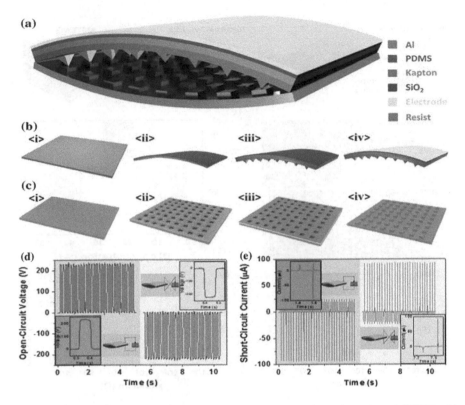

FIGURE 3.17 Structure and fabrication process of the PDMS–Al based TENG. (a) Schematic diagram of device structure; (b, c) fabrication flowchart of the top plate (b) and bottom plate (c) of the TENG; (d, e) the V_{oc} and I_{sc} of the TENG. (Reproduced with permission from American Chemical Society [16].)

PTFE–steel device: In our work, based on the spring steel and PTFE as paired triboelectric materials, a novel retractable spring-like-electrode TENG (SL-TENG) has been assembled for harvesting vibratory energy [18]. As shown in Figure 3.19, the device is a multilayer TENG with PTFE/steel/PET/steel repeated units, in which spring steel is employed as skeleton, tribolayer, and electrode. During the friction process, negative charges on the surface of PTFE film and equal positive charges on the surface of steel are induced due to the large difference in electron affinity between them. Later, under the periodic contact–separation by external force from a line motor, the mechanical energy is converted to electricity constantly. Experimental results indicate that due to the good elasticity of spring steel and the unique spring structure, the contact–separation synchronism of SL-TENG is greatly improved and further strengthened as frequency increases, which can be used to tandem stack the spring to efficiently convert vibration energy in a small volume, showing good integration. The three-layer SL-TENG in a volume of ~5 cm³ can reach the maximum negative current of 9.4 µA and positive current of 8 µA at the frequency of 7 Hz. Besides, the electrical output is applied to alternately light tens of commercial

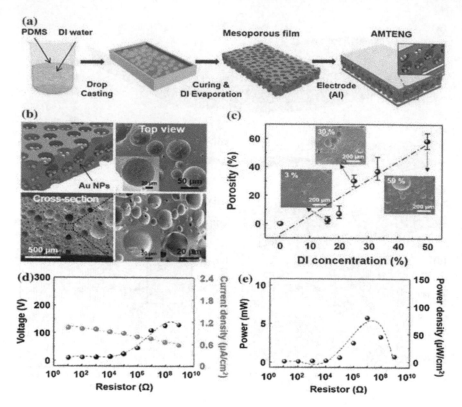

FIGURE 3.18 The structure and electrical output for the mesoporous PDMS–Au device. (a) Schematic diagrams of device, (b) top and cross-sectional SEM images of a mesoporous PDMS film with Au NPs, (c) porosity changes as a function of the deionized water (DI) water concentration, (d) the output voltage and current and (e) the output instantaneous power. (Reproduced with permission from Royal Society of Chemistry Ref. [17].)

LEDs, and the results demonstrate that SL-TENG is promising in the application of self-powered sensor for monitoring road potholes.

Kapton–Al device: In addition to the contact–separation dielectric-to-conductor TENG, some sliding devices have been suggested for harvesting sliding electrification between two sliding surfaces. For example, L. Lin et al developed a lateral sliding TENG by using a segmentally patterned disk structure for collecting energy from rotary motion [19]. As shown in Figure 3.20a–c, the device is composed of two disk-shaped components with four sectors each. At first, two acrylic sheets were processed by laser cutting to form the desired four-sector-structured templates for the effective contacting parts of TENG. Then, a layer of 50 μm thick Kapton film with Au electrode was manually attached on one template, while a piece of Al foil with the same shape was attached on the other template. The Al foil and Kapton film were brought to a face-to-face close contact, and they could spin around the common axis under external force. The working principle of TENG depends on rotary electrification and cyclic charge separation induced from relative motion between Al and Kapton, as shown in Figure 3.20d. In such a TENG, a V_{oc} of 230 V and $\Delta\sigma$ of

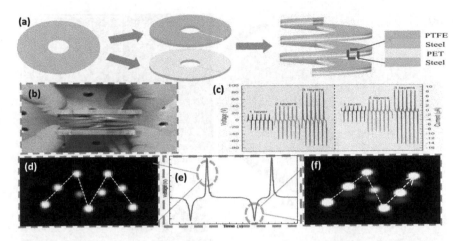

FIGURE 3.19 The spring-like PTFE–steel based TENG. (a) The structure and fabrication process of the TENG, (b) photograph of the TENG, (c) the output voltage, and (d) the output current for TENGs with one layer, two layers, and three layers. (d) "M" shaped LEDs are lit when positive current is generated. (e) The graph of tuned alternating current signal generated by the TENG. (f) The photograph of "N" shaped LEDs are lit when negative current flows. (Reproduced with permission from Royal Society of Chemistry Ref. [18].)

$40.8 \ \mu C \ m^{-2}$ are obtained. Also, the J_{sc} increases with increasing rotary speed, which shows that it has potential for application in automobile brakes and transmitters.

PTFE–copper device: P. Bai et al. further developed a coaxial cylindrically structured TENG for harvesting rotating mechanical energy based on copper (conductor) and polymer PTFE as triboelectric materials [20]. The schematic diagram of a TENG with six strip units is shown in Figure 3.21. The device is a core–shell structure which consists of a column linked to a rotatable motor and a stationary hollow tube. With the same central angle, the copper strip and PTFE film can be aligned completely. At the aligned position, electrons are transferred from copper to PTFE surface because of their opposite triboelectric polarities. Since positive charges on copper surface and negative charges on PTFE surface are equal, no electron flows in the external circuit. With a relative rotation occurring, applied by a rotary motor, triboelectric charges on the mismatched areas cannot be compensated. Consequently, the generated electric field will drive electron flow through the external circuit to produce current. For the parallel connection of six sliding units, output current from different units are synchronized. As a result, a high I_{sc} of 60 μA was achieved at a rotation speed of 1,000 r minute^{-1}. Besides, the V_{oc} reached a maximum value of 320 V as well.

3.2.3 Semiconductor Device Structure

3.2.3.1 The Role of Semiconductor for TENGs

From the viewpoint of triboelectric material science, dielectric-to-dielectric and dielectric-to-conductor types of TENGs are common approaches to achieve an effective triboelectrification conversion. This is because they generally have a larger

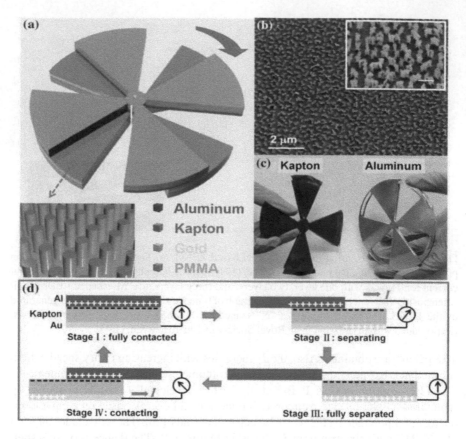

FIGURE 3.20 Basic structure and proposed working principle of the Kapton–Al based TENG. (a) A schematic structure of the TENG. (b) Top view SEM image of Kapton nanorods. (c) Photograph of the two parts of a real TENG. (d) The working principle of TENG with the electron flow diagram in four consecutive stages within a full cycle of electricity generation. (Reproduced with permission from American Chemical Society Ref. [19].)

difference in triboelectric polarities so as to increase charge transfer in triboelectrification process. Therefore, a greater tribocharge density can be produced on the contact surface, which determines the output performance of a TENG. However, with the development of TENG research, many semiconductor materials which not only have certain triboelectric properties, but also have other good electrical-conversion characteristics such as photoelectric properties and piezoelectric properties have been applied as functional layers in TENG designs for enhancing the electrical output through the combination of multienergy conversion mechanisms. Just based on the multifunctional nature of semiconductors, in recent researches, developing hybrid semiconductor-based TENGs that can simultaneously convert different forms of energy or realize multiconversion mechanisms has become an important direction to effectively enhance the device's output. According to the material paired with the semiconductor, semiconductor TENGs can be divided into two categories: semiconductor-to-dielectric device and semiconductor-to-conductor device.

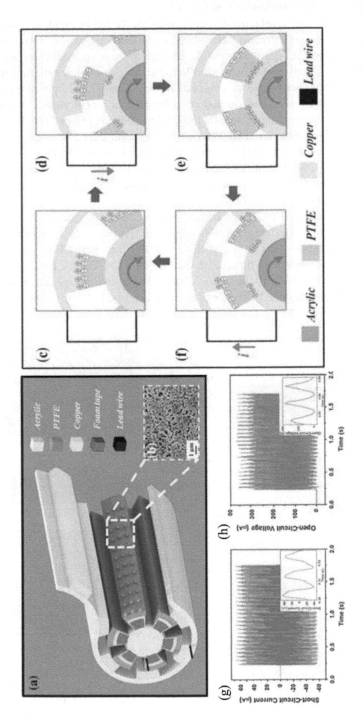

FIGURE 3.21 The rotating PTFE–copper based TENG. (a) Schematic of device with six strip units. (b) SEM image of PTFE surface. (c–f) The working principle of the rotating TENG. The output capability of this TENG: the I_{sc} (g) and the V_{oc} (h). (Reproduced with permission from American Chemical Society Ref. [20].)

3.2.3.2 Semiconductor-to-Dielectric Device

Recently, we demonstrated a dielectric-to-semiconductor TENG by employing semi-conductor organic–inorganic hybrid perovskite as the triboelectric layer and light absorber [21], as shown in Figure 3.22. With the high absorption coefficient across visible spectrum, low trap-state density, and long carrier diffusion length and life-time, organic–inorganic perovskite as a class of approximately ideal light absorbers have been extensively focused and studied on solar cells in recent years. Just over the past several years, perovskite solar cells (PSCs) have been developed rapidly, and their power conversion efficiency (PCE) has already exceeded 24%, showing vigor-ous development trend and potential for applications. In our device structure of FTO (Fluorine-doped SnO_2 transparent conductive electrode)/TiO_2/perovskite/pentacene-PTFE/Al, we introduced planar TiO_2 as electron transport layer (ETL) and ultrathin pentacene as hole transport layer (HTL).

Without illumination, a TENG can work and merely convert vibratory energy into electricity when a repeated external force is applied to the mover of a linear motor. At initial state, no charge exists on the surface of PTFE and perovskite. But after two friction layers are physically brought in contact for a few cycles, tribonegative charges accumulate on PTFE surface, while equal tribopositive charges accumulate on perovskite surface due to the difference of electron affinity. Later, the triboelectric

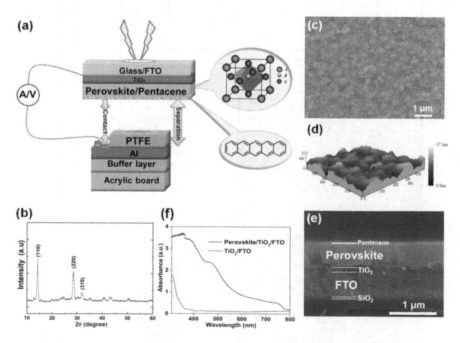

FIGURE 3.22 A semiconductor-to-conductor TENG with perovskite–PTFE structure. (a) Schematic diagram of the TENG, (b) XRD pattern, (c) surface morphology SEM and (d) roughness AFM of the perovskite film, (e) cross-section SEM of the FTO/TiO_2/perovskite/pentacene friction layer And (f) UV-vis absorbance of perovskite film on TiO_2/FTO. (Reproduced with permission from Springer Nature Ref. [21].)

charges will produce reciprocal inductive charges on the FTO and Al electrodes and induce an oscillating V_{oc} (i.e., electric potential difference) between them as the devices are in periodic contact–separation. In this case, the electrical output signals (V_{oc}, I_{sc}, and Q_{sc}) are just determined by charge densities of the triboelectric surface.

However, under illumination, based on the coupling of triboelectric and photo-electric conversion mechanisms, the as-prepared device can simultaneously harvest vibratory energy and solar energy. Compared to the dark condition, in a TENG with ~0.7 cm² effective area, the V_{oc}, I_{sc}, and maximum transfer charge (Q_{sc}) are increased by 55.7%, 50.8%, and 58.2% upon illumination, respectively, as shown in Figure 3.23a–f. In comparison with the dark condition shown in Figure 3.23g, the experimental results have indicated that the significant improvement of output performance under illumination is mainly attributed to two reasons: On one hand, a larger number of electron–hole pairs are generated within perovskite upon illu-mination, as shown in Figure 3.23h. For the introduction of TiO$_2$ ETL and penta-cene HTL distributes an optimized energy-level arrangement in device structure, the photo-induced electrons would be favorably transported away from perovskite to TiO$_2$ and then FTO. Meanwhile, the photo-induced holes are quickly captured by pentacene. Thus, the more effective charge separation and transfer in perovskite interface leads to a considerable increase of surface charge density of perovskite. On the other hand, the reduction of internal resistance in the device is another impor-tant source to improve the device output due to the improvement of conductivity in perovskite film under illumination.

Besides, the TENG shows a fast response on both the full-spectrum simulated sunlight and monochromatic light extending from ultraviolet to entire visible region which enhances its potential application in photodetection. This work presents a route to designing a high-performance TENG with good photoelectric materials as function layers via photoelectric–triboelectric hybridization.

3.2.3.3 Semiconductor-to-Conductor Device

In 2019, our group also developed a brand-new semiconductor-to-conductor TENG with a semiconductor TiO$_2$ and an ultrathin metal Au as the friction layers [22], as shown in Figure 3.24. The upper TiO$_2$ layer is a triboelectric layer and a good light absorber as well, while lower Au layer acts as another triboelectric layer and an electrode simultaneously. The two parts are fixed onto the stator and mover of a linear motor in the experimental measurement for harvesting the vibratory energy and solar energy.

In the dark condition, when TiO$_2$ and Au are fully in contact under mechanical force, the existing electrons on TiO$_2$ are transferred from its surface to Au. When TiO$_2$ and Au are separated, the positive charges on TiO$_2$ surface and negative charges on Au surfaces will induce potential difference in space, driving electron flow from Au electrode to FTO electrode. Thus, there is a positive current in the external cir-cuit. While they approach each other, the potential difference between FTO and Au decreases, and therefore electrons flow from FTO back to Au, generating the nega-tive current.

Compared with the dark condition, as shown in Figure 3.25, the current polarity is reversed instantaneously upon illumination with 12 and 2 times enhancement for

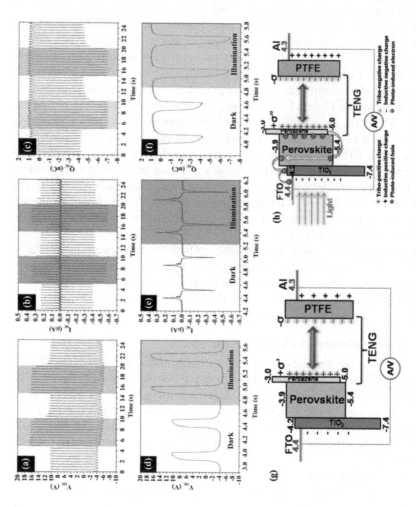

FIGURE 3.23 The performance of the optimal perovskite–PTFE based TENG under switchable illumination. (a–c) are figures of Voc, Isc, and Qsc, respectively. (d–f) are partial enlarged views of a–c, respectively. The working mechanisms of the TENG under darkness(g) and illumination (h). (Reproduced with permission from Springer Nature Ref. [21].)

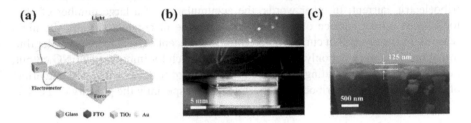

FIGURE 3.24 The structure and characterization of TiO_2–Au based TENG. (a) The schematic diagram of the TENG. (b) The photograph of the TENG under illumination. (c) The cross-sectional SEM image of the TiO_2/FTO segment. (Reproduced with permission from Elsevier Ref. [22].)

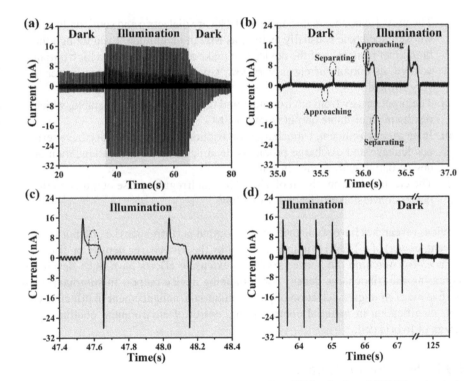

FIGURE 3.25 The comparison and enlarged details of TiO_2–Au based TENG under darkness and under illumination. (a) The output current of the TENG under darkness and illumination. The enlarged plot of the TENG corresponding to outputs at the moment of (b) switching the light on, (c) continuous light illumination, and (d) switching the light off. (Reproduced with permission from Elsevier Ref. [22].)

negative and positive short-circuit current peak, respectively. Moreover, the TENG can be transformed to a solar cell and generate a photocurrent plateau due to the formation of metal–semiconductor Schottky contact of TiO_2/Au in the device. By carefully analyzing the charge transfer in the dark and under illumination, it is found that the surface states play a key role in determining the magnitude and polarity of triboelectric current. In other words, the accumulation of a large number of photoelectrons on the surface of TiO_2 and the increase of conductivity are the main reasons for the reversal of current polarity and significant current increase under illumination. This work not only provides a new approach for improving TENG output, but also a new understanding about the mechanism of triboelectrification and effect of surface states on the triboelectric performance, especially the polarity of TENG.

3.3   THE OUTPUT ENHANCEMENT MECHANISM IN TENGs

However, the output capacity of a single TENG isn't high and stable and seriously hinders its commercial development. In general, there exist following factors influencing the output performance of a TENG.

i. The main functional materials of a TENG are dielectric and semiconductor, whose impedance is usually as high as MΩ in magnitude. Even though the high inner resistance of the device can make a TENG produce a high output voltage, its output current is very low. Consequently, the output power is very limited.
ii. The input energy from environment and organisms is quite unstable, which results in an unstable output from a TENG.
iii. In an air environment, tribocharges on frictional material surface have serious leakage and discharge problems, leading to an obvious attenuation of device output with change in time.
iv. The electrical output signal of a TENG is an irregular pulse output, and the power conversion efficiency is quite low.

Previous researches have confirmed that the output performance (i.e., output voltage, current, power) of a TENG is mainly decided by the tribocharge density on frictional surface. To overcome the above problems, extensive efforts have been devoted to increasing the tribocharge density for enhancing device output. In summary, there are five main strategies: selection of paired materials, enhancement in effective contact, modification in material composition, control of environment condition, and design of hybrid cell.

3.3.1   SELECTION OF PAIRED MATERIALS

The first strategy is to select the paired frictional materials with a large difference in triboelectric polarity as far as possible. The relevant methods have been emphasized in the previous discussions.

3.3.2   ENHANCEMENT IN EFFECTIVE CONTACT

The second strategy is to enhance the effective contact area, as shown in Figure 3.26. Due to surface roughness, the effective contact ratio of solid–solid materials is generally far less than 100%. Thus, introducing micro/nanostructures, such as nanorods,

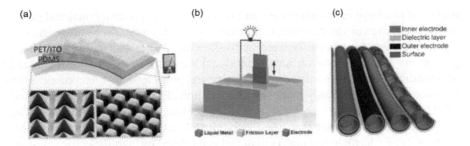

FIGURE 3.26 Strategy for enhancing the effective contact. (a) Surface microstructure. (b) Liquid/solid interface. (c) Soft contact. (Reproduced with permission from American Chemical Society ref. [23], Wiley Ref. [24] and Springer Nature Ref. [25].)

nanowires, and pyramid/cube-like arrays on triboelectric materials, has been confirmed as a straightforward and useful approach [23], as shown in Figure 3.26a. Further, W. Tang et al. [24] designed the first liquid-metal-based TENG in 2015, as shown in Figure 3.26b. In such a device structure, solid–liquid contact is used instead of solid–solid contact so as to enhance the effective contact greatly and thus achieve an output charge density as high as 430 μC m^{-2} with an instantaneous energy conversion efficiency up to 70.6%. Moreover, J. Wang et al. [25] presented a tube-like shape-adaptive TENG by using soft material silicone rubber as a friction layer, as shown in Figure 3.26c. Thanks to closer contact of triboelectric materials, the device obtained a high charge density of 250 μC m^{-2}.

3.3.3 MODIFICATION OF MATERIAL COMPOSITION

The third strategy is based on the modification of material composition, as shown in Figure 3.27. This method can be classified into two categories: surface functionalization and bulk modification. For surface functionalization, by exposing the chemical function groups on triboelectric material surface, the charge capture ability of materials is boosted obviously. As an example, a study by S. Wang et al. [26] utilized thiol and silane monolayers to modify Au film and SiO$_2$ film in a TENG, respectively, as shown in Figure 3.27a. It is found that the output of an Au-based TENG is

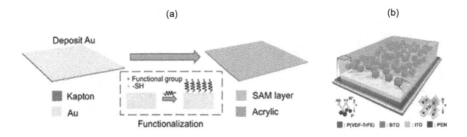

FIGURE 3.27 Strategy for modifying material composition. (a) Surface functionalization. (b) Bulk composition modification. (Reproduced with permission from Royal Society of Chemistry Ref. [26] and Wiley Ref. [27].)

improved to a large extent when the more triboelectrically positive functional group amine is attached on the Au surface. Similarly, the SiO_2-based TENG also achieved a great increase of output performance with silane molecules with amine as the head groups. For bulk modification, some functional nanoparticles are uniformly added into the dielectric films to enhance their charge-trapping capability. Accordingly, the output of a TENG obviously improves. For example, PDMS film using $BaTiO_3$ and $SrTiO_3$ with high dielectric constant as fillers can enhance TENG performance. Also, experiment has confirmed that when poly(vinylidenefluoride-*co*-trifluoroethylene) (P(VDF-TrFE)) ferroelectric matrix material is modified by high-dielectric-constant $BaTiO_3$ [27], both its charge-attracting and charge-trapping capabilities are boosted dramatically, as shown in Figure 3.27b. Therefore, the device's output power is increased by 150 times compared to a device without any modification.

3.3.4 CONTROL OF ENVIRONMENTAL CONDITIONS

The fourth strategy is to control the working environment conditions of a TENG, such as temperature and pressure, as shown in Figure 3.28. Many environmental factors affecting TENG output performance have been investigated. In 2017, C.X. Lu et al. [28] studied the temperature influences on the performance of a TENG, which is based on Al and PTFE as paired frictional materials, as shown in Figure 3.28a. Experimental results have shown that the device output decreased gradually with the temperature increasing from −20°C to 20°C, was a relatively stable value from 20°C to 100°C, but dropped rapidly over 100°C. The dependence of device output on temperature is mainly related to the decrease of material permittivity and surface defects induced by the increase of temperature, such as surface oxidation or defluorination. Xu et al. have also demonstrated that the ability to capture charges is weakened under high-temperature conditions. In addition, atmospheric pressure as another factor influencing the TENG output capability has been well studied by J. Wang et al. [29] as shown in Figure 3.28b. When operating in high vacuum with pressure of ~10^{-6} torr, a TENG based on Cu–PTFE achieved a record-high charge density of

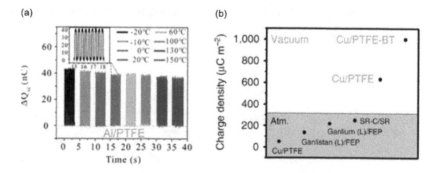

FIGURE 3.28 Strategies for controlling environmental conditions. (a) Changing temperature. (b) Changing pressure. (Reproduced with permission from Wiley Ref. [28] and Springer Nature Ref. [29].)

660 µC m^{-2}. These results have revealed that the higher the vacuum degree, the smaller the air breakdown effect, and the higher the device output.

3.3.5 Designs of Hybrid Cells

The fifth strategy is to design a hybrid TENG by integrating triboelectric conversion with other electrical conversion mechanisms, such as photoelectricity, piezoelectricity, and pyroelectricity. For example, a hybrid rotating-disk-based electromagnetic–triboelectric nanogenerator [30] has been proposed for converting rotary energy as a mobile power source, as shown in Figure 3.29a. The hybrid device has a planar structure with a diameter of 14 cm, which can produce an output power of 17 mW for TENG and 50 mW for electromagnetic generator (EMG) via harvesting mechanical energy from human motion. Also, this device shows a better charging capability than an individual harvesting unit in charging a capacitor of 6,600 µF. A commercial light bulb can be lit instantly by such a TENG, showing its potential for application in driving mobile electronic products. In addition, M. Han et al. [31] demonstrated a hybrid r-shaped piezoelectric–triboelectric nanogenerator for a more effective harvesting for the same mechanical input. The device is composed of an upper piezoelectric nanogenerator (PENG) and a lower TENG, whose schematic diagram is shown in Figure 3.29b. At the frequency of 5 Hz and periodic external force, the power density of PENG and TENG reach 10.95 and 2.04 mW cm^{-3}, respectively. The output of the hybrid device can power LEDs and LCDs, and the device has potential for applications in our daily life.

(a) Electromagnetic triboelectric hybridization

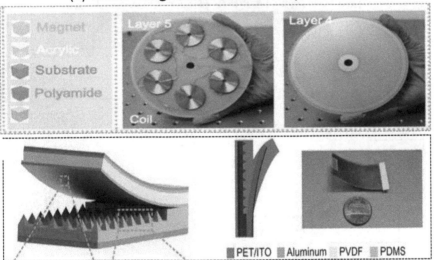

(b) Piezoelectric triboelectric hybridization

FIGURE 3.29 Strategies for designing hybrid cell. (a) Electromagnetic–triboelectric hybridization. (b) Piezoelectric–triboelectric hybridization. (Reproduced with permission from American Chemical Society [30,31].)

3.4   RECENT ADVANCEMENT OF TENGs FOR NANOENERGY

Due to many advantages such as wide range of material sources, multifunctional modes, simple preparation, low cost, and effective ability to harvest environmental mechanical energy, TENGs have attracted continuous attention from all over the world. At present, more than 40 countries, 400 organizations, and 3,000 scientists are devoted to studying TENGs. In recent years, TENGs have made many significant progresses in the fields of micro/nanoenergy, self-powered sensing, and "blue energy," as shown in Figure 3.30. First of all, TENG can convert almost all kinds of mechanical energy in our life, including human activities [32–34], vibration [35,36], mechanical trigger [37], tire rotation [38], wind energy [39,40], water energy [41], etc. into electrical energy for powering microelectronic devices, such as electric skin, implantable medical devices, and wearable electronic products. And the electrical signal (V_{oc}, I_{sc}, etc.) can be used as a sensing signal in many fields [4,42–44], such as the IoT, medical monitoring, environmental monitoring, infrastructure monitoring, and human–machine interaction. Besides, the network units integrated with several TENGs are expected to provide a brand-new technical solution for macroscale "blue ocean energy" [45,46].

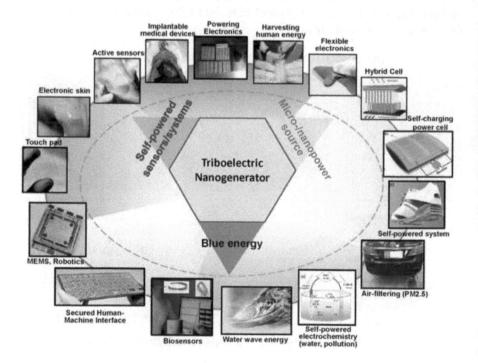

FIGURE 3.30   Summary of the three major fields of application of TENG: micro/nanopower sources, self-powered sensors/systems (or active sensors) and megascale blue energy. (Reproduced with permission from Elsevier Ref. [7].)

3.4.1 Micro/Nano Power Source

The recent key advances in TENG can be divided into two categories: one is shape adaptive devices, and the other is a fiber-based wearable power sources.

The shape-adaptive devices use elastic and deformed materials as functional materials, making them more natural and effective for human body. For example, a tube-like silicone TENG [47], which can be installed under shoes or clothes, has been developed to generate electricity from walking to power wearable electronic products such as watches and fitness trackers. Subsequently, Yi et al. [48] reported a highly stretchable TENG consisting of a conductive liquid and a soft rubber, as shown in Figure 3.31a. The stretchable device is very easy to adapt to uneven surface and is used in insoles and wearable bracelets to collect energy from walking and hand clapping. After a deformation period of 55,000, the device can withstand a tensile strain of up to 300% without obvious attenuation in electrical output. As shown in Figure 3.31b, Pu et al. [49] further improved the ability of device to withstand uniaxial strain to 1,160% based ionic conductor as electrode. In addition to energy capture in vitro, Zheng et al. [50] also reported a flexible and biodegradable TENG for the short-term conversion of energy in vivo, as shown in Figure 3.31c. It can capture energy from heartbeat, breathing exercises, and blood pressure for applications in transient, implantable, self-powered therapeutic or diagnostic devices, enabling the successful use of electricity-assisted neuronal cell localization.

The emergence of wearable electronic products and fiber-based power sources, such as batteries, supercapacitors, and nanogenerators, has drawn the attention of a lot of scientists. In 2014, Zhong et al. [51], for the first time, proposed a fiber-based TENG which can be woven into fabrics. As shown in Figure 3.32a and b, this device is composed of two kinds of core–shell fibers: one is carbon nanotubes coated with cotton threads, and the other is carbon nanotubes coated with PDMS. The fiber-based TENG coat can charge a commercial capacitor and trigger a wireless monitoring system for body temperature (in Figure 3.32c). Further, Dong et al. [52] firstly introduced a 3D woven structure into the design of fiber TENG by using stainless steel/polyester yarn instead of carbon nanotube, improving output performance of the device (in Figure 3.32d). The design can also be easily extended to a large area of wearable clothing. In addition, as shown in Figure 3.32e, it is demonstrated that silicone employed as core material in core–shell structure not only can make the fiber-based TENG with higher tensile strength [53], which can convert mechanical energy into electrical output more effectively. And based on a mixture of two polymer fibres, the reported hybrid power textile [54] introduces an innovative module fabricating strategy, which can harvest energy from both ambient sunlight and mechanical excitation simultaneously, capable of generating sufficient power for various practical applications, including continuously driving an electronic watch, directly charging a cell phone and driving the water splitting reactions (Figure 3.32f-h).

3.4.2 Self-Powered Sensors

TENG has shown a great potential in the fields of active sensing and self-powered sensors, because it can directly use electrical output signals for real-time sensing mechanical excitation signals without the need of extra converter. Many pioneering

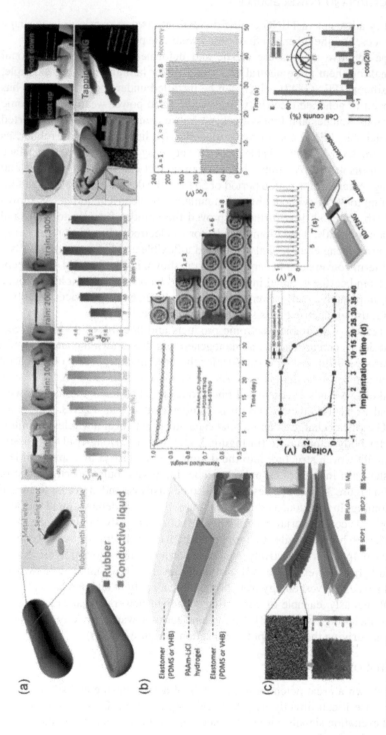

FIGURE 3.31 Shape-adaptive TENGs for conformal power sources. (a) A highly stretchable TENG composed of conductive liquid and soft rubber. (b) A skin-like TENG consisting of ionic hydrogel and elastomer films. (c) An implantable and biodegradable TENG. (Reproduced with permission from American Association for the Advancement of Science Refs. [48–50].)

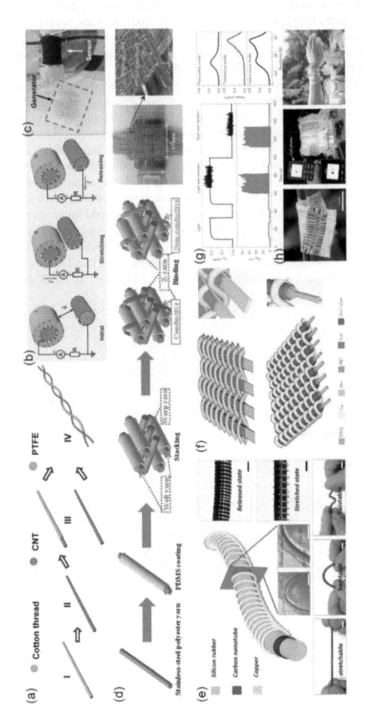

FIGURE 3.32 Fiber-based TENGs for wearable power sources. (a–c) The first fiber-based TENG and its fabrication process, working mechanism, and demonstration of a "power shirt," respectively. (d) The image of a 3D orthogonally woven TENG and its fabrication process. (e) A stretchable fiber-based TENG using silicone rubber as the core material. (f–h) A hybrid power textile's structural design (f), electrical outputs (g), and (h) demonstration of the power textile driving electronic devices. (Reproduced with permission from American Chemical Society Ref. [51], Wiley Refs. [52,53] and Royal Society of Chemistry Ref. [54].)

TENG designs including textile sensors [55–57], acoustic sensors [58,59], motion and acceleration sensors [60–62], and chemical sensors [63–66] have been demonstrated. With the rapid development of IoTs, advanced human–machine interaction has become more and more important for realizing convenient, safe, and novel communication between people and external devices [67,68].

Now, network security is getting increasing attention; therefore, it is very urgent to develop an effective and sustainable authentication solution. Previous researches have attested that the keyboard operation, i.e., a dynamic behavior with biological characteristic based on human input property, can improve network security level without interfacing with the user [69–71]. The concept to realize keyboard input using TENG sensor array was proposed by Chen et al. in 2015 [72]. Based on this work and the pioneering work by Li et al. [73] on TENG keyboard, a two-factor security dynamic system based on pressure-enhanced keyboard input was developed [74], as shown in Figure 3.33. The key for this system consists of a shielded electrode and a contact–separation TENG based on silica gel as the main structural material and ITO/PET as the electrode, which is able to authenticate and even identify users through their unique input behaviors with an accuracy rate of 98.7%. It is believed that TENG-based security system has potential for application in the computer and financial fields, which

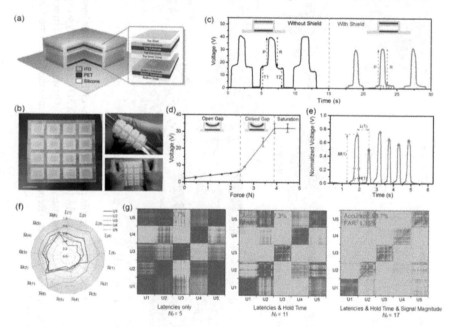

FIGURE 3.33 A keystroke-dynamics-enhanced security system with TENG array. (a) Schematic of a single TENG key. (b) Photographs of a triboelectric numeric keypad with 16 keys. (c) Comparison of electrical outputs from the triboelectric key and a reference key without shield electrode. (d) Relationship between the maximum output voltage and typing force. (e) Typical keystroke features from typical signals for constructing user profile models. (f) The radar plot of the normalized mean feature values of five users. (g) The difference score matrices across user inputs with different number of features. (Reproduced with permission from Elsevier Ref. [74].)

could promote cybersecurity to the next level. Moreover, by integrating with optical components such as LED, a single-electrode TENG key can serves as an optical wireless transmitter, which can be used to build a wireless touch-screen for security authentication and further for optical remote-control and pressure sensing [75].

Aside from the application of traditional human–machine interaction devices such as keyboard, TENG sensor was also used to construct a new type of hands-free human–machine interaction equipment that can convert real-time microactions of blinking into control commands [76]. As shown in Figure 3.34, TENG is a multilayer film structure based on tadpole-like PET film as substrate, FEP film and natural latex film as triboelectric functional layers, acrylic ring as spacer, ITO coated with FEP as a single electrode. By employing the system, blinking can work like a remote prosthetic hand to control the "on/off" state of appliances such as lamps, fans, and doorbells. These noninvasive and highly sensitive TENG sensors allow people to control and communicate with their devices via an unprecedented way.

Besides, acoustic human–machine interaction that converts sound into electrical signals is another important solution in the technological fields of information security and intelligent robotics. In 2014, Yang et al. [77] developed the first acoustic TENG device, which could be used as both a sustainable energy harvester and an active sensor. Subsequently, a paper-based TENG has been demonstrated with similar performance [78] and can also be used as sound source locators and self-powered microphones. Recently, a self-powered hearing sensor has been proposed for social robots and hearing aids [79]. The device is composed of an acrylic/Au substrate attached to FEP film, a spacer, and a Kapton film coated with Au, as shown in Figure 3.35. When the device is working, the deformation of the membrane caused by sound wave triggers the contact–separation of FEP film and Au electrode, thus generating electrical output. So, it has great potential for application in speech recognition safety system and hearing aids.

3.4.3 BLUE ENERGY

TENG which can almost harvest all kinds of mechanical energy in our environment, such as wind energy, rain-drop energy, ultrasonic energy, and water wave energy, has attracted many scientists' extensive research interest. Recent studies have shown that TENG is more effective in capturing the low-frequency energy (i.e., <5 Hz) than electromagnetic generator. Therefore, it is particularly important to use TENG to obtain the water wave energy in the vast ocean, namely "blue energy" [80]. Researchers have begun to study how to effectively connect multiple TENG units to achieve efficient large-scale array power generation networks.

There are many innovative device structures which have been designed for blue energy collection, such as parallel electrodes [81,82] and arch devices surrounding rolling ball [83]. Among these structures, the fully enclosed ball–shell structure [84] is considered to be most promising because it can obtain energy from all directions, as shown in Figure 3.36a–d. Further, soft silicone rubber used to prepare the dielectric layers on the ball and shell increases the effective frictional area due to the improvement of contact. Especially, when the ultraviolet treated silica balls and

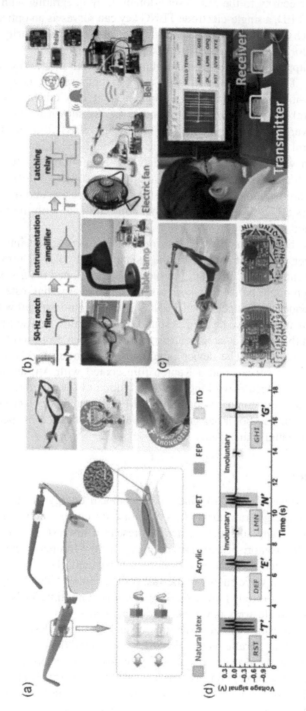

FIGURE 3.34 Eye-motion-triggered mechnosensational communication system based on TENG. (a) Schematic diagram and photographs of the TENG. (b) A smart home control system based on the TENG. (c) A hands-free typing system based on the TENG. (d) Voltage signals from voluntarily blinking the eye when typing the word "TENG." (Reproduced with permission from American Association for the Advancement of Science Ref. [76].)

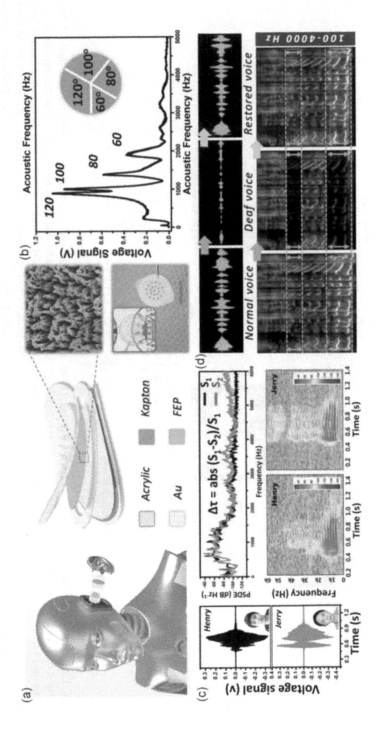

FIGURE 3.35 A TENG auditory sensor (TAS) for social robots and hearing aids. (a) Schematics and photographs of the TAS. (b) Broadband frequency response of the TAS with four sectorial boundaries. (c) Security application of the TAS in voice recognition. (d) Medical application of the TAS in hearing aids for restoring impaired frequency region. (Reproduced with permission from American Association for the Advancement of Science Ref. [79].)

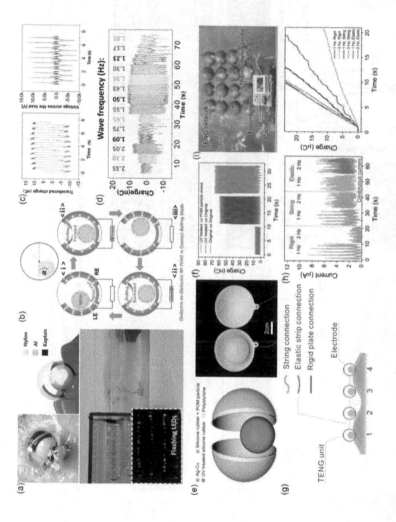

FIGURE 3.36 Ball–shell structured blue energy harvester using TENG. (a–d) A rolling-structured freestanding triboelectric-layer mode TENG. (a) Photographs, (b) schematic and working principle, (c) typical electrical outputs, and (d) output dependency on the wave frequency. (e–i) Coupled TENG network. (e) Schematic and photograph of silicone-based ball–shell TENG. (f) Comparison of output performance by modifying the silicone material. (g) Schematic of different types of connections with TENG units. (h) Comparison of output performance of different connections. (i) Demonstration of a string-connected TENG network to power a thermometer. Reproduced with permission from Wiley Ref. [84] and American Chemical Society Ref. [85].)

polyformaldehyde particles are added to the dielectric layer, the output performance is greatly improved, as shown in Figure 3.36e–i. In order to obtain blue energy on a large scale, Wang et al. [85] first proposed the concept of using a coupling network, but the optimal connection mechanism between TENG units is a key study. These results suggest considerable potential for TENG performance improvement and its potential for application in large-scale blue energy applications.

As shown in Figure 3.37, in recent research, depending on its high-voltage output character, TENG has also shown its potential for application in automobile exhaust treatment, air purification, miroplasma, etc.

3.5 CHALLENGE AND PERSPECTIVE

Indeed, TENG has made some considerably important progress in material development, structural designs, output capability, and practical application in recent years. However, the development of TENGs still faces the following critical problems at present, which may guide the development directions of TENGs in the near future.

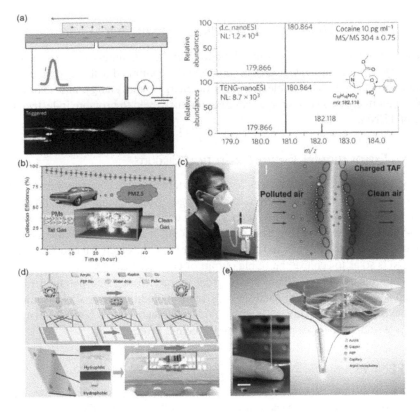

FIGURE 3.37 Application of TENGs as direct high-voltage power sources. (a) Sensitive nanocoulumb molecular mass spectrometry. (b) Particulate matter (PM) removal from automobile exhaust fumes. (c) Washable air filter. (d) Microfluidic transport system. (e) Triboelectric microplasma. (Reproduced with permission from Springer Nature Refs. [86,87], American Chemical Society Refs. [88,89], and Wiley Ref. [90].)

3.5.1 RESEARCH ON HIGH-PERFORMANCE SEMICONDUCTOR MATERIALS FOR TENGS

Due to the large internal resistance of dielectric materials, the output current of the reported TENGs is small generally. Therefore, the future research should focus on reducing the material's internal resistance and improving the performance indexes so as to enhance the device's output power and energy conversion efficiency.

3.5.2 IN-DEPTH STUDY ON WORKING PRINCIPLE AND SIMULATION MODELS

Up to now, the basic principle and simulation model of TENGs are still being improved, and new structures are also being proposed, but an authoritative theoretical system has not been established yet. Therefore, more research studies in the future need to focus on the working principles and simulation models. Then, the structures and parameters of TENGs will be optimized for the specific situations, so that TENGs can be used in actual applications more quickly.

3.5.3 RESEARCH ON HYBRID ENERGY UNITS

Each kind of nanogenerator has its advantages and disadvantages. If different kinds of nanogenerators are integrated or organically combined with other electricity-generation methods, such as electromagnetic energy, wind energy, and solar energy, to form an integrated generator, it can tremendously help to promote industrial progress of TENG.

3.5.4 EXPLORATION OF ENERGY COLLECTION CIRCUITS

The operation of electronic devices requires a stable electrical energy input. But the electrical energy output of TENGs often changes with the changes of the external excitations, so its output stability is poor, and it cannot provide stable input for electronic devices. Therefore, the next research should try to explore a suitable energy collection circuit for TENGs and convert and/or store the electrical energy generated by TENGs so as to achieve the stable output of electrical energy.

To summarize, TENGs have shown great research value and potential for commercial applications in energy and sensing fields, including personal electronic products, environmental monitoring, and biomedical detection. It is believed firmly that TENGs will appear in people's daily life in the near future.

REFERENCES

1. Z.L. Wang, W. Wu. Nanotechnology-enabled energy harvesting for self-powered micro-/nano systems. *Angew. Chem. Int. Ed.*, 2012, 51, 11700–11721.
2. F.R. Fan, Z.Q. Tian, Z.L. Wang. Flexible triboelectric generator. *Nano Energy*, 2012, 1, 328–334.
3. Z.L. Wang. Triboelectric nanogenerators as new energy technology and self-powered sensors-principles, problems and perspectives. *Faraday Discuss.*, 2015, 176, 447–458.

4. S. Wang, L. Lin, Z.L. Wang. Triboelectric nanogenerators as self-powered active sensors. *Nano Energy*, 2015, 11, 436–462.

5. G. Zhu, C.F. Pan, W.X. Guo, C-Y. Chen, Y.S. Zhou, R.M. Yu, Z.L. Wang. Triboelectric-generator-driven pulse electrodeposition for micropatterning. *Nano Lett.*, 2012, 12(9), 4960–4965.

6. C. Wu, A.C. Wang, W. Ding, H. Guo, Z.L. Wang. Triboelectric nanogenerator: A foundation of the energy for the new era. *Adv. Energy Mater.*, 2018, 9(1), 1802906.

7. Z.L. Wang. On Maxwell's displacement current for energy and sensors: The origin of nanogenerators. *Mater. Today*, 2017, 20(2), 74–82.

8. Z.L. Wang. On the first principle theory of nanogenerators from Maxwell's equations. *Nano Energy*, 2020, 68, 104272.

9. http://owlsmag.wordpress.com/2010/01/20/a-natural-history-devin-corbin/.

10. https://en.wikipedia.org/wiki/Van_de_Graaff_generator.

11. H. Zou, Y. Zhang, L. Guo, P. Wang, X. He, G. Dai, H. Zheng, C. Chen, A.C. Wang, C. Xu, Z.L. Wang. Quantifying the triboelectric series. *Nat Commun.*, 2019, 10, 1427. doi: 10.1038/s41467-019-09461-x.

12. S.H. Wang, L. Lin, Y.N. Xie, Q.S. Jing, S.M. Niu, Z.L. Wang. Sliding-triboelectric nanogenerators based on in-plane charge-separation mechanism. *Nano Lett.*, 2013, 13(5), 2226–2233.

13. P.-K. Yang, Z.-H. Lin, K.C. Pradel, L. Lin, X. Li, X. Wen, J.-H. He, Z.L. Wang. Paper-based origami triboelectric nanogenerators and self-powered pressure sensors. *ACS Nano*, 2015, 9(1), 901–907.

14. Y. Yang, H. Zhang, Z.-H. Lin, Y.S. Zhou, Q. Jing, Y. Su, J. Yang, J. Chen, C. Hu, Z.L. Wang. Human skin based triboelectric nanogenerators for harvesting biomechanical energy and as self-powered active tactile sensor system. *ACS Nano*, 2013, 7(10), 9213–9222.

15. S. Niu, S. Wang, L. Lin, Y. Liu, Y. Zhou, Y. Hu, Z.L. Wang. Theoretical study of contact-mode triboelectric nanogenerators as an effective power source. *Energy Environ. Sci.*, 2013, 6, 3576–3583.

16. S.H. Wang, L. Lin, Z.L. Wang. Nanoscale triboelectric-effect-enabled energy conversion for sustainably powering portable electronics. *Nano Lett.*, 2012, 12(12), 6339–6346.

17. J. Chun, J.W. Kim, W.S. Jung, C.Y. Kang, S.W. Kim, Z.L. Wang, J.M. Baik. Mesoporous pores impregnated with Au nanoparticles as effective dielectrics for enhancing triboelectric nanogenerator performance in harsh environments. *Energy Environ. Sci.*, 2015, 8(10), 3006–3012.

18. X.D. Yang, J.J. Han, F. Wu, X. Rao, G.D. Zhou, C.Y. Xu, P. Li, Q.L. Song. A novel retractable spring-like-electrode triboelectric nanogenerator with highly-effective energy harvesting and conversion for sensing road condition. *RSC Adv.*, 2017, 7, 50993–51000.

19. L. Lin, S.H. Wang, Y.N. Xie, Q.S. Jing, S.M. Niu, Y.F. Hu, Z.L. Wang. Segmentally structured disk triboelectric nanogenerator for harvesting rotational mechanical energy. *Nano Lett.*, 2013, 13(6), 2916–2923.

20. P. Bai, G. Zhu, Y. Liu, J. Chen, Q.S. Jing, W.Q. Yang, J.S. Ma, G. Zhang, Z.L. Wang. Cylindrical rotating triboelectric nanogenerator. *ACS Nano*, 2013, 7(7), 6361–6366.

21. X.D. Yang, J.J. Han, G. Wang, L.P. Liao, C.Y. Xu, G.D. Zhou, W. Hu, P. Li, B. Wu, Q.L. Song. Robust perovskite-based triboelectric nanogenerator enhanced by broadband light and interface engineering. *J. Mater. Sci.*, 2019, 54(12), 9004–9016.

22. J.J. Han, X.D. Yang, L.P. Liao, G.D. Zhou, G. Wang, C.Y. Xu, W. Hu, Q.L. Song. Photoinduced triboelectric polarity reversal and enhancement of a new metal/semiconductor triboelectric nanogenerator. *Nano Energy*, 2019, 58, 331–337.

23. F.-R. Fan, L. Lin, G. Zhu, W. Wu, R. Zhang, Z.L. Wang. Transparent triboelectric nano-generators and self-powered pressure sensors based on micropatterned plastic films. *Nano Lett.*, 2012, 12, 3109.

24. W. Tang, T. Jiang, F.R. Fan, A.F. Yu, C. Zhang, X. Cao, Z.L. Wang. Liquid-metal electrode for high-performance triboelectric nanogenerator at an instantaneous energy conversion efficiency of 70.6%. *Adv. Funct. Mater.*, 2015, 25, 3718.

25. J. Wang, S. Li, F. Yi, Y. Zi, J. Lin, X. Wang, Y. Xu, Z.L. Wang. Sustainably powering wearable electronics solely by biomechanical energy. *Nat.Commun.*, 2016, 7, 12744–12751.

26. S. Wang, Y. Zi, Y.S. Zhou, S. Li, F. Fan, L. Lin, Z.L. Wang. Molecular surface functionalization to enhance the power output of triboelectric nanogenerators. *J. Mater. Chem. A*, 2016, 4, 3728.

27. W. Seung, H.-J. Yoon, T.Y. Kim, H. Ryu, J. Kim, J.-H. Lee, J.H. Lee, S. Kim, Y.K. Park, Y.J. Park, S.-W. Kim. Boosting power-generating performance of triboelectric nanogenerators via artificial control of ferroelectric polarization and dielectric properties. *Adv. Energy Mater.* 2017, 7, 1600988.

28. C.X. Lu, C.B. Han, G.Q. Gu, J. Chen, Z.W. Yang, T. Jiang, C. He, Z.L. Wang. Temperature effect on performance of triboelectric nanogenerator. *Adv. Eng. Mater.*, 2017, 19, 1700275.

29. J. Wang, C. Wu, Y. Dai, Z. Zhao, A. Wang, T. Zhang, Z.L. Wang. Achieving ultrahigh triboelectric charge density for efficient energy harvesting. *Nat. Commun.*, 2017, 8, 88.

30. K. Zhang, X. Wang, Y. Yang, Z.L. Wang. Hybridized electromagnetic-triboelectric nanogenerator for scavenging biomechanical energy for sustainably powering wearable electronics. *ACS Nano*, 2015, 9(4), 3521–3529.

31. M. Han, X.-S. Zhang, B. Meng, W. Liu, W. Tang, X. Sun, W. Wang, H. Zhang. A r-shaped hybrid nanogenerator with enhanced piezoelectricity. *ACS Nano*, 2013, 7(10), 8554–8560.

32. T. Huang, C. Wang, H. Yu, H.Z. Wang, Q.H. Zhang, M.F. Zhu. Human walking-driven wearable all-fiber triboelectric nanogenerator containing electrospun polyvinylidene fluoride piezoelectric nanofibers. *Nano Energy*, 2015, 14, 226–235.

33. T.C. Hou, Y. Yang, H.L. Zhang, J. Chen, L.J. Chen, Z.L. Wang. Triboelectric nanogenerator built inside shoe insole for harvesting walking energy. *Nano Energy*, 2013, 2, 856–862.

34. B. Saravanakumar, R. Mohan, K. Thiyagarajan, S.J. Kim. Fabrication of a ZnO nanogenerator for eco-friendly biomechanical energy harvesting. *RSC Adv.*, 2013, 3, 16646–16656.

35. W.Q. Yang, J. Chen, Q.S. Jing, J. Yang, X.N. Wen, Y.J. Su, G. Zhu, P. Bai, Z.L. Wang. 3D stack integrated triboelectric nanogenerator for harvesting vibration energy. *Adv. Funct. Mater.*, 2014, 24, 4090–4096.

36. T. Quan, Y. Wu, Y. Yang. Hybrid electromagnetic-triboelectric nanogenerator for harvesting vibration energy. *Nano Res.*, 2015, 8(10), 3272–3280.

37. X. Wen, W. Yang, Q. Jing, Z.L. Wang. Harvesting broadband kinetic impact energy from mechanical triggering/vibration and water waves. *ACS Nano*, 2014, 8(7), 7405–7412.

38. H. Zhang, Y. Yang, X. Zhong, Y. Su, Y. Zhou, C. Hu, Z.L. Wang. Single-electrode-based rotating triboelectric nanogenerator for harvesting energy from tires. *ACS Nano*, 2014, 8(1), 680–689.

39. H.Y. Guo, J. Chen, L. Tian, Q. Leng, Y. Xi, C.G. Hu. Airflow-induced triboelectric nanogenerator as a self-powered sensor for detecting humidity and airflow rate. *ACS Appl. Mater. Interface*, 2014, 6, 17184–17189.

40. Y.N. Xie, S.H. Wang, S. Lin, Q.S. Jing, Z.H. Lin, S.M. Niu, Z.Y. Wu, Z.L. Wang. Rotary triboelectric nanogenerator based on a hybridized mechanism for harvesting wind energy. *ACS Nano*, 2013, 7, 7119–7125.

41. D. Choi, S. Lee, S.M. Park, H. Cho, W. Hwang, D.S. Kim. Energy harvesting model of moving water inside a tubular system and its application of a stick-type compact triboelectric nanogenerator. *Nano Res.*, 2015, 8(8), 2481–2491.

42. Z.L. Wang. Triboelectric nanogenerators as new energy technology for self-powered systems and as active mechanical and chemical sensors. *ACS Nano*, 2013, 7, 9533–9557.

43. Z.L. Wang, J. Chen. Progress in triboelectric nanogenerators as new energy technology and self-powered sensors. *Energy Environ. Sci.*, 2015, 8, 2250–2282.

44. A.S.M.I. Uddin, G.S. Chung. A self-powered active hydrogen sensor based on a high-performance triboelectric nanogenerator using a wrinkle-micropatterned PDMS film. *RSC Adv.*, 2016, 6, 63030–63036.

45. J. Chen, J. Yang, Z. Li, X. Fan, Y. Zi, Q. Jing, H. Guo, Z. Wen, K.C. Pradel. Networks of triboelectric nanogenerators for harvesting water wave energy: A potential approach toward blue energy. *ACS Nano*, 2015, 9, 3324–3331.

46. X. Wang, S. Niu, Y. Yin, F. Yi, Z. You, Z.L. Wang. Triboelectric nanogenerator based on fully enclosed rolling spherical structure for harvesting low-frequency water wave energy. *Adv. Energy Mater.*, 2015, 5, 1501467.

47. J. Wang, S. Li, F. Yi, Y. Zi, J. Lin, X. Wang, Y. Xu, Z.L. Wang. Sustainably powering wearable electronics solely by biomechanical energy. *Nat.Commun.*, 2016, 7, 12744–12751.

48. F. Yi, X. Wang, S. Niu, S. Li, Y. Yin, K. Dai, G. Zhang, L. Lin, Z. Wen, H. Guo, J. Wang, M.-H. Yeh, Y. Zi, Q. Liao, Z. You, Y. Zhang, Z.L. Wang. A highly shape-adaptive, stretchable and conductive liquid-electrode based triboelectric nanogenerator for energy harvesting and self-powered biomechanical monitoring. *Sci. Adv.*, 2016, 2, e1501624.

49. X. Pu, M. Liu, X. Chen, J. Sun, C. Du, Y. Zhang, J. Zhai, W. Hu, Z.L. Wang. Ultrastretchable, transparent triboelectric nanogenerator as electronic skin for biomechanical energy harvesting and tactile sensing. *Sci. Adv.*, 2017, 3, e1700015.

50. Q. Zheng, Y. Zou, Y. Zhang, Z. Liu, B. Shi, X. Wang, Y. Jin, H. Ouyang, Z. Li, Z.L. Wang. Biodegradable triboelectric nanogenerator as a life-time designed power source for implantable medical devices. *Sci. Adv.*, 2016, 2, e1501478.

51. J. Zhong, Y. Zhang, Q. Zhong, Q. Hu, B. Hu, Z.L. Wang, J. Zhou. Fiber-based generator for wearable electronics and mobile medication. *ACS Nano*, 2014, 8, 6273–6280.

52. K. Dong, J. Deng, Y. Zi, Y.-C. Wang, C. Xu, H. Zou, W. Ding, Y. Dai, B. Gu, B. Sun, Z.L. Wang. 3D orthogonal woven triboelectric nanogenerator for effective biomechanical energy harvesting and as self-powered active motion sensors. *Adv. Mater.*, 2017, 29, 1702648.

53. X. He, Y. Zi, H. Guo, H. Zheng, Y. Xi, C. Wu, J. Wang, W. Zhang, C. Lu, Z.L. Wang. A highly stretchable fiber-based triboelectric nanogenerator for self-powered wearable electronics. *Adv. Funct. Mater.*, 2017, 27, 1604378.

54. J. Chen, Y. Huang, N. Zhang, H. Zou, R. Liu, C. Tao, X. Fan, Z. L. Wang. Micro-cable structured textile for simultaneously harvesting solar and mechanical energy. *Nat Ener.*, 2016, 1, 16138.

55. Q.Z. Zhong, J.W. Zhong, X.F. Cheng, X. Yao, B. Wang, W.B. Li, N. Wu, K. Liu, B. Hu, J. Zhou. Paper-based active tactile sensor array. *Adv. Mater.*, 2015, 27, 7130–7136.

56. Y. Yang, H. Zhang, Z-H. Lin, Y.S. Zhou, Q. Jing, Y. Su, J. Yang, J. Chen, C. Hu, Z.L. Wang. Human skin based triboelectric nanogenerators for harvesting biomechanical energy and as self-powered active tactile sensor system. *ACS Nano*, 2013, 7, 9213–9222.

57. G. Zhu, W.Q. Yang, T. Zhang, Q. Jing, J. Chen, Y.S. Zhou, P. Bai, Z.L. Wang. Self-powered, ultrasensitive, flexible tactile sensors based on contact electrification. *Nano Lett.*, 2014, 14(6), 3208–3213.

58. A. Yu, M. Song, Y. Zhang, Y. Zhang, L. Chen, J. Zhai, Z. Wang. Self-powered acoustic source locator in underwater environment based on organic film triboelectrific nanogenerator. *Nano Res.*, 2015, 8, 765–773.

59. N. Arora, S.L. Zhang, F. Shahmiri, D. Osorio, Y-C. Wang, M. Gupta, Z. Wang, T. Starner, Z.L. Wang, G.D. Abowd. A thin and flexible self-powered microphone leveraging triboelectric nanogenerator. *Proc. ACM Interact. Mob. Wearable Ubiquitous Technol.*, 2018, 2(2) 1–28.

60. Y.S. Zhou, G. Zhu, S. Niu, Y. Liu, P. Bai, Q. Jing, Z.L. Wang. Nanometer resolution self-powered static and dynamic motion sensor based on micro-grated triboelectrification. *Adv. Mater.*, 2014, 26, 1719–1724.

61. F. Yi, L. Lin, S. Niu, J. Yang, W. Wu, S. Wang, Q. Liao, Y. Zhang, Z.L. Wang. Self-powered trajectory, velocity, and acceleration tracking of a moving object/body using a triboelectric sensor. *Adv. Funct. Mater.*, 2014, 24(47), 7488–7494.

62. C. Wu, X. Wang, L. Lin, H. Guo, Z.L. Wang. Paper-based triboelectric nanogenerators made of stretchable interlocking kirigami patterns. *ACS Nano*, 2016, 10(4), 4652.

63. Z.H. Lin, G. Zhu, Y.S. Zhou, Y. Yang, P. Bai, J. Chen, Z.L. Wang. A self-powered triboelectric nanosensor for mercury ion detection. *Angew. Chem. Int. Ed.*, 2013, 52(19), 5065–5069.

64. Z. Li, J. Chen, J. Yang, Y. Su, X. Fan, Y. Wu, C. Yu, Z.L. Wang. Beta-cyclodextrin enhanced triboelectrification for self-powered phenol detection and electrochemical degradation. *Energy Environ. Sci.*, 2015, 8(3), 887–896.

65. Z. Wen, J. Chen, M.-H. Yeh, H. Guo, Z. Li, X. Fan, T. Zhang, L. Zhu, Z.L. Wang. Blow-driven triboelectric nanogenerator as an active alcohol breath analyzer. *Nano Energy*, 2015, 16, 38–46.

66. H. Zhang, Y. Yang, Y. Su, J. Chen, C. Hu, Z. Wu, Y. Liu, C.P. Wong, Y. Bando, Z.L. Wang. Triboelectric nanogenerator as self-powered active sensors for detecting liquid/gaseous water/ethanol. *Nano Energy*, 2013, 2, 693–701.

67. S. Ornes. Core concept: The internet of things and the explosion of interconnectivity. *Proc. Natl. Acad. Sci. U. S. A.*, 2016, 113, 11059.

68. L. Atzori, A. Iera, G. Morabito. The internet of things: A survey. *Comput. Networks*, 2010, 54(15), 2787.

69. S.P. Banerjee, D.L. Woodard. Biometric authentication and identification using keystroke dynamics: A survey. *J. Pattern Recognit. Res.*, 2012, 7(1), 116.

70. F. Monrose, A.D. Rubin. Keystroke dynamics as a biometric for authentication. *Future Gener. Comput. Syst.*, 2000, 16(4), 351–359.

71. R.J. Spillane. Keyboard apparatus for personal identification. *Tech. Discl. Bull,* 1975, 17, 3346.

72. J. Chen, G. Zhu, J. Yang, Q. Jing, P. Bai, W. Yang, X. Qi, Y. Su, Z.L. Wang. Personalized keystroke dynamics for self-powered human–machine interfacing. *ACS Nano*, 2015, 9(1), 105–116.

73. S. Li, W. Peng, J. Wang, L. Lin, Y. Zi, G. Zhang, Z.L. Wang. All-elastomer-based triboelectric nanogenerator as a keyboard cover to harvest typing energy. *ACS Nano*, 2016, 10(8), 7973–7981.

74. C. Wu, W. Ding, R. Liu, J. Wang, A.C. Wang, J. Wang, S. Li, Y. Zi, Z.L. Wang. Keystroke dynamics enabled authentication and identification using triboelectric nanogenerator array. *Mater. Today*, 2018, 21, 216–222.

75. W. Ding, C. Wu, Y. Zi, H. Zou, J. Wang, J. Cheng, A.C. Wang, Z.L. Wang. Self-powered wireless optical transmission of mechanical agitation signals. *Nano Energy*, 2018, 47, 566–572.

76. X. Pu, H. Guo, J. Chen, X. Wang, Y. Xi, C. Hu, Z.L. Wang. Eye motion triggered self-powered mechnosensational communication system using triboelectric nanogenerator. *Sci. Adv.*, 2017, 3, e1700694.

77. J. Yang, J. Chen, Y. Liu, W. Yang, Y. Su, Z.L. Wang. Triboelectrification-based organic film nanogenerator for acoustic energy harvesting and self-powered active acoustic sensing. *ACS Nano*, 2014, 8(3), 2649–2657.
78. X. Fan, J. Chen, J. Yang, P. Bai, Z. Li, Z.L. Wang. Ultrathin, rollable, paper-based triboelectric nanogenerator for acoustic energy harvesting and self-powered sound recording. *ACS Nano*, 2015, 9(4), 4236–4243.
79. H. Guo, X. Pu, J. Chen, Y. Meng, M-H. Yeh, G. Liu, Q. Tang, B. Chen, D. Liu, S. Qi, C. Wu, C. Hu, J. Wang, Z.L. Wang. A highly sensitive, self-powered triboelectric auditory sensor for social robotics and hearing aids. *Sci. Robot.*, 2018, 3(20), eaat2516.
80. Z.L. Wang. Catch wave power in floating nets. *Nature*, 2017, 542, 159.
81. G. Zhu, Y. Su, P. Bai, J. Chen, Q. Jing, W. Yang, Z.L. Wang. Harvesting water wave energy by asymmetric screening of electrostatic charges on nanostructured hydrophobic thin-film surfaces. *ACS Nano*, 2014, 8, 6031–6037.
82. L. Xu, Y. Pang, C. Zhang, T. Jiang, X. Chen, J. Luo, W. Tang, X. Cao, Z.L. Wang. Integrated triboelectric nanogenerator array based on air-driven membrane structures for water wave energy harvesting. *Nano Energy*, 2017, 31, 351–358.
83. J. Chen, J. Yang, Z. Li, X. Fan, Y. Zi, Q. Jing, H. Guo, Z. Wen, K.C. Pradel. Networks of triboelectric nanogenerators for harvesting water wave energy: A potential approach toward blue energy. *ACS Nano*, 2015, 9, 3324–3331.
84. X. Wang, S. Niu, Y. Yin, F. Yi, Z. You, Z.L. Wang. Triboelectric nanogenerator based on fully enclosed rolling spherical structure for harvesting low-frequency water wave energy. *Adv. Energy Mater.*, 2015, 5, 1501467.
85. L. Xu, T. Jiang, P. Lin, J.J. Shao, C. He, W. Zhong, X.Y. Chen, Z.L. Wang. Coupled triboelectric nanogenerator networks for efficient water wave energy harvesting. *ACS Nano*, 2018, 12, 1849.
86. A. Li, Y. Zi, H. Guo, Z.L. Wang, F.M. Fernández. Triboelectric nanogenerators for sensitive nano-coulomb molecular mass spectrometry. *Nat. Nanotechnol.*, 2017, 12, 481–487.
87. J. Cheng, W. Ding, Y. Zi, Y. Lu, L. Ji, F. Liu, C. Wu, Z.L. Wang. Triboelectric microplasma powered by mechanical stimuli. *Nat. Commun.*, 2018, 9, 3733.
88. C.B. Han, T. Jiang, C. Zhang, X. Li, C. Zhang, X. Cao, Z.L. Wang. Removal of particulate matter emissions from a vehicle using a self-powered triboelectric filter. *ACS Nano*, 2015, 9, 12552–12561.
89. J. Nie, Z. Ren, J. Shao, C. Deng, L. Xu, X. Chen, M. Li, Z.L. Wang. Self-powered microfluidic transport system based on triboelectric nanogenerator and electrowetting technique. *ACS Nano*, 2018, 12(2), 1491–1499.
90. Y. Bai, C.B. Han, C. He, G.Q. Gu, J.H. Nie, J.J. Shao, T.X. Xiao, C.R. Deng, Z.L. Wang. Washable multilayer triboelectric air filter for efficient particulate matter PM2.5 removal. *Adv. Funct. Mater.*, 2018, 28, 1706680.

4 III-N Ultraviolet Light Emitters for Energy-Saving Applications

Dong-Sing Wuu
National Chung Hsing University

CONTENTS

4.1 Introduction 125
4.2 MOCVD Growth of GaN Templates for (Near) Ultraviolet LEDs 126
4.3 MOCVD Growth of AlN Template for Deep-UV LEDs 128
    4.3.1 Role of Alternating High and Low V/III Ratios on AlN Crystallinity 130
    4.3.2 Effect of Nanopatterned Sapphire Substrate on AlN Characteristics 141
    4.3.3 Effect of Defect Density of AlN Template on Deep-UV LEDs 152
4.4 Summary and Perspective 160
References.. 162

4.1 INTRODUCTION

This chapter elaborates the growth and device fabrication of III-N materials, which are considered to be ideal ultraviolet (UV) light-emitting diodes (LEDs) and have attracted considerable attention under the concept of saving energy and environment. A global treaty of the "Minamata Convention on Mercury" highlights the adverse effects of mercury on human health and environment. The import and export of certain mercury-based products will be altogether prohibited by 2020. Therefore, UV-LEDs are considered a suitable alternative to conventional UV lamps. The UV region includes the wavelength ranges 315–400 nm for UVA, 280–315 nm for UVB, and 200–280 nm for UVC (or deep UV). The wavelength of near-UV is 300–400 nm and that of DUV is 200–300 nm. The UVA range has many applications, including UV curing (for coatings, inks, and resins), lithography, sensing (fluorescent labels), medical field (e.g., blood gas analysis), and security detection (ID cards and banknotes). The UVB range has applications in UV curing, lithography, medical application (e.g., for psoriasis), and sensing (e.g., of gases). The deep UV range applications include water and air purification, sterilization (food and medical equipment), and sensing (e.g., DNA and gases). In this chapter, we firstly introduce the background, theory, and source of these III-N materials. Then, the fabrication issues, including epitaxial growth, structure design, and device fabrication, are described. In manufacturing UV-LEDs, the wide-bandgap material aluminum nitride (AlN) and its alloy aluminum gallium nitride ($Al_xGa_{1-x}N$) have been employed to prepare the multiquantum well structure. A metalorganic chemical vapor deposition (MOCVD) system is used for fabricating UV-LED epitaxial wafers with the newly designed GaN and AlN templates for emitting wavelengths from near to deep UV. The improved epitaxial quality and corresponding optical performance are discussed in detail. Finally, the challenges and prospects for UV LEDs are emphasized.

During the last decade, GaN, InGaN, and AlGaN have been used to develop high-performance optical devices such as blue, green, and UV LEDs and laser diodes [1–3]. These nitride-based LEDs are used widely as a mercury-free backlighting source in liquid-crystal displays, cell phone keypads, notebook lighting, indoor and outdoor lighting, and traffic light lamps. However, white-light LEDs have attracted considerable attention because of their application in solid-state illumination devices. Since blue LEDs are a commonly used pumping source for white-light LEDs, the recent progress on nitride-based LEDs such as patterned sapphire substrate [4,5], patterned sapphire templates [6], LEDs with embedded photonic crystals [7,8] and undercut sidewall LEDs [9] have received extensive attention, particularly regarding the blue emission region which is used to enhance the LEDs output power. For example, several approaches have been employed for improving the performance of UV LEDs. From the material aspect, more detail discussion about the growth of GaN and AlN templates by MOCVD for these UV LEDs will be described in Sections 4.2 and 4.3, respectively.

4.2 MOCVD GROWTH OF GAN TEMPLATES FOR (NEAR) ULTRAVIOLET LEDs

The near-UV LEDs still have a great potential to be fabricated as a pumping source for high color-rendering-index and highly stable white-light emission [10,11]. However, one problem associated with UV LEDs is their high sensitivity to threading dislocation density (TDD) [3]. It is well known that threading dislocations (TDs) which directly penetrate multiple quantum wells (MQWs) can work as nonradiative recombination centers, further degrading light emission efficiency as well as output power. Furthermore, these degradations in the UV LED samples are more pronounced than those in the blue LED ones [12] Therefore, several previous researches were proposed to reduce TDD effects in UV LEDs [13,14]. In particular, the epitaxial lateral overgrowth (ELOG) which uses the SiO_2 mask to block TDs was found to be efficient in reducing the TDD in the GaN epitaxial layers. It will be more attractive if a solution can not only reduce the TDD, but also enhance the light extraction.

The near-UV LED structure samples used in this study were grown on a (0001)-oriented sapphire substrate using the MOCVD system. The gallium, indium, aluminum, and nitrogen sources for the III-nitride epitaxial layer growth were trimethylgallium, trimethylindium, trimethylaluminum, and ammonia (NH_3), respectively. Biscyclopentadienyl magnesium and silane were used as the p-type and n-type doping sources, respectively. Prior to the epitaxial growth, the sapphire substrates were treated in a hydrogen ambient environment at 1,150°C to remove any surface contaminations, and a 30 nm GaN nucleation layer was then grown at 500°C. After the nucleation growth, the temperature was increased to 1,150°C and a 1–3 μm-thick n-type GaN:Si buffer layer was grown.

Typically, the surface morphology and roughness of the GaN buffer layers annealed under various temperature-ramping rates were analyzed by atomic force microscopy (AFM). The crystalline properties of the GaN buffer layers were analyzed by a double-crystal X-ray diffractometer. Scanning electron microscope was used to examine the overall surface morphology of the GaN epilayer. The electrical properties of the GaN samples were evaluated by van der Pauw–Hall measurement. The optical quality of the main GaN epilayer was characterized by photoluminescence spectra excited by a 325-nm He–Cd laser. After optimizing the parameters of the GaN template on sapphire, a MQW active layer, a thin p-AlGaN electron blocking layer and a very thin p⁺-GaN contact layer were grown successively. Various chip sizes of the LED samples used in this research were fabricated using standard photolithography and dry etch techniques. For the device testing, the current–voltage characteristics of the LEDs were measured using a semiconductor parameter analyzer at room temperature. The output power of the LED lamp was measured using an integrated sphere detector, and the measured deviation was around 5%.

The following are three examples for improving the near-UV and UVA LEDs using the GaN templates by MOCVD. First, we have proposed that the patterned distributed Bragg reflector (PDBR) can be used to replace the SiO$_2$ mask in an attempt to reduce the TDD in the epitaxial template and simultaneously enhance light extraction efficiency via the reflective behavior of the distributed Bragg reflector (DBR) mask [15]. The PDBR LED has an operating voltage very close to that of conventional LEDs, as well as a small leakage current. However, an increase in light output power of almost 39% was obtained by the PDBR LED at a forward current of 20 mA. The improvement in the light output power can be attributed to both an increase in light extraction by the periodically spaced hexagonal patterned DBR mask design as well as a decrease of TDD in the GaN template. This design can also be introduced into LEDs with an absorbing template or substrate, thus resulting in an improvement in light extraction, as well as better material quality.

Second, under high-current-density injection, the thermally induced effects usually cause the deteriorated carrier confinement in UV emitters, thus degrading their performance [16]. Hence, the vertical thin-film LEDs with great heat dissipation substrates [17,18] were developed to solve this urgent issue. Notably, the LED with flip-chip (FC) configuration [19–21] will also be an approach for better thermal management and high-power operating design. In order to solve the above-mentioned problems, we have developed high-power 380 nm flip-chip LEDs with embedded self-textured oxide mask (STOM-FCLED) apparatus, which is the new implement to reform the crystal quality of GaN epitaxial layers and intensify light output of LEDs simultaneously [22]. The STOM array structure fabricated by FC configuration can not only block the propagation of TDs but also intensify light extraction. The STOM-FCLED enhances the output power compared to LEDs without STOM structure. We attribute this enhancement to the boost of light extraction and the reduction of TDD as the STOM array can not only play a role of scattering center but also block the propagation of TDs.

In addition, the performance of UV-LEDs is always limited by the very high density (10^9–10^{10}cm^{-2}) of TDs that form when the nitride materials are grown on lattice mismatched substrates. Therefore, how to further reduce the dislocation density is an important issue for fabricating high-performance UV LEDs. Many different growth approaches have been proposed for TDD reduction such as epitaxial lateral overgrowth and patterned sapphire substrate [6,23,24]. However, both of these cases require additional etching process to generate a template for the subsequent MOCVD growth of GaN epilayers. To some extent, these complicated procedures do not avoid some negative effects on the as-grown samples. It is expected that the LED epitaxial structure can be grown through a single MOCVD process.

Finally, we have proposed a heavily Mg-doped GaN insertion layer (HD-IL) technique to improve crystal quality of the GaN layer and the rest of the required GaN-based LED structure to favor an overgrowth of GaN and blocking mechanism for TDs. The improvement in crystal quality of GaN could generate high-performance 380-nm GaN-based LEDs on sapphire substrates using a heavily Mg-doped GaN insertion layer (HD-IL) technique, which can greatly decrease the dislocation density [25]. This results in the improvement of the light output power. Experimental results indicated that the LED sample with HD-IL exhibited a 28% enhancement in light output power compared with that of the conventional LED sample. The improvement can be attributed to the reduction of nonradiative recombination centers from a reduced dislocation density in the active layer. In view of the fact that it only needs a single MOCVD growth process, the presented HD-IL technique can be used as a suitable growth template for high-quality UVA emitters.

Since the UV region includes the wavelength ranges 315–400 nm for UVA, 280–315 nm for UVB, and 200–280 nm for deep UV, the GaN template by MOCVD can only be applied for near-UV, UVA, and UVB LED structures. For the deep UV LED structure, the AlN template has to be used due to the absorbing issue of the GaN template. In the next section, we will focus on the MOCVD growth issues of deep-UV LEDs using the AlN template.

4.3   MOCVD GROWTH OF AlN TEMPLATE FOR DEEP-UV LEDs

There are several important factors to be considered for enhancing deep-UV LED properties. As illustrated in Figure 4.1, detailed descriptions are listed below:

i. A sapphire substrate has high transparency in the deep-UV region and low cost. However, the high lattice mismatch and the thermal expansion coefficient between the AlN epilayer and the sapphire substrate lead to excessive dislocation density and a thick epilayer with cracks.

ii. The AlN template should exhibit a low dislocation density (<10^8cm^{-3}). The internal quantum efficiency (IQE) of deep-UV LEDs can be increased by reducing the dislocation density of the AlN template.

iii. The structural design of the Al$_x$Ga$_{1-x}$N buffer can control the strain of the n-Al$_x$Ga$_{1-x}$N layer.

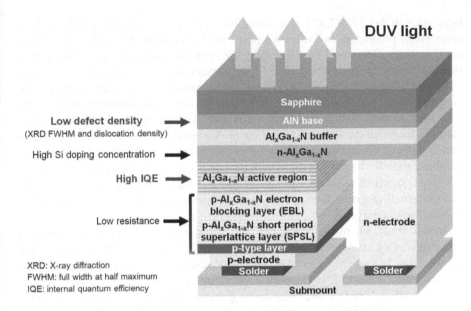

FIGURE 4.1 Illustration of key factors for improving deep-UV LED performance.

iv. The concentration of Si doping of n-Al$_x$Ga$_{1-x}$N can be increased to provide efficient current distribution.

v. The carrier confinement of MQWs can be enhanced to improve the IQE of deep-UV LEDs.

vi. Polarization of MQWs can be changed by controlling the strain and Al content of Al$_x$Ga$_{1-x}$N layers. The polarization degree mainly dominates the emission of MQWs for transverse electric (TE, E⊥c) or transverse magnetic (TM, E ∥ c) polarized light.

vii. The hole concentration of the p-Al$_x$Ga$_{1-x}$N layer can be enhanced to improve the hole injection. The hole concentration decreases with increasing Al content because of the high activation energy of Mg atoms in high-Al-content Al$_x$Ga$_{1-x}$N.

viii. High transparency in the deep-UV region and efficient hole injection in the p-Al$_x$Ga$_{1-x}$N layer can enhance the deep-UV LED performance.

ix. Using low-resistance, deep-UV-reflective p-layer improves the LED performance.

x. The light extraction efficiency of deep-UV LEDs can be enhanced using FCs in the device modules.

Aluminum nitride (AlN) with a wide bandgap (6.2 eV) has many fascinating features including high thermal stability, high electrical resistivity (~10^{13} Ω·cm), and high thermal conductivity (3.3 W·K^{-1}·cm^{-1}) [26]. Therefore, it is applied extensively in optoelectronic, high-power electronic, and piezoelectric devices, such as UV-LEDs, high-electron-mobility transistors, and surface acoustic wave devices [27]. Generally, SiC and sapphire are the most common growth templates for AlN. Unfortunately, the fabrication difficulty of SiC results in its high price, so it could be unadoptable in commercial production today. For sapphire, the lattice mismatch with AlN epilayer easily induces plenty of TDs. Moreover, the apparent mismatch of thermal expansion coefficient between sapphire and AlN causes tensile strain and then cracks epilayer with a higher thickness [28]. These defects possibly act as high-resistivity and nonradiative recombination centers to degrade the device performance. Besides, the high bond energy (2.88 eV/atom) of Al-N and the high sticking coefficient of Al adatoms facilitate the uneven dispersion of Al-related nucleation sites and formation of defects, especially at a low growth temperature (<1,300°C) [29]. Therefore, these methods for improving the AlN-based devices include the can generate more interfaces for effectively inhibiting dislocations in these stacking structures and the nanopatterned sapphire substrate (NPSS) for suppressing the lattice mismatch at AlN/substrate interface.

In this study, all the AlN epilayer structures were grown on 2-inch sapphire substrates using an Aixtron CCS 3″ by 2″ MOCVD system. NH_3 and trimethylaluminum (TMA) were supplied as N and Al sources, respectively. The substrate was preheated at 1,000°C to clean the surface. Before building epilayer structures, a ~27 nm-thick AlN buffer layer was firstly grown at 1,000°C on sapphire substrates while V/III (NH_3/TMAl) ratio and the growth pressure individually were 800 and 100 mbar. Then four distinct AlN layer structures with constant thicknesses (~1 μm) were prepared on the buffer layer at 1,100°C. Other layers were grown at 1,130°C under a pressure of 100 mbar.

4.3.1 ROLE OF ALTERNATING HIGH AND LOW V/III RATIOS ON ALN CRYSTALLINITY

Most researches report that a high growth temperature (≥1,300°C) could optimize the quality of AlN epilayer and get an atomically flat surface because of enhancing the surface migration of Al adatoms [30,31]. Other literatures have pointed out that the adjustment of growth program, the design of epilayer, and the surface treatment of growth substrate all have potential for ameliorating the AlN characteristics [28,32–34]. In the design of epilayer structure, the temperature alteration of each growth layer at high temperature is very common. Furthermore, the insertion of superlattice (SL) structure is also a general and useful process of restraining defects to achieve good crystallinity. For instance, Imura et al. have suggested that inserting the 10-period AlN/$Al_{0.6}Ga_{0.4}N$ SLs grown at 1,250°C between 2.5-μm-thick AlN and buffer layer not only improves the full width at half maximum (FWHM) of (0002) and ($10\overline{1}2$), but also reduces both screw and edge dislocations [35]. However, one can clearly understand that the methods mentioned above mostly work at high temperature. We suppose that if one can build an epilayer structure at a low growth temperature (<1,300°C) for improving the quality of AlN epilayer, the production cost and time both could be reduced, and the lifetime of heating system could be prolonged. Additionally, although the pairs and constituents of SL layer are often explored, the insertion site in the AlN epitaxial structure lacks related investigations. We consider that this site could also play an important role in affecting the AlN characteristics, so its insertion mechanism should be studied.

Consequently, we created a novel technique to prepare an AlN epilayer and a SL structure at a low growth temperature (1,100°C). The concept of this technique expresses that using the alternate high and low V/III ratios forms interfaces which can effectively inhibit dislocations in these stacking structures. Four different AlN epitaxial structures as shown in Figure 4.2 were designed in this study to verify this concept explicitly. We explored how these structures affected the characteristics of AlN epilayer as well as the role and growth mechanism of SL layer with distinct insertion sites. The study revealed that the crystallinity of AlN epilayer improved significantly after inserting the SL layers by X-ray diffraction (XRD) measurement. Moreover, it was worth noting that FWHM of (002) and (102) showed different sensitivity to the insertion sites of SL layer. Based on the detailed analyses by transmission electron microscopy (TEM) and XRD, the growth mechanisms of inserting the SL layers in different sites of the AlN epitaxial structure were studied.

In the growth of AlN epilayers, the V/III ratio strongly affects the surface mobility of Al adatoms that usually accompanies the transition of growth mode between longitudinal and transverse domains [36]. This relationship has to be realized in order to precisely control the growth characteristics. Thereby, at first, sample A with a simple epitaxial structure is used to understand the influence of V/III ratio on the AlN epilayer growth at V/III = 50–800. Figure 4.3 shows the surface morphology images of sample A at various V/III ratios. At V/III ratio = 50, only few small voids are present on the epilayer surface, as shown in Figure 4.3a. In Figure 4.3b–d, many voids with large size can be observed with V/III ratio increasing from 100 to 300. While increasing the V/III ratio from 400 to 800, the morphologies of AlN display loose appearance like islands instead of the dense surface. Particularly, most island crystals with ≤ 200 nm-diameter exist on the surface at V/III = 800. This indicates that a high V/III ratio should result in a three-dimensional (3D) growth mode in the shape of an island. Furthermore, a low V/III ratio should lead to a two-dimensional (2D) growth mode, which is a layer-by-layer mode. We consider that a strong parasitic gas reaction between Al source and ammonia gas is induced obviously by a high V/III ratio. As this parasitic gas reaction occurs, the diffusion length of Al adatoms can be shortened, which can then limit the ability of lateral growth [37]. Therefore, the island shape can dominate the crystal formation. On the other hand, a low V/III promotes lateral growth rather than longitudinal growth.

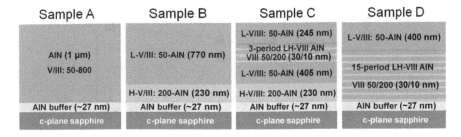

FIGURE 4.2 Illustration of samples with four distinct AlN layer structures. (From Wang, T.Y. et al. 2016. *CrystEngComm* 18:9152–9159. Reproduced by permission of The Royal Society of Chemistry.

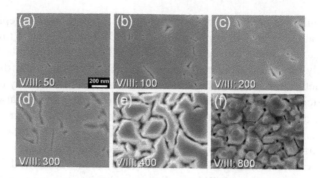

FIGURE 4.3 Surface morphology of sample A with V/III ratios of (a) 50, (b) 100, (c) 200, (d) 300, (e) 400, and (f) 800. (From Wang, T.Y. et al. 2016. *CrystEngComm* 18:9152–9159. Reproduced by permission of The Royal Society of Chemistry.)

The effect of V/III ratio on the AlN layer crystallinity in the sample A is attested by FWHM of AlN (0002) and (10$\bar{1}$2) as shown in Figure 4.4. The (0002) FWHM exhibits a gradually incremental trend from V/III = 50 to 800. The variation curve of (10–12) FWHM with V/III ratios is similar to "U" shape; i.e., it obviously decreases at first and slowly increases up to V/III = 800. This indicates that the (0002) FWHM is more sensitive to the V/III ratio than the (10$\bar{1}$2) FWHM. Moreover, one knows that the FWHM of symmetric AlN (0002) corresponds to the screw dislocation and that of asymmetric AlN (10$\bar{1}$2) corresponds to the edge and mixed dislocations [38]. It can be found that there is a serious reduction of the (10–12) FWHM with increasing the V/III ratio to 200 and the voids become large and many rather than tiny and few. As a result, we exclude the possibility that the voids are attributed to the edge and mixed dislocations. Since the (0002) FWHM and the scale and number of voids simultaneously increase with the V/III ratio, we consider that the screw dislocation

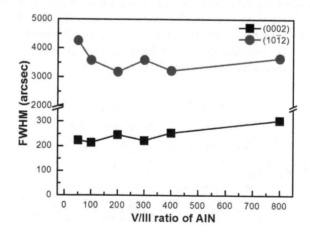

FIGURE 4.4 The FWHM data of AlN (0002) and (10$\bar{1}$2) for sample A with V/III ratios from 50 to 800. (From Wang, T.Y. et al. 2016. *CrystEngComm* 18:9152–9159. Reproduced by permission of The Royal Society of Chemistry.)

should be the cause for the existence of voids. As mentioned above, the function of alternating low and high V/III ratios for the epitaxial structure can be systematically concluded. At a low V/III ratio, the strong lateral growth ability and the voidless surface can be obtained. Introducing a high V/III ratio leads to high longitudinal growth rate forming island crystals to produce abundant crystal boundaries which are capable of suppressing the elongation of a dislocation. Hence, alternate stacking structures of high and low V/III ratios based on their features are employed in the samples B, C, and D. The V/III ratios of 50 and 200 are considered as the low and high ratios, respectively, because of the smooth surface and low ($10\bar{1}2$) and (0002) FWHMs at these two ratios. In the following sections, the benefit of using these specific structures in the improvement of AlN epilayers is analyzed and confirmed by comparing all samples.

From the oscillation of in-situ 405 nm reflectance curves in Figure 4.5, the curves of all samples show significant decrease at the time of initial growth due to the crystal nucleation. For sample A, the time for reflectance oscillations to become steady is lesser at V/III = 50 than V/III = 200. In addition to the time, the intensity of oscillations is the strongest at V/III = 50. As the discrepancy between these two V/III ratios

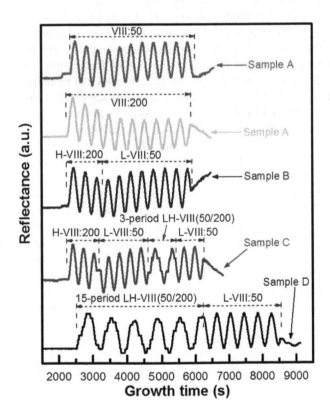

FIGURE 4.5 The 405-nm reflectance curves of samples A–D during the MOCVD growth. (From Wang, T.Y. et al. 2016. *CrystEngComm* 18:9152–9159. Reproduced by permission of The Royal Society of Chemistry.)

is little, it is determined that the growth mode is indeed dominated by the 3D island growth and the 2D layer-by-layer mode at V/III ratio = 200 (high ratio) and 50 (low ratio), respectively. For samples B to D, the reflectance curves all show the trend of drop and rise while the low V/III ratio layer is deposited on the high V/III ratio layer. This certifies our finding that primary growth of the high V/III ratio layer with abundant grain boundaries can retard dislocations, and then the low V/III ratio layer can strengthen the lateral growth to smooth the surface. Additionally, for samples C and D, after inserting the SL structure constructed by stacking the high and low V/III ratios layers to optimize the AlN layer quality, the following low V/III ratio layer shows a steady and continuous reflectance curve. This implies that the AlN layer could proceed to grow without the transition of growth mode. Furthermore, both the samples C and D with a SL structure have a smoother surface as shown in Figure 4.6b and c than samples without the SL structure like samples A and B as displayed in Figures 4.3a and c and 4.6a. Even the surfaces of samples C and D reveal an almost void-free state. Minor voids still can be observed in sample B, but they are smaller than those in sample A at V/III = 200. Obviously, this structure of alternate high and low V/III ratios assists in merging the AlN crystals and suppressing dislocations. The crystallinity of AlN layer should improve after using this structure.

Figure 4.7a shows the (0002) and (10$\bar{1}$2) FWHMs of AlN of samples A–D, including sample A at V/III = 50 and 200. Samples B–D have a lower FWHM of (0002) and (10–12) than sample A. In the previous observation of morphologies in samples A and B in Figures 4.3a and 4.6a, although sample B has more voids than sample A at V/III = 50, both the (0002) and (10–12) FWHMs can be reduced. Meanwhile, sample B shows a (10–12) FWHM similar to that of sample A with V/III = 200, but its (0002) FWHM slightly decreases. According to these results, we confirm that using the alternate high and low V/III ratios structure can improve the crystallinity of AlN layers. Interestingly, FWHM of the (10$\bar{1}$2) diffraction peak can dramatically decrease via inserting specific SL structures built by the alternate high and low V/III ratios in samples C and D. In addition, the improvement outcome of (0002) and (10$\bar{1}$2) FWHMs apparently depends on the insertion sites of SL structure. Comparing sample C with sample D, if the SL structure is situated in the middle layer, where the top AlN layer with the low V/III ratio follows, in sample C, the (10$\bar{1}$2) FWHM decreases by about 30%. If the SL structure firstly grows like that of sample D, the (0002) FWHM reduces by around 41% compared to that of sample C. We believe that the inserting sites of these SL structures certainly play an important role in achieving what defects are eliminated ffectively. The growth mechanisms are going to be further discussed later by detailed TEM analyses. One can observe Figure 4.7b

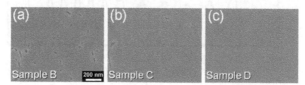

FIGURE 4.6 Surface morphologies of samples (a) B, (b) C, and (c) D. (From Wang, T.Y. et al. 2016. *CrystEngComm* 18:9152–9159. Reproduced by permission of The Royal Society of Chemistry.)

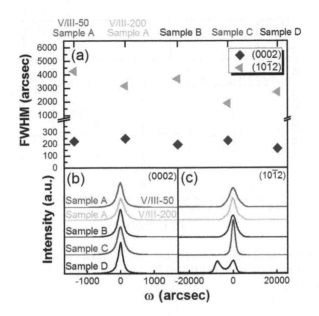

FIGURE 4.7 (a) AlN (0002) and (10$\bar{1}$2) FWHM and (b) and (c) individual rocking curves of all samples including sample A under V/III=50 and 200. (From Wang, T.Y. et al. 2016. *CrystEngComm* 18:9152–9159. Reproduced by permission of The Royal Society of Chemistry.)

that all samples display the purely single (0002) diffraction peak in the rocking curve of XRD spectra. In contrast to the (0002) diffraction peaks, Figure π.30(c) displays the (10$\bar{1}$2) diffraction peaks of all samples showing a unitary peak, except sample D. The (10$\bar{1}$2) diffraction peaks split into two peaks with the angle between 1° and 2°, suggesting that the crystals with (10$\bar{1}$2) have another slightly oblique orientation. In the phi-scan spectra of (10$\bar{1}$2) in Figure 4.8, the six-fold symmetry of AlN epilayers and the three-fold symmetry of the sapphire are clearly verified. This well consists with that the structures of AlN and sapphire are hexagonal wurtzite and trigonal corundum, respectively. An angular discrepancy of about 30° between AlN and sapphire exists and most possibly results from the relaxation of lattice mismatch by the 30° in-plane rotation [39].

Moreover, similar to Figure 4.7c, only sample D displays split peaks of (10$\bar{1}$2) determined by the phi scans in Figure 4.8. According to the researches of Ueno et al., the in-plane rotation domains which are separated by low-angle twist boundaries can induce the split of the AlN (10$\bar{1}$2) diffraction peaks into two peaks with an angle of around 1° [40]. This condition is in agreement with our study. As a result, we suggest that the sample D could produce other dislocations caused by the twist boundaries. In other words, despite the fact that primary growth of SL structure can inhibit the dislocations from the lattice and thermal mismatch (native dislocations) to optimize the AlN layer crystallinity, this structure simultaneously leads to the formation of other dislocations (external dislocations). It is worth noting that the total reduction in number of native dislocations is more than the increase in number of the external dislocations via the insertion of the SL structure at the early growth stage.

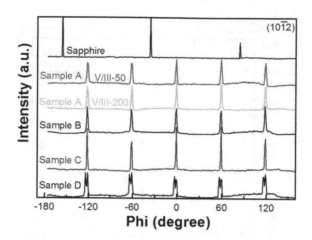

FIGURE 4.8 Phi scans of sapphire (10$\bar{1}$2) and AlN (10$\bar{1}$2) for samples A–D. (From Wang, T.Y. et al. 2016. *CrystEngComm* 18:9152–9159. Reproduced by permission of The Royal Society of Chemistry.)

For understanding the variation behaviors of dislocations in these four different structures of AlN layers in detail, the cross-sectional images are directly observed by TEM as shown in Figures 4.9 and 4.10. The diffraction patterns of all samples at the interface between sapphire and AlN layer are the same, so one of them is a reference in Figure 4.9g. It clearly reveals that the epitaxial relationship between AlN and sapphire shows (0002) AlN ∥ ∥(0006) sapphire, certifying the AlN layers grow along c-plane of sapphire and have a (0002) preferred orientation in the XRD spectrum. Two orthogonal orientations which are AlN [0002] and [$\bar{2}$110] are defined to help understand the dislocation behaviors. Comparing the bright-field image of sample A at V/III = 50 with that of sample B in Figures 4.9a and b, the distributions of dislocations are alike. From the two-beam condition with g = 0002 and $\bar{2}$110, both samples have few screw dislocations propagating from the substrate to the surface in Figures 4.9c and d based on the criterion of $g \cdot b$ = 0. In contrast to the screw dislocations, other numerous dislocations including edge and mixed dislocations significantly exist in Figures 4.9e and f. This is also consistent with the XRD results that the (0002) FWHM is certainly lower than (10$\bar{1}$2) FWHM. Additionally, the dislocation density close to the surface of sample A is about $7.8 \times 10^9 \, \text{cm}^{-2}$ and that of sample B is around $4.6 \times 10^9 \, \text{cm}^{-2}$ via our estimation. Thus, in fact, using the stacking structure of high and low V/III ratios can slightly reduce the dislocations. However, the annihilation of dislocations due to the insertion of SL layer built by the alternate high and low V/III ratios is evident in TEM images.

Regardless of the bright-field and two-beam condition images of samples C and D in Figures 4.10a–f, the numbers of dislocations near the surface are lower than those in samples A and B. In Figure 4.10a, c, and e, most of dislocations can be bended, form a closed loop, and combine to be restricted and eliminated beneath the top layer with the low V/III ratio in the sample C. The dislocation density of sample C is estimated to about $7.4 \times 10^8 \, \text{cm}^{-2}$, which is significantly lower than that of the samples A

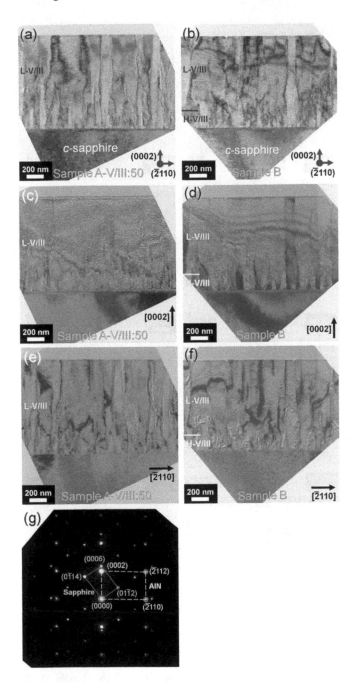

FIGURE 4.9 Bright-field images of (a) sample A with V/III = 50 and (b) sample B. The images under two-beam condition with g = 0002 of (c) sample A with V/III = 50 and (d) sample B and that with g = $\bar{2}$110 of (e) sample A with V/III = 50 and (f) sample B. (g) Diffraction pattern at AlN/sapphire interface. (From Wang, T.Y. et al. 2016. *CrystEngComm* 18:9152–9159. Reproduced by permission of The Royal Society of Chemistry.)

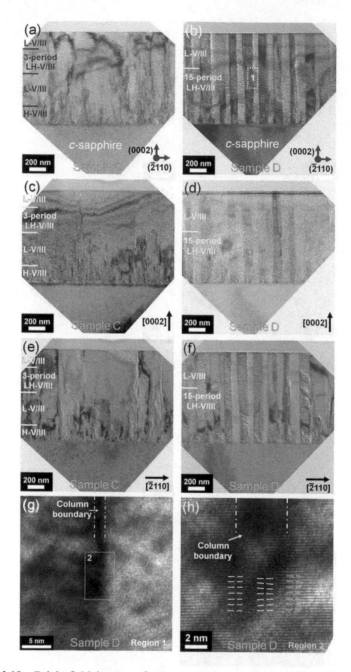

FIGURE 4.10 Bright-field images of (a) sample C and (b) sample D. The two-beam condition images of (c) sample C and (d) sample D along $g = 0002$ direction and those of (e) sample C and (f) sample D along $g = \bar{2}110$. (g) High-resolution cross-sectional TEM image at region 1 marked in Figure 4.10b. (h) High-resolution TEM image focusing on the column boundary noted as region 2 in Figure 4.10g. (From Wang, T.Y. et al. 2016. *CrystEngComm* 18:9152–9159. Reproduced by permission of The Royal Society of Chemistry.)

and B. Obviously, the insertion of this SL in the middle layer can greatly suppress the dislocations to avoid them propagating to the surface. For sample D in Figures 4.10b, d and f, interestingly, the first growth of the SL results in a column-like AlN layer. Similar to sample C, the sample D shows fewer dislocations than samples A and B.

Additionally, these columnar structures display the regular contrast. We consider that two possible factors cause the clear contrast. One is the nonuniform nucleation of AlN. As one knows, Al has a poor surface mobility at a low growth temperature (<1,300°C), so the Al-related nucleation sites could aggregate creating a regular or random pattern. Afterward, the AlN growth can follow in these nucleation sites to form crystals with a little different lattice spacing. But we carefully measure these column-like structures by the high-resolution images of TEM in Figures 4.10g and h and find that they have homogeneous lattice spacing. Consequently, the first factor can be excluded. The other reason is that the twist boundary occurs at the interface of the structures with the apparent contrast. We magnify the region 1 of Figure 4.10b and then observe there is a column boundary with ~2 nm width between the column structures in Figure 4.10g. From the magnification of the region 2 in Figure 4.10g, the clear lattice fringes are obtained in Figure 4.10h. We find that the right (green label) and the left (yellow label) areas both have identical lattice spacing in Figure 4.10h. However, at the column boundary, these two lattices have a misfit arrangement shown as the white marker. Most importantly, the misfit arrangement can create a great deal of stacking faults along nonlongitudinal orientation to induce the deterioration of the $(10\bar{1}2)$ FWHM. The occurrence of twist boundary also explains why the $(10\bar{1}2)$ of sample D has two peaks mentioned in the rocking curve in Figure 4.7c. As a result, this confirms the twist boundary indeed exists between the column structures and leads to the contrast. Moreover, since only the grain boundary can be noted by the two-beam condition alone $g = 0002$ in Figure 4.10d, we suppose that the screw dislocations could combine through the grain boundaries to the epilayer surface. This dislocation combination assists in reducing the dislocation density to improve the (0002) FWHM as displayed in Figure 4.7a. It is worth noticing that the whole AlN layer of growing the SL in the beginning are columnar structures, even the following layer has a powerful ability of lateral growth at a low V/III ratio. In summary, although the degeneration of lateral crystalline is ascribed to the formation of stacking faults by growing the SL at first, the threading dislocation along [0002] can be reduced by combining at the columnar grain boundaries.

As mentioned above, we select and build the growth mechanisms for these four AlN epilayer structures to help us understand the effect of alternate structure and SL on improving AlN quality grown at a low temperature (1,100°C). The growth mechanism of sample A is the simplest among all samples because of only one epilayer and a buffer layer. This mechanism is dominated by 3D and 2D modes according to the V/III ratios as discussed in the previous sections. For sample B, the grain boundaries induced by island crystals in the layer with high V/III ratio obstruct the dislocations. Afterward, the succeeding low V/III ratio is able to obtain a smooth AlN epilayer with better crystallinity than sample A. In addition to the stacking structure of the alternate high and low V/III ratios, samples C and D both have the SL with different insertion sites. It is very good that they show the significant improvement of crystallinity at a low growth temperature. Figure 4.11 shows the mechanism of growth for sample C. The stacking structure of alternate high and low V/III ratios at steps (i) and (ii) resembles sample B. The few dislocations denoted by the blue color can be suppressed. When the SL structure follows the preceding layers at steps (iii) and (iv), the dislocations form a closed loop and combined with each other to reduce their number. This is ascribed to the plentiful grain boundaries induced by alternately introducing the high and low V/III ratios. At the final stage of step (v), seldom dislocations propagate to the top AlN epilayer at the low V/III ratio. For sample D, the growth mechanism and dislocation behaviors are illustrated in Figure 4.12. The SL grows at first stage, which differs from other samples, as mentioned in steps (i) and (ii). We believe that since the thickness (30 nm) is insufficient to complete lateral growth at the low V/III ratio, this SL cannot afford to accommodate the island structure derived from the lower AlN buffer layer. Therefore, all grains grow along the longitudinal orientation to form the column morphologies. The junction of these column-like grains causes the formation of twist boundary, as shown in the magnified illustration of the red zone at step (iii). Although stacking faults can be generated in the twist boundary to damage the crystallinity, most TDs can be bended and

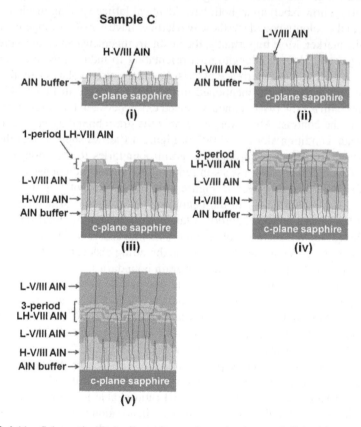

FIGURE 4.11 Schematic illustration of growth mechanism and dislocation behavior for sample C. (From Wang, T.Y. et al. 2016. *CrystEngComm* 18:9152–9159. Reproduced by permission of The Royal Society of Chemistry.)

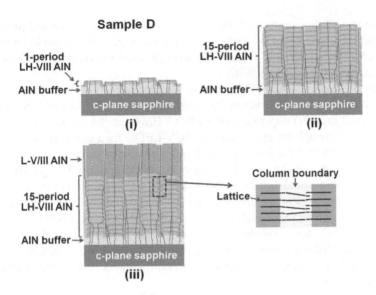

FIGURE 4.12 Schematic illustration of growth mechanism and dislocation behavior for sample D. (From Wang, T.Y. et al. 2016. *CrystEngComm* 18:9152–9159. Reproduced by permission of The Royal Society of Chemistry.)

can even integrate along the grain boundaries. Therefore, few dislocations occur in the top AlN layer at step (iii). In conclusion, different insertion sites of the SL can suppress distinct dislocations. We consider that if these two features of SL insertion sites can be combined together, the crystallinity of AlN epilayer grown at a low temperature could be greatly improved. This could be performed in future. Finally, we believe that the development of stacking structure based on the alternate high and low V/III ratios bring us a very useful method to obtain AlN epilayer at a low temperature.

4.3.2 EFFECT OF NANOPATTERNED SAPPHIRE SUBSTRATE ON AlN CHARACTERISTICS

The large mismatch of thermal expansion coefficient (TEC) between AlN and sapphire leads to the tensile stress and cracks in epilayer when layer thickness is higher than critical thickness [41]. Moreover, high lattice mismatch of AlN/sapphire causes a high density (10^9–10^{10}cm^{-2}) of TDs. Therefore, many literatures have reported the methods for improving epitaxial quality of AlN, including high temperature (1,700°C) annealing in N_2 + CO ambient [42], AlN grown at a high temperature (1,550°C) [39], enhancing the surface migration of Al adatoms [43], AlN with the SL structure by alternating the high and the low temperature [32], and AlN grown on patterned sapphire substrate (PSS) [44,45]. Among these methods, using PSS as a growth substrate is simple and direct to achieve high crystallinity AlN. Moreover, AlN epilayers grown on PSS with nanosize exhibit a coalescence thickness (<6 μm) lower than that of PSS with micron size [46]. Although high-crystal-quality AlN can be grown at high temperature ≥1,300°C, both of the wafer warpage and production cost could increase. Hence, we believe that the warpage and cost could be reduced when high-crystallinity AlN is grown by using a novel structure on NPSS at a low growth temperature below 1,150°C.

In our previous research [47], we have demonstrated the improvement of AlN crystallinity with a novel structure by alternating the high (200) and the low (50) V/III ratios at growth of 1,100°C. However, the crystal quality of AlN epilayer is not enough for application in DUV-LEDs. As a result, the optimization of the high (100) and the low (25) V/III ratios for the novel structure is achieved with NPSS to enhance the crystal quality of AlN in this study. The growth mechanism, residual stress, and dislocation behavior are systemically investigated in detail. Additionally, the growth mechanism and dislocation behavior of AlN/NPSS are explained. In contrast to the high-crystal-quality AlN grown on NPSS reported by literatures [44̄46], this study shows the lowest value for three features of AlN: the defect density (dislocation density of $1 \times 10^8 \, \text{cm}^{-2}$ and etching pit density (EPD) of $2.3 \times 10^5 \, \text{cm}^{-2}$), the coalescence thickness (total thickness of 2.55 μm), and growth temperature (1,130°C).

The AlN with a novel structure was grown on 2-inch flat sapphire substrate (FSS) and NPSS by employing Aixtron MOCVD system. The feature of NPSS is illustrated in Figure 4.13a; the NPSS has funnel-shaped patterns with diameters of 700 and 250 nm for top and bottom, respectively, and depth of 350 nm,. The substrates were preheated at 1,000°C to clear the surface before epilayer growth. Figure 4.13b shows the AlN with a novel structure on FSS and NPSS. The AlN structure was constructed with a ~27-nm-thick buffer layer; a 115-nm-thick high V/III ratio (H-V/III = 100) AlN, a 240-nm-thick low V/III ratio (L-V/III = 25) AlN, a three-period LH-V/III (alternation of H- and L-V/III ratios) AlN SL structure, and a 2.09-μm-thick L-V/III AlN top layer. To understand the growth mechanism of AlN grown on two types of substrates, the structure was split into five stages. The H-V/III AlN (115 nm), L-V/III AlN (240 nm), three-period LH-V/III AlN SLs, and L-V/III AlN top layer (2.09 μm) were denoted as stages 1–5 (S1–S5).

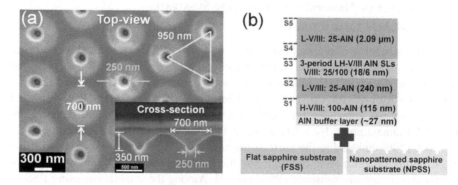

FIGURE 4.13 (a) Top view and cross-sectional FESEM images of NPSS. (b) A novel structure of AlN grown on FSS and NPSS. (From Wang, T.Y. et al. 2018. *Applied Surface Science* 455:1123–1130. Copyright 2018, with permission from Elsevier.)

To distinguish the growth mechanism of AlN grown on FSS and NPSS, the surface morphologies of AlN/FSS from S1 to S5 are shown in Figures 4.14 and 4.15. In the initial growth stage of AlN/FSS (Figure 4.14b), more AlN crystals are formed on the layer surface. For S2–S3 in Figure 4.14c and d, merging of crystals and decrease in void size can be observed. A flat surface with tiny voids is achieved when the AlN layer has fully coalesced at S4 (Figure 4.14e). At the growth's final stage (Figure 4.14f), plenty of cracks form on the L-V/III AlN layer surface. It could be ascribed to the high tensile stress caused by the TEC mismatch between AlN and sapphire. Meanwhile, the mismatch limits to a critical thickness of AlN grown on FSS. Once the layer thickness becomes higher than the critical thickness, the

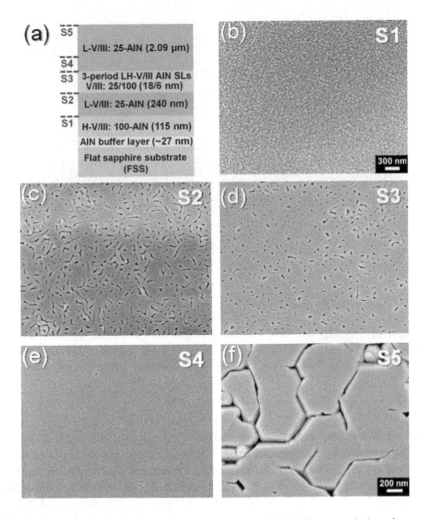

FIGURE 4.14 (a) Scheme of AlN structure grown on FSS. Surface morphology images of AlN at various growth stages for (b) S1, (c) S2, (d) S3, (e) S4, and (f) S5. (From Wang, T.Y. et al. 2018. *Applied Surface Science* 455:1123–1130. Copyright 2018, with permission from Elsevier.)

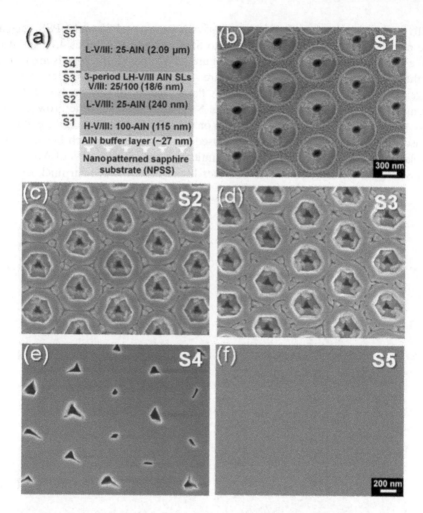

FIGURE 4.15 (a) Scheme of AlN structure grown on NPSS. Surface morphology images of AlN at various growth stages for (b) S1, (c) S2, (d) S3, (e) S4, and (f) S5. (From Wang, T.Y. et al. 2018. *Applied Surface Science* 455:1123–1130. Copyright 2018, with permission from Elsevier.)

cracks could be formed on the layer surface. Generally, tensile stress of epilayer could be released by creating dislocations. Therefore, the tensile stress of AlN is dependent on the both critical thickness and dislocation density. In our case, the dislocation density can be reduced by the formation of LH-V/III AlN SL structure, which increases the tensile stress and decreases the critical thickness of AlN. As a result, many cracks can be observed from the layer surface at the final stage. The residual stress of AlN would be discussed by Raman measurement in a later section. From the AFM measurement, the surface roughness is determined to be 1.6 nm with a scan area of 5 μm × 5 μm.

The surface morphology images of AlN/NPSS with growth stages 1–5 are presented in Figure 4.15. From S1 to S3 (Figure 4.15b–d), clear AlN crystals are formed on the layer surface. Meanwhile, the AlN crystals prefer to merge on the nonpatterned regions rather than on the patterned regions due to the differences in growth rates along the c-plane and the oblique plane. Therefore, the uncoalesced areas decrease gradually in S2–S3 (Figure 4.15c and d). The layer surface nearly fully coalesced with few voids at S4 (Figure 4.15e). Finally, the smooth layer surface can be achieved at S5, as shown in Figure 4.15f. The surface roughness is 0.7 nm with a scan area of 5 μm × 5 μm. To further understand the lateral growth mechanism of AlN/NPSS, the cross-sectional field emission scanning electron microscopy (FESEM) images are displayed in Figure 4.16. Before the AlN layer growth, the FESEM image of NPSS is presented in Figure 4.16a. The H-V/III AlN layer and the buffer layer are grown on the NPSS, as shown in Figure 4.16b, by following the funnel-shaped patterns. The AlN layer also grows along the oblique plane due to the high sticking coefficient of Al adatoms. For S2 and S3 (Figure 4.16c and d), an increase in the lateral growth leads to the formation of trench patterns after the LH-V/III AlN SL

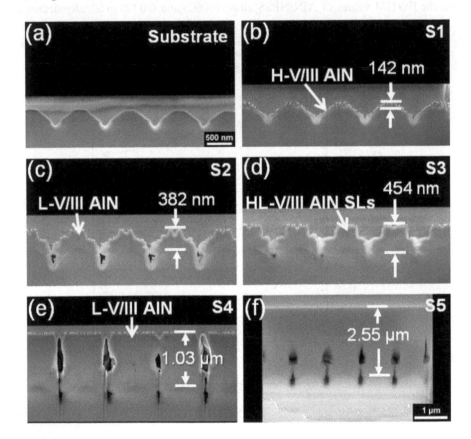

FIGURE 4.16 Cross-sectional FESEM images of (a) NPSS and AlN/NPSS with growth stages for (b) S1, (c) S2, (d) S3, (e) S4, and (f) S5. (From Wang, T.Y. et al. 2018. *Applied Surface Science* 455:1123–1130. Copyright 2018, with permission from Elsevier.)

structure is grown. Following the growth of the 576-nm-thick L-V/III AlN layer at S4 (Figure 4.16e), some key holes exist upon the patterned regions after the layer merges. In the final stage (Figure 4.16f), the AlN layer is fully coalesced after the growth of the 2.09-µm-thick L-V/III AlN top layer. These results suggest that the growth of AlN epilayer is dominated by epitaxial lateral overgrowth. In addition, it is observed that key holes are present on the patterned regions. In contrast to AlN/FSS, the surface morphology of AlN/NPSS at stage 5 reveals a flat surface without cracks as the tensile stress could be relaxed by the key holes.

To estimate the crystallinity of AlN layers, the FWHM values of AlN (0002) and $(10\bar{1}2)$ for AlN/FSS and AlN/NPSS are presented in Figure 4.17. The FWHM values of AlN/FSS along (0002) and $(10\bar{1}2)$, presented in Figure 4.17a, show a decreasing trend from S2 to S5. The FWHM values slightly decreased from 379 to 368 arcsec for AlN (0002) and significantly reduced from 6228 to 1640 arcsec for AlN $(10\bar{1}2)$. In addition, a drastic decrease in the FWHM value of AlN $(10\bar{1}2)$ can be observed from S3 to S4, which confirms that the dislocation density can be effectively reduced by employing the LH-V/III AlN SL structure. Figure 4.17b illustrates that the FWHM values of AlN/NPSS along (0002) and $(10\bar{1}2)$ gradually decrease

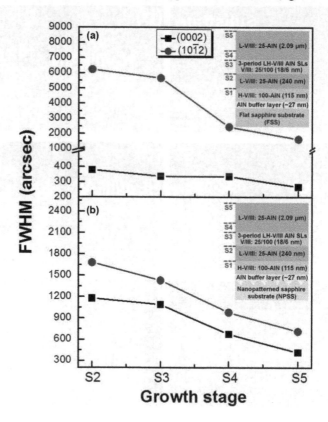

FIGURE 4.17 FWHM values of AlN at various stages grown on (a) FSS and (b) NPSS. (From Wang, T.Y. et al. 2018. *Applied Surface Science* 455:1123–1130. Copyright 2018, with permission from Elsevier.)

with growth stage 2 to stage 5. The FWHM value of (0002) peak reduced from 1178 to 415 arcsec and that of (10$\bar{1}$2) peak decreased from 1678 to 714 arcsec. Compared to AlN/FSS, the AlN/NPSS reveals a dramatic improvement of (10$\bar{1}$2) FWHM. This can be ascribed to the small area of the AlN layer grown on the nonpatterned regions, which can reduce the abundant dislocations induced by the lattice mismatch at the AlN/sapphire interface. Another factor is that growth of the AlN epilayer is dominated by the epitaxial lateral overgrowth due to the low surface migration of Al adatoms. Therefore, the dislocations can be effectively eliminated by bending or forming a closed loop.

Based on these two factors, the AlN layer grown on NPSS has a much lower (10$\bar{1}$2) FWHM than AlN/FSS. However, AlN/NPSS exhibits a slightly higher (0002) FWHM than AlN/FSS. According to the research result of Bai et al. [48], the AlN layer directly grown on FSS without any low-temperature buffer layer reveals a very low screw dislocation density. The screw dislocations are mainly related to the stacking disorder forming at the interface between the initial layer and sapphire. In addition, AlN has a higher formation energy of stacking faults compared to other nitride materials with wurtzite structure, which implies that the stacking fault is difficult to form [49]. Therefore, low density of screw dislocations and stacking faults can be observed in the AlN layer directly grown on sapphire. In our case, the (0002) FWHM of AlN/NPSS was slightly higher than that of AlN/FSS. It could be considered that the stacking disorder formed at the interface of the initial AlN layer and the patterned region of sapphire owing to the different growth rates of the AlN layer grown along (0002) and the oblique plane, followed by AlN layer grown along the funnel-shaped patterns. Although the stacking disorder can occur during the initial growth of the AlN layer, it can be effectively eliminated in the subsequent growth steps.

To estimate the stress of the AlN epilayer, the Raman spectra of AlN/FSS and AlN/NPSS were obtained and are presented in Figure 4.18. The peak location of E_2 (high) phonon mode of stress-free AlN is at 657.4 cm^{-1} [44,50]. Accordingly, the in-plane compressive stress (σ) of the AlN layer can be determined by the following equation [44,50]: $\omega_{E2(high)} - \omega_0 = C\sigma$, where $\omega_{E2(high)}$ is the peak position of E_2 (high) peak for the AlN epilayer, ω_0 is E_2 (high) peak at 657.4 cm^{-1} for stress-free AlN, and C is the biaxial strain coefficient, 3 cm^{-1}/GPa [51]. The position values of E_2 (high) peaks for AlN/FSS are presented in Figure 4.18a and Figure π.42; they are located at 658.6 cm^{-1} for S2, 657.4 cm^{-1} for S3, 654.4 cm^{-1} for S4, and 659.2 cm^{-1} for S5. From S2 to S4, the residual stress of AlN/FSS changed from compressive stress (0.4 GPa) to tensile stress (1 GPa). This corresponds to the coalescence evolution of the AlN layer observed from the surface morphology shown in Figure π.37. The complete coalescence of AlN induces a high tensile stress, which is a result of the TEC mismatch between AlN and sapphire. The increasing tensile stress with an increase in the layer thickness causes cracks on the surface. Hence, the stress of AlN/FSS at S5 is compressive stress (0.6 GPa). The peak locations of E_2 (high) phonon modes of AlN/NPSS for S2–S5 are presented in Figures 4.18b and 4.19; they are located at 659.2 cm^{-1} for S2 and 658 cm^{-1} for S3–S5. Meanwhile, the AlN/NPSS reveals compressive stress in S2–S5. From the surface morphology and the cross-sectional image presented in Figures 4.15 and 4.16, the layer surface of AlN/NPSS is observed to be fully coalesced at S5, and key holes exist on the patterned regions. A compressive stress of 0.2 GPa for AlN/NPSS at S5 is close to stress-free AlN. This result corresponds to the previous research results, which indicated that the flat AlN layers grown on patterned substrates have a compressive stress that is close to zero stress

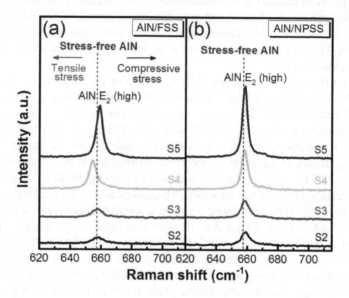

FIGURE 4.18 Raman spectra of AlN grown on (a) FSS and (b) NPSS. (From Wang, T.Y. et al. 2018. *Applied Surface Science* 455:1123–1130. Copyright 2018, with permission from Elsevier.)

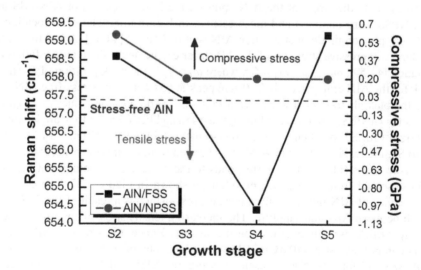

FIGURE 4.19 Raman shift and residual stress of AlN/FSS and AlN/NPSS at various growth stages. (From Wang, T.Y. et al. 2018. *Applied Surface Science* 455:1123–1130. Copyright 2018, with permission from Elsevier.)

[44,52]. As mentioned above, the AlN/NPSS epilayers reveal a compressive stress rather than tensile stress because the stress is relaxed by the key holes.

Further realizing the epitaxial relationship at AlN/sapphire interface and dislocation behaviour, the diffraction pattern and cross-sectional transmission electron microscopy (TEM) image of AlN are presented in Figures 4.20 and 4.21. Figure 4.20a shows the diffraction pattern at the AlN/sapphire interface; the epitaxial relationships of all samples are AlN (0002) ∥ sapphire (0006) and AlN (1$\bar{1}$00) ∥ sapphire (11$\bar{2}$0). The cross-sectional bright-field TEM image of AlN/FSS is presented in Figure 4.20b. Although many edge dislocations can be eliminated below the L-V/III AlN layer, edge dislocations propagated to the layer surface much faster than screw dislocations. The total dislocation density can be estimated to be 6.2×10^8 cm^{-2}. The TEM image of AlN/NPSS is shown Figure 4.21a. It clearly shows that only few dislocations propagated to the layer surface and the key holes presented on the patterned regions. The total dislocation density near the layer surface can be evaluated to approximately 1×10^8 cm^{-2}. Compared to AlN/FSS, AlN/NPSS reveals a reduction in the dislocation density. Figure 4.21b shows the TEM image of the AlN/NPSS interface in region 1 that is marked in Figure 4.21a. A clear contrast can be observed between the AlN grown along the c-plane and that grown on the oblique plane, as seen in region 2 marked in Figure 4.21b. The contrast could be ascribed to the different crystallinity caused by the growth directions. To further confirm this point, the high-resolution TEM image focused on region 2 is illustrated in Figure 4.21c. At the AlN/NPSS interface, the lattice fringes of the AlN layer include two regions, one for the stacking order and the other for the stacking disorder. The lattice fringes have different stacking (marked with red and blue lines) in the stacking disorder region. Clearer lattice fringes focused on the region of stacking disorder are presented in Figure 4.21d; the different stackings of lattice fringes are denoted by yellow and green lines. In addition, the clear contrast at region 2 can be considered to be due to the high density of stacking disorder of lattice fringes. According to the XRD result shown in Figure 4.17, AlN/NPSS has a slightly higher (0002) FWHM than

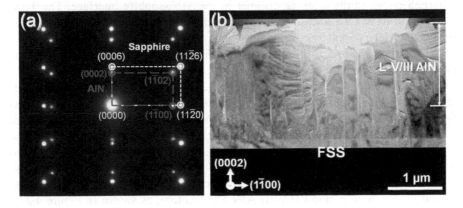

FIGURE 4.20 (a) Diffraction pattern at AlN/sapphire interface. (b) Cross-sectional bright-field TEM image of AlN/FSS. (From Wang, T.Y. et al. 2018. *Applied Surface Science* 455:1123–1130. Copyright 2018, with permission from Elsevier.)

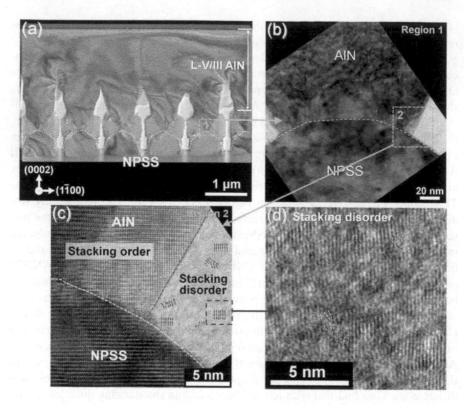

FIGURE 4.21 (a) Cross-sectional bright-field TEM image of AlN/NPSS. (b) The image focused on region 1 marked in Figure 4.21a. (c) High-resolution image focused on region 2 marked in Figure 4.21b. (d) Clear lattice fringes focused on stacking disorder area. (From Wang, T.Y. et al. 2018. *Applied Surface Science* 455:1123–1130. Copyright 2018, with permission from Elsevier.)

AlN/FSS owing to the stacking disorder at the AlN/sapphire interface. The TEM result confirms that the stacking disorder is formed at the interface between AlN and the oblique plane of patterned regions.

As mentioned above, to understand the role of NPSS on AlN crystallinity, the growth mechanism and dislocation behavior of AlN/NPSS are demonstrated in Figure 4.22. At initial growth stage in Figure 4.22a, the H-V/III AlN layer and buffer layer grown on NPSS are denoted in 3D. The AlN crystals with plenty of dislocations marked in blue color are formed due to the lattice mismatch between AlN and sapphire. When the L-V/III AlN is deposited on H-V/III AlN (Figure 4.22b), the lateral growth ability is increased. Meanwhile, the transition of growth mode from 3D to 2D results in some dislocations getting eliminated due to the combining of dislocations. In addition, it worth noticing that most of the dislocations are eliminated by inducing a three-period LH-V/III AlN SL structure, as shown in Figure 4.22c. Finally, very few dislocations propagate to layer surface by employing the thick L-V/III AlN to flatten the layer surface (Figure 4.22d). It can be found that the growth of AlN is dominated by epitaxial lateral overgrowth, while some key holes exist upon

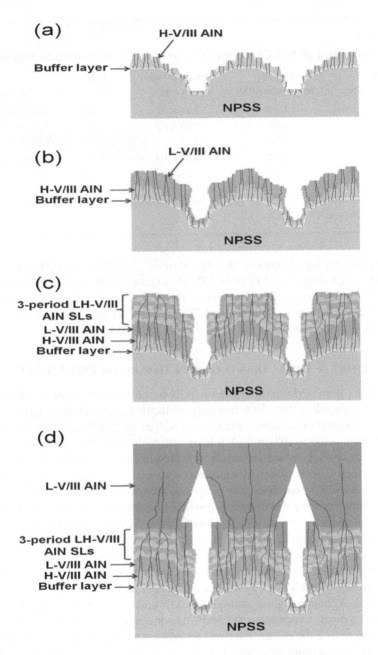

FIGURE 4.22 Schematic illustrations of growth mechanism and dislocation behavior for AlN grown on NPSS. (a) H-V/III AlN and buffer layer were firstly deposited on NPSS. (b) The L-V/III AlN grew on H-V/III AlN and reduced some dislocations. (c) The growth of three-period HL-V/III AlN can assist in the elimination for most of the dislocations. (d) Only few dislocations propagated to surface as L-V/III AlN was grown at the final stage. (From Wang, T.Y. et al. 2018. *Applied Surface Science* 455:1123–1130. Copyright 2018, with permission from Elsevier.)

TABLE 4.1

The Comparison of AlN Characteristics between Literatures and This Study

Total AlN Thickness (μm)	Growth Temperature (°C)	Total TD from TEM (cm^{-2})	EPD (cm^{-2})	References
8	1,200	1.25×10^8	3×10^6	Tran et al. [53]
6	1,250	3.83×10^8	2.7×10^8	Zhang et al. [45]
2.8	1,250	–	2×10^6	Kim et al. [35]
1.7	1,450	4.52×10^8	–	Kitagawa et al. [54]
2.55	1,130	1×10^8	2.3×10^5	This work

Source: From Wang, T.Y. et al. 2018. *Applied Surface Science* 455:1123–1130. Copyright 2018, with permission from Elsevier.

the patterned regions. Therefore, the crystal quality of AlN is effectively enhanced. The dislocation density and EPD of AlN in literatures and this study are listed in Table 4.1. In contrast to these literatures, this study reveals the lowest dislocation density for 1×10^8 cm^{-2} and ultra-low EPD for 2.3×10^5 cm^{-2}. It suggests that high feasibility of this technique makes it applicable to DUV-LEDs.

4.3.3 EFFECT OF DEFECT DENSITY OF AlN TEMPLATE ON DEEP-UV LEDs

To construct high-performance devices, AlN-based epilayers require a low defect density and smooth surface. Unfortunately, a mismatch between the high lattice constants and thermal expansion coefficients of AlN and sapphire leads to a high dislocation density ($\sim 10^{10}$ to 10^{11} cm^{-2}) [41]. These dislocation defects lead to the formation of a nonradiative recombination center, resulting in device degradation. Meanwhile, the dislocation defects of the AlN template strongly affect the IQE of MQWs [55]. Therefore, several researchers have proposed methods to improve the crystallinity of AlN, including the enhancement of surface migration for Al adatoms [42], use of AlN grown at high temperatures (≥1,300°C) [43], AlN SLs with the alternation of low and high temperatures as a buffer structure [32], the use of an AlGaN/AlN SL buffer structure [35], and an AlN epilayer grown on a nanopatterned sapphire substrate [45]. Besides these methods, Shatalov et al. have reported enhancement in the IQE of AlGaN-based MQWs by the use of an AlN template with a low dislocation density [56]. Although high-crystal-quality AlN epilayers can be obtained by utilizing a high growth temperature (≥1,300°C), a lower thermal budget (<1,150°C) is expected to reduce the production cost and wafer warpage, especially for the use of larger diameter sapphire substrates. Commonly, the production cost of the heater for a higher temperature (>1,300°C) is approximately three times more than that of the conventional heater (<1,200°C). Based on the estimation, for the same machine life and yield rate, the cost of LED wafer produced by MOCVD with the conventional heater is at least 70% less than the cost with the heater for a higher temperature.

In a study reported previously by our group [47], the AlN structure with three-period SLs achieved by the alternation of high and low V/III ratios can effectively eliminate the dislocation density for an AlN layer grown on flat sapphire at a low temperature of 1,100°C. However, the AlN crystal quality is not sufficient for application in DUV-LEDs. Therefore, to further enhance the AlN crystallinity, the V/III ratio of the AlN structure and period number (0–30) of the SLs grown on an NPSS are systematically optimized in this study. Meanwhile, the AlGaN MQW structure with an emission wavelength of 280 nm grown on an AlN template was utilized to confirm the epitaxial quality of AlN and the feasibility of device applications. To understand the effect of the defects of the AlN template on the MQW structure, the relative IQE of the MQWs can be obtained by photoluminescence (PL) measurements under room temperature (RT) and low-temperature (LT) conditions by the using the relative relationship IQE = I_{RT}/I_{LT}. This relationship assumes that the IQE of the MQWs is 100% at LT. Previously, Hirayama et al. have suggested that the highest relative IQE of 86% is achieved by using $In_xAl_yGa_{1-x-y}N/In_xAl_zGa_{1-x-z}N$ MQWs at a wavelength of 280 nm [57]. Banal et al. have also proposed that the use of $Al_xGa_{1-x}N/AlN$ MQWs with an emission wavelength of 247 nm results in a relative IQE of approximately 69% [58]. In this study, the 280 nm AlGaN-based MQWs grown on a low-defect-density AlN template exhibited a relative IQE as high as 85%. The result suggests that the high-IQE AlGaN MQW structure can be achieved using an engineered low-defect-density AlN template.

As shown in Figure 4.23a, the NPSS exhibited funnel-shaped patterns with top and bottom diameters of 700 and 250 nm, respectively, and a depth of 350 nm. Figure 4.23b shows the AlN structures with a total thickness of 2.55 μm, the structures of which have the same buffer layers consisting of an approximately 27-nm-thick AlN nucleation layer, a 115-nm-thick high V/III (H-V/III = 100) AlN, and a 240-nm-thick low V/III (L-V/III = 25) AlN. Following the 0–30 period, LH-V/III (alternation of H- and L-V/III) AlN SLs and an L-V/III AlN top layer were deposited onto the buffer layers, respectively. Meanwhile, the thickness of the L-V/III AlN top layer increased from 2.16 to 1.45 μm with an increase in the SL period from 0 to 30.

To understand the growth mechanism of the AlN structure deposited on NPSS, the cross-sectional FESEM images of AlN with three-period SLs recorded at various

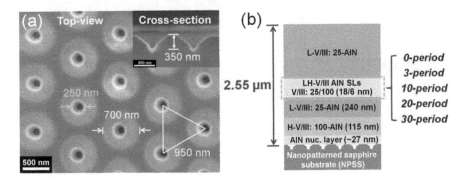

FIGURE 4.23 (a) Features of NPSS and (b) epitaxial structures of AlN with 0–30 period SLs. (From Wang, T.Y. et al. 2017. *Scientific Reports* 7:14422. Copyright 2017, Springer Nature.)

growth steps are shown in Figure 4.24. At the growth step-1 (Figure 4.24a), the AlN layer covered all of the NPSS, even the patterned regions, because of the high sticking coefficient of the Al adatoms [29]. From step-2 to step-3 (Figures 4.24b and c, respectively), the AlN epilayer was grown along the c-plane direction despite the increase in the lateral growth rate of the layer. At growth step-4 (Figure 4.24d), the epilayer coalesced, and some holes were observed on the patterned regions. These results indicated that the growth mechanism of the AlN epilayer is dominated by epitaxial lateral overgrowth. Finally, the flat AlN epilayer was observed at growth step-5 (Figure 4.24e). Notably, some key holes were formed on the patterned regions, which could decrease the tensile stress caused by the lattice mismatch between the AlN layer and sapphire. Figure 4.25 shows the further evaluation of the crystallinity of the AlN epilayers in terms of the FWHM values for (0002) and $(10\bar{1}2)$ for the AlN structure from the growth step-2 to step-5. The (0002) and $(10\bar{1}2)$ FWHM values decreased as a function of the growth steps. The FWHM value of (0002) decreased from 1178 to 415 arcsec, while that of $(10\bar{1}2)$ decreased from 1678 to 714 arcsec. The (0002) FWHM corresponded to screw dislocations, while the (10–12) FWHM corresponded to the edge and mixed dislocations [56]. In this study, the growth evolution of the epilayer was dominated by ELOG as shown in the FESEM images in Figure 4.24. The decrease in both (0002) and $(10\bar{1}2)$ FWHM values was related to the fact that ELOG can assist in the elimination of dislocations by bending or combining them.

Figure 4.26 shows the epitaxial relationship and dislocation behavior of AlN grown on NPSS. The diffraction pattern at the AlN/NPSS interface (Figure 4.26a), where the relationship is AlN (0002) ∥ sapphire (0006) and AlN $(1\bar{1}00)$ ∥ sapphire $(1\bar{1}20)$, indicating that the AlN epilayer exhibits a preferred (0002) orientation (along the c-plane direction). In our previous research, it is clearly indicated that the

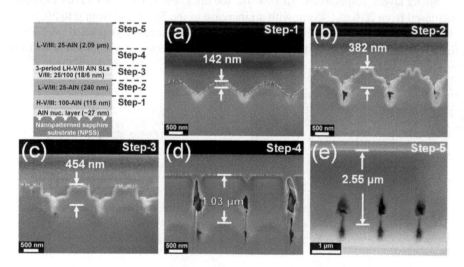

FIGURE 4.24 Cross-sectional FESEM images of the AlN structure grown at (a) step-1, (b) step-2, (c) step-3, (d) step-4, and (e) step-5. (From Wang, T.Y. et al. 2017. *Scientific Reports* 7:14422. Copyright 2017, Springer Nature.)

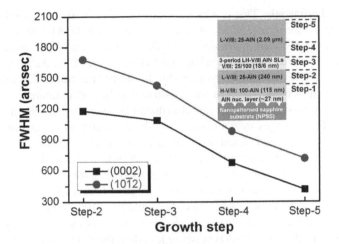

FIGURE 4.25 FWHM values of the AlN structure at various growth steps. (From Wang, T.Y. et al. 2017. *Scientific Reports* 7:14422. Copyright 2017, Springer Nature.)

insertion of AlN SLs into the AlN epilayer (on flat sapphire) is beneficial to decrease the dislocation density. Without inserting the AlN SLs, the dislocations can easily propagate from the AlN/sapphire interface to the surface of AlN epilayer. However, after inserting the AlN SLs into the AlN epilayer, the dislocation propagation can

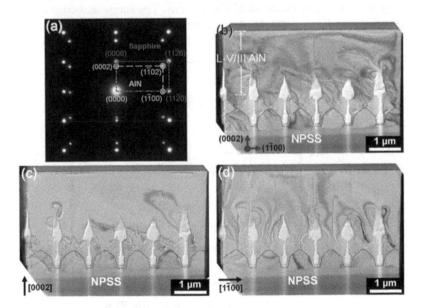

FIGURE 4.26 (a) Diffraction pattern of the AlN/sapphire interface. (b) Cross-sectional bright-field TEM image of the AlN structure/NPSS. TEM images with a two-beam condition along (c) $g = 0002$ and (d) $g = 1\bar{1}00$. (From Wang, T.Y. et al. 2017. *Scientific Reports* 7:14422. Copyright 2017, Springer Nature.)

be eliminated efficiently. This is why the insertion of AlN SLs was used during the growth of AlN epilayer on NPSS. Actually, the insertion of AlN SLs is also helpful to reduce the dislocation density of the AlN epilayer on NPSS, as discussed in Figure 4.27a. The cross-sectional bright-field TEM image of the AlN (Figure 4.26b) revealed only a few dislocations propagating to the layer surface. The total dislocation density near the layer surface was estimated to be approximately $1 \times 10^8 \text{cm}^{-2}$, suggesting that most dislocations are eliminated by inducing the SL structure and ELOG. The TEM images shown under a two-beam condition are utilized to distinguish the screw, edge, and mixed dislocations, as shown in Figures 4.26c and d. The Burgers vector (b) of the screw dislocation is $b = <0001>$ and that of edge dislocations is $b = 1/3<1\bar{1}20>$ [49]. Based on the invisible criterion $g \cdot b = 0$, the screw and edge dislocations were observed along $g = 0002$ and $g = 1\bar{1}00$, respectively. Meanwhile, mixed dislocations were observed at $g = 0002$ and $g = 1\bar{1}00$. Notably, the number of edge dislocations was greater than that of screw dislocations near the layer surface, corresponding to the (0002) FWHM being less than the $(10\bar{1}2)$ FWHM.

To enhance the crystal quality of the AlN structure grown on NPSS, the period number of the SLs was optimized. In Figure 4.27a, both (0002) and $(10\bar{1}2)$ FWHM values gradually reduced with the increase in the SL period from 0 to 30. Moreover, the decrease in the FWHM values attained stability at SL period greater than 10. Before inserting the AlN SLs, the (0002) and $(10\bar{1}2)$ FWHM values of the AlN epilayer were 500 and 792 arcsec, respectively. In the SL periods of 0–30, the lowest (0002) FWHM value of 331 arcsec and $(10\bar{1}2)$ FWHM value of 652 arcsec were obtained by utilizing the 20-period SLs. These results indicated that the SLs can effectively assist in the elimination of dislocations. The screw and edge dislocation densities (D_{screw} and D_{edge}) can be derived from the following equations: $D_{screw} =$

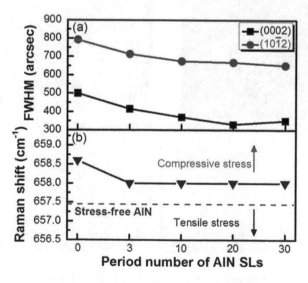

FIGURE 4.27 (a) FWHM values and (b) Raman shift of AlN with 0–30 period SL structures. (From Wang, T.Y. et al. 2017. *Scientific Reports* 7:14422. Copyright 2017, Springer Nature.)

$\beta_{(002)}/9b^2_{\text{screw}}$ and $D_{\text{edge}} = \beta_{(102)}/9b^2_{\text{edge}}$ [59]. In these two equations, $\beta_{(002)}$ and $\beta_{(102)}$ represent the FWHM values of AlN(0002) and AlN(10$\bar{1}$2), respectively. Additionally, b is the Burgers vector length. Here, b_{edge} and b_{screw} are 0.3189 and 0.5185 nm, respectively. Via our calculations, the screw and edge dislocation densities of the AlN epilayer without inserting the AlN SLs are 5.02×10^8 and $2.61 \times 10^{10}\,\text{cm}^{-2}$, respectively. With the insertion of the 20-period SLs, these two dislocation densities can be reduced to 2.20×10^8 and $2.01 \times 10^{10}\,\text{cm}^{-2}$, respectively. In comparison to the dislocation density ($1 \times 10^8\,\text{cm}^{-2}$) from the TEM image (Figure 4.26b), these calculated dislocation densities are higher. This is attributed that the estimation of dislocation density via the TEM result is observed near the epilayer surface. It can be found that the reduction of screw dislocation density is more obvious when the technique of SLs combined with NPSS is used. In addition to AlN crystallinity, the stress management of AlN is crucial for the fabrication of DUV-LEDs. Hence, the residual stress of AlN with 0–30 period SL structures is estimated by Raman measurement (Figure 4.27b). The frequency of the E_2 high-phonon mode was located at $658.6\,\text{cm}^{-1}$ for AlN with 0-period SLs and at $658\,\text{cm}^{-1}$ for AlN with 3–30 period SL structures. For stress-free AlN [44,50], the frequency was located at $657.4\,\text{cm}^{-1}$. Notably, the AlN epilayers with 0–30 period SL structures exhibited a higher frequency compared to the stress-free frequency, which is indicative of compressive stress. Moreover, the in-plane compressive stress (σ) of AlN can be evaluated by the following Eq. (4.1) [44,50]:

$$C\sigma = \omega_{E2\,(\text{high})} - \omega_0 \tag{4.1}$$

Here, C is the biaxial strain coefficient ($3\,\text{cm}^{-1}/\text{GPa}$) [51], and $\omega_{E2\,(\text{high})}$ and ω_0 are the frequencies of the E_2 high-phonon mode for AlN with SL structures and stress-free AlN, respectively. Hence, the compressive stress of AlN with 0 period and 3–30 period SL structures are 0.4 and 0.2 GPa, respectively. This result indicated that the compressive stress state is related to the existence of SLs on the patterned regions. A similar result has been reported by Dong et al. [44], wherein the AlN epilayer with a frequency of $658.7\,\text{cm}^{-1}$ and a compressive stress of 0.43 GPa are obtained for AlN grown on NPSS at a temperature of 1,200°C. In this study, the AlN with SL structures grown at a low temperature of 1,130°C revealed a lower compressive stress.

To realize the density of the dislocations that have propagated to the epilayer surface, the EPD of AlN with SL periods 0–30 was estimated (Figure 4.28). After etching in the KOH solution, the difference in the etching rates of the screw, edge, and mixed dislocations led to etching pits with different pit sizes. Kitagawa et al. have suggested that the smallest etching pits correspond to edge-type dislocations, while the largest hexagonal pits correspond to screw- or mixed-type dislocations [54]. In this study, edge-type dislocations decreased with the increase in the SL period from 0 to 30 owing to dislocation elimination (Figure 4.28a–e). Overall, the EPD decreased from 1.8×10^6 to $1 \times 10^5\,\text{cm}^{-2}$ with the increase in the SL period from 0 to 30. In contrast to the EPD of $2 \times 10^6\,\text{cm}^{-2}$ reported by Kim et al. [35] and the EPD of $1 \times 10^6\,\text{cm}^{-2}$ reported by Tran et al. [53], the AlN with 20–30 period SLs revealed an ultra-low EPD value of $1 \times 10^5\,\text{cm}^{-2}$ in this study, indicating that the increase in the SL period effectively assists in the decrease of the dislocation density. This result

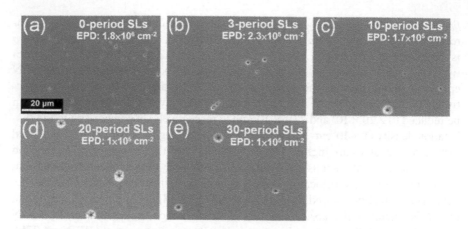

FIGURE 4.28 Surface morphology images of AlN with (a) 0, (b) 3, (c) 10, (d) 20, and (e) 30 period SL structures after etching in a KOH solution. (From Wang, T.Y. et al. 2017. *Scientific Reports* 7:14422. Copyright 2017, Springer Nature.)

corresponds to the FWHM values of AlN along (0002) and (10$\bar{1}$2). To achieve a high relative IQE of the AlGaN MQW structure, the AlN with 20-period SLs was utilized as a growth template.

To confirm the defect density of the AlN template on the relative IQE of AlGaN-based MQWs, 280-nm-MQW structures were prepared on AlN both without and with 20-period SLs, which are denoted as structures A and B, respectively. For structure A (shown in Figure 4.29a), the emission wavelength of the MQWs was observed at 276 nm for RT and 272 nm for LT (10 K). A clear shoulder peak at 260 nm for the LT was caused by the $Al_{0.6}Ga_{0.4}N$ barrier layer and buffer layer. If the IQE of the MQWs under the LT condition was assumed to be 100%, the relative IQE of structure A at RT can be evaluated as 22.4%. In contrast to structure A, structure B exhibited narrower emission peaks at 281 nm for RT and 279 nm for LT (Figure 4.29b). The relative IQE of structure B was estimated as 85%, indicating that the 280-nm-MQW structure grown on a low-defect-density AlN template effectively enhances the relative IQE. By comparing the RT-PL characteristics of AlGaN MQWs on AlGaN/AlN templates without and with the AlN SLs (Figures 4.29a and b), we can observe that the main emission wavelengths of these two MQWs were centered at 276 and 281 nm, respectively. The composition-pulling effect can be used to realize the result. Without the insertion of the AlN SLs, there existed a relatively larger compressive stress in the wells. This would result in the composition tendency of AlGaN wells toward higher Al content to minimize the lattice mismatch [60]. However, as the AlN SLs were inserted into the AlN epilayer, the compressive stress in the wells became smaller, leading to a lower Al content in AlGaN wells. This is why the AlGaN MQWs on AlGaN/AlN template without the insertion of the AlN SLs have a shorter emission wavelength.

In addition, another interesting phenomenon was also observed in PL results. Comparing the main RT-PL peaks of these two structures, the AlGaN MQWs on

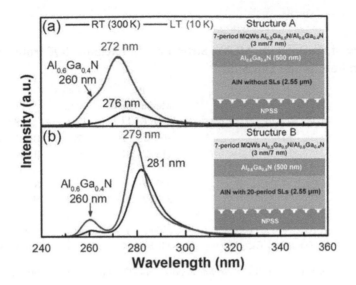

FIGURE 4.29 PL spectra of (a) structure A and (b) structure B under room temperature (300 K) and low temperature (10 K). (From Wang, T.Y. et al. 2017. *Scientific Reports* 7:14422. Copyright 2017, Springer Nature.)

AlGaN/AlN template without the insertion of the AlN SLs possessed an apparently larger FWHM value (149.9 nm) than that with the insertion of the AlN SLs (14.1 nm). There could be two reasons to explain it. The first reason is the formation of quantum-confined Stark effect (QCSE) in the structure without the AlN SLs, which causes the band tilt [61,62]. Second, due to the higher defect density of AlN epilayer without the insertion of AlN SLs, the quality of the AlGaN MQWs was lower. Thus, the FWHM value of RT-PL peak for the structure A was apparently large. In general, the QCSE is related to the red-shift of emission peak. However, compared to the PL characteristic of the structure B (with inserting the AlN SLs), the blue-shift phenomenon occurred in the structure A. We can speculate that this shift of emission peak is mainly dominated by the composition-pulling effect. Besides, the differences between emission peaks measured at RT and LT for the AlGaN MQWs on AlGaN/AlN templates without and with the insertion of the AlN SLs were measured to be 4 and 2 nm, respectively. The larger shift of PL peak in the structure A is possible owing to its higher compressive stress in the wells measured at RT. On the other hand, the relative IQE values obtained from previous studies as well as in this study are summarized in Table 4.2. As shown in Table 4.2, Hirayama et al. have obtained a high relative IQE of 86% using the $In_xAl_yGa_{1-x-y}N/In_xAl_zGa_{1-x-y}N$ MQWs. For the AlGaN-based MQWs, the relative IQE of the listed research results is shown to be around 43%–69%. In the previous studies, the AlN templates were prepared at a high growth temperature ($\geq 1,250°C$). Herein, a high relative IQE of 85% for 280 nm AlGaN-based MQWs was achieved by using an AlN template grown under a low growth temperature (1,130°C), and it has hardly ever been reported for AlGaN one. Therefore, the low-EPD AlN templates designed by defect reduction engineering demonstrate high potential for high-IQE deep-UV LED applications.

TABLE 4.2

Summary of Wavelength, MQWs Structure, and Relative IQE from Previous Research Results and This Study

Wavelength	MQW Structure (Well/Barrier)	PL Measurement at RT/LT	Relative IQE I_{RT}/I_{LT}	Reference
247 nm	$Al_xGa_{1-x}N/AlN$	300 K/8.5 K	69%	Banal et al. [58]
250 nm	$Al_{0.7}Ga_{0.3}N/AlN$	285 K/12 K	50%	Bhattacharyya et al. [63]
278 nm	$Al_xGa_{1-x}N/Al_yGa_{1-y}N$	300 K/14 K	55%	Shatalov et al. [56]
280 nm	$Al_xGa_{1-x}N/Al_yGa_{1-y}N$	300 K/77 K	50%	Hirayama et al. [57]
	$In_xAl_yGa_{1-x-y}N/In_xAl_zGa_{1-x-z}N$		86%	
283 nm	$Al_{0.4}Ga_{0.6}N/Al_{0.5}Ga_{0.5}N$	300 K/10 K	43%	Dong et al. [44]
281 nm	$Al_{0.5}Ga_{0.5}N/Al_{0.6}Ga_{0.4}N$	300 K/12 K	85%	This work

Source: From Wang, T.Y. et al. 2017. *Scientific Reports* 7:14422. Copyright 2017, Springer Nature.

4.4 SUMMARY AND PERSPECTIVE

In this chapter, the GaN and AlN templates on c-plane sapphire substrates using MOCVD for UV LEDs have been described. The performance of UV LEDs is always limited by the very high density (10^9–10^{10} cm^{-2}) of TDs that form when the nitride materials are grown on lattice mismatched substrates. Therefore, several approaches to further reduce the dislocation density have been realized in fabricating high-performance near-UV to 380 nm LEDs on the GaN templates. On the other hand, the 280-nm MQWs with high IQE have been successfully grown on low-defect-density AlN templates by employing the LH-V/III SL structure with alternation of high and low V/III ratios. The results show the feasibility for improving the crystallinity of AlN layers grown on NPSS at a low temperature (~1,100°C). The key points of this chapter are summarized below.

We succeeded in developing a novel structure by using alternate high and low V/III ratios for significantly improving the quality of AlN epilayers grown at a low temperature (1,100°C). This result was well confirmed by discussing four different structures of AlN epilayers including the stacking structure and the SL layers which were built by the alternate high and low V/III ratios. In addition, from the observation of FESEM, the growth mode of AlN epilayer changed from the layer-by-layer to the island growth while increasing the V/III ratios from low (50) to high (800). Actually, the design concept of this structure takes into account both features simultaneously at low and high V/III ratios. The low V/III ratio exhibited the strong lateral growth ability to smooth the epilayer and the high V/III ratio facilitated the formation of island crystals with abundant grain boundaries to obstruct the propagation of dislocations. The FWHM values of AlN (0002) and (10$\bar{1}$2) both were improved by employing the stacking structure constructed from the alternate high and low V/III ratios, especially the SL. After employing the SL, the dislocation density declined from $7.8 \times 10^9 \mathrm{cm}^{-2}$ to $7.4 \times 10^8 \mathrm{cm}^{-2}$. In addition, we also found that the FWHM values of (0002) and ($10\bar{1}2$) had different sensitivity to the insertion site of SL. For the SL inserted in the middle layer, the ($10\bar{1}2$) FWHM could be reduced by about 30% compared to the AlN layer with the SL firstly grown on the buffer layer. On the other hand, when the SL was deposited at first, the (0002) FWHM of AlN layer displayed a decrement of around 41% compared to that with a SL in the middle layer. Using TEM, the behavior of dislocations could be observed in detail. It showed that the dislocations could be bended to form a closed loop and combined with each other to be restricted and decreased. Moreover, it was notable that the columnar structure was formed as the SL grew initially. Meanwhile, most of the dislocations conjugated each other through the column boundaries. Although we also found that a great deal of stacking faults along nonlongitudinal orientation existed in the column boundary, their numbers are less than the reduced number of TDs. As a result, the whole crystallinity was significantly improved. Besides, we built the detailed growth mechanism of the SL inserted at various positions for assisting in realizing the benefit of using the stacking structure. We concluded that this novel structure had to be a good candidate for preparing the AlN epilayer at a low growth temperature if necessary.

Low-defect-density AlN epilayers on NPSS have successfully demonstrated at a low growth temperature of 1,130°C by using a novel structure with alternation of H- and L-V/III ratios. The serious cracks caused by the high tensile stress of AlN/FSS can be observed when layer thickness is higher than critical thickness. This result was in agreement with the change of residual stress of AlN from tensile stress to compressive stress. In contrast to the AlN/FSS, the AlN epilayer grown on NPSS was dominated by ELOG, while some key holes existed upon the patterned regions. Hence, the crack-free AlN/NPSS with a flat surface revealed compressive stress as tensile stress was reduced by the key holes. Besides, the FWHM value of AlN ($10\bar{1}2$) was significantly reduced from 1,640 to 714 arcsec when the AlN epilayer was grown on NPSS. In addition, lowest dislocation density of AlN/NPSS can be estimated to approximately $1 \times 10^8 \mathrm{cm}^{-2}$ from TEM analysis, with an ultra-low EPD of $2.3 \times 10^5 \mathrm{cm}^{-2}$. These results indicated that low-defect-density AlN can be built by employing the structure design and NPSS. Finally, both the growth mechanism and dislocation behavior of AlN/NPSS were systematically determined. We believe that this low-defect-density template technique has high potential in the deep-UV LED growth and related applications.

The low-defect-density AlN templates were successfully fabricated at a low growth temperature (1,130°C) to achieve high-IQE MQWs. From the FESEM analysis of the growth evolution of AlN with three-period SLs grown on NPSS, the growth of the AlN layers was dominated by ELOG. Meanwhile, some key holes were observed upon the pattern regions. These key holes could reduce the tensile stress related to the lattice mismatch between AlN and sapphire. Besides, the (0002) and ($10\bar{1}2$) FWHM values decreased with the increase in the growth steps because of the elimination of dislocations. From the TEM analysis, the ELOG and induction of SLs can assist in the elimination of most dislocations, with only a few of the dislocations propagating to the layer surface. As a result, the total dislocation density of AlN with three-period SLs is estimated to be approximately $1 \times 10^8 \text{cm}^{-2}$. Further enhancement of the AlN crystal quality was achieved by the increase in the SL period (0–30), and the AlN structure with 20-period SLs exhibited the lowest (0002) FWHM value of 331 arcsec and $(10\bar{1}2)$ FWHM value of 652 arcsec, as well as an ultra-low EPD of $1 \times 10^5 \text{cm}^{-2}$. Therefore, the AlN structure with 20-period SLs is selected as a growth template for preparing the AlGaN MQWs with an emission wavelength of 280 nm. Notably, the relative IQE was dramatically enhanced by approximately four times (from 22.8% to 85%) for the growth of 280-nm MQWs on the AlN template with 20-period SLs. These results show that by defect engineering, the presented the AlN templates can provide an alternate approach for developing high-efficient deep-UV LEDs.

REFERENCES

1. Schubert, E. F. 2003. *Light-Emitting Diodes.* Cambridge, U.K.: Cambridge University Press.
2. Zukauskas, A., M. S. Shur, and R. Gaska. 2002. *Introduction to Solid-State Lighting.* New York: Wiley.
3. Nakamura, S. and S. F. Chichibu. 2000. *Introduction to Nitride Semiconductor Blue Laser Diode and Light Emitters Diodes.* London, U.K.: Taylor & Francis.
4. Lee, J.-H., J. T. Oh, Y. C. Kim, and J. H. Lee. 2008. Stress reduction and enhanced extraction efficiency of GaN-based LED grown on cone-shape-patterned sapphire. *IEEE Photonics Technology Letters* 20: 1563–1565.
5. Lin, H. C., R. S. Lin, J. I. Chyi, and C. M. Lee. 2008. Light output enhancement of InGaN light-emitting diodes grown on masklessly etched sapphire substrates. *IEEE Photonics Technology Letters* 20: 1621–1623.
6. Chiu, C. H., H. H. Yen, C. L. Chao, Z. Y. Li, P. C. Yu, H. C. Kuo, T. C. Lu, S. C. Wang, K.M. Lau, and S. J. Cheng. 2008. Nanoscale epitaxial lateral overgrowth of GaN-based light-emitting diodes on a SiO_2 nanorod-array patterned sapphire template. *Applied Physics Letters* 93: 081108-1-3.
7. Park, J., J. K. Oh, K. W. Kwon, Y. H. Kim, S. S. Jo, J. K. Lee, and S. W. Ryu. 2008. Improved light output of photonic crystal light-emitting diode fabricated by anodized aluminum oxide nano-patterns. *IEEE Photonics Technology Letters* 20: 321–323.
8. Kwon, M. K., J. Y. Kim, I. K. Park, K. S. Kim, G. Y. Jung, S. J. Park, J. W. Kim, and Y. C. Kim. 2008. Enhanced emission efficiency of GaN/InGaN multiple quantum well light-emitting diode with an embedded photonic crystal. *Applied Physics Letters* 92: 251110-1-3.
9. Kuo, D. S., S. J. Chang, T. K. Ko, C. F. Shen, S. J. Hon, and S. C. Hung. 2009. Nitride-based LEDs with phosphoric acid etched undercut sidewalls. *IEEE Photonics Technology Letters* 21: 510–512.
10. Sheu, J. K., S. J. Chang, C. H. Kuo, Y. K. Su, L. W. Wu, Y. C. Lin, W. C. Lai, J. M. Tsai, G. C. Chi, and R. K. Wu. 2003. White-light emission from near UV InGaN–GaN LED chip precoated with blue/green/red phosphors. *IEEE Photonics Technology Letters* 15: 18–20.
11. Narukawa, Y., I. Niki, K. Izuno, M. Yamada, Y. Murazaki, and T. Mukai. 2002. Phosphor-conversion white light emitting diode using InGaN near-ultraviolet chip. *Japanese Journal of Applied Physics* 41: L371–L373.
12. Mukai, T., K. Takekawa, and S. Nakamura. 1998. InGaN-based blue light-emitting diodes grown on epitaxially laterally overgrown GaN substrates. *Japanese Journal of Applied Physics* 37: L839–L841.

13. Wuu, D. S., W. K. Wang, W. C. Shih, R. H. Horng, C. E. Lee, W. Y. Lin, and J. S. Fang. 2005. Enhanced output power of near-ultraviolet InGaN–GaN LEDs grown on patterned sapphire substrates. *IEEE Photonics Technology Letters* 17: 288–290.
14. Mukai, T., and S. Nakamura. 1999. Ultraviolet InGaN and GaN single-quantum-well-structure light-emitting diodes grown on epitaxially laterally overgrown GaN substrates. *Japanese Journal of Applied Physics* 38: 5735–5739.
15. Lin, W. Y., D. S. Wuu, S. C. Huang, and R. H. Horng. 2011. Enhanced output power of near-ultraviolet InGaN/AlGaN LEDs with patterned distributed Bragg reflectors. *IEEE Transactions on Electron Devices* 58: 173–179.
16. Huh, C., W. J. Schaff, L. F. Eastman, and S. J. Park. 2004. Temperature dependence of performance of InGaN/GaN MQW LEDs with different indium compositions. *IEEE Photonics Technology Letters* 25: 61–63.
17. Lin, W. Y., D. S. Wuu, K. F. Pan, S. H. Huang, C. E. Lee, W. K. Wang, S. C. Hsu, Y. Y. Su, S. Y. Huang, and R. H. Horng. 2005. High-power GaN–Mirror–Cu light-emitting diodes for vertical current injection using laser liftoff and electroplating techniques. *IEEE Photonics Technology Letters* 17: 1809–1811.
18. Chu, C. F., C. C. Cheng, W. H. Liu, J. Y. Chu, F. H. Fan, L. H. C. Cheng, T. Doan, and C. A. Tran. 2010. High brightness GaN vertical light-emitting diodes on metal alloy for general lighting application. *Proceedings of the IEEE* 98: 1197–1207.
19. Shatalov, M., A. Chitnis P. Yadav, Md. F. Hasan J. Khan, V. Adivarahan, H. P. Maruska, W. H. Sun, and M. Asif Khan. 2005. Thermal analysis of flip-chip packaged 280 nm nitride-based deep ultraviolet light-emitting diodes. *Applied Physics Letters* 86: 201109–1–201109–3.
20. Shen, C. F., S. J. Chang, W. S. Chen, T. K. Ko, C. T. Kuo, and S. C. Shei. 2007. Nitride-based high-power flip-chip LED with double-side patterned sapphire substrate. *IEEE Photonics Technology Letters* 19: 780–782.
21. Lee, C. E., Y. C. Lee, H. C. Kuo, T. C. Lu, and S. C. Wang. 2008. High-brightness InGaN-GaN flip-chip light-emitting diodes with triple-light scattering layers. *IEEE Photonics Technology Letters* 20: 659–661.
22. Shen, K. C., W. Y. Lin, D. S. Wuu, S. Y. Huang, K. S. Wen, S. F. Pai, L. W. Wu, and R. H. Horng. 2013. An 83% enhancement in the external quantum efficiency of ultraviolet flip-chip light-emitting diodes with the incorporation of a self-textured oxide mask. *IEEE Electron Device Letters* 34: 274–276.
23. Horng, R.H., W.K. Wang, S.C. Huang, S.Y. Huang, S.H. Lin, C.F. Lin, and D.S. Wuu. 2007. Growth and characterization of 380-nm InGaN/AlGaN LEDs grown on patterned sapphire substrates. *Journal of Crystal Growth* 298: 219.
24. Bohyama, S., H. Miyake, K. Hiramatsu, Y Tsuchida, and T. Maeda. 2005. Freestanding GaN substrate by advanced facet-controlled epitaxial lateral overgrowth technique with Masking Side Facets. *Japanese Journal of Applied Physics* 44: L24.
25. Huang, S. C., D. S. Wuu, P. Y. Wu, and S. H. Chan. 2009. Improved output power of 380 nm InGaN-based LEDs using a heavily Mg-doped GaN insertion layer technique. *IEEE Journal of Selected Topics in Quantum Electronics* 15: 1132–1136.
26. Claudel, A., V. Fellmann, I. Gélard, N. Coudurier, D. Sauvage, M. Balaji, E. Blanquet, R. Boichot, G. Beutier, S. Coindeau, A. Pierret, B. Attal-Trétout, S. Luca, A. Crisci, K. Baskar, and M. Pons. 2014. Influence of the V/III ratio in the gas phase on thin epitaxial AlN layers grown on (0001) sapphire by high temperature hydride vapor phase epitaxy. *Thin Solid Films* 573: 140–147.
27. Devillers, T., L. Tian, R. Adhikari, G. Capuzzo, and A. Bonanni. 2015. Mn as surfactant for the self-assembling of $Al_xGa_{1-x}N$/GaN layered heterostructures. *Crystal Growth & Design* 15: 587–592.
28. Banal, R. G., Y. Akashi, K. Matsuda, Y. Hayashi, M. Funato, and Y. Kawakami. 2013. Crack-free thick AlN films obtained by NH_3 nitridation of sapphire substrates. *Japanese Journal of Applied Physics* 52: 08JB21.

29. Sun, X., D. Li, Y. Chen, H. Song, H. Jiang, Z. Li, G. Miao, and Z. Zhang. 2013. In situ observation of two-step growth of AlN on sapphire using high-temperature metal–organic chemical vapour deposition. *CrystEngComm* 15: 6066–6072.

30. Balajia, M., R. Ramesh, P. Arivazhagan, M. Jayasakthi, R. Loganathan, K. Prabakaran, S. Suresh, S. Lourdudoss, and K. Baskar. 2015. Influence of initial growth stages on AlN epilayers grown by metal organic chemical vapor deposition. *Journal of Crystal Growth* 414: 69–75.

31. Imura, M., N. Fujimoto, N. Okada, K. Balakrishnan, M. Iwaya, S. Kamiyama, H. Amano, I. Akasaki, T. Noro, T. Takagi, and A. Bandoh. 2007. Annihilation mechanism of threading dislocations in AlN grown by growth form modification method using V/III ratio. *Journal of Crystal Growth* 300: 136–140.

32. Yan, J., J. Wang, Y. Zhang, P. Cong, L. Sun, Y. Tian, C. Zhao, and J. Li. 2015. AlGaN-based deep-ultraviolet light-emitting diodes grown on High-quality AlN template using MOVPE. *Journal of Crystal Growth* 414: 254–257.

33. Nakarmi, M. L., B. Cai, J. Y. Lin, and H. X. Jiang. 2012. Three-step growth method for high quality AlN epilayers. *Physica Status Solidi A* 209: 126–129.

34. Dong, P., J. Yan, J. Wang, Y. Zhang, C. Geng, T. Wei, P. Cong, Y. Zhang, J. Zeng, Y. Tian, L. Sun, Q. Yan, J. Li, S. Fan, and Z. Qin. 2013. 282-nm AlGaN-based deep ultraviolet light-emitting diodes with improved performance on nano-patterned sapphire substrates. *Applied Physics Letters* 102: 241113.

35. Kim, J., J. Pyeon, M. Jeon, and O. Nam. 2015. Growth and characterization of high quality AlN using combined structure of low temperature buffer and superlattices for applications in the deep ultraviolet. *Japanese Journal of Applied Physics* 54:081001.

36. Okada, N.; N. Kato, S. Sato, T. Sumii, T. Nagai, N. Fujimoto, M. Imura, K. Balakrishnan, M. Iwaya, S. Kamiyama, H. Amano, I. Akasaki, H. Maruyama, T. Takagi, T. Noro, A. Bandoh. 2007. Growth of high-quality and crack free AlN layers on sapphire substrate by multi-growth mode modification. *Journal of Crystal Growth* 298:349–353.

37. Uchida, T., K. Kusakabe, and K. Ohkawa. 2007. Influence of polymer formation on metalorganic vapor-phase epitaxial growth of AlN. *Journal of Crystal Growth* 304: 133–140.

38. Chierchia, R., T. Bottcher, H. Heinke, S. Einfeldt, S. Figge, and D. Hommel. 2003. Microstructure of heteroepitaxial GaN revealed by x-ray diffraction. *Journal of Applied Physics* 93: 8918.

39. Wu, P., M. Funato, and Y. Kawakami. 2015. Environmentally friendly method to grow wide-bandgap semiconductor aluminum nitride crystals: Elementary source vapor phase epitaxy. *Scientific Reports* 5: 1–9.

40. Ueno, K., J. Ohta, H. Fujioka, and H. Fukuyama. 2011. Characteristics of AlN films grown on thermally-nitrided sapphire substrates. *Applied Physics Express* 4: 015501.

41. Imura, M., K. Nakano, N. Fujimoto, N. Okada, K. Balakrishnan, M. Iwaya, S. Kamiyama, H. Amano, I. Akasaki, T. Noro, T. Takagi, and A. Bandoh. 2007. Dislocations in AlN epilayers grown on sapphire substrate by high-temperature metal-organic vapor phase epitaxy. *Japanese Journal of Applied Physics* 46: 1458.

42. Miyake, H., G. Nishio, S. Suzuki, K. Hiramatsu, H. Fukuyama, J. Kaur, and N. Kuwano. 2016. Annealing of an AlN buffer layer in N_2–CO for growth of a high-quality AlN film on sapphire. *Applied Physics Express* 9: 025501.

43. Hirayama, H., T. Yatabe, N. Noguchi, T. Ohashi, and N. Kamata. 2007. 231–261nm AlGaN deep-ultraviolet light-emitting diodes fabricated on AlN multilayer buffers grown by ammonia pulse-flow method on sapphire. *Applied Physics Letters* 91: 071901.

44. Dong, P., J. Yan, Y. Zhang, J. Wang, J. Zeng, C. Geng, P. Cong, L. Sun, T. Wei, L. Zhao, Q. Yan, C. He, Z. Qin, and J. Li. 2014. AlGaN-based deep ultraviolet light-emitting diodes grown on nano-patterned sapphire substrates with significant improvement in internal quantum efficiency. *Journal of Crystal Growth* 395: 9–13.

45. Zhang, L., F. Xu, J. Wang, C. He, W. Guo, M. Wang, B. Sheng, L. Lu, Z. Qin, X. Wang, and B. Shen. 2016. High-quality AlN epitaxy on nano-patterned sapphire substrates prepared by nano-imprint lithography. *Scientific Reports* 6: 35934.
46. Jain, R., W. Sun, J. Yang, M. Shatalov, X. Hu, A. Sattu, A. Lunev, J. Deng, I. Shturm, Y. Bilenko, R. Gaska, and M. S. Shur. 2008. Migration enhanced lateral epitaxial overgrowth of AlN and AlGaN for high reliability deep ultraviolet light emitting diodes. *Applied Physics Letters* 93: 051113.
47. Wang, T. Y., J. H. Liang, G. W. Fu, and D. S. Wuu. 2016. Defect annihilation mechanism of AlN buffer structures with alternating high and low V/III ratios grown by MOCVD. *CrystEngComm* 18: 9152–9159.
48. Bai, J., T. Wang, P.J. Parbrook, K. B. Lee, and A.G. Cullis. 2005. A study of dislocations in AlN and GaN films grown on sapphire substrates. *Journal of Crystal Growth* 282: 290–296.
49. Dovidenko, K., S. Oktyabrsky, and J. Narayan. 1997. Characteristics of stacking faults in AlN thin films. *Journal of Applied Physics* 82: 4296.
50. Prokofyeva, T., M. Seon, J. Vanbuskirk, M. Holtz, S. A. Nikishin, N.N. Faleev, H. Temkin, and S. Zollner. 2001. Vibrational properties of AlN grown on (111)-oriented silicon. *Physical Review B* 63: 125313.
51. Sarua, A., M. Kuball, and J. E. V. Nostrand. 2004. Phonon deformation potentials of the E_2(high) phonon mode of $Al_xGa_{1-x}N$. *Applied Physics Letters* 85: 2217.
52. Chen X., J. Yan, Y. Zhang, Y. Tian, Y. Guo, S. Zhang, T. Wei, J. Wang, and J. Li. 2016. Improved crystalline quality of AlN by epitaxial lateral overgrowth using two-phase growth method for deep-ultraviolet stimulated emission. *IEEE Photonics Journal* 8: 2300211.
53. Tran, B. T., N. Maeda, M. Jo, D. Inoue, T. Kikitsu, and H. Hirayama. 2016. Performance improvement of AlN crystal quality grown on patterned Si(111) substrate for deep UV-LED applications. *Scientific Reports* 6: 35681.
54. Kitagawa, S., H. Miyake, and K. Hiramatsu. 2014. High-quality AlN growth on 6H-SiC substrate using three dimensional nucleation by low-pressure hydride vapor phase epitaxy. *Japanese Journal of Applied Physics* 53: 05FL03.
55. Kneissl, M., T. Kolbe, C. Chua, V. Kueller, N. Lobo, J. Stellmach, A. Knauer, H. Rodriguez, S. Einfeldt, Z. Yang, N. M. Johnson, and M. Weyers. 2011. Advances in group III-nitride-based deep UV light-emitting diode technology. *Semiconductor Science and Technology* 26: 014036.
56. Shatalov, M., W. Sun, A. Lunev, X. Hu, A. Dobrinsky, Y. Bilenko, J. Yang, M. Shur, R. Gaska, C. Moe, G. Garrett, and M. Wraback. 2012. AlGaN deep-ultraviolet light-emitting diodes with external quantum efficiency above 10%. *Applied Physics Express* 5: 082101.
57. Hirayama, H., S. Fujikawa, N. Noguchi, J. Norimatsu, T. Takano, K. Tsubaki, and N. Kamata. 2009. 222–282 nm AlGaN and InAlGaN-based deep-UV LEDs fabricated on high-quality AlN on sapphire. *Physica Status Solidi A* 206: 1176–1182.
58. Banal, R. G., M. Funato, and Y. Kawakami. 2011. Extremely high internal quantum efficiencies from AlGaN/AlN quantum wells emitting in the deep ultraviolet spectral region. *Applied Physics Letters* 99: 011902.
59. Wang, T. Y., S. L. Ou, R. H. Horng, and D. S. Wuu. 2014. Growth evolution of Si_xN_y on the GaN underlayer and its effects on GaN-on-Si (111) heteroepitaxial quality. *CrystEngComm* 16: 5724–5731.
60. Li, X., S. Sundaram, P. Disseix, G. Le Gac, S. Bouchoule, G. Patriarche, F. Réveret, J. Leymarie, Y. El Gmili, T. Moudakir, F. Genty, J-P. Salvestrini, R. D. Dupuis, P. L. Voss, and A. Ougazzaden. 2015. AlGaN-based MQWs grown on a thick relaxed AlGaN buffer on AlN templates emitting at 285 nm. *Optical Materials Express* 5: 380–392.
61. Tamulaitis, G. 2011. Ultraviolet light emitting diodes. *Lithuanian Journal of Physics* 51: 177–193.

62. Li, L., Y. Miyachi, M. Miyoshi, and T. Egawa. 2016. Enhanced emission efficiency of deep ultraviolet light-emitting AlGaN multiple quantum wells grown on an N-AlGaN underlying layer. *IEEE Photonics Journal* 8: 1601710.

63. Bhattacharyya, A., T. D. Moustakas, L. Zhou, D. J. Smith, and W. Hug. 2009. Deep ultraviolet emitting AlGaN quantum wells with high internal quantum efficiency. *Applied Physics Letters* 94: 181907.

5 *In-situ* Growth of Spherical Graphene Films on Cemented Carbide for Spatial Sensor Matrix

Xiang Yu, Zhen Zhang, Jing-xuan Pei, Jian-kang Huang, and Xiao-yong Tian
China University of Geosciences (Beijing)

CONTENTS

5.1 Introduction 168
5.2 Space Sensors and SGF 168
    5.2.1 Research Summary of SGF 169
        5.2.1.1 Performance Advantage 170
        5.2.1.2 Preparation Method 171
    5.2.2 Preparation of SGF by Metal Catalysis 172
        5.2.2.1 Growth Mechanism of Graphene by Metal Catalysis........ 173
        5.2.2.2 Selection of the Carbon Source 174
        5.2.2.3 The Catalytic Metal 174
5.3 Synthesis of SGF on Cemented Carbide 176
    5.3.1 Experimental Procedure 176
        5.3.1.1 Experimental Equipment 177
        5.3.1.2 Preparation Process... 178
        5.3.1.3 Characterization of the Samples 179
    5.3.2 *In situ* Growth of Graphene on Cemented Carbide Surface............ 179
        5.3.2.1 Annealing vs. Graphene Structure....................... 179
        5.3.2.2 Influence of Annealing Temperature 182
        5.3.2.3 Action Mechanism of Si Atoms in Co-Catalyzed Graphene Generation.................. 186
    5.3.3 Controlled Preparation of Graphene Defects and Number of Layers by C_2H_2 Gas Flow Rate... 188
        5.3.3.1 C_2H_2 Gas Flow vs. Atomic Content of Films 189
        5.3.3.2 C_2H_2 Gas Flow Rate vs. Phase of Interfacial Layer.......... 189

5.3.3.3   C₂H₂ Gas Flow Rate vs. Layers and Lattice
            Defects of Graphene 190
5.3.3.4   C₂H₂ Gas Flow vs. Graphene Morphology 191
5.3.4   Reaction Mechanism of Co-Catalyzed SGF Growth 193
5.4   Summary and Perspectives 194
References 195

5.1   INTRODUCTION

Sensing is a unique approach to obtain important data in different fields. Among this, a sensor modified by SGF (spherical graphene film) represents a first choice applicable for the space sensing. Absence of feasible method to achieve the *in-situ* and controllable preparation of SGF on the targeted substrate is in the way of the application benefits. SGF can provide more chemical modification sites due to the large specific surface area [1,2]. SGF can not only improve the sensitivity of the sensor, but also extend the detection range of the sensor to meet the lightweight and multifunctional requirements of the space sensors. Available methods such as chemical or chemical vapor deposition (CVD) techniques, not to mention the synthesis of SGFs. Consequently, the *in-situ* and controllable growth of SGF on the requested substrate has drawn a lot of attention.

SGF has begun to attract the attention of experts, but the preparation and application of graphene pellets is still in the air. In this chapter, the progress in preparation and application of SGF is reviewed, and an innovative method for *in-situ* and controlled growth of SGF on cemented carbide base surface is proposed.

5.2   SPACE SENSORS AND SGF

Space exploration enables humans to explore space resources, develop space boundaries, and realize sustainable and long-term development of the Earth's homeland. The devices for the space exploration are commonly subjected to operate at a far distance from the earth, exhibit a long duration, and work in harsh environments (high vacuum, alternating temperature, and high radiation). In order to achieve a variety of space exploration missions, space detectors also need to carry a large number of sensing load instruments. As a result, a material, with resistance to space harsh environment, miniaturization, light weight and low energy consumption, becomes a good candidate for space exploration missions.

In addition, the friction and wear problems in the high vacuum of space environment are more serious than on the earth. The detector components, especially the metal parts, cannot form an oxide film on the surface for protection. In vacuum environment, there is a stronger interaction between the atoms at the contact sites, and they tend to stick together. This limits the proper operation of the space detecting device. The common fluid lubrication technology is not suitable for space environment due to its easy volatilization and decomposition under vacuum environment. Solid lubricant material is the main solution for space lubrication [3]. Consequently, it is the primary task to seek sensor materials with good electrical conductivity and excellent tribological properties for the space exploration.

Graphene is a two-dimensional (2D) material of a single-layered sheet structure composed of sp^2-C. The extremely strong σ bond and the π electrons freely moving in the sheet confer superior electrical conductivity, high strength, and ultra-lightweight property. Graphene can adapt to the high radiation and harsh environment of space and has a unique spatial lubrication mechanism, making it shine in space detection applications. The high radiation and extremely high and low temperatures in deep space contribute to the reduced sensitivity and shortened lifetime of the traditional semiconductor sensors [4]. Graphene-based sensors have the characteristics of high sensitivity and long life operation. Due to the large specific surface area, SGF can provide more chemically active sites [5,6], which in turn improve the sensor sensitivity and extend its detection range to meet multifunctionality requirements of the space sensor [1,2]. Various sensors based on SGF have been successfully developed through functional modification. This functional modification technology has matured to be used in the field of space detection. According to literature reports, NASA's graphene space detection sensor can not only detect trace elements in the space environment, but also detect structural defects on spacecraft. SGF-based sensor is the first choice for space detection. It has been shown that a variety of SGF-based sensors have been successfully developed by surface functionalization. However, the *in-situ* growth of SGF on different substrates still has great challenges.

There are two main issues in the surface growth of graphene spheres on the carbide substrate: (1) the generation of spherical graphene structures on the basis of *in-situ* growth and (2) the control of the number of layers and defects of the synthesized material. For the preparation of SGF membranes, spheroidal catalytic metal particles are applied as templates, and CVD method is used to grow graphene on the surface with subsequent etching to form a hollow SGF, which will be finally transferred to the substrate of interest [7,8]. However, the transfer process readily produces defects and impurities, which not only reduces the mechanical and electrical properties of the graphene matrix, but also decreases the binding ability [9,10]. As a common material for space detection devices, *in-situ* growth of graphene film as a sensor matrix on the surface of cemented carbide can avoid the problem of easy falling off and attenuation of detection accuracy caused by secondary transfer. At the same time, the morphology, number of layers, and defects of graphene films are the key factors affecting the performance of graphene. In this way, *in-situ* growth of SGF on the carbide, controllable preparation of graphene layers, and defects are three main issues in the way of SGF application in space exploration.

5.2.1 RESEARCH SUMMARY OF SGF

The excellent properties of graphene benefit from its special structure. However, the 2D structure of graphene is prone to agglomeration, resulting in a decrease in specific surface area and uneven layer number. Due to this phenomenon the electrical conductivity and lubrication performance of graphene cannot be fully developed for application in space exploration. And van der Waals forces between the graphene layers cause this agglomeration and overlap in an irreversible way. The solution to this disadvantage is to further increase the specific surface area of graphene through the preparation of three-dimensional (3D) graphene or adjust its morphology to a spherical shape [5,6]. Among them, the SGF has the largest specific surface area and the lowest density, and it can maintain its layered structure well. Consequently, SGF has great advantages in space detection and sensing through surface modification technology [11].

5.2.1.1 Performance Advantage

SGF has the largest specific surface area and the lowest density; maintaining the properties of graphene can reduce the stacking effect and increase the adsorption and resilience of graphene materials.

The spherical structure also imparts special properties to the graphene. SGF is a strong light absorber in the visible and near-infrared regions. Figure 5.1 shows the light absorption mechanism of SGF. Hao et al. found that, in the wavelength range of 350~2,500 nm, the average absorption rate of the shrinking graphene sphere can reach 97.4%. The shrinking graphene sphere is used as the absorber of the interface evaporation system. At ambient pressure, the evaporation efficiency is 84.6% under one sun. Enhanced solar absorption shrinking graphene balls, coupled with their aggregation and ubiquitous solution processing capabilities, make them promising for solar heating/distillation applications [12].

Since a large specific surface area can provide more chemical sites, SGF can realize high-sensitivity detection, thus meeting the requirements of light and multifunctional detection of space exploration [5,6]. At present, a variety of sensors have been successfully prepared using a graphene sphere as a matrix by surface modification technology [13,14]. Huang et al. successfully prepared 3D redox graphene (RGO) using SiO_2 microspheres as a support structure and achieved high-sensitivity detection of NH_3 by chemical modification. SiO_2 microspheres, as a support structure, provide more chemical modification sites for SGF. This 3D SiO_2–RGO framework exhibited superior response (31.5%) toward 50 ppm NH_3 in 850 s to that of unmodified 2D plane-stacked RGO network sensor (1.5%). And the selective detection of NH_3 by chemical modification has also been enhanced [1].

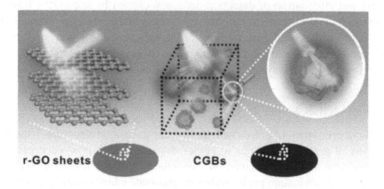

r-GO sheets   **CGBs**

FIGURE 5.1   Schematic drawings illustrating that flake-like graphene layers become reflective due to their planar microstructure, while layers of crumpled graphene balls are nonreflective due to extensive scattering within or between the particles [12]. (Reprinted with permission from Royal Society of Chemistry.)

Wang et al. synthesized a uniform 3D graphene nanodoped porous gold electrode by ion beam sputtering deposition (IBSD) and chemical etching techniques. Through the surface modification of pyrene functionalities, the prepared enzyme electrode has the advantages of good repeatability, high sensitivity, inherent selectivity, and wide detection range. As for glucose analysis, a broad linear range from 0.05 to 100 mM was obtained, and the linear range for hydrogen peroxide was 0.005 to 4 mM. Detection limits of 30 μM for glucose and 1 μM for hydrogen peroxide were achieved ($S/N = 3$), respectively [11]. Various other biosensors can be prepared using the same porous electrode and the same enzyme memory method.

At the same time, the SGF also has excellent electrochemical properties. Son *et al.* prepared graphene–SiO_2 spheres by CVD. The 3D graphene–silica assembly can be uniformly coated on a nickel-rich layered cathode. It improves battery cycle life and provides fast chargeability by suppressing harmful side reactions and providing an effective conductive path. SGF is also used as anode material with a high specific capacity of 716.2 mAh g^{-1}. A cell containing SGF increased the volumetric energy density by 27.6% compared to a control cell without graphene, showing the possibility of achieving 800 Wh L^{-1} in a commercial battery environment along with a high cyclability of 78.6% capacity retention after 500 cycles at 5°C and 60°C [15].

It can be seen that the special 3D structure of SGF has great advantages and application prospects in the field of sensors and electrochemistry. Moreover, the surface functionalization modification technology based on SGF has matured and can be used in space exploration [16]. However, the *in-situ* synthesis of SGF on desired substrate and the control of the number of layers and lattice defects of the SGF pose greater challenges for the current redox or CVD methods. Therefore, the *in-situ* controllable preparation of SGF is also a key challenge that restricts its further applications.

5.2.1.2 Preparation Method

At present, the mainstream preparation methods of graphene include stripping, CVD, redox, epitaxial growth, and metal-catalyzed growth. Among them, the stripping method allows to separate graphite into high-quality single layer or a few layers graphene, but it is not possible to regulate its morphology and size. Epitaxial growth method consists of processing SiC crystal under high temperature and vacuum, so that Si atoms volatilize and C atoms rearrange to form graphene. This method has high cost and many requirements on equipment, and the size and morphology of graphene crystals cannot be controlled.

Sohn et al. used polystyrene capsules as a template to mix graphene oxide with ultrasonic spray to form nanodroplets. After drying, spherical graphene oxide grows on the surface of the droplet. The polystyrene is then pyrolyzed, and the graphene oxide is reduced to graphene oxide spheres [17]. It was subsequently discovered that reduced graphene oxide spheres can also be formed using other nanospherical templates [18].

Yoon et al. used self-assembled 3D SiO_2 microspheres as a template, immersed them in polyvinyl alcohol–ferric chloride, and then fed them into a CVD tube furnace. At 1,000°C in Ar/H_2 gas atmosphere, iron ions were reduced to elemental iron

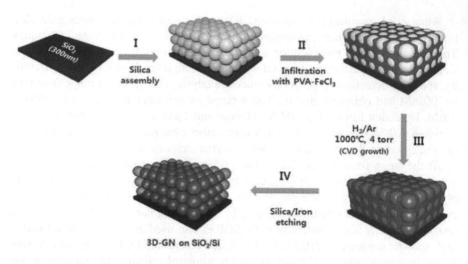

FIGURE 5.2   Preparation process of SGF by CVD [19].

as catalytic metal, and polyvinyl alcohol was used as carbon source to grow graphene on the surface of SiO_2 microspheres. Finally, SiO_2 and residual iron ions were removed by using an acid solution. Figure 5.2 shows the growth process diagram [19]. Subsequently, Lu et al. also applied SiO_2 spheres as a template and methane gas as a carbon source to directly form an SGF [20].

It has been found that SGF can be synthesized by both redox and CVD methods. However, the graphene prepared by the redox method has a large amount of impurities. The CVD method requires the help of spherical substrate template, and the technique is not easy to control the gaseous carbon source and catalytic reaction process, which will affect the quality of SGF. The most critical is that both methods require the transfer of the prepared graphene to the desired substrate and cannot achieve *in-situ* growth [21]. On the basis of the CVD method, the *in-situ* growth of graphene on the desired substrate can be realized by the metal-catalyzed method using a more easily controllable solid carbon source. In addition, a large specific surface area SGF can be obtained by controlling the morphology of the catalytic metal. The preparation of SGF is currently achieved using different catalytic metals [16,22]. Therefore, the preparation of SGF membrane by metal catalysis is our research focus.

5.2.2   PREPARATION OF SGF BY METAL CATALYSIS

Metal catalysis is the most promising method for the *in-situ* and controllable growth of SGF on different substrates. It utilizes a carbon source and a metal to catalyze the *in-situ* growth of C atoms on the substrate surface at a suitable temperature and pressure. The controllable preparation of graphene film layers and defects was achieved by controlling the carbon source and catalytic reaction conditions. Large specific surface area SGF can be obtained by controlling the morphology of catalytic metals. In addition, the catalytic metal is easily etched away with an acid solution to achieve *in-situ* growth of graphene. Therefore, the metal catalysis method has received a lot of attention.

5.2.2.1 Growth Mechanism of Graphene by Metal Catalysis

The mechanism of metal-catalyzed growth of graphene is similar to that of CVD. Current research indicates that the mechanism is of two types: one is the dissolution–precipitation mechanism of carbon atoms inside the metal, and the other is the atomic adsorption mechanism of carbon atoms on the metal surface [23,24]. Figure 5.3 depicts the schematic diagram of the mechanism for the metal-catalyzed growth of graphene. Dissolution–precipitation utilizes the difference in solubility of carbon atoms in metals at different temperatures. At high temperatures, the carbon atoms will dissolve into the interior of the metal due to the high solubility; the solubility will then decrease during the cooling process, so the C atoms will precipitate onto the surface, thereby nucleating and growing to form graphene. This mechanism is mostly applicable to the metals having a higher ability to dissolve carbon, such as Co and Ni. The surface adsorption mechanism is mainly directed to Cu, a catalytic metal with low solubility to carbon atoms. At high temperatures, due to the low solubility, carbon atoms are directly adsorbed on the surface of the metal to nucleate and grow into graphene. At the same time, when a layer of graphene is formed on the surface, self-limiting effect occurs, which makes it difficult for C atoms to continue to grow, and a single layer of graphene can be formed.

It is concluded that the key to metal catalysis mainly includes three points. One is the energy for the reaction and growth of graphene, the second is the carbon source, and the third is the catalytic metal. Current methods of providing energy include laser irradiation, electron irradiation, and high-temperature annealing. Among them, high-temperature annealing is the most common method currently used due to its simplicity and low cost. The conversion of carbon atoms to a graphene film can

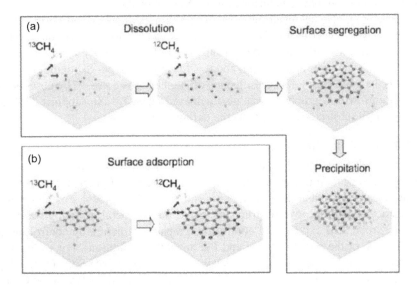

FIGURE 5.3 Schematic diagram of the mechanism of graphene growth by metal catalysis: (a) dissolution–precipitation mechanism and (b) surface adsorption mechanism [24]. (Reprinted with permission from American Chemical Society.)

be achieved by rapid temperature rise, but the setting of the annealing parameters needs to be adjusted according to the catalytic metal and the type of carbon source.

5.2.2.2 Selection of the Carbon Source

The carbon source of the metal-catalyzed method mainly includes a gas carbon source and a solid carbon source. The CVD method uses a gas carbon source such as methane. However, solid carbon is becoming the preferred choice for metal cataly-sis, because it is easier to control than gas sources. The current solid carbon sources include amorphous carbon, graphite-like carbon, diamond, single-crystal SiC, and organic carbon.

Liu et al. applied a magnetic filter cathode vacuum arc composite magnetron sputtering equipment to deposit amorphous carbon film as a carbon source and Ni as a catalytic metal to grow graphene on the surface of a silicon wafer. It was found that the quality of graphene obtained under vacuum conditions is better than that of Ar gas during annealing. Sun et al. used polymethyl methacrylate (PMMA) as a car-bon source on Cu substrate to obtain graphene with less defects [25]. Liu's team of Peking University used the carbon atoms dissolved in the metal to realize the growth of graphene through a simple vacuum annealing process.

On the basis of SiC epitaxy, SiC can be used as a carbon source to produce high-quality graphene at lower annealing temperatures and shorter annealing times. Metal-catalyzed SiC can form a metal silicide at the interface, which can further reduce the friction coefficient of the surface, and control the graphene structure by controlling the phase of the metal silicide.

Juang et al. first discovered that by depositing a Ni metal layer on the surface of SiC, during rapid heating, carbon atoms of SiC diffuse and dissolve into the Ni layer, and after cooling, graphene is formed on the Ni surface [26].

Machač et al. also achieved the formation of graphene by annealing the Ni/SiC composite layer. The effects of Ni layer thickness and Ni atom content on the struc-ture of graphene were studied. Annealing of Ni/6H-SiC (0001) system at 800°C yields the best quality graphene film, and it is found that the Ni–Si compound on the surface is the main reason for the influence of graphene structure [27]. Subsequently, the *in-situ* growth of graphene was also achieved on SiO_2 [28], sapphire [29], and Cu_2O (111) [8] substrates using SiC as the carbon source.

5.2.2.3 The Catalytic Metal

Studies revealed that metals such as Ni, Cu, and Co can effectively catalyze the growth of graphene from carbon sources. The preparation of graphene from Ni and Cu has been mentioned above. Another transition metal Co is also a catalytic metal of interest for *in-situ* growth of graphene. Since the Co metal has a high ability to dissolve carbon, it is expected to control the graphene structure by controlling the concentration of the carbon source and the annealing conditions. For example, Li *et al.* reported that by depositing a Co layer on the surface of 6H-SiC, a single layer and a few layers of graphene are grown by a selective reaction between the Co film and the SiC substrate at a high temperature [30].

Controlling the structure of the catalytic metal to regulate the morphology of the graphene is the key for the *in-situ* SGF growth. For example, Marchena et al.

used different densities of nano-Cu particles to prepare different sizes of globular graphene [16], as shown in Figure 5.4. Yoon et al. investigated Ni nanoparticles as a template to achieve *in-situ* growth of carbon atoms to form a 3D SGF [22], as shown in Figure 5.5. These methods use a spherical metal or metal oxide as a template to grow an SGF.

At present, it has been found that SGF can be grown on the surface of SiO_2 substrate even if the spherical metal template is not applied [28,31]. A Ni metal layer is deposited on the surface of SiO_2 substrate and the metal layer shrinks after annealing; then the C atoms grow on the surface nuclei, and the resulting graphene is spherical, as shown in Figure 5.6.

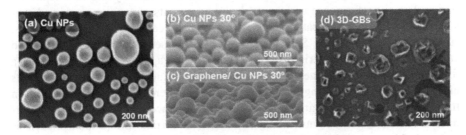

FIGURE 5.4   (a–d) SGF prepared using nano-Cu particles [16].

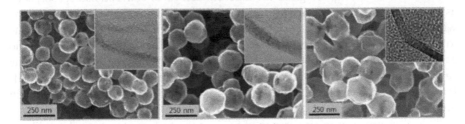

FIGURE 5.5   3D SGF prepared using a Ni metal template [22]. (Reprinted with permission from American Chemical Society.)

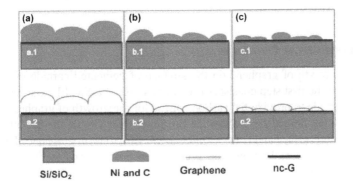

FIGURE 5.6   (a–c) Growth mechanism of SGF prepared by non-spherical metal template [28].

In summary, the metal catalysis method is the best way to achieve *in-situ* and controllable growth of SGF. Among them, the use of metal Co to catalyze SiC is more advantageous. At present, mostly single-crystal SiC is used as the carbon source, but the material is relatively expensive. Replacing single-crystal SiC with amorphous SiC (a-SiC) is an effective method to reduce cost. However, the composition of a-SiC, the catalytic metal, the annealing process parameters, and the growth mechanism still need to be further investigated.

5.3   SYNTHESIS OF SGF ON CEMENTED CARBIDE

From the above analysis, SGF has good electrical conductivity and excellent friction and wear performance and meets the requirements of space sensors and sensor conductive space lubrication materials. The metal catalysis method is the most suitable way for preparing SGF for space detecting sensor substrates. Cemented carbide is a common material used in space drilling applications for providing a first-hand insight into the geological structure and mineral distribution of planets and is also used as a cutting tool material. Development of a suitable material on the surface of cemented carbide tools could enable acquisition of drilling information and detection and automated correction of certain failure modes, such as auger choking, auger jamming, and bit wear. Graphene-based composites due to their remarkable mechanical, electrical, piezo-resistive and other physical properties become the best option given the inherent defects of conventional sensor, such as low sensitivity, short service life, and rapid attenuation. Especially, 3D graphene with hollow spheres on the surface displays enhanced sensing performance due to its large surface area that provides more binding sites. However, there are no acceptable methods to synthesize *in-situ* a graphene layer on the cemented carbide surface. The graphene with different layers and lattice defects is required for different sensing applications, but the control over these parameters is a bottleneck in graphene preparation [32,33].

This work proposes *in-situ* growth of SGF on the surface of cemented carbide by the method of metal catalysis. The simple thermal annealing used induces an *in-situ* transformation of magnetron-sputtered a-SiC films into graphene matrix [34]. And by controlling the experimental parameters, the number of layers and defects can be controlled [35].

5.3.1   Experimental Procedure

The *in-situ* growth of graphene on the surface of cemented carbide was achieved *via* two steps: the first step consists of the deposition of a-SiC film on the surface of cemented carbide to provide the carbon source for the growth of graphene. The second step is to use annealing to allow the binding phase (Co) in the cemented carbide to diffuse to the surface and react with the a-SiC film to form graphene. This chapter introduces experimental equipment, preparation process, and testing methods for graphene growth.

5.3.1.1 Experimental Equipment

1) Medium-Frequency Magnetron Sputtering Coating System

As one of physical vapor deposition (PVD) techniques, magnetron sputtering technology allows to deposit a variety of thin film materials such as metals, semiconductors, and insulators, with the advantages of fast coating rate, large and uniform coating area, high film-based bonding force, and low deposition temperature. The medium-frequency magnetron sputtering technique for deposition of a-SiC film on the surface of cemented carbide can avoid thermal stress and interface roughness caused by high-temperature deposition. Compared with radio-frequency magnetron sputtering, it has the advantages of low equipment cost and low radiation; in addition, it can avoid the target poisoning phenomenon often caused by direct current sputtering.

This experiment used SP0806AS multifunctional medium-frequency magnetron sputtering coating system developed by Beijing Power Source Technology Development Co., Ltd. Ltd. Figure 5.7 shows an external view of the equipment and a schematic structural view.

The system uses a twin target non-equilibrium magnetron sputtering to deposit a-SiC film. The ultimate vacuum of the equipment can reach 5.0×10^{-4} Pa; the sample is fixed on the vertical turret, about 20 cm away from the target. There are two sets of heating rods, and the maximum heating temperature can reach 400°C.

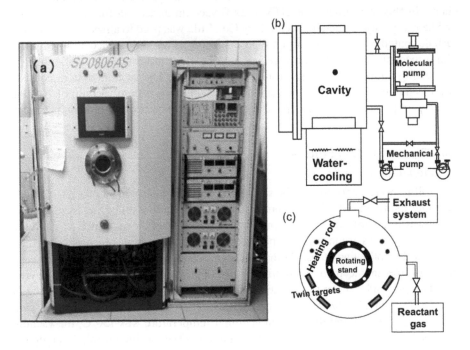

FIGURE 5.7 SP0806AS multifunctional magnetron sputtering coating system: (a) equipment appearance; (b) equipment schematic structure; (c) top view of equipment cavity.

FIGURE 5.8   Appearance of HTVO-2000 vacuum annealing furnace equipment.

2)   Vacuum Annealing Furnace

The annealing process is a key step to achieve Co-catalyzed a-SiC formation of graphene. In this experiment, the HTVO-2000 vacuum annealing furnace, developed by Shenyang Keyou Vacuum Technology Co., Ltd., was used to anneal the cemented carbide coated with the a-SiC layer. Figure 5.8 depicts the appearance of the furnace. The furnace uses molybdenum wire as a heating element for rapid temperature rise; the heating temperature ranges from 150 to 2,000°C. The vacuum system consists of a mechanical pump and a molecular pump with a maximum vacuum of 10^{-5} Pa. The cooling system is a split-type air-cooled chiller. The cooling process is to turn off the heating and cool with the furnace.

5.3.1.2   Preparation Process

1)   Preparation of a-SiC Layer

a-SiC films were deposited on the surface of cemented carbide YG 8 (WC-Co 8%) substrate using an SP0806 medium-frequency magnetron sputtering system. The target was a Si substrate having a purity of 99.96%, reaction gas was C_2H_2, and the sputtering gas was argon (Ar) with purity of 99.99%. The sample preparation method included the following steps: (1) the cemented carbide substrate was cleaned with both gasoline and acetone for 10 minutes under ultrasonication, followed by drying with nitrogen and transfer to a sputtering furnace for deposition; (2) glow cleaning of the substrate for 20 minutes before deposition. When the vacuum chamber pressure reached 3×10^{-3} Pa and the vacuum chamber temperature was 150°C, the vacuum chamber was filled with argon gas to 0.50 Pa; and (3) deposition using Si target and acetylene gas. The main deposition parameters are temperature = 150°C, Si target constant current = 22 A, duty ratio = 50%, bias = −100 V, and acetylene gas flow (15/20/25/30/35 sccm) as the experimental variables. The deposition time was 120 minutes, and the film thickness was $1.00 \pm 0.04\,\mu m$.

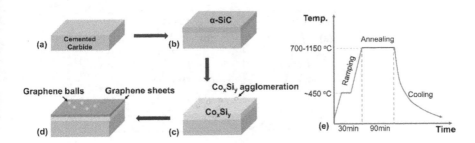

FIGURE 5.9 (a–e) Schematic illustrations of the graphene matrix growth process and annealing process.

2) Vacuum Annealing to Grow Graphene

Thermal annealing was used to generate graphene. The cemented carbide after deposition of the a-SiC film was placed in a HTVO-1200 vacuum high-temperature furnace for heating, annealing vacuum was kept below 5×10^{-4} Pa, and the heating rate was 20°C/min. Annealing temperature (700/850/1,000/1,150°C) and annealing time (30/50/70/90/110 minutes) were experimental variables. After annealing, the samples were cooled with the furnace in vacuum. Figure 5.9 summarizes the steps required for the growth of the graphene matrix.

5.3.1.3 Characterization of the Samples

The *in-situ* generation and number of graphene layers on the carbide were characterized using Raman spectrometer (HOEIBA Jobin Yvon, France). The structural changes of the graphene precipitates were measured by X-ray diffractometer (XRD; Bruker D8 Advance, CuKα, 40 kV, 40 mA, 2θ range 20°–80° and step size of 0.02°). The morphology of the cemented carbide surface was characterized by scanning electron microscopy (SEM, JEOL JSM 6301F). The surface film chemical composition was analyzed by EDS, and the atomic composition was obtained by taking the average of six measurements on the surface of an individual sample.

5.3.2 *IN SITU* GROWTH OF GRAPHENE ON CEMENTED CARBIDE SURFACE

In this study, Co from cemented carbide was used to catalyze *in-situ* the graphene formation from magnetron-sputtered a-SiC films by an annealing process. The annealing is an important parameter for the *in-situ* growth of SGF on cemented carbide substrate, since the diffusion of Co and the provision of kinetic energy can only be achieved when the temperature reaches a certain level [36]. So, we firstly deposited a-SiC films on the surface of cemented carbide under C_2H_2 gas flow rate of 30 sccm. Then, the annealing temperature was optimized considering the formation of favorable SGF layer.

5.3.2.1 Annealing vs. Graphene Structure

The Raman spectra of the deposited a-SiC films annealed at five different temperatures are shown in Figure 5.10. The Raman spectra of non-annealed and annealed samples at temperatures of 700°C and 850°C comprise diffuse peaks, typical of

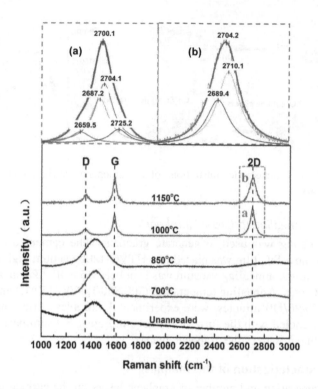

FIGURE 5.10 Raman spectra of Co-catalyzed a-SiC films at five temperatures. Inset 10-a (top-left corner) is the enlarged 2D peak obtained at 1,000°C, and inset 10-b (top-right corner) is the enlarged 2D peak obtained at 1,150°C.

amorphous structures, indicating that within this temperature range, crystallisation of the a-SiC film is not prominent and the films retain their amorphous structure. When the annealing temperature was increased to 1,000°C and 1,150°C, D, G, and symmetric 2D peaks appeared in the Raman spectra. The G peak indicates the presence of graphite phase, the D peak reveals the chaos level of phase structure, and the double-resonance 2D peaks suggest that a graphene phase was generated [37].

The characteristics of the D, G, and 2D peaks of the graphene obtained at 1,000°C and 1,150°C are listed in Table 5.1. Raman spectroscopy can clearly distinguish the

TABLE 5.1

Peak Positions, Peak Intensity Ratios, and Graphene Grain Sizes

	D Band		G Band		2D Band				
T (°C)	Position (cm⁻¹)	FWHM (cm⁻¹)	Position (cm⁻¹)	FWHM (cm⁻¹)	Position (cm⁻¹)	FWHM (cm⁻¹)	I_{2D}/I_G	I_D/I_G	L_a (nm)
1,000	1,353	45	1,585	31	2,700	44	1.304	0.339	56.710
1,150	1,357	46	1,583	34	2,704	72	0.964	0.407	47.235

number of graphene layers from a single layer to five layers and indicate the defects in graphene structure. The main indicators are intensity ratio of 2D and G peaks (I_{2D}/I_G), full width at half maximum (FWHM) of the peak, and the number of fitted Lorentzian peaks [38]:

$$L_a(nm) = 2.4 \times 10^{-10} \lambda^4 \left(I_D/I_G\right)^{-1} \tag{5.1}$$

where $\lambda = 532$ nm. The Raman spectrum at annealing temperature of 1,000°C showed that I_{2D}/I_G value was higher than 1, the FWHM was 45 cm^{-1}, and the 2D peak was at 2,700 cm^{-1}. For a single layer, the intensity of the 2D peak was four times higher than that of the G peak, whilst the intensity of the 2D peak was 1–2 times higher than that of the G peak for two layers [39]. The FWHM of the 2D peak for a single layer of graphene was in the range of 30–45 cm^{-1}, for two layers in the range of 45–60 cm^{-1}, and for three layers in the range of 65–70 cm^{-1}, whilst the FWHM of the 2D peak of graphite was lower than 30 cm^{-1} [39,40]. The Inset 10-a in Figure 5.10 indicates that the 2D peak can be deconvoluted into four peaks with maxima at 2,659, 2,687, 2,704, and 2,725 cm^{-1}. In general, a 2D peak of a single layer of graphene can be fitted by a single Lorentzian profile, a 2D peak of two layers of graphene can be fitted by four Lorentzian profiles, and a 2D peak of three layers of graphene can be fitted into six Lorentzian profiles, whilst a 2D peak of four or more graphene layers can be fitted only with two Lorentzian profiles [41]. Therefore, it can be concluded that by annealing at 1,000°C, high-quality graphene with two layers was successfully grown on the cemented carbide.

The intensity ratio of the 2D and G peaks at annealing temperature of 1,150°C is lower than that at 1,000°C, whilst the FWHM of the former is greater than of the latter; thus, it is clear that graphene is generated, but the number of layers produced at 1,150°C is obviously higher. The 2D peak may be fitted by two Lorentzian profiles with maxima at 2,689 and 2,710 cm^{-1} (Inset 10-b in Figure 5.10). It is suggested that the number of graphene layers was increased. In addition, the crystallite size of graphene (L_a) decreased and the I_D/I_G ratio increased, indicating that the number of film defects increased with the annealing temperature [28,42]. The reason for the increase of the number of defects with temperature is that the temperature increase leads to an increase in the number of graphene layers. This induces the formation of boundaries among the graphene layers and the increase of the ratio of I_D/I_G.

Figure 5.11 displays the I_{2D}/I_G ratio and FWHM of Raman mapping of the graphene film synthesised at two annealing temperatures, 1,000°C and 1,150°C, with the size of $100 \times 100\,\mu m^2$. As shown in Figure 5.11a, at 1,000°C, the I_{2D}/I_G ratio is in the range of 1.1–1.5, and FWHM, Figure 5.11c, is in the range of 45–57 cm^{-1}. At 1,150°C, Figure 5.11b, the I_{2D}/I_G ratio value drops and ranges from 0.7 to 1.2, whilst the FWHM value, Figure 5.11d, ranges partly from 65 to 80 cm^{-1} but also less than 30 cm^{-1}. From these findings, it can be concluded that the graphene film has high quality and consists of two layers at 1,000°C, whilst at 1,150°C, the film comprises several layers of graphene and graphite phases.

Thus, in the suitable annealing temperature range (1,000°C–1,150°C), the graphene film was formed on the cemented carbide by Co-catalyzed a-SiC transformation. When the annealing temperature was increased from 1,000°C to 1,150°C, the

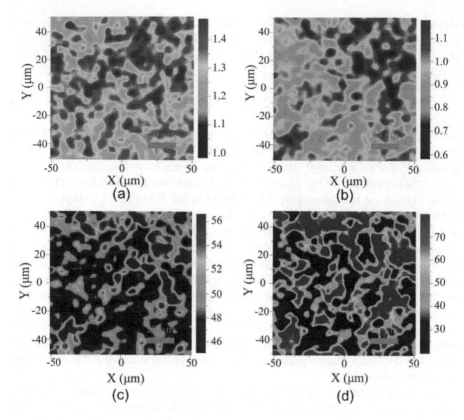

FIGURE 5.11   Raman mapping of the graphene film grown on the cemented carbide surface by annealing of a-SiC (size of $100 \times 100\,\mu m^2$): (a) I_{2D}/I_G peak ratio at 1,000°C; (b) I_{2D}/I_G peak ratio at 1,150°C; (c) FWHM of 2D peak at, 1,000°C; (d) FWHM of 2D peak at 1,150°C.

number of graphene layers, as well as the number of defects of the graphene films, was increased, and the quality of the graphene layer was degraded.

5.3.2.2   Influence of Annealing Temperature

The number of graphene layers is an important parameter to distinguish between graphene and graphite phases and to gauge the quality of graphene. Therefore, it was attempted to investigate how the annealing temperature influences the number of graphene layers. The following approaches were applied: (1) probing the impact of the annealing temperature on the graphene layer structure by XRD, (2) SEM monitoring of morphological changes of the graphene layer upon the annealing process, and (3) analysis of the annealing temperature impact on the graphene film composition by EDS.

Figure 5.12 displays the XRD patterns of a-SiC films precipitated on cemented carbide after annealing at five different temperatures. The results of XRD analyses reveal that only the basal peak (WC) appeared without annealing and even after annealing at 700°C or 850°C. These peaks are not very sharp below 850°C,

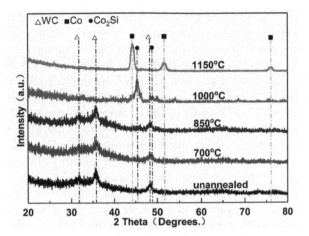

FIGURE 5.12 XRD patterns of a-SiC films recorded at five annealing temperatures.

because of the effect of a-SiC on the surface. This indicates the selected temperatures are inadequate for a-SiC film crystallization. The results are consistent with the results of Raman spectroscopy. When the annealing temperature was increased to 1,000°C and 1,150°C, Co_2Si and Co phases appeared in the XRD spectra, denoting that within this temperature range Co diffused from cemented carbide into the SiC film and reacted with SiC, thus preventing the crystallization of a-SiC. However, the diffraction peak of β-SiC appeared on the surface of the Si slice at 1,000°C and 1,150°C, (Figure 5.13), indicating that the a-SiC film was definitely engaged in the crystallization process and, thus, confirming the beneficial role of Co.

Different products were generated at different annealing temperatures by the catalytic reaction of Co and a-SiC. The structure of precipitated graphene layer also changed with temperature. Systematic research on the impact of the annealing

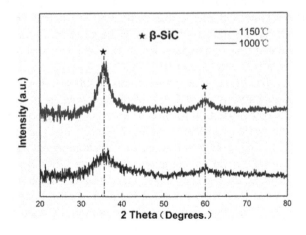

FIGURE 5.13 XRD patterns of a-SiC films on the surface of Si slices at annealing temperatures of 1,000°C and 1,150°C.

temperature on the structure of precipitated graphene layer could assist in revealing the mechanism by which the temperature causes the change in the number of graphene layers. Upon heating, Co and SiC interact, generating a Co–Si compound saturated with C atoms [43,44]. The Gibbs free energy of the Co–Si compound is lower than that of SiC, so the reaction (Co + SiC → Co_xSi_y + C) is thermodynamically a spontaneous reaction. If the reaction temperature is higher than 600°C, Co_2Si is generated (2Co + SiC → Co_2Si + C), and if the temperature is higher than 800°C, the materials further react to form CoSi (Co_2Si + SiC → 2CoSi + C) [36]. The formation of graphene is based on the reaction mechanism where C is isolated on the surface of the Co/Co_xSi_y film, owing to the decrease in solubility of C with temperature decrease [45]. As for the annealing temperature of 1,000°C, the Raman analysis shows that graphene was synthesized from the a-SiC film, which is the process catalyzed by Co. Because the number of graphene layers was low, the graphite phase was not detected by XRD at approximately 26.5° (2θ). The Co reacts with the a-SiC film to form Co_2Si, and the Co/Si atomic ratio after reaction may contribute to this reaction product [46]. When the Co/Si atomic ratio increases, a compound phase in the Co–Si bi-component system approaches Co, and the CoSi compound transforms to Co_2Si [42]. Thus, at the annealing temperature of 1,000°C, Co sufficiently reacts with the a-SiC film to catalyze graphene formation, generating carbon-saturated Co_2Si.

Unlike the reaction at 1,000°C, Co reacts with a-SiC at 1,150°C, and there is only the Co phase without Co_2Si compound. As mentioned before, Co reacts with the a-SiC film at 800°C to preferentially generate CoSi. At the annealing temperature of 1,150°C, the CoSi compound in the film disappears, and the Co phase appears. Meanwhile, the number of the generated graphene layers increases, and the graphite phase could be also detected.

Therefore, in the suitable annealing temperature range (1,000°C–1,150°C), Co diffuses in a-SiC film, catalyzing graphene formation. At 1,000°C, Co reacts with a-SiC to form Co_2Si compound saturated with C atoms, and a graphene layer with low I_D/I_G (0.339) and high I_{2D}/I_G (1.304) ratio is obtained. When the annealing temperature is further increased, the Co_2Si compound transforms into Co, and the amount of precipitated C is increased, as well as the number of graphene layers.

The impact of the annealing temperature on graphene morphology is depicted in Figure 5.14. The dendritic defects and large block regions, generated during deposition, remain in the a-SiC film even at temperatures of 700°C or 850°C, indicating the insufficient atomic activity. At 1,000°C, white particles covered by a transparent graphene film appear on the a-SiC film surface, and the morphology of graphene films is wrapped around the particles with 3D spheres (Figure 5.14d). The magnified area in the circle in Inset 14-1 in Figure 5.14 reveals the morphology and particle size of the transparent graphene film, and typical graphene film wrinkles can be observed. If white particles disappear, the transparent graphene film forms hollow graphene balls (as shown by the dotted line in Inset 14-1 in Figure 5.14). Compared with the SEM micrograph at 1,000°C, the white particles almost completely disappear at 1,150°C, and the hollow graphene balls are formed (as shown by the circle in Figure 5.14e), whilst the transparency of the film is relatively poor, implying the increased thickness of the graphene film (Figure 5.14e). Moreover, the wrinkles on the surface of

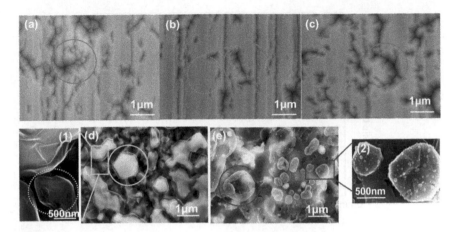

FIGURE 5.14 SEM micrographs of (a) non-annealed a-SiC films and films annealed at five temperatures: (b) 700°C, (c) 850°C, (d) 1,000°C, and (e) 1,150°C. Inset 14-1 (on the left) is the enlarged area obtained at 1,000°C, and inset 14-2 is the same for the micrograph obtained at 1,150°C.

the graphene balls are deeper, and the transparency reduces, so the film of the hollow graphene balls becomes thicker (as shown in Inset 14-2 in Figure 5.14).

The presented results suggest that at annealing temperatures of 1,000°C and 1,150°C, Co diffused into a-SiC film and catalyzed the formation of graphene thereon. The metal reacted with a-SiC film, and metal silicide spontaneously agglomerated to reduce the surface energy, which is in agreement with previously reported results that graphene film is generated preferentially at areas where the metal disappeared after contraction [47].

Thus, it can be concluded that at the annealing temperature of 1,000°C, the graphene balls precipitated from Co_2Si particles agglomerated on cemented carbide; hollow graphene balls were generated, and graphene sheets with graphene balls were obtained. When the annealing temperature was increased to 1,150°C, Co_2Si particles disappeared, the film of hollow graphene balls became thicker, and the number of graphene layers increased. Such graphene sheets, with hollow graphene balls that meet the requirement of high specific surface area for an ideal graphene film, can be beneficial for improvement of the efficiency of strain sensors by the reduction of size and power consumption [41].

Influence of annealing temperature on the chemical composition of graphene film was monitored by EDS. In the temperature range of 1,000–1,150°C, when the annealing temperature was increased, the Co_2Si compound disappeared, so the Co phase appeared in the film, and the number of generated graphene layers increased. In the Co–Si system, different Co/Si ratios resulted in different Co–Si phases [36]. The mechanism by which the precipitation layer structure of segregated C was altered with temperature may be revealed only by a systematic research of the impact of the annealing temperature on the atomic content during the synthesis of graphene. Figure 5.15 exhibits the EDS results of the a-SiC film at five annealing temperatures. The contents of Si and C remained essentially unchanged, and Co particles did not

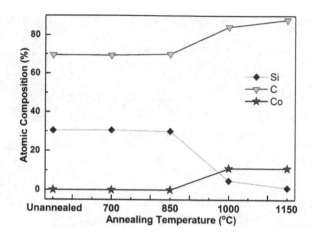

FIGURE 5.15   EDS results of a-SiC films annealed at five different temperatures.

exist up to 850°C (Figure 5.15). This indicates that in the temperature range of 700–850°C, the Co binding phase in cemented carbide is unable to diffuse into the a-SiC film. The Co can only react with Si to generate Co_2Si, and the reaction with SiC is not possible, so it does not appear in the a-SiC film. When the annealing temperature was increased, Co atoms appeared in the film, the content of Si atoms decreased, and the content of C atoms became higher, which implies that Co diffused into the a-SiC film.

Unlike conventional graphene epitaxial growth by SiC that requires high temperatures (>1,200°C) and high vacuum (10^{-4} Pa) conditions for Si sublimation from SiC film to enable restructuring of C atoms and graphene formation, graphene can be synthesized at low temperature using a SiC film with Co catalyst [48]. This study shows that at 1,000°C, Co diffused into the a-SiC film, and at Co concentration of 11.43%, the content of Si dropped from 30.5% to 4.22%, whilst the amount C atoms increased to 84.35%. It suggests that Co catalyzed the sublimation of Si and the restructuring of C atoms, decreasing the content of Si and rearranging C atoms to generate graphene. Compared to that, at 1,150°C, the content of Co atoms remained essentially unchanged, 10.83%, the amount of Si was even lower (0.93%), and the amount of C atoms increased to 88.24%. Having in mind the aforementioned results of Raman and XRD measurements, it can be stated that at the annealing temperature of 1,150°C, the Co-catalyzed sublimation of Si was accelerated, inducing Co_2Si cracking and the increase in the number of segregated carbon atoms in Co layer, as well as the number of graphene layers.

5.3.2.3   Action Mechanism of Si Atoms in Co-Catalyzed Graphene Generation

At 1,000°C, the sublimation of Si atoms was accelerated by Co catalysis, so the content of Si in the film decreased, and C was segregated to generate graphene. When the annealing temperature was increased to 1,150°C, the Co-catalyzed sublimation of Si atoms was enhanced, decreasing the content of Si, which induced Co_2Si

compound disappearance and the increase in the number of formed graphene layers. Therefore, the change of Si content in the film has a direct impact on graphene generation and the number of layers. Thus, the role of Si atoms is very important for a proper understanding of Co-catalyzed graphene formation from a-SiC and the change in the number of graphene layers as well.

Figure 5.16 shows the proposed mechanism describing the role of Si atoms in both Co-catalyzed graphene generation from a-SiC film and change in the number of layers. The microstructure and changes in the film are demonstrated by the ball-and-stick structure. A variation in the content of Si atoms in the film may alter the structure of Co metal and cause a change in graphene formation and the number of graphene layers. Three different parts of Si atom behavior can be observed: (1) Si reacts with Co, and C particles are freed (black represents Si particles in the film, light color represents Co particles in the film, and dark gray corresponds to C particles, Figure 5.16b); (2) the content of Si particles decreases and C particles precipitate to generate graphene (C particles are combined to generate the FLG, as shown in Figure 5.16e); and (3) Si particles disappear, the number of graphene layers increases, and clusters are gathered to generate the MLG, as shown in Figure 5.16g).

The proposed three-step mechanism agrees well with the experimental findings.

1. *Si reacted with Co to generate the compound saturated with C phase.* During annealing, Co diffused into and reacted with a-SiC film (Figure 5.16a), particularly with Si particles to generate CoSi, and free C was distributed in the product (Figure 5.16b). XRD results in Figure 5.12 illustrate that if the annealing temperature is higher than 850°C, Co reacts with the a-SiC film to generate the C-saturated CoSi compound.

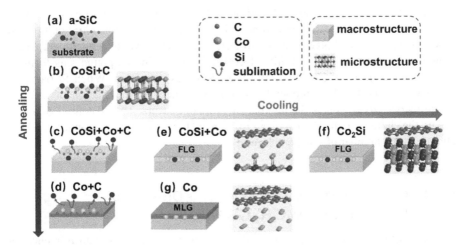

FIGURE 5.16 (a–g) The schematic illustration of graphene formation mechanism by annealing of a-SiC film on the surface of cemented carbide: the changing trend of temperature is demonstrated by the arrow; colors from shallow to deep mean that temperature changes from low to high for annealing and from high to low for cooling.

2. *If the content of Si atoms decreases, C atoms segregate in Co layer to generate graphene.* As shown in Figures 5.16c–g, the Si content in the film decreased as Si atoms sublimated, and the CoSi compound cracked to generate free Co metal (Figures 5.16c and d). As the temperature decreases (during cooling), the solubility of C in the Co metal phase would decline and C atoms in the Co layer would precipitate to form graphene (Figures 5.16e and g). Raman (Figure 5.11) and EDS (Figure 5.15) analysis showed that when the annealing temperature was 1,000°C, Co catalyzed Si sublimation and transport of C atoms through free Co metal to generate the graphene layer.

3. *If Si particles disappear, the Co–Si compound is not generated, and the number of graphene layers increases.* As shown in Figure 5.16e, during cooling, C particles passed through Co layer and precipitated to generate graphene; remnant Si particles reacted with Co to generate Co_2Si, and free Co phase disappeared, so C particles stopped precipitating and forming graphene, producing only a few graphene layers (Figure 5.16f). When the annealing temperature was increased, Si atoms in the film disappeared (Figure 5.16d); during cooling Co metal was not engaged in the reaction, and the Co_2Si compound was not generated, so the precipitation of C atoms was actually enhanced, and the number of graphene layers was increased (Figure 5.16g). EDS results in Figure 5.15 indicate that with further increase in the annealing temperature, the Si content in the film significantly decreased and almost disappeared below the detection limit; at the annealing temperature of 1,000°C, the Co_2Si compound disappeared, and only the free Co layer existed in the film, whilst the number of graphene layers increased.

Summarizing the aforementioned findings, the influence of Si particles is a synergistic effect of three different actions: Si particles' sublimation, CoSi compound cracking, and C atoms' precipitation through a free metal layer to generate graphene. When the content of Si in the film decreased, the Co_2Si compound was formed, the free metal layer disappeared, and the FLG was generated. When Si particles disappeared, the Co–Si compound was not generated, so the amount of free Co phase increased, as well as the number of graphene layers. The content of Si determined the changes in the Co phase and influenced the generation of graphene and the change in the number of graphene layers.

5.3.3 CONTROLLED PREPARATION OF GRAPHENE DEFECTS AND NUMBER OF LAYERS BY C_2H_2 GAS FLOW RATE

As a constituent element of graphene, the C content has an important influence on the number of layers and defects of graphene and is a key factor for achieving controlled growth of graphene. We control the content of C atoms by adjusting the gas flow rate of acetylene during the deposition process of a-SiC film. In order to analyze the effect of carbon content on the defects and number of layers of graphene, five acetylene gas flow rates (15/20/25/30/35 sccm) were used, accompanied by annealing at 1,000°C for 90 minutes.

5.3.3.1 C_2H_2 Gas Flow vs. Atomic Content of Films

The variation trend of atomic content on the film surface before and after annealing under five acetylene flows during the deposition of a-SiC was determined by EDS. Figure 5.17 displays the EDS results of the relative contents of three atoms (C, Si, Co) at five C_2H_2 gas flow rates before and after annealing.

It can be seen that different C_2H_2 gas flow rates have a significant influence on the relative content of specific atomic species. Before annealing, the film did not contain Co atoms, and with the increase of gas flow, the content of C atoms increased, and Si atom content decreased in the film. After annealing, Co atoms appeared in the film, and the relative content of Co atoms decreased with the increase of C_2H_2 gas flow. However, the content of C and Si atoms decreased with the increase of gas flow, which is the same trend as showed before for the annealing process. It can be concluded that the C content in the interfacial layer can be directly controlled by the flow of acetylene gas during the deposition of a-SiC films.

5.3.3.2 C_2H_2 Gas Flow Rate vs. Phase of Interfacial Layer

The atomic content may also affect the crystal phase of the precipitate, and we need to determine the variation of the interfacial layer phase with the atomic content. Figure 5.18 shows the XRD patterns of samples after annealing under five C_2H_2 gas flows during the deposition of a-SiC films. The results reveal that under C_2H_2 gas flow of 15 and 20 sccm, the mixed phase of Co_2Si and $CoSi$ compounds appeared in the XRD pattern. The diffraction peaks of $CoSi$ compound decreased, while the diffraction peaks of Co_2Si compound were enhanced with the increase of C_2H_2 gas flow. And starting from C_2H_2 gas flow of 25 sccm, the interfacial layer only has pure Co_2Si phase. This suggests that with the increase of C_2H_2 gas flow, the phase of interfacial

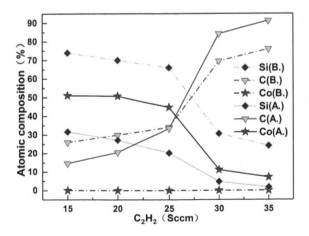

FIGURE 5.17 EDS results of atomic content before and after annealing of SiC films deposited under five C_2H_2 flow rates. (Reprinted with permission from Elsevier.)

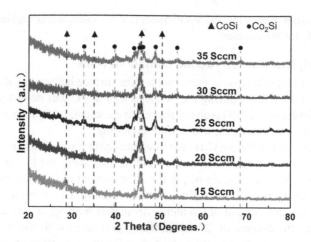

FIGURE 5.18 XRD pattern of a-SiC films deposited under five C_2H_2 flow rates after annealing. (Reprinted with permission from Elsevier.)

layer was converted from CoSi to Co_2Si after annealing. The main reason is that with the increase of C_2H_2 gas flow, in the SiC film, the content of C atom increased, and the content of Si atom decreased, resulting in an increase of the Co/Si ratio after annealing. So the cobalt–silicon compounds got converted from CoSi to Co_2Si; finally, only pure Co_2Si was found in the interfacial layer. It can be concluded that although the flow of C_2H_2 will also affect the phase of the interfacial layer, when the acetylene gas flow exceeds 25 sccm, the phase of interfacial layer is only composed of Co_2Si phase. The effect of atomic content on the formation of graphene can be discussed in the situation of same phase.

5.3.3.3 C_2H_2 Gas Flow Rate vs. Layers and Lattice Defects of Graphene

The influence of C atomic content changes on the number of layers and defects of graphene layers was explored by Raman. Figure 5.19 displays the Raman spectra of graphene formed by five C_2H_2 gas flow rates from a-SiC films deposited on cemented carbide.

As can be seen, the D, G, and 2D peaks appeared at the five C_2H_2 gas flow rates, indicating that the graphene is generated. The ratios of D to G and 2D to G peaks have changed significantly with the increase of C_2H_2 gas flow, indicating that the C_2H_2 gas flow rate has an obvious influence on the number of layers and defects of graphene.

Table 5.2 depicts the Raman peak information of graphene formed under five C_2H_2 gas flow rates from SiC films deposited on cemented carbide surface. We can consider the change of graphene defects by the value of I_D/I_G. We can see that the I_D/I_G of graphene decreases continuously with the increase of C_2H_2 gas flow. This indicates that the lattice defects of graphene decreased with the increase of C atom in the interfacial layer. When the gas flow rate is 15 and 20 sccm, the I_D/I_G ratio is larger, probably because the D peak is strong due to more amorphous carbon on the surface at this time. At 25 sccm, the I_D/I_G ratio decreased obviously, suggesting that the lattice defects of graphene were reduced. This is because the interfacial

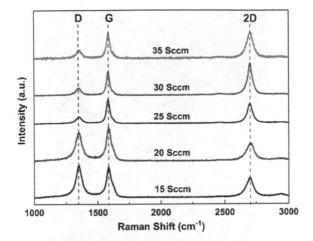

FIGURE 5.19   Raman spectra of graphene grown at five C_2H_2 gas flow rates from a-SiC films deposited on cemented carbide. (Reprinted with permission from Elsevier.)

TABLE 5.2
Raman Peak Information of Graphene at Different C_2H_2 Flow Rates Used during Deposition of SiC Thin Films

C_2H_2 (Sccm)	D Peak		G Peak		2D Peak				
	Position (cm^{-1})	Intensity	Position (cm^{-1})	Intensity	Position (cm^{-1})	Intensity	FWHM	I_{2D}/I_G	I_D/I_G
15	1,349.7	127.8	1,585.2	116.8	2,700.8	74.4	71.1	0.637	1.094
20	1,351.3	570.3	1,587.3	656.0	2,704.8	345.7	69.6	0.527	0.869
25	1,353.4	48.6	1,583.6	133.4	2,699.4	84.8	52.8	0.636	0.364
30	1,349.3	213.2	1,585.3	720.2	2,701.5	906.9	44.7	1.259	0.296
35	1,352.7	74.0	1,585.3	255.0	2,704.2	247.3	63.4	0.970	0.290

layer was just Co_2Si, and the increase of C content in the film provided a sufficient C content for the nucleation and growth of graphene. In addition, the number of layers of graphene can be considered by the FWHM of 2D peak and the I_{2D}/I_G ratio [40]. When the flow rate of acetylene is 25 sccm, the I_{2D}/I_G ratio is 0.636, and the FWHM of 2D peak is 52.8, which indicates that the SGF comprises a few layers. When the acetylene flow rate is 30 sccm, the I_{2D}/I_G ratio is about 1.26, the number of layers is two. And when the gas flow rate is 35 sccm, the I_{2D}/I_G decreased, so the number of graphene layers increased with the increase of C atom content.

5.3.3.4 C_2H_2 Gas Flow vs. Graphene Morphology
Finally, the surface morphology before and after annealing of SiC films deposited under five C_2H_2 gas flows is exhibited in Figure 5.20. It can be seen that after annealing of SiC films, deposited at different acetylene gas flows, Co_2Si white particles

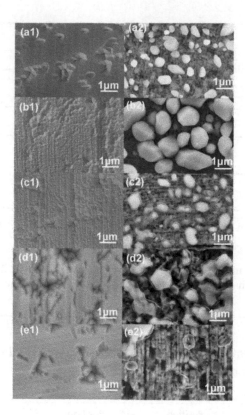

FIGURE 5.20 SEM images of graphene films grown at five C_2H_2 flow rates during a-SiC films' deposition. (a1–e1) 15 sccm – 35 sccm deposition of SiC films and (a2–e2) 15 sccm – 35 sccm deposition of SiC films after annealing. (Reprinted with permission from Elsevier.)

appeared on the surface of all samples. At the C_2H_2 gas flow of 15 and 20 sccm, the graphene films did not appear on the surface of Co_2Si white particles, because the content of C atom in the interfacial layer was low and the formation of the graphene nucleus was limited. When the gas flow increased to 25–35 sccm, the precipitated C atoms increased, forming a transparent SGF wrapped on the surface of the Co_2Si particles. When the flow rate of C_2H_2 was 35 sccm, after annealing, due to the decrease of Si atoms, the Co_2Si particles disappeared, and the C atom content continues to increase, resulting in the formation of a transparent SGF without Co_2Si particles.

According to the above analysis, the C content of the precipitated layer determines the defects formation and number of layers of SGF. Under the annealing condition of 1,000°C for 90 minutes, the structure of SGF can be controlled by adjusting C_2H_2 gas flow during the deposition of SiC films. The C_2H_2 gas flow first affects the atomic content in the deposited SiC film. As the gas flow rate increases, the C atom content in the SiC increases and the Si content decreases. After annealing, the interfacial layer phase is converted from CoSi to Co_2Si compound, and finally the carbon atom is nucleated on the surface of the Co_2Si compound particle to form a SGF. And as the content of C atoms in the interfacial layer increases, the defects of graphene are continuously reduced; simultaneously the number of layers becomes uniform, reaching the lowest number at 30 sccm, and then the number of layers increases with the increase of the C_2H_2 gas flow rate. Due to the continuous reduction of the initial Si atom content, the amount of formed Co_2Si compound particles reduces, and finally, the hollow SGF forms on the surface of cemented carbide.

5.3.4 REACTION MECHANISM OF CO-CATALYZED SGF GROWTH

In view of the fact that SGF is expected to offer space sensors the required detection performance, a-SiC was deposited on the surface of the cemented carbide and by controlling the annealing process, and the acetylene gas flow during deposition of a-SiC film, *in-situ* growth of SGF with controllable number of layers and lattice defects was achieved. During the growth of graphene spheres, Co and SiC react to form the interfacial layer of cobalt silicon compounds, and C atoms form graphene spheres on the surface. The phase of interfacial layer and the content of C atoms in the SiC film directly affect the formation of the SGF. Here, we propose the reaction mechanism of *in-situ* and controllable growth of SGF by metal Co-catalyzed SiC.

Figure 5.21 summarizes the schematic diagram of Co-catalyzed a-SiC-controlled growth of SGF. The red arrow coordinates in the figure indicate the acetylene gas flow rate, and the acetylene gas flow rate increases in the direction of the arrow. The gray arrow indicates the C atom content, and the direction of the arrow indicates the increase of the C atom content in the film, which is consistent with the change trend of the acetylene gas flow rate. (Reprinted with permission from *Elsevier* publisher)

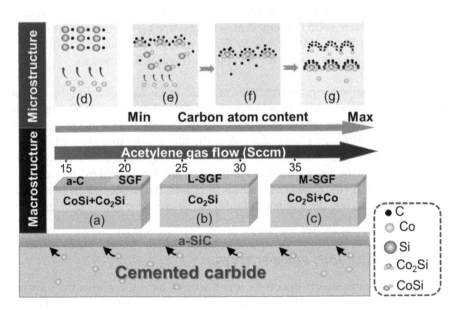

FIGURE 5.21 (a–g) Schematic diagram of controllable growth of SGF by Co-catalyzed a-SiC. (Reprinted with permission from Elsevier.)

Figure 5.21 can be divided into top and bottom parts by the middle arrow. The bottom part shows the macrostructure of the sample: The Co atom (light-colored sphere) in the cemented carbide diffuses to surface and reacts with SiC film during the annealing process; then the three-stage macroscopic phase of the sample interfacial layer and the graphene changes as the flow rate of the acetylene gas increases (Figures 5.21a–c). The top part is the microscopic reaction mechanism. With the increase of C atom content in the film, microscopic change of the precipitated phase and the SGF can be seen in Figures 5.21e–g.

Therefore, along the direction of the arrow in Figure 5.21, the *in-situ* controllable growth mechanism of the SGF can be divided into three stages as the flow of acetylene gas increases:

i. When the content of acetylene gas is 15–20 sccm, the deposited SiC film has less C atom content and more Si atom content relatively, resulting in the interfacial layer being a mixed phase of CoSi and Co_2Si after annealing. And there are not enough C atoms to nucleate and grow on the surface, so the defects in SGF are higher (Figures 5.21a and e).

ii. When the acetylene gas content is increased to 25–30 sccm, as the acetylene gas flow rate increases, the C atom content in the film increases, and the Si atom content decreases. After annealing, the interfacial layer is a pure Co_2Si phase, and the C atom begins to grow on the surface of Co_2Si. An F-SGF (few-layer spherical graphene film) is formed (Figure 5.21b and f), and the defects are reduced.

iii. When the acetylene gas content is increased to 35 sccm, the C atom content in the film continues to increase, and an M-SGF (multilayer spherical graphene film) is formed, and the defects of SGF are further reduced. And due to the decrease in the content of Si atoms, free Co is formed at the same time as the formation of Co_2Si in the interfacial layer, and a transparent SGF containing no white particles appears (Figures 5.21c and g).

Through the above analysis, the mechanism of Co-catalyzed a-SiC and the controllable *in-situ* growth of SGF can be determined: After depositing a-SiC film on the surface of cemented carbide, controllable *in-situ* growth of SGF is realized upon annealing at 1,000°C; the C atom content in the film can be directly controlled by adjusting the flow rate of acetylene gas when depositing the a-SiC film, while the lattice defects and the number of layers of the SGF are determined by the content of surface C atoms. As the content of C atoms increases, the lattice defects of the SGF tend to decrease, while the number of layers of the graphene sphere decreases first and then increases.

5.4 SUMMARY AND PERSPECTIVES

SGF is an admirable matrix option for space sensors but suffers from the substrate transfer and the difficulty to control the number of layers and lattice defects. This work summarizes the research progress in the preparation of SGF and the applications in sensor and provides a new idea and solution for *in-situ* growth of graphene spheres with controllable structure on a commonly used cement carbide substrate. The a-SiC film was deposited on the cemented carbide substrate, followed by vacuum annealing to achieve the *in-situ* growth of the SGF. The control of layer number and lattice defects was achieved by adjusting the C_2H_2 gas flow during the deposition of a-SiC films. And the reaction mechanism of controllable growth of SGF by Co-catalyzed SiC approach was revealed.

The main conclusions include the following:

1. Graphene spherical films can be grown on spherical metal or metal oxide templates by metal catalysis. Moreover, graphene matrix sensor prepared by surface functionalization is applied in space exploration.
2. A suitable annealing temperature is the key factor for the *in-situ* growth of graphene. The *in-situ* growth of the SGF was achieved in the temperature range of 1,000°C–1,150°C, after the a-SiC film was deposited on cemented carbide. Annealing provides kinetic energy for diffusion of the cohesive phase Co and reaction. Therefore, when the annealing temperature is lower than 1,000°C, graphene cannot be grown.
3. The number of layers and the lattice defects of the SGF are mainly affected by the C atom content on the surface of the a-SiC film. The relative content of C atoms can be directly controlled by adjusting the C_2H_2 gas flow rate during deposition of a-SiC. With the increase of the content of C atoms, the lattice defects of the SGF tend to decrease, while the number of layers tends to increase.

This research study provides a new deep insight into the transfer-free direct growth of graphene on the surface of cemented carbide, explaining the mechanism of chemical processes behind especially the structural, chemical, and morphological aspects of the graphene layer formation.

REFERENCES

1. D. Huang. 2017. Three-dimensional chemically reduced graphene oxide templated by silica spheres for ammonia sensing. *Sensors and Actuators B-Chemical* 252: 956–64.
2. N. Baig. 2018. Electrodes modified with 3D graphene composites: A review on methods for preparation, properties and sensing applications. *Microchimica Acta* 185: 283.
3. S. Houri. 2017. Direct and parametric synchronization of a graphene self-oscillator. *Applied Physics Letters* 110: 073103.
4. J.P. Cheng. 2017. Hybrid nanomaterial of alpha-Co(OH)$_2$ nanosheets and few-layer graphene as an enhanced electrode material for supercapacitors. *Journal of Colloid and Interface Science* 486: 344–50.
5. L. Embrey. 2017. Three-dimensional graphene foam induces multifunctionality in epoxy nanocomposites by simultaneous improvement in mechanical, thermal, and electrical properties. *ACS Applied Materials & Interfaces* 9: 39717–27.
6. S. Kumar. 2014. Radiation stability of graphene under extreme conditions. *Applied Physics Letters* 105: 133107.
7. M. Qi. 2014. Improving terahertz sheet conductivity of graphene films synthesized by atmospheric pressure chemical vapor deposition with acetylene. *Journal of Physical Chemistry C* 118: 15054–60.

8. J.W. Liu. 2015. Direct graphene growth on (111) Cu$_2$O templates with atomic Cu surface layer. *Carbon* 95:608–15.

9. A.M. Alexeev. 2017. A simple process for the fabrication of large-area CVD graphene based devices via selective in situ functionalization and patterning. *2D Materials* 4: 1.

10. X.Y. Ren. 2013. A review of cemented carbides for rock drilling: An old but still tough challenge in geo-engineering. *International Journal of Refractory Metals & Hard Materials* 39: 61–77.

11. J.M. Wang. 2015. Graphene nanodots encaged 3-D gold substrate as enzyme loading platform for the fabrication of high performance biosensors. *Sensors and Actuators B-Chemical* 220: 1186–95.

12. W. Hao. 2018. Crumpled graphene ball-based broadband solar absorbers. *Nanoscale* 10: 6306–12.

13. B. Cao. 2011. The ripple's enhancement in graphene sheets by spark plasma sintering. *AIP Advances* 1: 032170.

14. J. Bao. 2019. 3D graphene/copper oxide nano-flowers based acetylcholinesterase biosensor for sensitive detection of organophosphate pesticides. *Sensors and Actuators B-Chemical* 279: 95–101.

15. I.H. Son. 2017. Graphene balls for lithium rechargeable batteries with fast charging and high volumetric energy densities. *Nature Communications* 8: 1561.

16. M. Marchena. 2017. Direct growth of 2D and 3D graphene nano-structures over large glass substrates by tuning a sacrificial Cu-template layer. *2D Materials* 4: 025088.

17. K. Sohn. 2012. Oil absorbing graphene capsules by capillary molding. *Chemical Communications* 48:5968–70.

18. Z.G. An. 2016. Glass-iron oxide, glass-iron and glass-iron-carbon composite hollow particles with tunable electromagnetic properties. *Journal of Materials Chemistry C* 4: 7979–88.

19. J.C. Yoon. 2013. Three-dimensional graphene nano-networks with high quality and mass production capability via precursor-assisted chemical vapor deposition. *Scientific Reports* 3: 1788.

20. Y.Q. Lu. 2016. Direct fabrication of metal-free hollow graphene balls with a self-supporting structure as efficient cathode catalysts of fuel cell. *Journal of Nanoparticle Research* 18: 160.

21. L. Li. 2016. Solid-phase coalescence of electrochemically exfoliated graphene flakes into a continuous film on copper. *Chemistry of Materials* 28: 3360–6.

22. S.M. Yoon. 2012. Synthesis of multilayer graphene balls by carbon segregation from nickel nanoparticles. *ACS Nano* 6:6803–11.

23. J.A. Rodriguez-Manzo. 2011. Graphene growth by a metal-catalyzed solid-state transformation of amorphous carbon. *ACS Nano* 5:1529–34.

24. X.S. Li. 2009. Evolution of graphene growth on Ni and Cu by carbon isotope labeling. *Nano Letters* 9:4268–72.

25. Z.Z. Sun. 2010. Growth of graphene from solid carbon sources. *Nature* 468: 549–52.

26. Z.Y. Juang. 2009. Synthesis of graphene on silicon carbide substrates at low temperature. *Carbon* 47: 2026–31.

27. P. Machac. 2012. Synthesis of graphene on SiC substrate via Ni-silicidation reactions. *Thin Solid Films* 520: 5215–8.

28. G.H. Pan. 2013. Transfer-free growth of graphene on SiO$_2$ insulator substrate from sputtered carbon and nickel films. *Carbon* 65: 349–58.

29. M. Miyoshi. 2015. Transfer-free graphene synthesis on sapphire by catalyst metal agglomeration technique and demonstration of top-gate field-effect transistors. *Applied Physics Letters* 107: 073102.

30. C. Li. 2011. Preparation of single- and few-layer graphene sheets using Co deposition on SiC substrates. *Journal of Nanomaterials* 2011: 1–7.

31. Y.B. Dong. 2018. Transfer-free, lithography-free, and micrometer-precision patterning of CVD graphene on SiO_2 toward all-carbon electronics. *APL Materials* 6: 026802.
32. J. Wu. 2018. Synthesis of hollow fullerene-like molybdenum disulfide/reduced graphene oxide nanocomposites with excellent lubricating properties. *Carbon* 134: 423–30.
33. H.C. Lin. 2015. The essential role of Cu vapor for the self-limit graphene via the Cu catalytic CVD method. *Journal of Physical Chemistry C* 119: 6835–42.
34. X Yu, 2018. Synthesis of transfer-free graphene on cemented carbide surface. *Scientific Reports* 8: 1–10.
35. Z Zhang, 2019. Adjusting acetylene gas flow to grow a spheroidal graphene film with controllable layer number and lattice defects. *Surface and Coatings Technology* 364: 416–421.
36. L. Zhang. 2006. Experimental investigation and thermodynamic description of the Co–Si system. *Calphad-Computer Coupling of Phase Diagrams & Thermochemistry* 30: 470–81.
37. X.H. Cao. 2014. Three-dimensional graphene materials: Preparation, structures and application in supercapacitors. *Energy & Environmental Science* 7: 1850–65.
38. M.A. Pimenta. 2007. Studying disorder in graphite-based systems by Raman spectroscopy. *Physical Chemistry Chemical Physics* 9: 1276–91.
39. K. Grodecki. 2016. SEM and Raman analysis of graphene on SiC(0001). *Micron* 80: 20–3.
40. Z.W. Peng. 2011. Direct growth of bilayer graphene on SiO_2 substrates by carbon diffusion through nickel. *ACS Nano* 5: 8241–7.
41. L.M. Malard. 2009. Raman spectroscopy in graphene. *Physics Reports-Review Section of Physics Letters* 473: 51–87.
42. G. Rius. 2012. Synthesis of patterned nanographene on insulators from focused ion beam induced deposition of carbon. *Journal of Vacuum Science & Technology B* 30: 03D113.
43. S.W. Park. 1997. Investigation of Co/SiC interface reaction. *Journal of Electronic Materials* 26: 172–7.
44. C.Y. Kang. 2012. Few-layer graphene growth on 6H-SiC(0001) surface at low temperature via Ni-silicidation reactions. *Applied Physics Letters* 100: 251604.
45. K.S. Kim. 2014. Fabrication and characterization of hydrogen sensors based on transferred graphene synthesized by annealing of Ni/3C-SiC thin films. *Surface Review and Letters* 21: 1450050.
46. M. Hasegawa. 2015. In situ SR-XPS observation of Ni-assisted low-temperature formation of epitaxial graphene on 3C-SiC/Si. *Nanoscale Research Letters* 10: 421.
47. A.A. Woodworth. 2010. Surface chemistry of Ni induced graphite formation on the 6H-SiC (0001) surface and its implications for graphene synthesis. *Carbon* 48: 1999–2003.
48. M. Valakh. 2014. Free-standing graphene monolayers in carbon-based composite obtained from SiC: Raman diagnostics. *Physica Status Solidi a-Applications and Materials Science* 211: 1674–8.

6 Membrane Materials for Vanadium Redox Flow Battery

Jiaye Ye and Lidong Sun
Chongqing University

CONTENTS

6.1 Introduction 199
6.2 Working Principle of RFB.. 201
    6.2.1 Working Principle... 201
    6.2.2 Membrane Requirement 203
    6.2.3 Membrane Characterization 205
6.3 Membrane Materials for VRFB 207
    6.3.1 Perfluorinated Sulfonic Acid Membranes 207
        6.3.1.1 *In-situ* Sol-Gel Modification.. 208
        6.3.1.2 Surface Modification... 210
        6.3.1.3 Solution Recasting 212
        6.3.1.4 Other Modification Methods... 213
    6.3.2 Partially Fluorinated Membranes... 214
    6.3.3 Non-Fluorinated Membranes.. 216
        6.3.3.1 Poly(ether ether ketone) Based Membrane 216
        6.3.3.2 Polybenzimidazole-Based Membrane 218
        6.3.3.3 Polyimide-Based Membrane... 221
        6.3.3.4 Other Hydrocarbon Membranes 222
    6.3.4 Porous Membranes 223
        6.3.4.1 Solvent-Template Method 224
        6.3.4.2 Phase Inversion Method.. 225
6.4 Summary and Perspectives.. 230
References.. 231

6.1 INTRODUCTION

To reduce the carbon footprint induced by consuming the fossil fuels, renewable energy (e.g., solar and wind energy) has generated worldwide interest as an alternative candidate. However, the intermittent nature of the renewable energy requires a reliable, efficient, and cost-effective energy storage technique to match with the grid. In the meantime, an efficient and stable operation of the grid also needs a large-scale energy storage system to perform different functions, as shown in Figure 6.1. Among various electrochemical energy storage systems, redox flow batteries (RFBs) exhibit a number of merits, particularly the decoupled power and capacity, and have been regarded as promising candidates for large-scale energy storage. The membrane is one of the key components in a battery. It prevents the electrodes from short circuit and stops the electrolytes from crossover and meanwhile allows the proton transportation to complete the circuit. Figure 6.2 summarizes the number of literatures

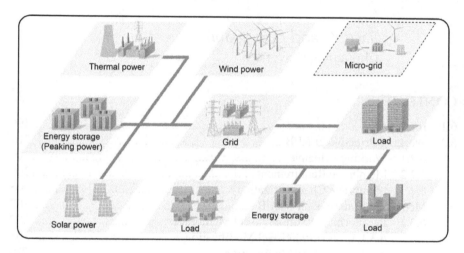

FIGURE 6.1   An illustration showing a simple power grid.

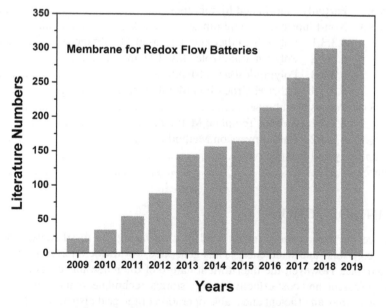

FIGURE 6.2   The number of literatures on membranes for RFBs.

on membranes for RFBs in the last decade. These data were extracted from Web of Science on December 31, 2019, with the keywords "membrane" and "redox flow batteries." Obviously, the number gradually increases year by year, suggesting that the study on membranes has attracted growing attention. As such, the development of the membranes and their working principle, requirement, modification, and performance are herein discussed.

6.2 WORKING PRINCIPLE OF RFB

6.2.1 WORKING PRINCIPLE

RFB is an electrochemical energy storage system, in which the electrical energy is stored in two separate solutions with different redox couples. The most prominent characteristic of RFB is the decoupled configuration of power and capacity. As such, the RFBs can be flexibly designed by adjusting the size of cell stack and the volume of electrolyte solution.

Figure 6.3 shows a single-cell structure of an RFB. It consists of membrane, gaskets, electrode carbon felts, flow frames, and graphite bipolar flat plates, with the membrane being in the center and the others stacked at its two sides. The membrane is the key component, which prevents the electrodes from short circuit and separates

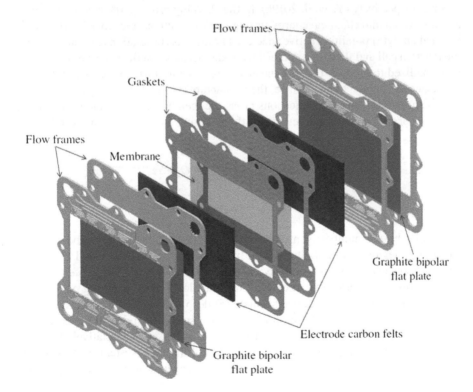

FIGURE 6.3   The configuration of an RFB cell. (Reprinted after reference Guarnieri et al. (2018) with permission, © Elsevier, 2018.)

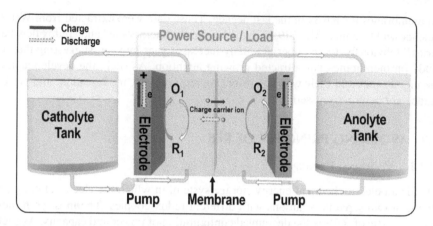

FIGURE 6.4 The working principle of the RFB. (Redrawn after reference Ye et al. (2019b) with permission, © IOP Publishing, Ltd, 2019.)

the catholyte and anolyte in two compartments. Figure 6.4 illustrates the working principle of the battery. The redox reactions can be described using Equations 6.1 and 6.2, where the O_1/R_1 and O_2/R_2 represent the redox couples in the catholyte and anolyte, respectively (Ye et al. 2019b). In the charging process, the electrical energy is transformed into chemical energy through redox reactions and stored in the catholyte and anolyte containing active species. The electrolyte solutions are pumped into the reaction cell and flow through the electrode surfaces. On the cathode surface, the R_1 is oxidized to produce O_1 with the electrons passing through the external circuit to the anode. On the anode surface, the O_2 captures the electrons to generate R_2. In the meantime, the charge carrier ions cross the membrane to complete the circuit. The discharging process is a reverse of the charging as described in the following equations.

$$\text{Cathode:} \quad R_1 \xleftarrow[\text{discharge}]{\text{charge}} O_1 + xe^- \tag{6.1}$$

$$\text{Anode:} \quad O_2 + ye^- \xleftarrow[\text{discharge}]{\text{charge}} R_2 \tag{6.2}$$

The RFBs have a number of merits, including (1) decoupled power and capacity, enabling a flexible configuration design by adjusting the cell stack and electrolyte volume (concentration), respectively; (2) high reliability; (3) long lifetime over 10 years; (4) high energy efficiency, with the chemical reactions taking place in liquid state between different active species; (5) deep discharging ability, the energy being stored in the electrolyte solutions instead of solid electrodes, in the absence of phase change and electrode collapse; (6) rapid response; and (7) relatively low cost. However, the low energy density is a short slab of the RFB system. In this regard, enhancing the solubility of active species in solvent or improving the work potential window are effective strategies to boost the energy density.

Based on the category of the supporting electrolyte, the RFBs can be classified into aqueous, non-aqueous, and hybrid systems. The aqueous RFBs employ the acid (e.g., sulfuric acid, hydrochloric acid, or mixed acid), neutral (e.g., sodium chloride or potassium sulphate), or base (e.g., sodium hydroxide or potassium hydroxide) solutions as supporting electrolytes to dissolve the active species. The non-aqueous RFBs generally use organic solvents. Some additives are usually added into the electrolytes to enhance the conductivity and/or stability. The hybrid ones are composed of an aqueous and non-aqueous half-cell.

6.2.2 Membrane Requirement

To date, the vanadium redox flow battery (VRFB) is one of the most promising systems for commercialization. The redox couples are all based on the vanadium element that solves the electrolyte crossover contamination induced by different elements (e.g., Fe–Cr, V–Fe, and Zn–Fe systems). The working principle of the VRFB is shown in Figure 6.5. Initially, the redox active species in the catholyte and anolyte are the VO^{2+} and V^{3+} ions, respectively. During charge process, the electrolyte is pumped into the cell, and the VO^{2+} is oxidized on the cathode surface to VO_2^+, while the electrons flow through external circuit to the anode where the V^{3+} is reduced to V^{2+}. In the meantime, the protons (charge carrier ions) are transported through the ion exchange membrane to ensure the electrical neutrality of system. And the reactions are reversed during the discharge process as follows (Leung et al. 2012; Sum et al. 1985; Sum and Skyllas-Kazacos 1985).

$$\text{Cathode:} \quad VO^{2+} + H_2O \leftrightarrow VO_2^+ + 2H^+ + e^- \left(E^0 = 1.00\,V\,vs.\ SHE\right) \quad (6.3)$$

$$\text{Anode:} \quad V^{3+} + e^- \leftrightarrow V^{2+} \left(E^0 = -0.26\,V\,vs.\ SHE\right) \quad (6.4)$$

The ion exchange membrane (IEM) is a key component in the VRFB. It is used to separate the catholyte (cathode) and anolyte (anode) in the cell, and meanwhile

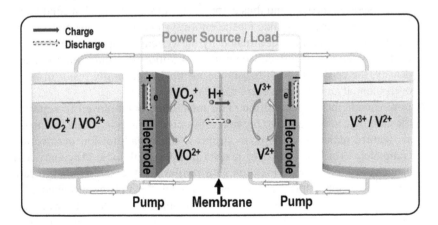

FIGURE 6.5 Illustration of the working principle of VRFB. (Redrawn after reference Ye et al. (2019b) with permission, © IOP Publishing, Ltd., 2019.)

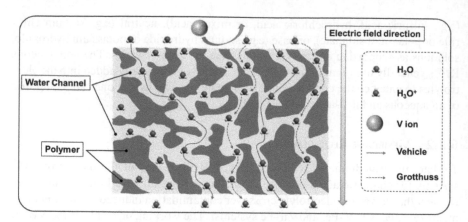

FIGURE 6.6   The transport mechanism of protons in IEM.

it allows the protons to get transported through the membrane. Figure 6.6 shows the transport of protons in the membrane via vehicle and Grotthuss mechanisms (Ye et al. 2019a). In the vehicle mechanism, the protons are transported by hydronium. The Grotthuss mechanism is also named as "hopping mechanism" in which the protons hop from the hydronium donor sites to the adjacent acceptors (Amjadi et al. 2010). The proton transport direction is the same as the electric field direction in cell interior.

An ideal proton exchange membrane for VRFB should possess high proton conductivity, low vanadium ion permeation, good chemical and mechanical stability, and low cost. As previously mentioned, the membrane significantly affects the cell performance. Hence, a competent membrane is expected to meet the following requirements.

1. High ion conductivity. The ion exchange membrane allows the protons to get transported through and thus complete the circuit. A high ion conductivity could lower the resistance for the entire system. The voltage and energy efficiency of the cell are dominated by the system resistance, which leads to the ohmic loss during cell operation. Additionally, under high current density, the cells with high-resistance membrane generally result in large overpotential and thereby severe capacity loss.
2. Low vanadium ion permeation. The membrane is employed to separate the catholyte and anolyte to prevent cross-contamination, while its intrinsic feature makes it difficult to suppress the vanadium ion permeation completely. The ion crossover results in poor coulombic efficiency and capacity loss. A low permeation rate minimizes the self-discharge rate (or open-circuit voltage decay) and improves the voltage efficiency. To evaluate the overall performance of the membrane, the ratio between the proton conductivity and the ion permeability is defined as the ion selectivity. Generally, the higher the ion selectivity, the better the membrane performance.

3. Good chemical and mechanical stability. The membrane works under a harsh environment containing acid solutions of high concentration (generally 2–3 mol L^{-1}). Moreover, in the charge state, the VO_2^+ ions in catholyte exhibit a strong oxidizing feature. The membrane also suffers from large stress during the cell assembly and the liquid flow. The commercial VRFBs are commonly required to operate for a decade. Hence, the chemical and mechanical stabilities of the membrane are critical factors determining the battery performance and lifetime.

4. Low cost. The materials in a VRFB system mainly include membranes, electrodes, pumps, bipolar plates, current collectors, vanadium electrolytes, cell frames and gaskets, and control system. The membrane is one of the core components. The Nafion series membranes (DuPont) have been widely used in the VRFBs, which account for nearly 40% of the total cost for the cell stack (Doetsch and Burfeind 2016). Therefore, exploring membranes of low cost is one of the major issues for the development and application of the VRFBs.

6.2.3 MEMBRANE CHARACTERIZATION

Generally, the characterization of the membrane for VRFB includes determining the ion exchange capacity (IEC), VO^{2+} permeability, proton conductivity, water uptake, swelling ratio, tensile strength, chemical stability, cell performance, and so on.

The IEC represents the mole ratio of exchange groups (e.g., sulfonic acid groups) per gram of dried membranes. The IEC can be measured by the traditional titration. The dried membrane (cation exchange membrane) is first cut into small pieces and soaked into saturated sodium chloride aqueous solution for 24 hours. After that, the solution is titrated with 0.01 mol L^{-1} sodium hydroxide solution, and the IEC can be calculated using Eq. 6.5:

$$IEC = \frac{C_{NaOH} \times V_{NaOH}}{W_{dry}} \quad (6.5)$$

where C_{NaOH}, V_{NaOH}, and W_{dry} are the concentration of NaOH solution (mol L^{-1}), the consumed volume of NaOH solution (mL), and the mass of the dry membrane (g), respectively.

The vanadium ion permeability is measured through the following steps. An H-shaped apparatus is used, with the electrolyte solutions being separated by a membrane. To reduce the influence of osmotic pressure, $VOSO_4$ in H_2SO_4 and $MgSO_4$ in H_2SO_4 solutions of the same concentration and volume are adopted to fill in the H-shaped apparatus. Continuous magnetic stirring is applied during the whole testing process to avoid concentration polarization at the membrane surfaces. The solution is sampled from the $MgSO_4$ compartment at a fixed time interval to examine the VO^{2+} concentration by the UV-vis spectrometer. The permeability (P) of vanadium ions can be calculated by the following equation:

$$V\frac{dC_t}{dt} = A\frac{P}{L}(C_0 - C_t) \quad (6.6)$$

where V, A, and L represent the volume (mL) of the solution in both reservoirs, the active area of the membrane (cm^2), and the thickness of the membrane (cm), respectively; C_0 is the initial concentration of VO^{2+} in the VOSO$_4$ compartment (mol L^{-1}), and C_t is the vanadium concentration in the MgSO$_4$ compartment at particular time of t (min).

The membrane resistivity can be examined by electrochemical impedance spectroscopy or a resistance tester. The membrane is assembled into the cell with two electrodes. With the resistance, thickness, and active area of the membrane, its conductivity (σ) can be calculated using Eq. 6.7:

$$\sigma = \frac{L}{AR} \tag{6.7}$$

where, L, A, and R are the thickness (cm), active area (cm^2), and resistance (Ω) of the membrane. The membrane's resistivity is inversely proportional to its conductivity.

The water uptake (WU) and swelling ratio (SR) of the membrane can be determined by measuring the mass and dimension changes upon soaking in water. The dried membrane is cut into identical rectangles and immersed in the deionized water over 24 hours. After that, the membrane is taken out from the water and cleaned by wiping out the residual water quickly. The mass and length of the soaked membrane are recorded. The WU and SR are computed by the equations below:

$$WU = \frac{W_{wet} - W_{dry}}{W_{dry}} \times 100\% \tag{6.8}$$

$$SR = \frac{L_{wet} - L_{dry}}{L_{dry}} \times 100\% \tag{6.9}$$

where the W_{dry} and W_{wet} are the masses of the membranes in dry and wet state, respectively, and the L_{dry} and L_{wet} represent the lengths of the membranes in dry and wet states, respectively.

The tensile strength can be examined by an electromechanical universal testing machine. The tensile strength (T) can be obtained using Eq. 6.10:

$$T = \frac{F_m}{WL} \tag{6.10}$$

where the F_m is the maximum strength and the W and L are the width and thickness of the membrane, respectively.

During the VRFB's operation, the VO$_2^+$ ions of strong oxidizing nature would degrade the membrane, with the VO$_2^+$ ions being reduced to VO^{2+} ions in the meantime. Hence, the chemical stability can be evaluated by immersing the membrane into VO$_2^+$ solution and measuring the concentration change of the VO$_2^+$ ions by UV-vis spectrometry.

The single-cell VRFB is usually assembled by sandwiching the membrane between two electrodes and clamped by two bipolar flat plates, as shown in Figure 6.3. The properties of charge–discharge, open-circuit voltage decay, rate performance, and polarization can be measured by a battery test system. The coulombic efficiency (CE), energy efficiency (EE), and voltage efficiency (VE) of the cell are computed by the following equations:

$$CE = \frac{\int I_{dc}\, dt}{\int I_c\, dt} \times 100\% \tag{6.11}$$

$$EE = \frac{\int V_{dc} I_{dc}\, dt}{\int V_c I_c\, dt} \times 100\% \tag{6.12}$$

$$VE = \frac{EE}{CE} \times 100\% \tag{6.13}$$

where the I_c, I_{dc}, V_c, and V_{dc} represent the charging current, the discharging current, the charging voltage, and the discharging voltage, respectively.

6.3 MEMBRANE MATERIALS FOR VRFB

The study on IEMs can be in retrospect to the research about electrical properties of semipermeable membrane by Ostwald in 1890. In 1950, the Ionics Inc. developed a high-selectivity and low-resistance IEM. In 1973, the Nafion membrane (a cation ion exchange membrane) was developed by DuPont company, with excellent chemical stability (Grot 1973). After that, many types of IEMs were investigated and used in different fields such as water treatment, fuel cells, and RFBs. In general, the IEMs can be classified into three types: anion exchange membrane, cation exchange membrane, and amphoteric membrane. The anion and cation exchange membranes are polymers with the backbones and/or side chains containing cationic and anionic function groups, respectively. The amphoteric membrane consists of both cationic and anionic function groups.

Due to the poor antioxidation behaviors, most of the earlier membranes are unsuitable for application in VRFBs, e.g., Selemion CMV, DMV (Asahi Glass Co., Japan) (Li et al. 2013b). Perfluorinated sulfonic acid membranes (such as Nafion, DuPont) have been widely used in VRFBs, because of their high proton conductivity, good chemical, and mechanical stability. However, the severe ion crossover and high cost impede the further development of VRFBs. Modifying the Nafion membranes or exploring non-perfluorinated alternatives are two major strategies currently. Here, the commonly used membranes and their application for VRFBs are discussed.

6.3.1 PERFLUORINATED SULFONIC ACID MEMBRANES

The copolymerization of tetrafluoroethylene and sulfonyl vinyl ether can result in the perfluorinated sulfonic acid polymer membranes. The Nafion membrane is a typical one produced by DuPont company, and its chemical structure is shown below (Mauritz and Moore 2004).

$$----[(CFCF_2)(CF_2CF_2)_m]----$$
$$|$$
$$OCF_2CFOCF_2CF_2SO_3H$$
$$|$$
$$CF_3$$

The m represents the number of replications of $-CF_2CF_2-$. The weight of dry per-fluorinated sulfonic acid polymers corresponding to a mole of sulfonic acid groups is defined as an equivalent weight. It can be examined by acid–base titration, analysis of atomic sulfur, and Fourier-Transform Infrared (FT-IR) spectroscopy (Mauritz and Moore 2004).

The Nafion membranes have been widely used in VRFBs since 1985, in light of their high proton conductivity, good chemical, and mechanical stability. The high proton conductivity is mainly attributed to the membrane hydrophilicity and the intrinsic channels (~2–4 nm) larger than the proton size (<0.24 nm) (Yuan et al. 2016c). However, the water channels are also larger than the vanadium ions (>0.6 nm) (Yuan et al. 2016c). As a consequence, the Nafion membranes exhibit large vanadium ion permeability and poor ion selectivity, leading to low coulombic efficiency, large energy loss, and low cyclic stability. As such, many efforts have been devoted to improving the Nafion-based membranes by blocking and/or prolonging the channels. One of the efficient strategies is to modify the membranes with additives.

The membranes used for VRFBs usually serve under the harsh environment, so the modification materials should possess good chemical stability. The commonly adopted materials include graphene, graphene oxide, SiO_2, WO_3, and so on, bearing the morphology of nanoparticles, nanosheets, or nanorods. Additives are introduced into the polymer matrix to reduce the vanadium ion crossover. It can be achieved by blending with polymers, surface modification, and *in-situ* reactions. Facile physical means, e.g., stretching, annealing, and immersing in boiling water, can also change the distribution and/or direction of the channels. Generally, the modified membranes are capable of suppressing the vanadium ion permeation and altering the water transport property, thereby improving the battery performance (e.g., coulombic efficiency, energy efficiency). On the other hand, they usually exhibit a low proton conductivity compared to the original ones, resulting in low voltage efficiency.

6.3.1.1   *In-situ* Sol-Gel Modification

The sol-gel technique is primarily based on the polymerization and hydrolysis reactions of active compound precursors. The precursors form homogeneous sol system, slowly change into gel counterpart by reaction, and develop into molecular or nanoscale materials after drying. The method can be used to incorporate inorganic fillers into a polymer matrix. The sol-gel technique is a facile and mild method to modify the Nafion membrane for VRFB application. The Nafion/SiO_2 hybrid membranes were prepared by *in-situ* growth of SiO_2 nanoparticles in hydrated Nafion for fuel cells (Mauritz and Warren 1989). Xi and co-workers further adopted the Nafion/SiO_2 hybrid membranes for VRFBs (Xi et al. 2007). Figure 6.7 illustrates a typical process to prepare the Nafion/SiO_2 membranes. The polar clusters (pores) are formed by the conductive groups in the Nafion. First, the original Nafion membrane is pretreated and soaked overnight in H_2O/MeOH. This aims to expand the channels in the matrix and

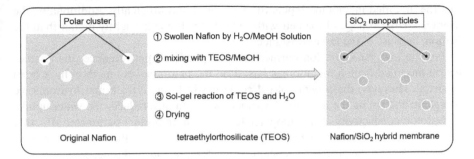

FIGURE 6.7 The preparation of Nafion/SiO₂ hybrid membrane by *in-situ* sol-gel method. (Redrawn after Xi et al. (2007) with permission, © Elsevier, 2007.)

facilitate the infiltration of the reaction solutions. The tetraethylorthosilicate/MeOH solution is subsequently introduced into the above solution and is stirred. After the hydrolysis reaction, the membrane is taken out and soaked in MeOH for 1–2 seconds to remove the residual reactants from the surface and then dried in vacuum to obtain the Nafion/SiO₂ hybrid membrane. The *in-situ*-formed SiO₂ nanoparticles fill into the clusters to reduce the size of the water channels, so that the vanadium ion permeation is retarded efficiently.

Thereafter, the sol-gel technique has been widely used to modify the Nafion membranes for VRFBs, such as Nafion/Si/Ti hybrid membrane (Teng et al. 2009). Figure 6.8a shows the changes of the VO^{2+} concentration across the Nafion and Nafion/Si/Ti hybrid membranes. Apparently, under the same conditions, the Nafion/Si/Ti hybrid membrane renders a smaller permeation than the Nafion counterpart. It is attributed to the organic-silica-modified TiO_2 nanoparticles that fill into the cluster of the Nafion matrix. Figure 6.8b are the charge–discharge curves of the VRFB single cell with Nafion and Nafion/Si/Ti hybrid membranes. The discharge capacity of the cell with the hybrid membrane is higher than that with the Nafion one, indicating

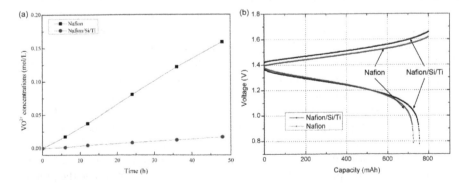

FIGURE 6.8 (a) Concentration variation of VO^{2+} in the permeation cell with Nafion and Nafion/Si/Ti membrane; (b) Charge–discharge curves for VRFB single cell with Nafion and Nafion/Si/Ti membrane. (Reprinted from Teng et al. (2009) with permission, © Elsevier, 2009.)

that the modified membrane repels the vanadium ions efficiently. Additionally, the average charge voltage of the cell with hybrid membrane is slightly higher than the Nafion, because of the large resistance.

For the low vanadium ion permeation and good cell performance, many studies on modified Nafion membranes based on the *in-situ* sol-gel method have been reported. However, the *in-situ* sol-gel route easily forms a thick and hydrophobic inorganic coating on the membrane surface, which in turn reduces the water uptake and proton conductivity. The nanoparticle size is an important parameter affecting the membrane property. Generally, the nanoparticle size and the coating thickness are dependent on the precursor concentration, reaction rate and time, etc.

6.3.1.2 Surface Modification

Surface modification aims to alter the surface property (such as wettability and adhesion) while maintaining the feature of the polymer matrix. The surface modification is an important method to enhance the Nafion membrane performance in VRFB. It gives rise to dense surface, positively charged groups, appropriate hydrophobicity, and sometimes enhanced proton conductivity. The commonly used methods include the interfacial polymerization, oxidation polymerization, grafting, electrodeposition, interface cross-linking, and surfactant treatment.

The cationic layer on surface can reduce the permeability of multivalent vanadium ions compared to the monovalent cation (e.g., the proton), due to the Donnan exclusion effect. To this end, the coatings usually contain plenty of positively charged groups. Luo and co-workers employed the polyethyleneimine (PEI) to modify the Nafion surface by interfacial polymerization (Luo et al. 2008a). The PEI coatings with lots of amino groups efficiently suppressed the vanadium ion permeation by charge repulsion. Hence, the modified membranes exhibited low permeability. On the other hand, the charge repulsion also retarded the proton transport and therefore increased the membrane resistance. The polypyrrole (PPY) with positive functional groups is also used to modify the Nafion membrane for VRFB. Zeng and co-workers prepared the PPY coatings by electrolyte soaking, oxidation polymerization, and electrodeposition (Zeng et al. 2008). In particular, the electrodeposition showed the best comprehensive performance in terms of the water uptake, vanadium ion permeation, and area resistance. However, the proton conductivity was also reduced in comparison to the original Nafion. Schwenzer and co-workers also synthesized the PPY coatings by chemical polymerization and compared the performance with the polyaniline (PANI) modified ones (Schwenzer et al. 2011). Both of the PPY and PANI modifications exhibited lower permeability of vanadium ions than the original Nafion membrane, with PANI coatings performing better. The hydrophobic PPY tended to aggregate in the center of the Nafion channels rather than on the side chains with hydrophilic sulfonic groups, while the PANI polymerized more slowly and preferred to bind to the side walls of the channels (Schwenzer et al. 2011). It is demonstrated that the performance of the modified membranes relies on the preparation condition and the property of the cationic polymers.

A few other modification methods are listed below. Ma et al. (2013) prepared the Nafion-g-DMAEMA composite membranes by radiation-induced graft copolymerization of the N, N-dimethylaminoethylmethacrylate (DMAEMA). The Nafion-g-PSSA was synthesized via the oxygen plasma induced grafting with poly(styrenesulfonic acid) (PSSA) (Yang et al. 2017). The Nafion-g-PSBMA was produced by grafting the zwitterionic sulfobetaine methacrylate (SBMA) onto Nafion via surface-initiated atom transfer radical polymerization (SI-ATRP) technique (Dai et al. 2018). Zhang and co-workers fabricated the Nafion/GO hybrid membranes by spin-coating the cross-linked GO layer onto the Nafion (Zhang et al. 2019). The membranes with angstrom-scaled channels could efficiently repel the vanadium ion permeability. In addition, surfactant treatment is also an effective technique to enhance the Nafion membrane. Teng and co-workers compared three kinds of surfactants for treating Nafion membranes for the VRFBs, i.e., tetramethylammonium bromide (cationic), sodium dodecyl sulfate (anionic), and Triton X-100 (non-ionic) (Teng et al. 2014). The micellar pore structure can be changed by the surfactants and further influence the transport of redox couples through the membrane. The Triton X-100-modified Nafion membranes exhibited the best performance among them.

The layer-by-layer self-assembly is another important surface modification method, first developed by Decher and co-workers in the 1990s (Decher et al. 1992). It is a powerful means to build multilayer thin films of controlled architecture and composition. The principle is based on the electrostatic attraction to form adsorption layer. Accordingly, the Nafion membrane can be modified by the method because of the changed groups. Xi and co-workers exploited the poly(diallyldimethylammonium chloride) (PDDA, polycation) and poly(sodium styrene sulfonate) (PSSS, polyanion) to prepare Nafion/[PDDA–PSSS]$_n$ membrane by layer-by-layer self-assembly method (Xi et al. 2008). The vanadium ion permeability of Nafion/[PDDA–PSSS]$_n$ membrane decreased with the increase in deposition layer. This was attributed to the PDDA–PSSS multilayer that covered or sealed the clusters (sulfonic group) and repelled the multivalent vanadium ions by the Donnan effect. As the PSSS contained –SO$_3^-$ groups, the Nafion/[PDDA–PSSS]$_n$ membrane showed a small decrease in proton conductivity, compared with the original Nafion membrane.

To reduce the vanadium ion crossover and meanwhile maintain the proton conductivity, the anionic layers should possess good proton conductivity. Accordingly, Lu and co-workers employed the phosphotungstic acid (PAW, inorganic compound) of strong acidity and high proton conductivity and polycation chitosan (CS, polycation) to prepare Nafion/[CS–PAW]$_n$ membrane (Lu et al. 2014). Because of the good conductivity of the PWA, the CS–PWA bilayers not only reduced the vanadium ion permeability, but also minimized the impact on the proton conductivity of the Nafion membrane. A few other examples of modified Nafion membranes for VRFBs by layer-by-layer self-assembly are as follows: Nafion/[PDDA–ZrP]$_n$ (ZrP: zirconium phosphate) (Zhang et al. 2017), Nafion/[PDDA-CNC]$_n$ (CNC: cellulose nanocrystals) (Kim et al. 2019), and so on. The open-circuit voltage decay and cell efficiency with the Nafion and Nafion/[PDDA–ZrP]$_3$ membranes are compared in Figure 6.9. Generally, the vanadium ion permeation leads to self-discharge and the open-circuit voltage decay. A long retention time and slow decay suggests that the Nafion/[PDDA–ZrP]$_3$ membrane performs better in suppressing the ion permeation. Eventually, higher coulombic efficiency and energy efficiency are achieved with the modified Nafion membrane, because of the low vanadium ion permeability and high proton conductivity.

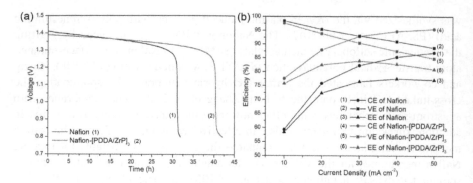

FIGURE 6.9 VRFB single cell with Nafion and Nafion–[PDDA/ZrP]$_3$ membranes: (a) open-circuit voltage decay curves at 50% state of charge and (b) different efficiencies. (Reprint after Zhang et al. (2017) with permission, © Elsevier, 2017.)

Although the method could achieve high battery performance, the chemical stability or durability of the resulting membranes is the key concern. As the polyelectrolytes or inorganic compounds are deposited on Nafion surface by electrostatic force, the modified membranes usually exhibit relatively poor stability under the harsh operation conditions of strong acid, oxidizing species, flow electrolyte, and electric field. Hence, the adhesion strength of the multilayers is a crucial factor to be considered for the layer-by-layer self-assembly method.

6.3.1.3   Solution Recasting

The recasting method is a technique to dissolve the Nafion polymer with an organic solvent, incorporate some organic and/or inorganic additives, and then recast to obtain a hybrid membrane dispersed with strengthening materials in the form of nanoparticles, nanotubes, nanosheets, etc.

Similar to the *in-situ* sol-gel technique, the incorporation of nanoscaled materials into the polymer matrix can adjust the size of the water channels to reduce vanadium ion permeability. A few additives containing conductive groups are capable of further enhancing the membrane and battery performance. Sun and co-workers adopted tungsten oxide (WO$_3$) nanoparticles to modify the Nafion membrane by solution recasting method (Sun et al. 2018). The Nafion/WO$_3$ hybrid membrane (1 wt% WO$_3$ ratio of Nafion polymer) exhibited high thermal stability, good mechanical property, reduced water uptake, and low vanadium permeability, compared with the original one. It was attributed to the cross-linking interactions between sulfonate groups and WO$_3$ nanoparticles (Sun et al. 2018). Aziz and Shanmugam employed the ZrO$_2$ nanotubes to disperse in the Nafion matrix and obtained the Nafion/ZrO$_2$-nanotube hybrid membrane (Aziz and Shanmugam 2017). The nanotubes embedded in the matrix partially blocked, elongated, and reduced the size of water channels, due to the electrostatic interaction between ZrO$_2$ and sulfonic groups. Hence, the vanadium ion permeability of the hybrid membrane was much lower than that of the original one. The ZrO$_2$ nanotubes exhibited good chemical stability and ensured the stable performance of the hybrid membrane.

The additives could enhance the dimensional stability of the membrane, which on the other hand could reduce the proton conductivity. Compared with the Nafion polymer, the inorganic materials exhibit low water uptake, which is a critical factor for proton transport through membrane by vehicle or Grotthuss mechanism. However, too many additives also lead to the membrane brittleness and instability caused by aggregation. Besides, additives of high conductivity (e.g., some carbon materials) are undesirable, which may result in short circuit between the cathode and anode.

Blending with organic materials is another way to tailor the chemical and physical properties of the membrane. Unlike the inorganic materials, the blending of polymers usually gives rise to good miscibility for the composite membrane. The polyvinylidene fluoride (PVDF) is a semi-crystalline polymer with good chemical stability that can blend with ionomers (ion containing polymers, e.g., Nafion). Zhang's group (Mai et al. 2011) prepared the Nafion/PVDF composite membrane for VRFB. The highly crystalline and hydrophobic PVDF reduced the water uptake and vanadium ion permeation. The polytetrafluoroethylene (PTFE) is another chemical stable polymer. A Nafion/PTFE composite membrane was prepared by solution casting for VRFB (Teng et al. 2013b).The addition of hydrophobic PTFE reduced the water uptake, ion exchange capacity, and proton conductivity of the membrane but enhanced the crystallinity, thermal stability, and antipermeability capacity for vanadium ions. As shown in Figure 6.10a, the vanadium ion permeation declines gradually with increased amount of PTFE additives. The cell with the optimized Nafion/ PTFE membrane (i.e., $N_{0.7}P_{0.3}$) shows a relatively higher average charge voltage and lower charge capacity, as displayed in Figure 6.10b. This is due mainly to the high resistance and large hydrophobicity of PTFE. However, the modified one exhibits a higher discharge capacity, indicating the good blocking effect for the vanadium ion permeation.

6.3.1.4 Other Modification Methods

Pretreatment, such as annealing at 120°C and immersion in boiling water and sulfuric acid, can also change the ion transport behavior in the Nafion membrane (Xie et al. 2015). Jiang and co-workers investigated the effect of the wet, boiled, and

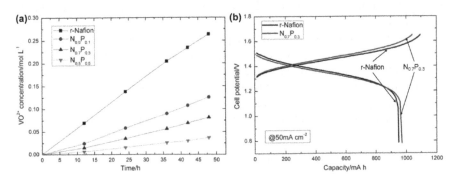

FIGURE 6.10 (a) Concentration variation of VO^{2+} ions in the permeation cell with recasted Nafion membrane (r-Nafion) and modified Nafion membrane (containing PTFE of various ratios) and (b) charge–discharge curves for VRFB single cell with the recasted Nafion membrane and modified Nafion membrane (70 wt% Nafion and 30 wt% PTFE). (Reprinted from Teng et al. (2013b) with permission, © Elsevier, 2013.)

boiled and dried pretreatment processes on Nafion properties for the VRFB (Jiang et al. 2016). Immersion in the water or acid solution could swell the micromicelle structure of the Nafion membrane. This increased the channel size and improved the proton conductivity but accelerated the crossover of vanadium ions compared to the original membrane. The boiled membrane exhibited high water uptake and swelling ratio. This was mainly due to the boiling temperature being close to the glass transition temperature of the dry Nafion ionomer (i.e., 110°C–120°C). When the membrane was boiled in acid solution, the isolated free volume voids in the Nafion would be connected by tiny channels and hence result in severe vanadium ion crossover when applied for VRFB (Jiang et al. 2016). The boiled and dried membrane displayed area shrink, which reduced vanadium ion permeability and enhanced the VRFB performance. The order of vanadium ion permeation is as follows: boiled > wet > as-received > boiled and dried.

To reduce the use of high-cost Nafion membrane, a thin Nafion layer is usually combined with other polymer layers to obtain multilayered composite membrane. The membrane reduces the permeability of vanadium ions and possesses good mechanical stability. Luo and co-workers fabricated the Nafion/sulfonated poly(ether ether ketone) (SPEEK) double-layered composite membrane, where the diamine was used to cross-link the sulfonic acid groups of Nafion and SPEEK (Luo et al. 2008b). The SPEEK has the advantages of low cost, low vanadium ion permeation, controllable proton conductivity, and good mechanical stability. Hence, the Nafion/SPEEK composite membrane exhibited low ion permeation and high battery performance. Besides, the inorganic layer can also enhance the Nafion properties for VRFB application, such as double-layered Nafion/silicate composite membrane (Kim et al. 2014). It was prepared by hot pressing the silicate/Nafion solution onto the Nafion membrane. A double-layered Nafion/zeolite composite membrane was prepared by a similar route, in which the zeolitic layer was coated on pure Nafion matrix with the Nafion solution as the adhesive (Yang et al. 2015). The sub-nanometer-sized zeolitic pores repelled the hydrated vanadium ions and permitted the hydrated protons. It is noteworthy that if the multilayered composite membrane is developed by only physical adsorption, delamination phenomena, high resistance, and poor stability usually arise.

The polytetrafluoroethylene is also frequently used. However, the PTFE contains no ion conductive groups (e.g., sulfonic groups). A thick PTFE layer shows large area resistance and leads to low-voltage efficiency for VRFB application. In addition, the PTFE layer is porous and inefficient for repelling the vanadium ion permeation. Teng et al. (2013a) impregnated the porous PTFE membrane with Nafion solution to prepare PTFE/Nafion composite membrane, which exhibited low vanadium ion permeability.

6.3.2 Partially Fluorinated Membranes

The perfluorinated sulfonic acid membranes (e.g., Nafion) exhibit high proton conductivity and good chemical and mechanical stability, which are however of high cost and involve complicated preparation and severe vanadium ion permeation. The partially fluorinated membranes (e.g., PVDF and ETFE) display comparable chemical stability and are of low cost and attractive for VRFB application.

The PVDF is a representative partially fluorinated material, which exhibits semi-crystallinity, good stability, and competitive cost. Similar to the PTFE, the PVDF contains no ion conductive groups and has small pores. As such, the PVDF is required to be modified before use. The modification methods described above can also be applied to PVDF. Hietala and co-workers modified the commercial PVDF membrane by irradiated graft with styrene and further sulfonated to prepare PVDF-g-PSSA composite membrane (Hietala et al. 1997). The sulfonation reaction took place mainly at the *para* position of the phenyl rings. The membrane displayed a good proton conductivity comparable to the Nafion (Hietala et al. 1997). Inspired by this work, Luo et al. fabricated the PVDF-g-PSSA membrane for VRFB application, which exhibited low vanadium ion permeability and high proton conductivity (Luo et al. 2005). Qiu and co-workers pre-irradiated the styrene and maleic anhydride (MAn) to graft them on PVDF polymer to produce the PVDF-g-PS-co-PMAn membrane and obtained the PVDF-g-PSSA-co-PMAn membrane upon further sulfonation (Qiu et al. 2008). Similarly, they modified the PVDF by grafting with styrene and dimethylaminoethyl methacrylate (DMAEMA) using γ-irradiation techniques and subsequent sulfonation and protonation and eventually generated a PVDF-based amphoteric ion exchange membrane for VRFB (Qiu et al. 2009). It is noted that the cation exchange membranes usually have high proton conductivity, while the anion counterparts suppress the proton transport. Consequently, although the amphoteric ion exchange membranes exhibited low vanadium ion permeability, they would cause some loss in proton conductivity and voltage efficiency.

The poly(ethylene–tetrafluoroethylene) (ETFE) is another important partially fluorinated polymer. Similarly, the ETFE possesses good chemical stability in the absence of ion conductive groups, and its main chain bears good thermal and mechanical stability. As such, the ETFE membrane is usually modified by introducing ion conductive groups by means of radiation grafting. Qiu and co-workers used the γ-radiation technique to prepare the ETFE-g-PMAOEDMAC (poly(methacryloxyethyl dimethyl ammonium chloride)) membrane with low vanadium ion permeation (Qiu et al. 2007). Similarly, the 2-hydroxyethyl methacrylate (HEMA) and glycidyl methacrylate (GMA) were grafted on ETFE by pre-irradiation to produce the ETFE-g-p(GMA-co-HEMA) composite membrane for VRFB (Li et al. 2017).

Besides, the non-fluorinated polymers are employed to synthesize the partially fluorinated membrane of high stability by introducing fluorine or fluorine-containing groups. For example, the sulfonated poly(fluorenyl ether ketone) (SPFEK) was directly fluorinated with fluorine gas and composited with the 3-aminopropyltriethoxylsilane (APTES) to fabricate the F-SPFEK-APTES composite membrane. This increased the oxidation stability and suppressed vanadium permeation without a large decrease in proton conductivity (Chen et al. 2012). The oxidation stability of semi-fluorinated sulfonated polyimide (SPI) membrane was higher than that of the non-fluorinated SPI counterparts, while the conductivity of the semi-fluorinated SPI could be controlled by adjusting the sulfonation degree (Li et al. 2016).

The non-fluorinated monomers can also be exploited to generate the fluorinated polymer membrane upon fluorination reaction and polyreaction. For example, the fluorinated SPI membranes were synthesized by typical polycondensation reaction with side chains containing $-CF_3$ groups (Yang et al. 2018).

6.3.3 Non-Fluorinated Membranes

The non-fluorinated membranes, especially those based on the aromatic polymers, have attracted a lot of interest, because of their high chemical stability, low cost, and simple preparation. A few examples include the poly(ether ether ketone), polyimide, polybenzimidazole, poly(arylene ether), poly(ether sulfone), and poly(phthalazinone ether ketone). Such kinds of aromatic polymers cannot be directly used as IEMs in the absence of conductive groups with strong hydrophobicity. However, they can be modified by introducing the desirable functional groups. The sulfonation is a simple way to enhance the conductivity, hydrophilicity, solubility in solvents, and membrane-forming ability. The non-fluorinated aromatic ring membrane is a promising alternative to the Nafion membrane. Some common non-fluorinated membranes are discussed here for the VRFB application, such as the sulfonated poly(ether ether ketone) with negatively charged groups ($-SO_3^-$), the polybenzimidazole with positively charged N-containing group, the amphoteric sulfonated polyimide, and other polymers.

6.3.3.1 Poly(ether ether ketone) Based Membrane

Poly(ether ether ketone) (PEEK) is a semi-crystalline polymer with good chemical, mechanical, and thermal stability. The PEEK is converted into sulfonated poly(ether ether ketone) upon sulfonation by concentrated sulfuric or chlorosulfuric acid. In particular, the concentrated sulfuric acid is usually employed since the chlorosulfuric acid is too strong to control the side reactions. The repeated units of PEEK and SPEEK are shown in Figure 6.11.

The proton conductivity of SPEEK can be adjusted by the proportion of introduced sulfonic acid groups. The structure of SPEEK consists of PEEK and PEEK–SO_3H units, and the x in Figure 6.11 represents the ratio of the PEEK–SO_3H units (sulfonation degree). Hence, the sulfonation degree (SD) can be described as follows (Huang et al. 2001):

$$SD = \frac{M_1}{M_1 + M_2} \tag{6.14}$$

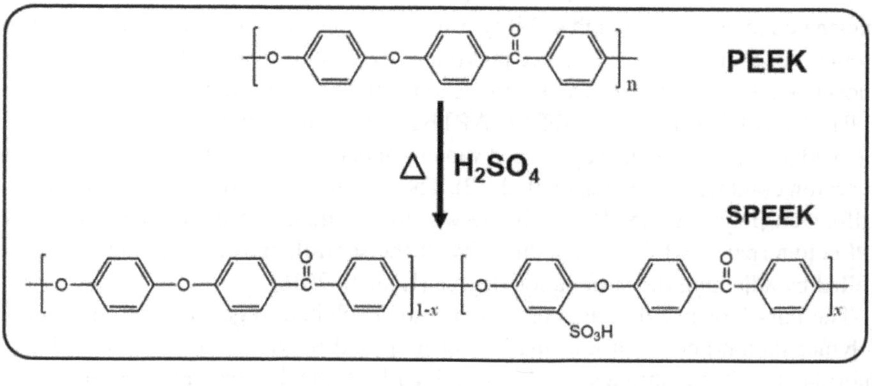

FIGURE 6.11   Structure and synthesis procedure of the SPEEK.

M_1—the molar number of PEEK–SO$_3$H units

M_2—the molar number of PEEK units.

The SD can be examined by nuclear magnetic resonance spectra and acid–base titration, which is closely related to the temperature and reaction time of sulfonation. A high SD generally gives rise to high proton conductivity of the SPEEK. However, the high SD on the other hand increases the vanadium ion permeability and reduces the mechanical stability. And the oxidizing VO$_2^+$ is easy to degrade the SPEEK of high SD value during cell operation, thus influencing the cyclic stability of VRFB. Therefore, it is also necessary to modify the SPEEK-based membranes. The modification techniques are similar to those used for Nafion membranes, including *in-situ* sol-gel growth, solution recasting, and surface modification.

The graphene oxides are widely used to modify the SPEEK membrane, in view of the two-dimensional structure, abundant functional groups, high surface area, and chemical stability. For example, the hybrid membrane of SPEEK/GO nanosheets was prepared by solution casting method (Dai et al. 2014). The vanadium ion permeability decreased with the increase in GO content, as the GO nanosheets embedded in the SPEEK matrix served as a blocking barrier for ion permeation. The oxygen-containing species on GO surface enhanced the hydrophilicity and dispersity of GOs and thus the mechanical stability of the membrane. Further, the p-phenylene diamine-functionalized GO (PPD-GO, amine-functionalized GO) and phenyl isocyanate treated sulfonated GO (isGO) were blended with SPEEK to obtain the SPEEK/PPD-GO membrane (Kong et al. 2016) and SPEEK/isGO membrane (Park and Kim 2016), respectively. Both membranes exhibited low vanadium ion permeability and high single-cell performance.

The inorganic materials of high stability, simple preparation, and low cost are commonly used to modify the SPEEK membranes. The mesoporous silica (diameter: 9–10 nm) is a typical example, with good chemical and mechanical stability. The hydroxyl groups on the surface further assisted to reduce the crossover and enhance the stability by forming hydrogen bonds with sulfonic groups. The VRFB single cells with the hybrid membranes exhibited high coulombic efficiency and good energy efficiency, compared with the Nafion-based ones.

The organic materials of good inoxidizability, hydrophilicity, and proton conductivity are usually used to modify SPEEK membrane. Inspired by the poly(vinylidene fluoride-co-hexafluoropropylene) (P(VDF-co-HFP)) that exhibited good mechanical and electrochemical stability for fuel cell and Li-ion battery, Li and co-workers employed the P(VDF-co-HFP) blending with high-SD SPEEK to obtain the SPEEK/P(VDF-co-HFP) composite membrane (Li et al. 2014d). The membrane with the hydrophobic P(VDF-co-HFP) units reduced the water uptake and swelling ratio and therefore suppressed the vanadium ion permeation and improved the mechanical and chemical stability. The polyetherimide was previously exploited to modify the Nafion with high performance in VRFB (Luo et al. 2008a). Similarly, it was also blended with SPEEK to prepare the composite membrane (Liu et al. 2014b). The membrane reduced the swelling ratio and improved the dimension stability, because of the cross-linking and hydrogen bonds between the N-containing groups (polyetherimide) and sulfonic acid groups (SPEEK). The amphoteric membrane repelled the vanadium ion permeation efficiently in light of the charge exclusion (protonation of the N-containing groups). In general, the cross-linking can be done in two ways: the polymer cross-linked with SPEEK membrane directly and the polymer cross-linked with the monomer of SPEEK to form the SPEEK-based membrane. Accordingly, the SPEEK was blended with quaternized poly(ether imide) (QAPEI) to obtain another amphoteric SPEEK-based composite membrane (Liu et al. 2015a). The QAPEI had a similar effect as the polyetherimide that improved the thermal and oxidative stability and reduced the water uptake and vanadium ion permeability. The inferior miscibility between the main chains of SPEEK and QAPEI lead to a high degree of microphase separation that is beneficial to the proton transport. The stable organic materials for high-ion-selectivity SPEEK-based hybrid membranes include PVDF, triphenylamine, and lignin.

Besides, the surface modification can also achieve outstanding performance for the SPEEK-based membranes. In dopamine solution, at pH = 8.5, the monomers are easily self-polymerized and oxygenated to form a thin polydopamine (PDA) layer on the surface of a wide range of inorganic and organic solid materials (Liu et al. 2014b). Accordingly, the SPEEK membrane surface was coated with a PDA thin layer (as shown in Figure 6.12) to enhance its property (Xi et al. 2015a). The hydrogen bonds between the rich hydroxyls of PDA and the sulfonic acid groups of SPEEK improved the mechanical stability of the membrane. The PDA as a preselecting layer reduced the vanadium ion permeability, water uptake, and swelling ratio and enhanced the chemical stability of SPEEK. Additionally, the SPEEK blended with other materials, such as the SPEEK/PVDF/poly(ether sulfone) (Fu et al. 2015), SPEEK/PANI-GO (polyaniline-functionalized graphene oxide) amphiprotic membrane (Zhang et al. 2018a), and SPEEK/MWCNTs@PDA (polydopamine-decorated multiwalled carbon nanotubes) (Zhang et al. 2018b), also exhibited low vanadium ion permeability and high VRFB-single-cell performance.

6.3.3.2 Polybenzimidazole-Based Membrane

Polybenzimidazole (PBI) is an aromatic polymer of low cost, which is resistant to strong acids and bases and has good mechanical property. It is synthesized in the molten state or in solution from the condensation reaction of aromatic bis-o-diamines

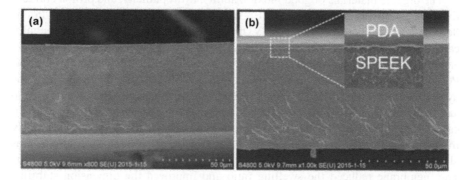

FIGURE 6.12 The cross-sectional SEM images of (a) SPEEK and (b) PDA/SPEEK. (Reprinted from Xi et al. (2015a) with permission, © Royal Society of Chemistry, 2015.)

and dicarboxylates. The chemical structure of the PBI is shown below, poly[2,2'-(*m*-phenylene)-5,5'-bibenzimidazole] (Zhou et al. 2015).

Similar to other pristine hydrocarbon polymers, the PBI is an ionic insulator (~10^{-12} S cm^{-1}) (Pohl and Chartoff 1964). The ionic conductivity is inversely proportional to the anion size, indicating that the protonated PBI is inert and the mobile anion plays a dominant role in the conduction process (Wainright et al. 1995). As a base polymer, the PBI can be protonated through the hydrogen bond under strong acid (Peng et al. 2016). The conductivity increases with the concentration of immersion acid, and it is also described as the acid-doped PBI. The doping level is defined as the number of moles of acid per mole repeat unit of PBI (Jang et al. 2016) and is dependent on the microstructure (dense or loose) of the PBI. Accordingly, much efforts have been devoted to developing high-performance VRFB by improving the acid-doped PBI. A typical acid-doping process is described as follows. The PBI membrane is immersed in an acid solution, and then the acid infiltrates into the PBI matrix. The acid-doping process generates free volumes (pores) (Zhou et al. 2015). The imidazole nitrogen with negative charge can be protonated in acidic solution forming the acid–base complexes. The porous PBI matrix with the acid solution conducts ions under electric filed. Therefore, the supporting electrolyte of sulfonic acid in VRFBs can improve the conductivity of the PBI membranes to enhance the voltage efficiency.

Zhao's group investigated the commercial PBI membranes for VRFBs and compared them with the Nafion counterparts (Zhou et al. 2015). The as-received PBI presented nearly no pores and low conductivity. The conductivity was enhanced significantly after immersing in high concentration sulfonic acid to form continuous proton conductive channels. The vanadium ion permeability of PBI membrane was two orders of magnitude smaller than that of the Nafion counterpart, which was assigned to the size sieving and charge repulsion effect. Enhancing the acid-doping level is an efficient way to improve the conductivity of PBI membrane. Peng and co-workers used the phosphoric acid pre-swelling strategy to boost the acid-doping level (Peng et al. 2016). The membrane was soaked in phosphoric acid first to expand the PBI matrix and then immersed in high concentration sulfonic acid. The pretreatment formed large free volumes in the PBI matrix. It absorbed more sulfonic acid with high proton conductivity, compared to the direct sulfonic acid treatment process. Although the conductivity of the PBI membrane was improved, the vanadium ion permeability was maintained.

Based on PBI membrane, Jang and co-workers synthesized the poly[2,2'-(2-benzimidazole-p-phenylene)-5,5'-bibenzimidazole] (BI*p*PBI) containing two benzimidazole groups (Jang et al. 2016). The BI*p*PBI membrane exhibited an amorphous structure because of the benzimidazole side groups and possessed high absorptivity for sulfonic acid and water and thus facilitated the proton transport. It also suppressed the vanadium ion permeation because of the strong Donnan effect induced by the positively charged bibenzimidazole groups. To further improve the chemical stability, a Nafion layer was spray-coated onto the BI*p*PBI membrane (Ahn et al. 2018). The sulfonic acid groups of Nafion formed the acid–base complex with the nitrogen of benzimidazole, enhancing the adhesion strength. Such a combination made the membrane achieve high ion selectivity and battery performance. Liu and co-workers fabricated a dense PBI-based membrane cross-linked with partially quaternized poly(1-vinylimidazole) (QPVI, anionic polymer) (Liu et al. 2017). Because of the enhanced Donnan exclusion effect induced by the quaternized and bibenzimidazole groups, the modified membrane exhibited nearly four orders of magnitude lower vanadium ion permeability than the Nafion membrane. At low current density, the VRFB single cell with the modified membrane also showed higher coulombic efficiency and energy efficiency than that with the Nafion membrane. A cross-linked and methylated PBI membrane was synthesized by adding the dibromoxylene into the PBI casting solution, with the resultant membrane being subsequently soaked in the iodomethane and NaOH solution (Chang et al. 2017; Chen et al. 2018). The membrane exhibited reinforced chemical and mechanical stability. The positively charged backbone (nitrogenous heterocyclic ring) could reduce the vanadium ion permeability. To further enhance the mechanical property and chemical resistance, the covalently cross-linked sulfonated membrane was fabricated by solution casting method using bisphenol epoxy resin as a cross-linker (Xia et al. 2017). The membrane showed lower vanadium ion permeability (3–4 orders of magnitude), as well as 6–30 times higher ion selectivity, than the Nafion membrane.

Chen and co-workers prepared the PBI-based membrane of high proton conductivity by introducing the pyridine groups into the PBI backbone (Chen et al. 2019). The pyridine group is similar to the benzimidazole group. Hence, the membrane exhibited low vanadium ion permeability and good proton conductivity, compared with the pristine PBI membrane. It is worth noting that, similar to the polymers with sulfonic acid groups, the proton is transported through hopping from the protonated =N–H or doped acid to neighboring non-protonated molecules (Bouchet and Siebert 1999; Glipa et al. 1999). As such, it enhances the acid-doping level of the PBI membrane to improve its proton conductivity by adding the imidazole groups into the PBI backbone or shrinking the interspacing between nitrogen atoms along the backbone (Ding et al. 2018; Guan et al. 2012). The PBI with a bulky group, such as the naphthalene group, can enhance the rigidity of the chains to improve the acid-doping level (Geng et al. 2019). The conductivity can be enhanced by alleviating the tortuosity of the proton transport channels in the PBI membrane (Lee et al. 2019).

The PBI conductivity cannot be increased by direct sulfonation reactions. Alternatively, condensation polymerization with the monomers is another effective strategy. Xia and co-workers employed the 4,4′-dicarboxydiphenyl ether (can be sulfonated), 5-aminoisophthalic acid, and 3,3′-diaminobenzidine in polyphosphoric acid for condensation polymerization and subsequent sulfonation to obtain sulfonated PBI-based polymer (Xia et al. 2017). Similarly, the sulfonate (5-sulfoisophthalic acid monosodium salt) was introduced into the PBI to enhance the proton conductivity and control its SD (Ding et al. 2019). The proton conductivity increased with the SD, but too high a sulfonation degree would form hydrogen bonds between the –N= of imidazole rings and sulfonate, thereby impeding the proton transport (Ding et al. 2019).

6.3.3.3   Polyimide-Based Membrane

Polyimide possesses good mechanical and chemical stability. It forms a network, and its proton conductivity can be enhanced by introducing sulfonic acid groups into the chains. The sulfonated polyimide can be synthesized by direct sulfonation or polymerization of its sulfonated monomers. A typical chemical structure of the sulfonated polyimide is shown below (Yue et al. 2012).

The repeated units with bulky aromatic groups improve its chemical stability. The SPI contains the sulfonic acid groups and the N-containing groups, with controllable proton conductivity and charge repulsion for vanadium ions. As a result, it is a suitable separator for VRFB and promising candidate for the Nafion membrane.

The sulfonated polyimide was employed as a proton conductivity membrane for VRFB by Yue and co-workers (Yue et al. 2011). Further, it was self-assembled with the chitosan to form SPI/CS composite membrane (Yue et al. 2012). The CS with positive charges promoted the Donnan exclusion effect of the SPI membrane. The modification method of SPI also contains blending with inorganic and/or polymer materials, cross-linking with the polymer, modification of the monomer, and copolymerization to obtain novel SPI-based membrane.

The inorganic materials, such as TiO_2, ZrO_2, AlOOH, and MoS_2, are widely used to modify the SPI membrane. The SPI was blended with TiO_2 nanoparticles to reduce water uptake and enhance dimension stability (Li et al. 2013a). The ZrO_2 of high hydrophilicity was adopted to further improve the membrane stability, because of the good antioxidation ability and acid resistance (Li et al. 2014a). The ZrO_2 nanoparticles possess high specific surface area and hydrophilicity and promote the proton conductivity of SPI. In general, the materials with hydroxyl groups can form hydrogen bonds with the sulfonic acid groups. This can enhance the stability and reduce vanadium ion permeation of the polymer matrix. The AlOOH is rich with hydroxyl groups on the surface, thereby enhancing the ion selectivity of the membrane when incorporated into the SPI (Zhang et al. 2014). To reduce the vanadium ion permeability and maintain the high proton conductivity of the membrane, the functionalized filler is preferable. The MoS_2 could be sulfonated to achieve s-MoS_2 nanosheet with sulfonic acid groups and used to modify the SPI membrane with high proton conductivity and good performance for VRFB (Li et al. 2015).

Formation of composite membrane by blending with other polymers can also obtain high ion selectivity. In general, the microphase separation promotes the proton conductivity and enhances the performance of VRFB. The hydrophilic polyvinyl alcohol (PVA) was blended with the SPI by means of microphase separation, because of the aggregation of hydrophilic PVA chains and hydrophobic SPI main chains (Liu et al. 2015b). The chain of the SPI could also introduce the hydrophilic and/or hydrophobic segment to facilitate the microphase separation for high-performance membranes in VRFB (Yu et al. 2019). To enhance the proton conductivity and chemical and mechanical stability of the SPI, it is effective to convert the linear structure to a branched structure. The branched SPI has more free volumes for water adsorption, which in turn facilitates the proton transport; and the 3D crosslinked structure exhibits high chemical and dimension stability (Zhang et al. 2016). A moderate branching degree is necessary, since too many free volumes deteriorate the vanadium ion permeation. It is reported that the branched SPI blending with the molybdenum disulfide nanosheets could further enhance the membrane performance (Pu et al. 2018).

6.3.3.4   Other Hydrocarbon Membranes

A few other hydrocarbon membranes for VRFB application are briefly introduced as follows.

For the cation exchange membranes, the polymers usually contain bulky aromatic rings and cation exchange groups (e.g., sulfonic acid group). Typical examples include the sulfonated poly(fluorenyl ether ketone sulfone), sulfonated poly(ether sulfone), sulfonated poly(phthalazinone ether sulfone), sulfonated poly(arylene thioether), sulfonated poly(arylene ether), sulfonated polyimide, sulfonated poly (fluorenyl ether ketone), sulfonated poly(fluorenyl ether thioether ketone), and their derivatives, to name but a few.

The anion exchange membranes can reduce the vanadium ion crossover induced by the Donnan exclusion effect, while they permit the proton transport. As such, the flow batteries with anion exchange membranes generally exhibit relatively high coulombic efficiency. On the other hand, the affixed positively charged groups hamper the proton transport to some extent, lowering the voltage efficiency. The anion exchange membranes usually possess quaternary ammonium groups and/or pyridinium groups (Zhang et al. 2015). For the strong oxidizing VO_2^+ species, the pyridinium groups are more stable than the quaternary ammonium counterparts and facilitate the proton transport by the interaction with acid (Zhang et al. 2013a, b). The pyridinium groups were introduced in poly(phenyl sulfone) to obtain anion exchange membrane of low vanadium ion permeation, low self-discharge rate, and high coulombic efficiency, compared with the Nafion membrane (Zhang et al. 2015). The polysulfone is a chemically stable and cost-effective polymer and is usually synthesized for anion exchange membrane with quarternized ammonium groups. For example, Jung and co-workers prepared a polysulfone-based anion exchange membrane with quaternary benzyl trimethylammonium groups, exhibiting high chemical stability and high battery performance (Jung et al. 2013). It was demonstrated that the high oxidizing environment (e.g., VO_2^+) primarily degraded the polymer backbones in hydrocarbon cation exchange membranes, ascribed to quaternary ammonium groups (Jung et al. 2013). Additionally, the quarternized ammonium groups are incorporated into the anion membranes by the chloromethylation reactions with carcinogenicity (Luo et al. 2016).

6.3.4 POROUS MEMBRANES

The porous membranes can cut down the use of raw materials and thus reduce the overall cost of the VRFB system. The connective pores in the membrane are filled with electrolyte under cell operation, improving the proton conductivity and cell performance. The commercial Daramic microporous membranes were initially investigated for VRFBs. They comprise polyethylene with a special mineral oil as a stabilizer against oxidation (Mohammadi and Skyllas-Kazacos 1995). The membrane was usually used in lead-acid batteries in view of the high stability in sulfonic acid. However, it is unsuitable for VRFB application because of severe vanadium ion permeation and instability under VO_2^+ strong oxidizing environment, compared with the membranes of aromatic hydrocarbon polymers and Nafion counterparts. In 2011, Zhang's group proposed a polyacrylonitrile porous membrane by phase inversion method for VRFB application, as shown in Figure 6.13 (Zhang et al. 2011). The concept demonstrates that the decrease in pore size can enhance the ion selectivity of the membrane (size sieving effect), which provides a new way to prepare membranes for VRFBs.

The solvent-template and phase inversion methods are commonly used for porous membrane preparation (Jang et al. 2016). The porous membranes can be classified as those containing ionic groups and non-ionic groups. The sieving action and/or Donnan effect are the basis for the porous membrane. The porous membranes are also modified by various techniques to tune the size and density of the pores.

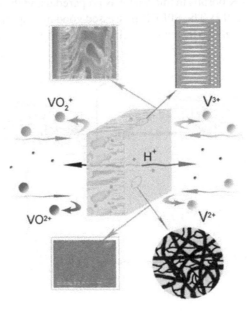

FIGURE 6.13 Schematic of porous membrane prepared by phase inversion for VRFB. (Reprinted from Zhang et al. (2011) with permission, © Royal Society of Chemistry, 2011.)

6.3.4.1 Solvent-Template Method

The solvent-template method usually adds porogen in the casting solution to form a uniform mixture. The porogen should be easily dissolved in the polymer solution and uniformly distributed in the polymer matrix. After forming the membrane, the porogen should be easily removed from the polymer matrix to obtain porous structure.

Maurya and co-workers fabricated a nanoporous PBI membrane by adding polyethylene glycol (PEG) in casting solution (Maurya et al. 2016). The PEG porogen was easily dissolved in water, resulting in the formation of nanopores in the PBI matrix. The pore size and amount were dependent on the PEG addition. Little porogen leads to less connected pores that would suppress the vanadium ion permeation with large membrane resistance. Excessive porogen generated large pores with high proton conductivity but rendered severe vanadium permeation and poor mechanical stability of the membrane. As aforementioned, the nanoporous structure of the PBI membrane can be connected upon acid doping for high proton conductivity. In addition, the low vanadium ion permeation of the PBI membrane is attributed to the sieving effect of the nanopores and the Donnan effect of the positively charged N-containing groups. Using a similar strategy, Wang and co-workers fabricated the SPFEK porous membrane using the imidazole as porogen and subsequently modified the membrane by layer-by-layer self-assembly technique to obtain the SPFEK–[PDDA–PSSS]$_n$ porous membrane (as shown in Figure 6.14) (Wang et al. 2014). Compared with the pristine membrane, the SPFEK–[PDDA–PSSS]$_n$ porous membrane exhibited high ion selectivity because of the [PDDA–PSSS] self-assembled layers. The poly(ether sulfone) (PES)/SPEEK porous membrane was prepared using the phenolphthalein as porogen (Chen et al. 2017). The SPEEK possessed good hydrophilicity and enhanced the conductivity for high-performance VRFBs. The concentration of VO^{2+} diffusion

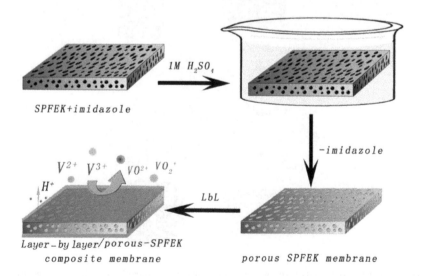

FIGURE 6.14 The preparation process of SPFEK–[PDDA–PSSS]$_n$ membrane for VRFB. (Reprinted from Wang et al. (2014) with permission, © Elsevier, 2014.)

through the membrane increased with the porogen content, as shown in Figure 6.15a. The membrane matrix free of porogen exhibited the lowest vanadium ion permeation. However, the cell was unable to complete the charge–discharge process in light of the high resistance, as displayed in Figure 6.15b. The voltage efficiency was improved with the increased amount of pores in the membrane, accompanied by an accelerated vanadium ion permeation. The energy efficiency of the cell with 10% P membrane was higher than that of the cell with 20% P and 30% P membranes (Figure 6.15b, d). In addition, the charge capacity and discharge energy of cell with the 10% P membrane were higher than that of the Nafion 115 (Figure 6.15c). There are other commonly used porogens for porous membranes, such as dibutyl phthalate for porous PBI membrane, sulfolane for porous PES/SPEEK membrane, and so on.

6.3.4.2   Phase Inversion Method

Phase inversion is a facile and cost-effective method to prepare porous polymer membranes and was developed by Loeb and Sourirajan in 1962. The phase inversion represents the conversion of liquid polymer solution into solid polymer membrane. In particular, it is an immersion–precipitation induced phase separation by soaking the casting film in a non-solvent to obtain a solid membrane (Vandezande et al. 2008), as illustrated in Figure 6.16. The preparation process of porous membrane

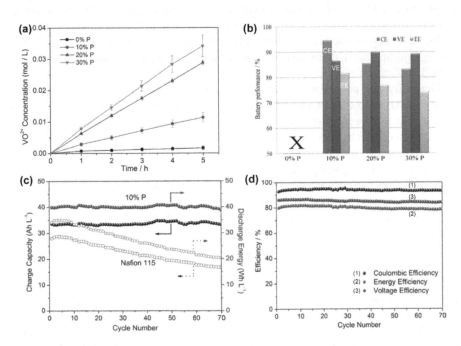

FIGURE 6.15   (a) The change of VO^{2+} concentration at the deficiency side of the diffusion cell with different phenolphthalein contents for PES hierarchical porous membranes, (b) the performance of a VRFB with PES hierarchical porous membranes at 80 mA cm^{-2}, (c) charge capacity and discharge energy over cycling at 80 mA cm^{-2}, and (d) cycle performance of PES porous membrane (10% phenolphthalein) in VRFB single cell at 80 mA cm^{-2}. (Reprinted from Chen et al. (2017) with permission, © Elsevier, 2017.)

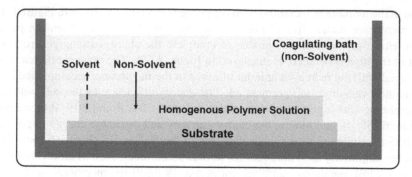

FIGURE 6.16 Illustration of the preparation process for porous membrane by immersion–precipitation induced phase separation.

is briefly described as follows. A homogeneous polymer solution (single phase) is in thermodynamic equilibrium. Upon soaking in the non-solvent, the stable state is broken with the polymer solution transforming into polymer-rich and polymer-lean phases. The solvent in the two phases mixes with the non-solvent in the bath. The phase-separation occurs via nucleation and growth mechanism. Eventually, the polymer-rich phase develops into the matrix, while the polymer-lean phase is removed to form pores in the matrix, giving rise to the porous membrane (Vandezande et al. 2008). The porous membranes were initially applied for water treatment. After Zhang's group (2011), many researchers investigated various porous membranes for VRFB application.

Generally, the porous membrane developed by phase inversion is asymmetric with a porous support layer and skin layer (Li et al. 2014b). The skin layer determines the ion selectivity and resistivity of the membrane. It is also a general way to enhance the ion selectivity by modifying the membranes with other materials. The membrane morphology is an important factor influencing the battery performance and can be tailored by adjusting the parameters during the phase separation, such as the composition of casting solution, the composition and concentration of non-solvent, the additives, and the coagulating bath (Li et al. 2013b). The polymer matrices for porous membranes usually have good mechanical and chemical stability, as well as high hydrophobicity (e.g., PBI, PVDF, PES, and so on). The hydrophobicity of the polymer can limit the water transport and prevent the vanadium ion permeation to some extent.

The PVDF porous membranes were prepared by immersion–precipitation induced phase separation for VRFBs (Wei et al. 2013). The mass exchange rate decreased with the increase in PVDF content in solvent. A low exchange rate facilitated the crystallization of PVDF and the formation of spherical particles instead of open sponge-like structure, as displayed in Figure 6.17a–d (Wei et al. 2013). The denser sphere structure is larger than that in the pore size in open sponge-like structure and is efficient to suppress vanadium ion permeation (Figure 6.17e).

The PES porous membranes were prepared by phase inversion, and the porous morphology was adjusted by the hydrophilic poly(vinyl pyrrolidone) (PVP) (Li et al. 2013b). The PES had strong hydrophobicity and led to a high-resistance membrane.

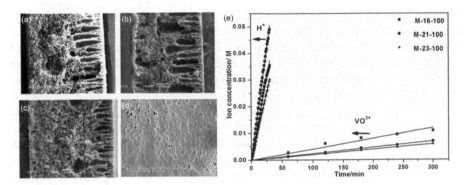

FIGURE 6.17 The cross-sectional morphology (a–c) of the PVDF porous membranes (a: 16%, b: 21%, c: 23%), as well as the corresponding top-view surface (d: 23%). (e) The ion concentration in the acceptor phase of the dialysis set-up vs diffusion time for the PVDF membranes prepared from solutions with different polymer concentrations. (Reprinted from Wei et al. (2013) with permission, © Royal Society of Chemistry, 2013.)

The addition of the PVP decreased the void size in the support layer but enlarged the pore size in the skin layer for improved proton conductivity, as shown in Figure 6.18. Such a porous structure resulted in a large vanadium ion permeability. To enhance the ion selectivity of the PES-based porous membranes, Li and co-workers coated a thin Nafion layer (~1 μm) on the PES/SPEEK porous membrane (Li et al. 2014b). The hydrophilic SPEEK was used to adjust the porous structure of the PES membrane. The Nafion layer served as a permselective layer to enhance the ion selectivity

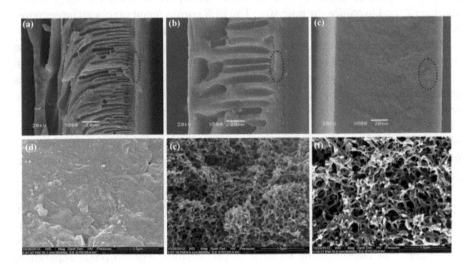

FIGURE 6.18 Cross-sectional morphology of PES membranes with different PVP proportions: (a) 1:0.18, (b) 1:0.33, and (c) 1:0.54 and the corresponding morphology of skin layers (d) 1:0.18, (e) 1:0.33, and (f) 1:0.54. (Reprinted from Li et al. (2013b) with permission, © Elsevier, 2013.)

together with the skin layer. The stable inorganic materials, combining with polymer matrix, can also act as the permselective layer to enhance the ion selectivity of the porous membrane. Yuan and co-workers fabricated an ultra-thin zeolite-coated PES porous membrane for VRFB (Yuan et al. 2016c). Particularly, the zeolites with micropores (~0.5 nm) and well-defined channels can efficiently select the protons (<0.24 nm) and vanadium ions (>0.6 nm).

The ion selectivity of the porous membrane can be improved by introducing the positively charged groups or polymers on the surface, based on the Donnan exclusion effect. Yuan and co-workers fabricated the PPY on the skin layer of porous PES membrane by *in-situ* polymerization (Yuan et al. 2016a). The PES/PPY composite membrane was obtained with high ion selectivity. When the positively charged imidazole groups were cross-linked on the pore walls of polysulfone membranes, the ion selectivity was also higher than that of the pristine one (Zhao et al. 2016). Similarly, the agents or polymers possessing positively charged groups, such as pyridine, butanediamine, and trimethylamine, are adopted to improve the ion selectivity of the porous membrane.

A small pore size suppresses the vanadium ion permeation, but the proton conductivity declines as well. As such, the pores are usually filled with organic and/or inorganic materials to reject the bulky vanadium ions whereas allow the tiny protons. This produces membranes of high ion selectivity. For example, the sulfonic silica was filled in the porous PVDF membrane via sol-gel process (Ling et al. 2019). The PVDF rendered good stability, and the sulfonic silica in the pores repelled the vanadium ion permeation and provided new pathways for proton transport. Hence, the hybrid membrane exhibited good ion selectivity and high performance in VRFB. Similarly, the silica was used to modify the PES/SPEEK porous membrane for high ion selectivity (Xi et al. 2015b). The solvent-induced polymer chain rearrangement is another method to reduce the pore size. The PES porous membrane with larger pores was modified by isopropanol, and the pore size was adjusted by subsequently controlling the isopropanol evaporation (Lu et al. 2016). The morphology changes of the porous membrane before and after solvent treatment are compared in Figure 6.19a–f. Upon treatment, the pore size on the skin layer decreased significantly, which is beneficial for the suppression of vanadium ion permeation. The change of pore size with the treatment can be described as follows (Lu et al. 2016), as illustrated in Figure 6.19g. The membrane swells and allows the rearrangement of polymer chains when immersed in a proper solvent. The swelling force declines while the cohesive force increases during the solvent evaporation process. Because of the chain reorganization, the membrane shrinks from the initial state under large cohesive force.

The pore morphology is diverse in the porous membrane, and particularly the sponge-like pores are desirable for the VRFB application because of the high ion selectivity (Qiao et al. 2019). The morphology can be tuned by changing the constituent of the coagulating bath, as discussed above and illustrated in Figure 6.16. For nanoporous membrane by immersion–precipitation induced phase separation, the non-solvent in coagulating bath is generally the deionized water. In this case, finger-like voids are usually developed because of the instantaneous demixing in water (Qiao et al. 2019). To obtain a membrane with sponge-like pores, Qiao and co-workers added inorganic salts into the non-solvent. This delayed the demixing

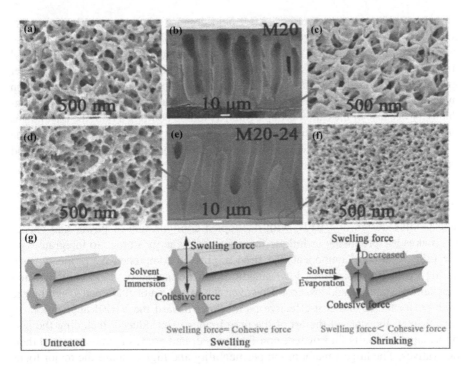

FIGURE 6.19 Cross-sectional morphology of the PES membranes (b and e) and the corresponding magnified morphology of the pore wall of the macrovoid (a and d) as well as the skin-layer. (c and f) The influence of the solvent treatment on the morphology of the porous membrane (g). (M20: the membrane with 20 wt% PVP and not treated by isopropanol; M20–24: the membrane with 20 wt% PVP and immersed in isopropanol for 24 hours). (Reprinted from Lu et al. (2016) with permission, © Royal Society of Chemistry, 2016.)

process and thus achieved high ion selectivity with the sponge-like pores in the membrane (Qiao et al. 2019). In order to enhance the hydrophilicity for high proton conductivity, the sodium p-styrene sulfonate (NaSS) was grafted on the porous polysulfone membrane by UV-initiated vinyl monomer polymerization (Li et al. 2014c). The grafted degree of NaSS adjusted the hydrophilicity of the membrane and thereby controlled the battery performance.

Apart from the porogen addition and immersion–precipitation induced phase inversion techniques, the porous membranes can also be achieved by vapor-induced phase inversion and hydrolysis in a polymer. Zhang's group fabricated a sponge-like porous PBI membrane for VRFB by vapor-induced phase inversion method (Yuan et al. 2016b). The porous membrane contained many micron-sized cells with ultra-thin walls and plenty of positively charged groups in PBI. The VRFB single cells with the membranes exhibited outstanding performance, i.e., with no obvious efficiency decay even after more than 13,000 cycles. Luo and co-workers prepared the porous PBI membrane with dense skin layer using the same method (Luo et al. 2016). The membrane had symmetrical sponge-like cross-sections and closed cells to resist water and vanadium ion permeation. The repulsion of vanadium ions was improved with the Donnan effect. After acid doping, the proton conductivity also improved. Ma et al. synthesized the NaSS–ST–VBC–BMA copolymer (by free radical polymerization of NaSS, styrene (ST), 4-chloromethylstyrene (VBC), and butyl methacrylate (BMA)) and further cross-linked the polymer with the bulky ester groups at the side chains (Ma et al. 2018). A porous membrane was obtained upon hydrolyzing. The cross-linking restricted the swelling and enhanced the proton conductivity because of the cation groups. Such a method combined both the advantages of the dense and porous membranes with high VRFB performance. To ensure the stability of the membrane, the hydrolysis should be conducted at the side chains and under mild conditions.

6.4 SUMMARY AND PERSPECTIVES

The flexible design of VRFB, with decoupled configuration of power and capacity, makes it a promising technique for large-scale energy storage to integrate with the grid. As a core component, the IEM requires a comprehensive property, such as high proton conductivity, low vanadium ion permeability, good chemical and mechanical stability, and low cost. However, it is still a major challenge to synthesize high-performance and cost-effective membranes toward the VRFB application. In this chapter, the state-of-the-art of the membranes is introduced, including the perfluorinated, partially fluorinated, non-fluorinated, and porous membranes and their derivatives. The large vanadium ion permeability and high cost are the major focus for the development of the commercial Nafion membranes, a typical perfluorinated counterpart. It is effective to modify the Nafion membranes by incorporating inorganic and organic additives to tailor the intrinsic channels and reduce the use of raw materials. As an alternative to the Nafion membrane, the non-fluorinated hydrocarbon membranes are developed, with low vanadium ion permeability and competitive cost. Nonetheless, proton conductivity and chemical stability are the challenging issues. The same strategy has been widely used to enhance the membranes with the inorganic and organic species. The porous membranes are further explored based on the size sieving effect. The ion selectivity, a key factor affecting the membrane performance, is improved by adjusting the pore or channel size. This includes filling the matrix, modifying the surface, turning the component of coagulating bath (for porous membrane), and so on.

The hydrocarbon polymer is a promising candidate for VRFB application, in view of its high ion selectivity and low cost. However, its chemical stability is unsatisfactory. Further development of the membranes should be focused on tailoring the polymers toward high ion selectivity, long-term chemical and mechanical stability, and low cost. First, the hydrocarbon polymers can be improved by introducing the fluorine-containing groups or stable species to enhance the stability and proton conductivity and meanwhile maintaining the low cost. Second, a judicious combination of different polymers, various ion exchange groups, and modification or preparation methods could also be a promising way for membranes of high ion selectivity, good chemical stability, and low cost. Finally, the degradation mechanism in the cell operation process should be further investigated for better understanding.

REFERENCES

Ahn S., H. Jeong, J. Jang, J. Lee, S. So, Y. Kim, Y. Hong and T. Kim. 2018. Polybenzimidazole/ Nafion hybrid membrane with improved chemical stability for vanadium redox flow battery application. *RSC Advances* 8: 25304–12.

Amjadi M., S. Rowshanzamir, S. Peighambardoust, M. Hosseini and M. Eikani. 2010. Investigation of physical properties and cell performance of Nafion/TiO$_2$ nanocomposite membranes for high temperature PEM fuel cells. *International Journal of Hydrogen Energy* 35: 9252–60.

Aziz M. and S. Shanmugam. 2017. Zirconium oxide nanotube-Nafion composite as high performance membrane for all vanadium redox flow battery. *Journal of Power Sources* 337: 36–44.

Bouchet R. and E. Siebert. 1999. Proton conduction in acid doped polybenzimidazole. *Solid State Ionics* 118: 287–99.

Chang Z., D. Henkensmeier and R. Chen. 2017. One-step cationic grafting of 4-hydroxy-tempo and its application in a hybrid redox flow battery with a crosslinked PBI membrane. *ChemSusChem* 10: 3193–97.

Chen D., M. Hickner, S. Wang, J. Pan, M. Xiao and Y. Meng. 2012. Directly fluorinated polyaromatic composite membranes for vanadium redox flow batteries. *Journal of Membrane Science* 415–416: 139–44.

Chen D., D. Li and X. Li. 2017. Hierarchical porous poly (ether sulfone) membranes with excellent capacity retention for vanadium flow battery application. *Journal of Power Sources* 353: 11–18.

Chen D., H. Qi, T. Sun, C. Yan, Y. He, C. Kang, Z. Yuan and X. Li. 2019. Polybenzimidazole membrane with dual proton transport channels for vanadium flow battery applications. *Journal of Membrane Science* 586: 202–10.

Chen R., D. Henkensmeier, S. Kim, S. Yoon, T. Zinkevich and S. Indris. 2018. Improved all-vanadium redox flow batteries using catholyte additive and a cross-linked methylated polybenzimidazole membrane. *ACS Applied Energy Materials* 1: 6047–55.

Dai J., Y. Dong, C. Yu, Y. Liu and X. Teng. 2018. A novel Nafion-g-PSBMA membrane prepared by grafting zwitterionic SBMA onto Nafion via SI-ATRP for vanadium redox flow battery application. *Journal of Membrane Science* 554: 324–30.

Dai W., Y. Shen, Z. Li, L. Yu, J. Xi and X. Qiu. 2014. SPEEK/graphene oxide nanocomposite membranes with superior cyclability for highly efficient vanadium redox flow battery. *Journal of Materials Chemistry A* 2: 12423–32.

Decher G., J. Hong and J. Schmitt. 1992. Buildup of ultrathin multilayer films by a self-assembly process: III. Consecutively alternating adsorption of anionic and cationic polyelectrolytes on charged surfaces *Thin Solid Films* 210–211: 831–35.

Ding L., X. Song, L. Wang and Z. Zhao. 2019. Enhancing proton conductivity of polybenzimidazole membranes by introducing sulfonate for vanadium redox flow batteries applications. *Journal of Membrane Science* 578: 126–35.

Ding L., X. Song, L. Wang, Z. Zhao and G. He. 2018. Preparation of dense polybenzimidazole proton exchange membranes with different basicity and flexibility for vanadium redox flow battery applications. *Electrochimica Acta* 292: 10–19.

Doetsch C. and J. Burfeind. 2016. Chapter 12: Vanadium redox flow batteries. *Storing Energy*. Elsevier, Amsterdam. pp. 227–46.

Fu Z., J. Liu and Q. Liu. 2015. SPEEK/PVDF/PES composite as alternative proton exchange membrane for vanadium redox flow batteries. *Journal of Electronic Materials* 45: 666–71.

Geng K., Y. Li, Y. Xing, L. Wang and N. Li. 2019. A novel polybenzimidazole membrane containing bulky naphthalene group for vanadium flow battery. *Journal of Membrane Science* 586: 231–39.

Glipa X., B. Bonnet, B. Mula, D.J. Jones and J. Roziere. 1999. Investigation of the conduc-
    tion properties of phosphoric and sulfuric acid doped polybenzimidazole. *Journal of
    Materials Chemistry* 9: 3045–49.
Grot W. 1973. Laminates of support material and fluorinated polymer containing pendant
    side chains containing sulfonyl groups. US Patent 3770, 1973-11-6.
Guan Y., H. Pu, M. Jin, Z. Chang and A. Modestov. 2012. Proton conducting membranes
    based on poly(2,2′-imidazole-5,5′-bibenzimidazole). *Fuel Cells* 12: 124–31.
Guarnieri M., A. Trovò, A. D'anzi and A. Piergiorgio. 2018. Developing vanadium redox flow
    technology on a 9-kw 26-kwh industrial scale test facility: Design review and early
    experiments. *Applied Energy* 230: 1425–34.
Hietala S., S. Holmberg, M. Karjalainen, J. Nasman, M. Paronen, R. Serimaa, F. Sundholm
    and S. Vahvaselka. 1997. Structural investigation of radiation grafted and sulfonated
    poly(vinylidene fluoride), PVDF, membranes. *Journal of Materials Chemistry* 7:
    721–26.
Huang R., P. Shao, C. Burns and X. Feng. 2001. Sulfonation of poly(ether ether ketone)(peek):
    Kinetic study and characterization. *Journal of Applied Polymer Science* 82: 2651–60.
Jang J., T. Kim, S. Yoon, J. Lee, J. Lee and Y. Hong. 2016. Highly proton conductive, dense
    polybenzimidazole membranes with low permeability to vanadium and enhanced
    H_2SO_4 absorption capability for use in vanadium redox flow batteries. *Journal of
    Materials Chemistry A* 4: 14342–55.
Jiang B., L. Yu, L. Wu, D. Mu, L. Liu, J. Xi and X. Qiu. 2016. Insights into the impact of the
    Nafion membrane pretreatment process on vanadium flow battery performance. *ACS
    Applied Materials & Interfaces* 8: 12228–38.
Jung M., J. Parrondo, C. Arges and V. Ramani. 2013. Polysulfone-based anion exchange
    membranes demonstrate excellent chemical stability and performance for the all-
    vanadium redox flow battery. *Journal of Materials Chemistry A* 1: 10458–64.
Kim J., J. Jeon and S. Kwak. 2014. Nafion-based composite membrane with a permse-
    lective layered silicate layer for vanadium redox flow battery. *Electrochemistry
    Communications* 38: 68–70.
Kim M., D. Ha and J. Choi. 2019. Nanocellulose-modified Nafion 212 membrane for improv-
    ing performance of vanadium redox flow batteries. *Bulletin of the Korean Chemical
    Society* 40: 533–38.
Kong L., L. Zheng, R. Niu, H. Wang and H. Shi. 2016. A sulfonated poly(ether ether ketone)/
    amine-functionalized graphene oxide hybrid membrane for vanadium redox flow bat-
    teries. *RSC Advances* 6: 100262–70.
Lee Y., S. Kim, A. Maljusch, O. Conradi, H. Kim, J. Jang, J. Han, J. Kim and D.
    Henkensmeier. 2019. Polybenzimidazole membranes functionalised with 1-methyl-
    2-mesitylbenzimidazolium ions via a hexyl linker for use in vanadium flow batteries.
    Polymer 174: 210–17.
Leung P., X. Li, C. De León, L. Berlouis, C. Low and F. Walsh. 2012. Progress in redox flow
    batteries, remaining challenges and their applications in energy storage. *RSC Advances*
    2: 10125–56.
Li J., S. Liu, Z. He and Z. Zhou. 2016. Semi-fluorinated sulfonated polyimide membranes
    with enhanced proton selectivity and stability for vanadium redox flow batteries.
    Electrochimica Acta 216: 320–31.
Li J., Y. Zhang and L. Wang. 2013a. Preparation and characterization of sulfonated polyimide/
    TiO_2 composite membrane for vanadium redox flow battery. *Journal of Solid State
    Electrochemistry* 18: 729–37.
Li J., Y. Zhang, S. Zhang and X. Huang. 2015. Sulfonated polyimide/s-MoS_2 composite mem-
    brane with high proton selectivity and good stability for vanadium redox flow battery.
    Journal of Membrane Science 490: 179–89.

Li J., Y. Zhang, S. Zhang, X. Huang and L. Wang. 2014a. Novel sulfonated polyimide/ZrO$_2$ composite membrane as a separator of vanadium redox flow battery. *Polymers for Advanced Technologies* 25: 1610–15.

Li X., A. Dos Santos, M. Drache, X. Ke, U. Gohs, T. Turek, M. Becker, U. Kunz and S. Beuermann. 2017. Polymer electrolyte membranes prepared by pre-irradiation induced graft copolymerization on ETFE for vanadium redox flow battery applications. *Journal of Membrane Science* 524: 419–27.

Li Y., X. Li, J. Cao, W. Xu and H. Zhang. 2014b. Composite porous membranes with an ultrathin selective layer for vanadium flow batteries. *Chemical Communications* 50: 4596–99.

Li Y., H. Zhang, X. Li, H. Zhang and W. Wei. 2013b. Porous poly (ether sulfone) membranes with tunable morphology: Fabrication and their application for vanadium flow battery. *Journal of Power Sources* 233: 202–08.

Li Y., H. Zhang, H. Zhang, J. Cao, W. Xu and X. Li. 2014c. Hydrophilic porous poly(sulfone) membranes modified by UV-initiated polymerization for vanadium flow battery application. *Journal of Membrane Science* 454: 478–87.

Li Z., L. Liu, L. Yu, L. Wang, J. Xi, X. Qiu and L. Chen. 2014d. Characterization of sulfonated poly(ether ether ketone)/poly(vinylidene fluoride-co-hexafluoropropylene) composite membrane for vanadium redox flow battery application. *Journal of Power Sources* 272: 427–35.

Ling L., M. Xiao, D. Han, S. Ren, S. Wang and Y. Meng. 2019. Porous composite membrane of PVDF/sulfonic silica with high ion selectivity for vanadium redox flow battery. *Journal of Membrane Science* 585: 230–37.

Liu G., Z. Xia, S. Jin, X. Guo and J. Fang. 2017. Preparation and properties of polybenzimidazole/quaternized poly(1-vinylimidazole) cross-linked blend membranes for vanadium redox flow battery applications. *High Performance Polymers* 30: 612–23.

Liu S., L. Wang, Y. Ding, B. Liu, X. Han and Y. Song. 2014b. Novel sulfonated poly (ether ether ketone)/polyetherimide acid-base blend membranes for vanadium redox flow battery applications. *Electrochimica Acta* 130: 90–96.

Liu S., L. Wang, D. Li, B. Liu, J. Wang and Y. Song. 2015a. Novel amphoteric ion exchange membranes by blending sulfonated poly(ether ether ketone)/quaternized poly(ether imide) for vanadium redox flow battery applications. *Journal of Materials Chemistry A* 3: 17590–97.

Liu S., L. Wang, B. Zhang, B. Liu, J. Wang and Y. Song. 2015b. Novel sulfonated polyimide/ polyvinyl alcohol blend membranes for vanadium redox flow battery applications. *Journal of Materials Chemistry A* 3: 2072–81.

Liu Y., K. Ai and L. Lu. 2014a. Polydopamine and its derivative materials: Synthesis and promising applications in energy, environmental, and biomedical fields. *Chemical Reviews* 114: 5057–115.

Loeb S. and S. Sourirajan. 1962. *Sea Water Demineralization by means of an Osmotic Membrane. Saline Water Conversion—II Chapter*, ACS publications, Washington, DC. pp. 117–32.

Lu S., C. Wu, D. Liang, Q. Tan and Y. Xiang. 2014. Layer-by-layer self-assembly of Nafion–[CS–PWA] composite membranes with suppressed vanadium ion crossover for vanadium redox flow battery applications. *RSC Advances* 4: 24831–37.

Lu W., Z. Yuan, Y. Zhao, X. Li, H. Zhang and I.F.J. Vankelecom. 2016. High-performance porous uncharged membranes for vanadium flow battery applications created by tuning cohesive and swelling forces. *Energy & Environmental Science* 9: 2319–25.

Luo Q., H. Zhang, J. Chen, P. Qian and Y. Zhai. 2008a. Modification of Nafion membrane using interfacial polymerization for vanadium redox flow battery applications. *Journal of Membrane Science* 311: 98–103.

Luo Q., H. Zhang, J. Chen, D. You, C. Sun and Y. Zhang. 2008b. Preparation and characterization of Nafion/SPEEK layered composite membrane and its application in vanadium redox flow battery. *Journal of Membrane Science* 325: 553–58.

Luo T., O. David, Y. Gendel and M. Wessling. 2016. Porous poly(benzimidazole) membrane for all vanadium redox flow battery. *Journal of Power Sources* 312: 45–54.

Luo X., Z. Lu, J. Xi, Z. Wu, W. Zhu, L. Chen and X. Qiu. 2005. Influences of permeation of vanadium ions through PVDF-g-PSSA membranes on performances of vanadium redox flow batteries. *The Journal of Physical Chemistry B* 109: 20310–14.

Ma J., S. Wang, J. Peng, J. Yuan, C. Yu, J. Li, X. Ju and M. Zhai. 2013. Covalently incorporating a cationic charged layer onto Nafion membrane by radiation-induced graft copolymerization to reduce vanadium ion crossover. *European Polymer Journal* 49: 1832–40.

Ma Y., N. Qaisrani, L. Ma, P. Li, L. Li, S. Gong, F. Zhang and G. He. 2018. Side chain hydrolysis method to prepare nanoporous membranes for vanadium flow battery application. *Journal of Membrane Science* 560: 67–76.

Mai Z., H. Zhang, X. Li, S. Xiao and H. Zhang. 2011. Nafion/polyvinylidene fluoride blend membranes with improved ion selectivity for vanadium redox flow battery application. *Journal of Power Sources* 196: 5737–41.

Mauritz K. and R. Moore. 2004. State of understanding of Nafion. *Chemical Reviews* 104: 4535–85.

Mauritz K. and R. Warren. 1989. Microstructural evolution of a silicon oxide phase in a perfluorosulfonic acid ionomer by an in situ sol-gel reaction. 1. Infrared spectroscopic studies. *Macromolecules* 22: 1730–34.

Maurya S., S. Shin, J. Lee, Y. Kim and S. Moon. 2016. Amphoteric nanoporous polybenzimidazole membrane with extremely low crossover for a vanadium redox flow battery. *RSC Advances* 6: 5198–204.

Mohammadi T. and M. Skyllas-Kazacos. 1995. Use of polyelectrolyte for incorporation of ion-exchange groups in composite membranes for vanadium redox flow battery applications. *Journal of Power Sources* 56: 91–96.

Park S. and H. Kim. 2016. Preparation of a sulfonated poly(ether ether ketone)-based composite membrane with phenyl isocyanate treated sulfonated graphene oxide for a vanadium redox flow battery. *Journal of the Electrochemical Society* 163: A2293–98.

Peng S., X. Yan, D. Zhang, X. Wu, Y. Luo and G. He. 2016. A H_3PO_4 preswelling strategy to enhance the proton conductivity of a H_2SO_4-doped polybenzimidazole membrane for vanadium flow batteries. *RSC Advances* 6: 23479–88.

Pohl H. and R. Chartoff. 1964. Carriers and unpaired spins in some organic semiconductors. *Journal of Polymer Science Part A: General Papers* 2: 2787–806.

Pu Y., S. Zhu, P. Wang, Y. Zhou, P. Yang, S. Xuan, Y. Zhang and H. Zhang. 2018. Novel branched sulfonated polyimide/molybdenum disulfide nanosheets composite membrane for vanadium redox flow battery application. *Applied Surface Science* 448: 186–202.

Qiao L., H. Zhang, W. Lu, Q. Dai and X. Li. 2019. Advanced porous membranes with tunable morphology regulated by ionic strength of nonsolvent for flow battery. *ACS Applied Materials & Interfaces* 11: 24107–13.

Qiu J., M. Li, J. Ni, M. Zhai, J. Peng, L. Xu, H. Zhou, J. Li and G. Wei. 2007. Preparation of ETFE-based anion exchange membrane to reduce permeability of vanadium ions in vanadium redox battery. *Journal of Membrane Science* 297: 174–80.

Qiu J., J. Zhang, J. Chen, J. Peng, L. Xu, M. Zhai, J. Li and G. Wei. 2009. Amphoteric ion exchange membrane synthesized by radiation-induced graft copolymerization of styrene and dimethylaminoethyl methacrylate into PVDF film for vanadium redox flow battery applications. *Journal of Membrane Science* 334: 9–15.

Qiu J., L. Zhao, M. Zhai, J. Ni, H. Zhou, J. Peng, J. Li and G. Wei. 2008. Pre-irradiation grafting of styrene and maleic anhydride onto PVDF membrane and subsequent sulfonation
for application in vanadium redox batteries. *Journal of Power Sources* 177: 617–23.

Schwenzer B., S. Kim, M. Vijayakumar, Z. Yang and J. Liu. 2011. Correlation of structural
differences between Nafion/polyaniline and Nafion/polypyrrole composite membranes
and observed transport properties. *Journal of Membrane Science* 372: 11–19.

Sum E., M. Rychcik and M. Skyllas-Kazacos. 1985. Investigation of the v(v)/v(iv) system for
use in the positive half-cell of a redox battery. *Journal of Power Sources* 16: 85–95.

Sum E. and M. Skyllas-Kazacos. 1985. A study of the v(II)/v(III) redox couple for redox flow
cell applications. *Journal of Power Sources* 15: 179–90.

Sun C., A. Zlotorowicz, G. Nawn, E. Negro, F. Bertasi, G. Pagot, K. Vezzù, G. Pace, M.
Guarnieri and V. Di Noto. 2018. [Nafion/(WO$_3$)x] hybrid membranes for vanadium
redox flow batteries. *Solid State Ionics* 319: 110–16.

Teng X., J. Dai and J. Su. 2014. Effects of different kinds of surfactants on Nafion membranes for all vanadium redox flow battery. *Journal of Solid State Electrochemistry*
19: 1091–101.

Teng X., J. Dai, J. Su, Y. Zhu, H. Liu and Z. Song. 2013a. A high performance polytetra-
fluoroethene/Nafion composite membrane for vanadium redox flow battery application.
Journal of Power Sources 240: 131–39.

Teng X., C. Sun, J. Dai, H. Liu, J. Su and F. Li. 2013b. Solution casting Nafion/polytetra-
fluoroethylene membrane for vanadium redox flow battery application. *Electrochimica
Acta* 88: 725–34.

Teng X., Y. Zhao, J. Xi, Z. Wu, X. Qiu and L. Chen. 2009. Nafion/organic silica modified
TiO$_2$ composite membrane for vanadium redox flow battery via in situ sol–gel reactions. *Journal of Membrane Science* 341: 149–54.

Vandezande P., L. Gevers and I. Vankelecom. 2008. Solvent resistant nanofiltration:
Separating on a molecular level. *Chemical Society Reviews* 37: 365–405.

Wainright J., J. Wang, D. Weng, R. Savinell and Litt M. 1995. Acid-doped polybenzimid-
azoles: A new polymer electrolyte. *Journal of the Electrochemical Society* 142:
L121–L23.

Wang Y., S. Wang, M. Xiao, D. Han and Y. Meng. 2014. Preparation and characterization of
a novel layer-by-layer porous composite membrane for vanadium redox flow battery
(VRB) applications. *International Journal of Hydrogen Energy* 39: 16088–95.

Wei W., H. Zhang, X. Li, H. Zhang, Y. Li and I. Vankelecom. 2013. Hydrophobic asymmetric
ultrafiltration PVDF membranes: An alternative separator for VFB with excellent stability. *Physical Chemistry Chemical Physics* 15: 1766–71.

Xi J., W. Dai and L. Yu. 2015a. Polydopamine coated SPEEK membrane for a vanadium
redox flow battery. *RSC Advances* 5: 33400–06.

Xi J., Z. Wu, X. Qiu and L. Chen. 2007. Nafion/SiO$_2$ hybrid membrane for vanadium redox
flow battery. *Journal of Power Sources* 166: 531–36.

Xi J., Z. Wu, X. Teng, Y. Zhao, L. Chen and X. Qiu. 2008. Self-assembled polyelectrolyte
multilayer modified Nafion membrane with suppressed vanadium ion crossover for
vanadium redox flow batteries. *Journal of Materials Chemistry* 18: 1232–38.

Xi X., C. Ding, H. Zhang, X. Li, Y. Cheng and H. Zhang. 2015b. Solvent responsive silica
composite nanofiltration membrane with controlled pores and improved ion selectivity
for vanadium flow battery application. *Journal of Power Sources* 274: 1126–34.

Xia Z., L. Ying, J. Fang, Y.-Y. Du, W.-M. Zhang, X. Guo and J. Yin. 2017. Preparation of
covalently cross-linked sulfonated polybenzimidazole membranes for vanadium redox
flow battery applications. *Journal of Membrane Science* 525: 229–39.

Xie W., R. Darling and M. Perry. 2015. Processing and pretreatment effects on vanadium
transport in Nafion membranes. *Journal of the Electrochemical Society* 163: A5084–89.

Yang M., C. Lin, J. Kuo and H. Wei. 2017. Effect of grafting of poly(styrenesulfonate) onto Nafion membrane on the performance of vanadium redox flow battery. *Journal of Electroanalytical Chemistry* 807: 88–96.

Yang P., S. Xuan, J. Long, Y. Wang, Y. Zhang and H. Zhang. 2018. Fluorine-containing branched sulfonated polyimide membrane for vanadium redox flow battery applications. *ChemElectroChem* 5: 3695–707.

Yang R., Z. Cao, S. Yang, I. Michos, Z. Xu and J. Dong. 2015. Colloidal silicalite-Nafion composite ion exchange membrane for vanadium redox-flow battery. *Journal of Membrane Science* 484: 1–9.

Ye J., Y. Cheng, L. Sun, M. Ding, C. Wu, D. Yuan, X. Zhao, C. Xiang and C. Jia. 2019a. A green SPEEK/lignin composite membrane with high ion selectivity for vanadium redox flow battery. *Journal of Membrane Science* 572: 110–18.

Ye J., L. Xia, C. Wu, M. Ding, C. Jia and Q. Wang. 2019b. Redox targeting-based flow batteries. *Journal of Physics D: Applied Physics* 52: 443001.

Yu L., L. Wang, L. Yu, D. Mu, L. Wang and J. Xi. 2019. Aliphatic/aromatic sulfonated polyimide membranes with cross-linked structures for vanadium flow batteries. *Journal of Membrane Science* 572: 119–27.

Yuan Z., Q. Dai, Y. Zhao, W. Lu, X. Li and H. Zhang. 2016a. Polypyrrole modified porous poly(ether sulfone) membranes with high performance for vanadium flow batteries. *Journal of Materials Chemistry A* 4: 12955–62.

Yuan Z., Y. Duan, H. Zhang, X. Li, H. Zhang and I. Vankelecom. 2016b. Advanced porous membranes with ultra-high selectivity and stability for vanadium flow batteries. *Energy & Environmental Science* 9: 441–47.

Yuan Z., X. Zhu, M. Li, W. Lu, X. Li and H. Zhang. 2016c. A highly ion-selective zeolite flake layer on porous membranes for flow battery applications. *Angewandte Chemie International Edition* 55: 3058–62.

Yue M., Y. Zhang and Y. Chen. 2011. Preparation and properties of sulfonated polyimide proton conductive membrane for vanadium redox flow battery. *Advanced Materials Research* 239: 2779–84.

Yue M., Y. Zhang and L. Wang. 2012. Sulfonated polyimide/chitosan composite membrane for vanadium redox flow battery: Influence of the infiltration time with chitosan solution. *Solid State Ionics* 217: 6–12.

Zeng J., C. Jiang, Y. Wang, J. Chen, S. Zhu, B. Zhao and R. Wang. 2008. Studies on polypyrrole modified Nafion membrane for vanadium redox flow battery. *Electrochemistry Communications* 10: 372–75.

Zhang B., E. Zhang, G. Wang, P. Yu, Q. Zhao and F. Yao. 2015. Poly(phenyl sulfone) anion exchange membranes with pyridinium groups for vanadium redox flow battery applications. *Journal of Power Sources* 282: 328–34.

Zhang D., Q. Wang, S. Peng, X. Yan, X. Wu and G. He. 2019. An interface-strengthened cross-linked graphene oxide/Nafion 212 composite membrane for vanadium flow batteries. *Journal of Membrane Science* 587: 117189.

Zhang H., H. Zhang, X. Li, Z. Mai and J. Zhang. 2011. Nanofiltration (NF) membranes: The next generation separators for all vanadium redox flow batteries (VRBS)? *Energy & Environmental Science* 4: 1676–79.

Zhang H., H. Zhang, F. Zhang, X. Li, Y. Li and I. Vankelecom. 2013a. Advanced charged membranes with highly symmetric spongy structures for vanadium flow battery application. *Energy & Environmental Science* 6: 776–81.

Zhang L., L. Ling, M. Xiao, D. Han, S. Wang and Y. Meng. 2017. Effectively suppressing vanadium permeation in vanadium redox flow battery application with modified Nafion membrane with nacre-like nanoarchitectures. *Journal of Power Sources* 352: 111–17.

Zhang S., B. Zhang, D. Xing and X. Jian. 2013b. Poly(phthalazinone ether ketone ketone) anion exchange membranes with pyridinium as ion exchange groups for vanadium redox flow battery applications. *Journal of Materials Chemistry A* 1: 12246–54.

Zhang Y., J. Li, L. Wang and S. Zhang. 2014. Sulfonated polyimide/alooh composite membranes with decreased vanadium permeability and increased stability for vanadium redox flow battery. *Journal of Solid State Electrochemistry* 18: 3479–90.

Zhang Y., H. Wang, W. Yu and H. Shi. 2018a. Structure and properties of sulfonated poly(ether ether ketone) hybrid membrane with polyaniline-chains-modified graphene oxide and its application for vanadium redox flow battery. *Chemistry Select* 3: 9249–58.

Zhang Y., H. Wang, W. Yu, J. Shi and H. Shi. 2018b. Sulfonated poly(ether ether ketone)-based hybrid membranes containing polydopamine-decorated multiwalled carbon nanotubes with acid-base pairs for all vanadium redox flow battery. *Journal of Membrane Science* 564: 916–25.

Zhang Y., S. Zhang, X. Huang, Y. Zhou, Y. Pu and H. Zhang. 2016. Synthesis and properties of branched sulfonated polyimides for membranes in vanadium redox flow battery application. *Electrochimica Acta* 210: 308–20.

Zhao Y., M. Li, Z. Yuan, X. Li, H. Zhang and I.F.J. Vankelecom. 2016. Advanced charged sponge-like membrane with ultrahigh stability and selectivity for vanadium flow batteries. *Advanced Functional Materials* 26: 210–18.

Zhou X., T. Zhao, L. An, L. Wei and C. Zhang. 2015. The use of polybenzimidazole membranes in vanadium redox flow batteries leading to increased coulombic efficiency and cycling performance. *Electrochimica Acta* 153: 492–98.

7 Thin-Film Solid Oxide Fuel Cells

Jong Dae Baek
Yeungnam University

Ikwhang Chang
Wonkwang University

Pei-Chen Su
Nanyang Technological University

CONTENTS

List of Acronyms 240
7.1 Introduction 241
    7.1.1 Brief Introduction to SOFCs 243
    7.1.2 Fuel Cell Losses 244
7.2 TF-SOFCs 245
    7.2.1 Definition of TF-SOFCs 245
    7.2.2 Development of TF-SOFCs 246
        7.2.2.1 Silicon-Based TF-SOFCs 246
        7.2.2.2 Porous Substrates with Nanoscale Surface Pores 247
    7.2.3 Scaling Up and System Level Work 248
7.3 Nano Thin-Film Electrolyte Materials and Deposition Methods 249
    7.3.1 Electrolyte 249
        7.3.1.1 Oxygen Ion Conductor for TF-SOFCs 250
        7.3.1.2 Proton Conducting Electrolyte for TF-SOFCs 251
    7.3.2 Cathode and Cathode Interlayer in TF-SOFCs 252
    7.3.3 Anode 252
    7.3.4 Deposition Method for TF-SOFC Components 253
        7.3.4.1 Electrolyte 253
7.4 Supporting Substrates 255
    7.4.1 Free-Standing MEA Configuration 255
        7.4.1.1 Silicon Wafers 255
        7.4.1.2 Glass Ceramic 257
    7.4.2 Porous-Substrate-Supported MEA 258
        7.4.2.1 Anodic Aluminum Oxide 258
        7.4.2.2 Conventional Ni–YSZ Supporting Substrate 258
        7.4.2.3 Other Porous Substrates 259
    7.4.3 Scaling-Up of the Single Cells on Membrane 260

      7.4.3.1 Enlarging the Lateral Dimension of Free-
              Standing Membranes 263
      7.4.3.2 3D Corrugated Electrolyte Membrane 267
      7.4.3.3 Large Membrane Arrays with Mechanical
              Supporting Layer 269
7.5   Issues of Thin-Film Stress..272
    7.5.1   Residual Stress...272
    7.5.2   Thermal Stress...273
    7.5.3   Membrane Buckling 274
7.6   Conclusions and Perspectives 276
References...277

LIST OF ACRONYMS

AAO	Anodic Aluminum Oxide
AFC	Alkaline Fuel Cell
ALD	Atomic Layer Deposition
ASR	Area-Specific Resistance
BS	Black Silicon
BYZ	Yttria-doped Barium Zirconate
BCY	Yttria-doped Barium Cerate
CVD	Chemical Vapor Deposition
DC	Direct Current
DRIE	Deep Reactive Ion Etching
EDX	Energy Dispersive X-Ray Spectroscopy
EIS	Electrochemical Impedance Spectroscopy
FCC	Face-Centered Cubic
FESEM	Field Emission Scanning Electron Microscopy
GDC	Gadolinium-Doped Ceria
HF	Hydrofluoric acid
HOR	Hydrogen Oxidation Reaction
KOH	Potassium hydroxide
LPCVD	Low-Pressure Chemical Vapor Deposition
MCFC	Molten Carbonate Fuel Cell
MEMS	Microelectromechanical system
MEA	Membrane Electrode Assembly
MIEC	Mixed Ionic–Electronic Conductor
MTS	Microthermal Stage
NSL	Nanosphere Lithography

OCV	Open Circuit Voltage
OM	Optical Microscope
ORR	Oxygen Reduction Reaction
PAFC	Phosphoric Acid Fuel Cell
PEMFC	Polymer Electrolyte Membrane Fuel Cell
PEN	Positive electrode-Electrolyte-Negative electrode
PLD	Pulsed Laser Deposition
PVD	Physical Vapor Deposition
RF	Radio Frequency
SDC	Samarium-Doped Ceria
SiN	Silicon Nitride
SOFC	Solid Oxide Fuel Cell
TF-SOFC	Thin-Film Solid Oxide Fuel Cell
LT-SOFC	Low-Temperature Solid Oxide Fuel Cell
IT-SOFC	Intermediate-Temperature Solid Oxide Fuel Cell
HT-SOFC	High-Temperature Solid Oxide Fuel Cell
TEC	Thermal Expansion Coefficient
TPB	Triple Phase Boundary
XPS	X-Ray Photoelectron Spectroscopy
XRD	X-Ray Diffraction
YSZ	Yttria-Stabilized Zirconia

7.1 INTRODUCTION

Solid oxide fuel cells (SOFCs) are efficient electrochemical devices that directly convert the chemical energy into electrical energy, and continuous electricity can be produced as long as fuel is consistently fed. The fuel is supplied into the anode side where it undergoes oxidation process, and the released electrons are transferred to an external circuit. Most SOFC systems do not require an external reformer and can directly utilize hydrocarbon fuels because of their high operating temperature, typically close to 1,000°C.

Fuel cell researches have been active since the first invention of hydrogen fuel cell by William Grove using sulfuric acid solution and zinc and platinum electrodes in 1839. Following Grove, Mond and Langer in 1889 reported fuel cell performance enhancement by considering a porous electrode structure which became the electrode structure of modern fuel cell (Mond and Langer 1890). After that, the use of fuel cell continued to create new milestones, and most scientists attempted to develop fuel cells with various fuels and electrolytes throughout the remainder of the century.

Of many types of fuel cells, SOFC was originally invented to replace a commercial light source by using solid ion conductor which is made of 85% zirconia and 15%

yttria. But it disappeared after tungsten lamp was introduced in 1905. In 1937, Baur and Preis first operated SOFCs at 1,000°C using coke as the fuel and magnetite as the oxidant with the knowledge of previous research works regarding SOFCs (Baur and Preis 1937). Until the 1960s, fundamentals of SOFC in empirical phase were developed, and after 1960, the number of patents regarding SOFC technology rapidly increased. The first paper "A solid electrolyte fuel cell" was published in English by Weissbart and Ruka in 1962 (Weissbart and Ruka 1962). They used $Zr_{0.85}Ca_{0.15}O_{1.85}$ as an oxygen ion conductor and revealed that the cell output is essentially limited by the resistance of the electrolyte and incompact electrolyte structure. A remarkable step forward from 1970 was the development of the EVD (electrochemical vapor deposition) method for perfect closing of pores of the solid electrolyte (Feduska and Isenberg 1983). Conventional SOFC systems have been researched for more than a couple of decades. In 1999, the Solid State Energy Conversion Alliance (SECA) by US Department of Energy initiated the development of SOFC stacks and systems for quick commercialization at low cost. From 2003, US government started the hydrogen fuel initiative program to develop hydrogen fuel cell vehicles and infrastructures for commercialization.

Yttria-stabilized zirconia (YSZ), which is an oxygen ion conductor, is the most typical electrolyte material for SOFCs. Typical operating temperatures for current SOFCs are between 800°C and 1,000°C. A high operating temperature is necessary in order to activate the ion transportation process across the electrolyte, as well as for higher electrochemical reaction kinetics on electrodes. Therefore, researches on how to lower operating temperatures without sacrificing high fuel cell performance have been active. Reducing operating temperature has a great potential in minimizing interfacial diffusion between electrode and electrolyte, simplifying integration of components, alleviating material degradation, and improving thermal cycling capabilities. Most importantly, it offers an opportunity for SOFCs to be a mobile power source rather than only a stationary power plant. However, lowering operating temperature results in inevitable performance drop from both the slow ionic transportation and sluggish electrode reaction processes. Since the ionic transportation across the electrolyte is a process following Ohm's law, the resistance is proportional to the distance the ion travels. It is therefore intuitive to minimize the electrolyte thickness in order to have lower ohmic resistance.

As such, using a thinner electrolyte for lower ohmic resistance and therefore lower operating temperature has been a trend for the past two decades in SOFC research. The thickness of the typical YSZ electrolyte ranges from micrometers using powder-based processes to only 10 nm using vacuum-based thin-film deposition (Baek, Liu, and Su 2017). The cell configuration also varies from the "electrolyte support," having a thick YSZ as the main mechanical support of the entire cell, to the "free-standing configuration without support," having a completely free-hanging membrane. Instead of using conventional powder or slurry followed by sintering process, the thin membranes are fabricated using thin-film deposition techniques, so that a very thin film can be fabricated. The challenge will be to ensure the membrane is completely gas-tight to prevent gas leakage and pinhole-free to prevent current leakage. In addition, scaling up the cells with nanoscale thin-film membranes is another major challenge, and several solutions suggesting new cell configurations instead of conventional SOFCs have been published.

This chapter will discuss the configurations that have been used for TF-SOFCs and review the relevant literature to explore their pros and cons. The context concerns only SOFCs using thin-film electrolytes of sub-micrometer thickness and the supporting structure specifically designed for it. The typical "thin-film electrolyte" referred to in many literatures that has thickness in the range of tens of micrometers is not within the scope of the discussion here. Specifically, this chapter will review the following:

1. Overview of TF-SOFCs operating below 500°C including their fabrication methods, cell configurations, electrochemical performance, and technical issues.
2. State-of-the-art cell configurations to improve fuel cell performances and membrane stability against thermo-mechanical stresses.
3. Methodology of scaling up the membrane electrode assembly (MEA) of TF-SOFCs to achieve higher total power output within a confined reaction area.

7.1.1 BRIEF INTRODUCTION TO SOFCs

Figure 7.1 schematically shows the basic working principles of SOFCs with different ion conducting electrolytes. The three main components of an SOFC include a cathode, an electrolyte as an oxygen ion (O^{2-}) conductor (Figure 7.1a) or a proton (H^+) conductor (Figure 7.1b), and an anode. Electrochemical reaction occurs between reactants (gaseous fuels such as H_2 and hydrocarbons) and oxidants (air). The electrolyte layer should be dense and gas-tight without pinholes to avoid electrical short circuit and with high fracture toughness to maintain a robust solid electrolyte membrane. The electrode layers need to be highly porous structures to facilitate efficient gas diffusion.

In an oxygen-ion-conducting SOFC, oxygen is reduced at the cathode to oxygen ions and then transported through the electrolyte, reacts with the hydrogen at the anode, and finally forms water and releases electricity. Oppositely, for a

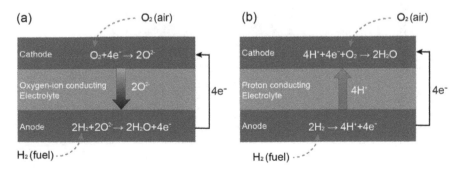

FIGURE 7.1 Schematic diagrams of SOFCs with the associated electrochemical reactions: (a) oxygen-ion conductor. (b) Proton conductor.

proton-conducting SOFC, protons migrate through the electrolyte from the anode to the cathode side, where they react with oxygen ions to form water (H_2O). A fully dense and solid-oxide electrolyte layer is spatially sandwiched between the nanoporous cathode and anode, and electrons are directly released to the external circuit by electrochemical reaction from fuels.

7.1.2 FUEL CELL LOSSES

The performance of an SOFC can be characterized by its current–voltage (I–V) curve in Figure 17.2, showing the voltage output of the fuel cell with respect to a given output. The fuel cell performance is typically evaluated with this curve, and the maximum ideal voltage can be directly determined by thermodynamics and can be obtained when the fuel cell is operated under the thermodynamically reversible condition. However, the actual voltage of a real fuel cell is always less than the thermodynamically estimated voltage. As the current is drawn from the fuel cell, the output voltage is immediately dropped from the reversible cell voltage. This voltage drop characterizes the irreversible losses in a fuel cell operation, and the more the current drawn, the greater these losses.

As mentioned above, three major types of fuel cell losses, which give a fuel cell I–V curve its characteristic shape can be defined as follows:

- **Activation losses**: η_{act} (losses due to electrochemical reaction)
- **Ohmic losses**: η_{Ohmic} (losses due to ionic and electronic conduction)
- **Concentration losses**: η_{conc} (losses due to mass transport).

Therefore, the actual voltage output for a fuel cell is expressed as the ideal voltage subtracted by the three main losses:

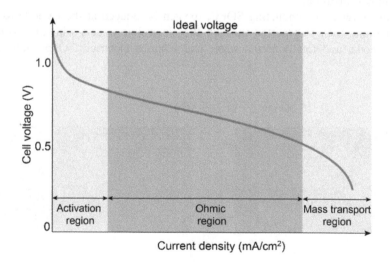

FIGURE 7.2 Schematic of current–voltage characteristics of an SOFC. The voltage drop results from three major losses, which are activation loss, ohmic loss, and concentration loss.

$$V = V_{\text{thermo}} - \eta_{\text{act}} - \eta_{\text{Ohmic}} - \eta_{\text{conc}} \tag{7.1}$$

where V_{thermo} represents open circuit voltage (OCV) determined by thermodynamics. In terms of electrochemical reaction in SOFCs, cathode kinetics for O_2 reduction is significantly slower than anode kinetics for H_2 oxidation (Adler 2004). Therefore, the activation loss and ohmic loss are critical losses that affect performance of low-temperature solid oxide fuel cells (LT-SOFCs), and especially, ohmic loss is the most obvious source of loss in TF-SOFCs.

The ohmic loss is mainly due to the resistance of ionic charge transport through the electrolyte layer. Since the electric conductivity is significantly higher than ionic conductivity, ohmic loss mostly results from ionic transportation inside electrolyte membrane. They are simply governed by Ohm's law:

$$V = i \cdot R = i \cdot \left(\frac{L}{A\sigma} \right) \tag{7.2}$$

where A is the electrochemically active area, L is the length of the ionic transport path (namely electrolyte thickness), and σ is the ionic conductivity of the electrolyte material. The voltage V represents the voltage, which must be applied to transport a charge at a rate given by i.

Since fuel cells are generally compared on a per-unit-area basis using current density instead of current, area-normalized fuel cell resistance, which is area-specific resistance (ASR) with units of $\Omega \cdot \text{cm}^2$ is reasonable. By using current density, $j = i \cdot A^{-1}$ and ASR, ohmic losses are expressed as

$$\eta_{\text{ohmic}} = j \cdot \text{ASR} = j \cdot \left(\frac{L}{\sigma} \right) \tag{7.3}$$

Based on Eq. 7.3, ohmic losses can be decreased either by reducing the electrolyte thickness (a dimensional property) or by using an electrolyte with higher ionic conductivity (a material property). With the help of MEMS-fabrication processes and vacuum-based thin-film deposition methods, the electrolyte thickness can be minimized to sub-micrometer scale. As shown in Figure 7.3, the electrolyte thickness of a typical TF-SOFC fabricated using powder process (Figure 7.3a) is around 50 μm which is rather a thick film, while the MEMS-based TF-SOFC using electrolyte deposited by atomic layer deposition (ALD) is only 70 nm in thickness (Figure 7.3b). This drastic decrease in thickness allows the operating temperature of the cell in Figure 7.3b to be below 500°C with decent output power density.

7.2 TF-SOFCs

7.2.1 Definition of TF-SOFCs

The electrolyte thicknesses in most works developed for TF-SOFCs are typically in the range of tens of micrometers. The electrolytes are either oxygen ion conductors or proton conductors and are usually fabricated with powder-based processes with

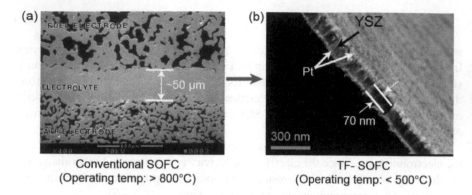

FIGURE 7.3 Electrolyte thickness reduction. (a) Conventional SOFC with electrolyte thickness of approximately 50 μm operating above 800°C. (Reprinted with the permission from Singhal (2000) © 2000 Elsevier Science B.V.) (b) TF-SOFC with electrolyte thickness of 70 nm operating below 500°C. (Reprinted with the permission from Su et al. (2008) © 2008 American Chemical Society.)

high-temperature sintering. The supporting structure is usually a porous anode substrate. The operating temperature for such TF-SOFCs reported are between 550°C to 750°C.

In this chapter, we focus the discussion on TF-SOFCs operated at low temperature below 500°C having sub-micrometer-thick electrolyte films. As the realization of such TF-SOFCs is largely reported on unconventional cell configurations such as using silicon wafer supports or anodic aluminum oxide (AAO) supports, they are still in the development stage, and the real product has not been realized to date. In most of the literature reporting the TF-SOFCs in this category, it is more often that the term "micro-SOFCs" is used due to their miniature cell size and the sub-micrometer MEA thickness.

7.2.2 DEVELOPMENT OF TF-SOFCs

7.2.2.1 Silicon-Based TF-SOFCs

Initially, the development of a nanoscale thin-film electrolyte was for the study of nanoionics within a typical oxygen ionic conductor by shifting bulk electrolyte to thin-film electrolytes. YSZ was the first material fabricated as a free-standing electrolyte for such purpose (Bruschi et al. 1999) since its bulk property is most well-studied and understood for comparison of nanoscale behavior. The work by Bruschi et al. was the first to demonstrate the use of silicon bulk micromachining techniques to fabricate a free-standing YSZ electrolyte. Silicon wafer has superior surface smoothness, and therefore using silicon as a supporting substrate would allow synthesis of ultra-thin YSZ with minimum occurrence of pinholes. YSZ was deposited with thickness of 300 nm by RF sputtering on <100> silicon wafers, where a silicon dioxide layer was first thermally grown on the wafer surface as an etch stop for the subsequent through-wafer etching in ethylenediamine-pyrocatechol (EDP) solution at 115°C. The resulting free-standing YSZ membrane has a lateral dimension of

170 μm. In 2004, Baertsch et al. demonstrated the successful fabrication of such free-standing electrolytes that further shrunk the thickness of YSZ and gadolinia-doped ceria (GDC) to only 100 nm, with a purpose to study membrane stress and geometric design criteria for micro-SOFCs (Baertsch et al. 2004). The work of using silicon micromachining processes to make YSZ membranes became popular after that. Since YSZ electrolyte film was usually deposited by RF sputtering with the thickness similar to the grain size, the film usually carries a high density of nanoscale features including high-density grain boundaries, nanovoids, interfaces, and dislocations, which trigger the study of nanoionics on such free-standing membranes (Tuller 2000). Despite the consistent success in YSZ membrane fabrication and surface potential study, it was only until 2006 that the fuel cell performance was successfully measured with high OCV and impressive (Huang et al. 2007). Afterward, the high performance has triggered the interest in pursuing higher power density. Various MEAs of TF-SOFCs have been competitively reported with enhanced power densities. Especially, to enlarge the reaction area within confined space, Su et al. reported OCV of 1.1 V and power density of 861 mW cm^{-2} at 450°C with three-dimensional (3D) corrugated MEA structures (Su et al. 2008). Many researches were conducted to break the barrier of power density of 1,000 mW cm^{-2} with flat TF-SOFCs. Kerman et al. first recorded 1,037 mW cm^{-2} at 500°C with flat TF-SOFCs (Kerman, Lai, and Ramanathan 2011), and then An et al. reported 1,300 mW cm^{-2} at 450°C with nano-structured and bi-layered TF-SOFCs in 2013 (An et al. 2013).

7.2.2.2 Porous Substrates with Nanoscale Surface Pores

Another well-studied approach of TF-SOFC is the depositing of the thin-film electrolyte on a porous substrate with nanoscale surface pores. This allows the thin-film electrolyte to form a gas-tight film to avoid leakage of gas and pinholes for electrical leakage. AAO is the most popular substrate, and other substrates include porous nickel, stainless steel, and a multiscale supporter. The fabrication process of the porous substrate-based TF-SOFC is much simpler than that for other TF-SOFCs because it does not require additional processes of lithography and etching. The porous substrate-based TF-SOFC strongly depends on its physical characteristics such as pore diameter, surface roughness, and porosity. The AAO which is one of the commercial nanoporous templates has very uniform and highly well-defined pore structure (typically >100-μm thickness and 20–100 nm diameter) (Lee et al. 2018). It is significantly rigid to control for hands-on processes. The AAO is also a cost-effective product because its parameters can easily be controlled using an electrochemical method. Unfortunately, the TF-SOFC using AAO substrate is susceptible to the electrical short due to the inherent columnar defects within the electrolyte. Kwon et al. reported a void within YSZ electrolyte grown on the AAO template as shown in Figure 7.4 (Kwon et al. 2011). These voids can be easily observed between two columnar YSZ grown on the AAO template. As the columns within the YSZ electrolyte gradually merge, the thickness of the deposited thin film increases. It is believed that the electrolyte thickness should be optimized between the minimum thickness to avoid the electrical short and the thinner thickness to reduce ohmic polarization.

The iron-based substrate (i.e. stainless steel, STS) for metal-supported TF-SOFCs are also attractive due to their easy fabrication, high thermal/electrical conductivity,

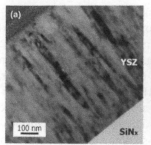
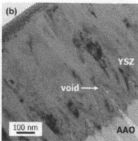

FIGURE 7.4 Cross-sectional images of 600-nm-thick YSZ films deposited by PLD on (a) a flat silicon nitride substrate and on a porous AAO substrate with 40-nm pores (b) and 60-nm pores (c). (Reprinted with the permission from Kwon et al. (2011) © 2011 WILEY-VCH Verlag GmbH & Co.)

and low material cost (Lee et al. 2018). As STS is more resistive to mechanical impacts and thermal shocks, the metal-supported TF-SOFCs are suitable for various power applications. Kim et al. demonstrated a TF-SOFC using a STS-based substrate and prepared using a tape-casting process (Kim et al. 2016). A dual-layer substrate of a buffer layer and an STS substrate are prepared using conventional tape-casting and co-firing processes. Noh et al. reported an electrode consisting of nanoscale grains near the TPB fabricated using a thin-film process (pulsed laser deposition, PLD) and a current collector layer with centimeter-scale grains grown by controlling background pressure during an electrode deposition (Noh et al. 2009). Therefore, the lateral dimension varies from nanometer to centimeter, which is called "multiscale architectured TF-SOFC." The peak power density and thermal cycle were 1,400 mW cm^{-2} at 600°C and 50 tests, respectively, between 400°C and 600°C.

7.2.3 Scaling Up and System Level Work

For the unconventional supporting substrates like silicon wafers or AAOs, new methods to scale up the single cell are necessary for further system development. As one of the modified free-standing TF-SOFCs, Tsuchiya et al. developed the 25 mm^2 MEA by depositing metallic grids to enhance mechanical robustness (Tsuchiya, Lai, and Ramanathan 2011). A power density of 155 mW cm^{-2} and a total power output of ~20 mW from a single cell were measured at 510°C. Chang et al. fabricated AAO-based TF-SOFC of Pt (anode)/YSZ-GDC-YSZ/Pt (cathode) using sputtering and ALD (Chang et al. 2016). The active area of a single cell was 2.56 cm^2. The maximum absolute power at 500°C was measured to be 44 mW. Also, they successfully demonstrated that the combined system with a TF-SOFC and a catalytic burner could be heated up from room temperature to 319°C without any other initial heating sources. Scherrer et al. demonstrated the assembly consisting of a free-standing SOFC and a microreactor is thermally stable and can be operated above 470°C by internal refining of butane (Scherrer et al. 2014). The electrochemical performance of their system was indicated by an OCV of 1.0 V and a maximum power density of 47 mW cm^{-2} at 565°C.

7.3 NANO THIN-FILM ELECTROLYTE MATERIALS AND DEPOSITION METHODS

7.3.1 ELECTROLYTE

Currently, YSZ is the most favorably used electrolyte material for TF-SOFC due to its superior chemical stability in both oxidizing and reducing environments. Despite the good chemical stability, YSZ has low ionic conductivity, about 0.02 S cm^{-1} at 800°C and 0.1 S cm at 1,000°C. This is one of the main reasons that SOFCs have such high operating temperatures. To make YSZ suitable for intermediate (500°C–700°C) and low (300°C–500°C) operating temperature regimes, there have been efforts to decrease the electrolyte thickness because ohmic loss is a function of ion-conducting membrane thickness. Especially, Y-doped BaZrO$_3$ (BYZ) has been one of the most promising proton-conducting electrolyte materials for its high proton conductivity and excellent chemical stability (Iwahara et al. 1993). The ion conductivities of various electrolytes which are representatively used in thin films are compared as shown in Figure 7.5. Various alternative electrolytes have been investigated to date, among which aliovalent-doped ceria and isovalent-cation-stabilized bismuth oxides are especially attractive because of their superior ionic conductivity at lower temperatures (Wachsman and Lee 2011).

One of the major requirements for the electrolyte is higher conductivity for oxygen ions over a wide range of O$_2$ partial pressures without electronic conductivity. Perhaps, the most conventional materials that conduct oxygen ions fast have crystal structures of fluorite type AO$_2$, where A is a tetravalent cation (Malavasi, Fisher, and Islam 2010). Another requirement is chemical stability because the electrolytes are

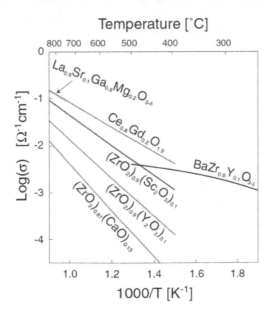

FIGURE 7.5 Comparison of ionic conductivity of various solid oxide electrolytes. (Reprinted with the permission from (Haile 2003). © Elsevier Science Ltd 2003.)

exposed to strongly oxidizing and reducing environments. In addition, mechanical stability of the membrane over high operating temperatures is required.

7.3.1.1 Oxygen Ion Conductor for TF-SOFCs

The most standard oxygen ion conductor in SOFCs is YSZ as mentioned above. YSZ is created by doping ZrO_2 with the acceptor Y_2O_3. ZrO_2 itself is generally not considered as a good electrolyte for SOFCs due to its low ionic conductivity. ZrO_2 has a crystal structure that varies with temperature and pressure. In its bulk form, ZrO_2 is typically stable in a monoclinic structure from room temperature up to 1,170°C, then it becomes tetragonal from 1,170°C to 2,370°C, and finally changes to cubic fluorite at temperatures greater than 2,370°C (Ishihara, Sammes, and Yamamoto 2003). The cubic fluorite structure of ZrO_2 has a face-centered cubic (FCC) zirconia lattice, and a cubic oxygen lattice is placed in the FCC lattice. The Zr^{4+} cations fully occupy the tetrahedral sites, and the O^{2-} anions and Zr^{4+} cations occupy the octahedral sites as schematically shown in Figure 7.6.

The doping of Y_2O_3 is for stabilizing the ZrO_2 but more importantly to introduce oxygen vacancies in order to improve oxygen ion conductivity. When doping ZrO_2 with Y_2O_3, the Zr^{4+} cations in the ZrO_2 lattice are substituted by the Y^{3+} cations, and subsequently, oxygen vacancies are created to maintain charge neutrality in the lattice. The oxygen vacancies provide a chance for ionic transportation through the electrolyte by hopping from vacancy to vacancy. Increasing dopant concentration leads to increased vacancy concentration, causing improved ionic conductivity (Fergus 2006). However, all ceramic electrolyte materials have their own optimal dopant concentrations to maximize ionic conductivity because of the formation of dopant-oxygen vacancy clusters which can reduce the mobility of the oxygen vacancies as charge carriers (Kilner and Waters 1982). For YSZ, maximum ionic conductivity is shown in 8–9 mol% Y_2O_3 doping (Pornprasertsuk et al. 2005).

Despite the expected low ionic conductivity of ZrO_2, when making the electrolyte with nanometer thickness, ion conduction was observed. TF-SOFC with pure ZrO_2 electrolyte is discussed in detail in Ko, Kerman, and Ramanathan (2012). ZrO_2 electrolytes of 50 nm thickness were grown with ultraviolet oxidation from precursor Zr films at room temperature, and porous platinum electrodes were applied. The

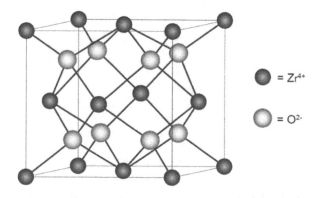

FIGURE 7.6 The cubic fluorite structure of ZrO_2.

maximum power density and open circuit voltage were measured to be 33 mW cm^{-2} and 0.91 V at 450°C, respectively.

Another popular oxygen ion conducting electrolyte material is doped ceria (CeO_2). Typical dopants include gadolinia (Gd_2O_3) and yttria (Y_2O_3), producing GDC and yttria-doped ceria (YDC), respectively. These doped ceria have much higher ionic conductivities than YSZ, especially at lower temperature regimes and lower activation energies. Kharton et al. reported that 10 mol% doped GDC ($Ce_{0.9}Gd_{0.1}O_{1.95}$) has an ionic conductivity of 0.01 S cm^{-1} at 500°C (Kharton, Marques, and Atkinson 2004). This is why the doped ceria materials are much attractive for use in lower operating temperature SOFC electrolytes.

However, doped ceria materials have a major disadvantage in SOFC electrolyte applications. At high temperatures above 600°C and low oxygen partial pressure environment, the conductivity is not purely ionic. Under reducing conditions, Ce^{4+} partially reduces to Ce^{3+}, which may lead to internal electronic shortening and also induces membrane instability due to volumetric change by increase in Ce radius. As the operating temperature is increased, this issue increases. Therefore, single-layer GDC is hardly used as an electrolyte film for TF-SOFCs. In response to this stability issue of doped ceria materials in a reducing environment, composite electrolytes having a ceria-based layer and a zirconia-based layer (YDC/YSZ, GDC/YSZ) have been investigated. Huang et al. first introduced 50-nm-thick GDC as an interlayer on YSZ electrolyte of TF-SOFCs and reported 400 mW cm^{-2} with OCV of 1.1 V at 400°C (Huang et al. 2007). Afterward, Kerman et al. considered optimally composite GDC–YSZ electrolyte in TF-SOFCs and achieved 665 mW cm^{-2} with OCV of 0.85 V at 520°C (Kerman, Lai, and Ramanathan 2012b).

7.3.1.2 Proton Conducting Electrolyte for TF-SOFCs

Although the current SOFC technology based on oxide ion conducting electrolytes is well established, it is still problematic with material incompatibility, low tolerance regarding the operation conditions, and fuel efficiencies. Proton transport in several members of acceptor-doped perovskites of the general formula ABO_3 is reported to be fast, with activation energies of about 0.45 eV, which makes them interesting as potential solid electrolytes for next generation protonic devices such as fuel cells, hydrogen sensors, and gas refiners (Bonano, Ellis, and Mahmood 1991).

Initially, the proton conductivity in doped perovskites was investigated with alkaline earth cerates and zirconates where the trivalent Ce^{4+} sites were partially substituted by trivalent dopant cations such as Y^{3+}, Gd^{3+}, or Nb^{3+}. Despite the high ionic conductivity of doped cerate perovskites, they suffer from significant chemical instability in CO_2 environments. Kreuer reported that $SrCeO_3$ and $BaCeO_3$ easily turn into $SrCO_3$ and $BaCO_3$ through reaction, even with a small amount of CO_2 (Kreuer 2003). Also, it is reported that $BaCeO_3$ easily decomposes into $Ba(OH)_2$ in the presence of water. In contrast, especially Y-doped $BaZrO_3$ (BYZ) has been one of the most promising proton conducting electrolyte materials for their high proton conductivity and excellent chemical stability (Iwahara et al. 1993).

Proton conducting electrolyte thin films were considered as TF-SOFCs only after the year of 2010. In 2010, Pergolesi et al. first demonstrated the usefulness of BYZ electrolyte on TF-SOFCs by depositing the electrolyte using PLD (Pergolesi et al. 2010). Afterward, free-standing BYZ thin-film electrolytes fabricated as cup-shaped arrays and as crater-patterned film were considered as working TF-SOFCs (Kim et al. 2011, Su and Prinz 2011). Bae et al. also demonstrated a bilayer electrolyte using the chemically more stable BYZ capping on BCY electrolyte as a working TF-SOFC with fuel cell performance (peak power density of 145 mW cm^{-2}, OCV of 0.98 V at 400°C). The work also observed the OCV stability over time and significant degradation in power density and drop in OCV, which was caused by the formation of barium carbonate (Bae et al. 2014). Li et al. subsequently addressed the issue of expansion of lattice parameter of BCY electrolyte and confirmed the formation of barium carbonate during fuel cell reaction, which caused poor fuel cell performance and induced cracks within the 300 nm thickness membrane (Li et al. 2014). In the same work, Li replaced unstable BCY with BZCY (ceria and yttria-doped barium zirconate) later.

7.3.2 CATHODE AND CATHODE INTERLAYER IN TF-SOFCs

For SOFCs operating at very low temperatures, the major limiting factor to govern overall cell performance is the polarization resistance on the cathode from the complex oxygen reduction reaction (ORR). The exact mechanism of ORR on metal surfaces still remains unclear, and rate-limiting reaction is not always predictable and depends on local conditions such as temperature, oxygen partial pressure, or micro-/nanostructures (Sun, Hui, and Roller 2009). Nevertheless, it is generally agreeable that for TF-SOFCs operating below 550°C, improving cathode reactions kinetics is the most critical in order to obtain high cell performance.

The search for high-performance cathodes for TF-SOFCs is not active, and Pt remains the most common material for TF-SOFCs. Porous Pt thin films with thickness around 100 nm deposited using sputtering methods are typically seen as a facile choice for both cathodes and anodes in most published works. The Pt thin film is made porous by controlling Ar base pressure in RF or DC sputtering chambers to obtain sufficient triple phase boundaries (TPBs), where gas, ion, and electron meet in specific spots. However, the sputtered Pt particles are thermally unstable, and morphological degradation of Pt is significant as soon as current starts to flow. Several works have put efforts to stabilize or delay the agglomeration of sputtered Pt film by considering capping layers on Pt surface. (Liu et al. 2018, Lee et al. 2016, Liu et al. 2015, Chang et al. 2015).

On the other hand, the interfacial kinetics between cathode and electrolyte are also known to be critical for improving fuel cell performance at low operating temperatures. It can be achieved by improving the oxygen ion incorporation into the electrolyte surface with a cathode interlayer. The effect of the surface exchange on the cathode performance has been well highlighted by Adler (2004) and Steele (1995).

7.3.3 ANODE

Anode should be a porous structure for effective gas exchange at the anode/electrolyte interface to reduce mass loss in transport and should be catalytically active for fuel oxidation. In addition, anode should not react with substrate. There are several criteria which must be met to function efficiently: (1) high electronic and ionic conductivity, (2) catalytic activity for fuel oxidation reaction (i.e. hydrogen oxidation reaction, HOR), and (3) chemical and thermal stability under reducing environment.

Noble metal catalysts, such as Pt, Pd, and Ru, have been utilized for high electronic conductivity and low polarization resistance for operation at <500°C (Jeong et al. 2015, Li et al. 2017). Kang et al. showed that the TF-SOFC using a porous Pt layer on the Pd anode is superior to that using pure Pt (Kang et al. 2011). It implies that the optimization of the Pt/Pd anode leads to the improvement of HORs. Recent researches of the anode are the direct operation of hydrocarbon fuels such as methane and butane. Takagi et al. reported that only Ru anode is utilized for direct methane oxidation at the operating temperature of 500°C (Takagi et al. 2013). The electrochemical outputs of 0.71V (OCV) and 450 mW cm^{-2} (peak power density) were measured without noticeable carbon decomposition. Also, this group reported that Ru–GDC composite thin films are utilized for direct methane refining in the TF-SOFC through a co-sputtering method (Takagi, Adam, and Ramanathan 2012). They reported that the peak power density of 275 mW cm^{-2} was measured at 485°C.

7.3.4 Deposition Method for TF-SOFC Components

7.3.4.1 Electrolyte

The most commonly used thin-film deposition methods for fabrication of TF-SOFC electrolytes and electrodes are vacuum-based processes, including RF or DC sputtering, PLD, and ALD. The requirement of having the electrolyte to be gas-tight and pinhole-free under sub-micrometer thickness makes it very challenging to adopt conventional powder-based methods (spin coating, spray pyrolysis, tape casting, sol-gel process, etc.) to fabricate the thin-film electrolyte.

Sputtering system as shown in Figure 7.7a can be classified as direct current (DC) sputtering and radio frequency (RF) sputtering depending on the power supply. In DC sputtering, 3–5 kV of voltage is biased to the target, while the chamber and the substrate are grounded. To induce ionization and plasma, Ar gas is supplied into the chamber, and the positive ions are moved toward the negatively charged target surface and sputtered on the substrate (Thiele et al. 1991). In case of a dielectric target, RF source is required. RF sputtering shares many characteristics of DC sputtering, with the major difference being the use of an RF power source at 13.56 MHz that applies an AC voltage between the electrodes (Smeacetto et al. 2010).

PLD is a physical vapor deposition (PVD) technique suitable for the preparation of oxide films in vacuum conditions and has gained considerable attention for the fabrication of TF-SOFCs due to excellent control of the crystallinity and morphology of the deposited material and preservation of the required stoichiometry (Pergolesi et al. 2010). A typical set-up for PLD is schematically shown in Figure 7.7b. In the PLD chamber, a target on which the material is to be deposited is ablated with a pulsed laser beam of wavelength in the range between 200 and 400 nm. The target

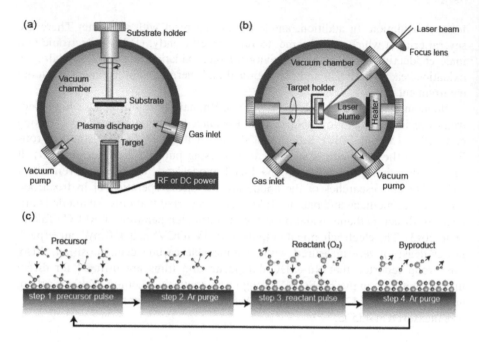

FIGURE 7.7 Schematic of thin-film growth techniques. (a) RF or DC sputtering. (b) PLD. (c) ALD.

is usually a small ceramic disc prepared by powder sintering. It should be dense to avoid droplet ejection from the target surface and have small grains to enhance the film stoichiometry. Excimer laser (Nd:YAG) is focused on the target at short pulse durations which transfers the heat energy to vaporize the target material to form the thin film. The deposition uniformity of the thin film depends on the spot size of the laser, plume temperature, and target-to-the-substrate distance (TSD). Reviews about detailed PLD mechanisms and feasible applications can be found elsewhere (Beckel et al. 2007, Ashfold et al. 2004, Krebs et al. 2003).

ALD is a versatile deposition technique which is able to meet the needs of atomic-level control of thickness and conformal deposition using self-limiting chemical reactions of gas-phase precursor molecules (George 2010).

Film growth by ALD is in a cyclic manner, and reactants are changed at each cycle. For this reason, a layer created during a cycle is different from one created during the next cycle since the chemical reaction on the surface at each cycle is different. The reactants are individually exposed to the substrate surface, allowing for a sequential layering process. ALD cycle generally consists of four steps as illustrated in Figure 7.7c: Step 1. Pulse of the first precursor into the process chamber and chemisorption of the precursors on the substrate. Step 2. Purge of the process chamber to remove reacted ligands and unreacted precursors. Step 3. Pulse of the second precursor followed by surface reaction to form the desired film layer. Step 4. Purge of by-products. The growth cycle is repeated to reach the desired film thickness. The growth rate in most cases is 0.1–3 Å/cycle (Shim et al. 2007).

7.4   SUPPORTING SUBSTRATES

TF-SOFCs can be classified into two different types according to its supporting substrate: free-standing type (non-supporting substrate) and porous-substrate type (supporting substrate), as illustrated in Figure 7.8. The substrate selection for supporting active membrane closely concerns the thermal and mechanical stability, chemical compatibility during fabrication, fuel cell operation, and electrochemical performance. For the free-standing MEA, the design considerations will be membrane residual stress, membrane lateral dimension, electrolyte deposition temperature, substrate surface finishing, and MEMS-process compatibility. For porous supporting substrates, the dimensions of surface pores will determine the minimum thickness of the MEA that can achieve a gas-tight membrane. Other considerations for porous supporting type include the deposition of conducting catalytic surface layer over the anode surface pore if the substrate is non-conductive and the reasonable deposition time of a gas-tight electrolyte film that is thick enough to fully cover the anode surface pores. Materials, geometry, and dimensions of TF-SOFCs reported to date are summarized in Table 7.1.

7.4.1   FREE-STANDING MEA CONFIGURATION

7.4.1.1   Silicon Wafers

Using silicon wafers as the supporting substrate has several advantages. First, the surface finishing of a standard silicon wafer is superior, which makes deposition of a fully dense and pinhole-free thin film of tens of nanometers thickness possible. In addition, it takes full advantage of using silicon bulk micromachining processes or MEMS process, which is compatible with most thin-film deposition techniques for electrolytes and electrolyte films. Subsequently, the scaling-up of the miniature single cell is straightforward utilizing the silicon foundry process as the process is based on photolithography and chemical etching, which can easily scale up with bulk batch processes.

Baertsch et al. first demonstrated the feasibility of TF-SOFC structure having free-standing membranes by silicon micromachining techniques (Baertsch et al. 2004). Even though no electrochemical results were reported, self-supporting thin-film electrolyte deposition and incorporation into a silicon processing platform were well demonstrated. Silicon nitride (Si_3N_4) was deposited on a silicon substrate which served as a wet etching mask layer, and after 100-nm-thick electrolyte deposition by e-beam

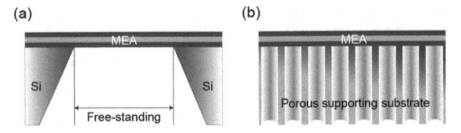

(a)   **(b)**

FIGURE 7.8   Cell configuration for TF-SOFCs: (a) Free-standing TF-SOFC structure. (b) Porous-substrate-supported TF-SOFC structure.

TABLE 7.1

Materials and Geometry of Substrate and MEA for TF-SOFCs

Geometry	Substrate	A*	E*	C*	A	E	C	Ref.
		Material			**Thickness (nm)**			
Square	Silicon	Pt	YSZ	Pt	10–30	100	80	(Jiang et al. 2008)
Square	Silicon	Pt	YSZ	Pt	80	50	80	(Huang et al.
				GDC–Pt	80	50	130	2007)
Square	Silicon	Pt	YSZ	Pt	120	70	120	(Su et al. 2008)
Square	Silicon	Pt	YSZ	Pt	60	80	60	(Chao et al. 2011)
Square	Silicon	Pt	YSZ	Pt	80	100	80	(Kerman, Lai, and Ramanathan 2011)
Circular	Silicon	Pt-Ni	YSZ	Pt	100/4–6 µm	750	50	(Rey-Mermet and Muralt 2008)
Square	Silicon	Pt	BCY	Pt	50	300	50	(Li et al. 2014)
Square	Foturan®	Pt	YSZ	Pt	100	600	100	(Tolke et al. 2012)
Square	Silicon	Pt	YSZ	La0.6Sr0.4 Co0.8 Fe0.2O3	30	54	47/1 µm	(Tsuchiya, Lai, and Ramanathan 2011)
Square	Silicon	Pt	YSZ	Pt LSCF	40–80	75–150	130 15–150	(Johnson et al. 2009)
Square	Silicon	LSCF	YSZ	LSCF	65	60	65	(Lai, Kerman, and Ramanathan 2011b)
Square	Silicon	Pt	YSZ	LSCF	120	60	67	(Lai, Kerman, and Ramanathan 2010)
Square	Silicon	Pt–Ni	YSZ	LSCF	80/5 µm	80	20	(Johnson et al. 2010)
Square	Silicon	Pd	YSZ	Pt	50	100	70	(Lai, Kerman, and Ramanathan 2011a)
Square	Silicon	Pt	YSZ	Pt–YSZ	50–60	100	100	(Yan et al. 2012)
Square	Silicon	Pt	YSZ	Pt	80	60	80	(Shim et al. 2007)
Square	AAO	Pt	YSZ	Pt	300–380	240	200	(Ha, Su, and Cha 2013)
Square	Silicon	Pt	BYZ	Pt	70	120	70	(Kim et al. 2011)
Square	Silicon	Pt	BYZ	Pt	120	300	120	(Su and Prinz 2011)
Hexagon	Silicon	LSC	YSZ	LSC	350	500	350	(Garbayo et al. 2014)
Square	Silicon	Ru	YSZ	Pt	50	110	70	(Takagi et al. 2013)
Square	AAO	Pt	BYZ	Pt	300	930–1,340	200	(Park et al. 2013)

(Continued)

TABLE 7.1 (*Continued*)
Materials and Geometry of Substrate and MEA for TF-SOFCs

Geometry	Substrate	Material			Thickness (nm)			Ref.
		A*	E*	C*	A	E	C	
Square	AAO	Pt	BYZ	Pt	300	900	200	(Ha et al. 2013)
Square	AAO	Pt	BYZ	Pt	350	1,000	200	(Chang, Heo, and Cha 2013)
Square	AAO	Pt	YSZ	Pt	320	70	150	(Ji et al. 2015)
Square	AAO	Pt	YSZ	Pt	80	300–900	80	(Kwon et al. 2011)
Circular	Porous STS	NiO–YSZ	YSZ	LSC	600	2,000	700	(Kim et al. 2016)
Square	Silicon	Pt	YDC–YSZ	Pt	80	10/50	30–80	(An et al. 2013)
Square	Silicon	Pt	YSZ	Pt	120	70	120	(Su and Prinz 2012)
Square	Silicon	Pt	ZrO_2	Pt	40	50	55	(Ko, Kerman, and Ramanathan 2012)
Square	Silicon	Pt	YDC–YSZ	Pt	80	70	80	(Fan et al. 2012)
Square	Silicon	LSC	YSZ	Pt	200	300	80	(Evans et al. 2015)

*A, *B, *C represent anode, electrolyte, and cathode, respectively.

evaporation, Si_3N_4 underneath electrolyte layer was removed by using plasma etching to release free-standing YSZ and GDC electrolyte membranes with lateral dimension as large as 1mm. The silicon nitride passivation layer on (100) silicon subsequently became a standard process for making free-standing TF-SOFC MEAs.

7.4.1.2 Glass Ceramic

Photo-structurable glass ceramic was once a choice for a TF-SOFC substrate. Gauckler group in ETH Zürich first demonstrated TF-SOFCs using silica-based glass ceramic which can be etched by hydrofluoric acid (HF) (Muecke et al. 2008). Double polished glass wafers (Foturan®, Mikroglas, Germany) were used, and a selectively crystalized region by UV-exposure and annealing can be etched by hydrofluoric acid (HF) to produce a free-standing membrane. The main advantage of glass ceramic is its thermal expansion coefficient (TEC) (at 20°C) (Tolke et al. 2012) that matches that of ceramic electrolyte like YSZ (at 20°C) as well as the potentially rapid fabrication process that generates free-standing membranes (Bieberle-Hutter et al. 2008). However, for larger aspect ratios, the etching rates are very different due to the mass transport limitation of fresh etchant to the etch front. Also, pinhole issues on the electrolyte membrane by HF etching should be solved as the HF is also an effective etchant to YSZ.

7.4.2 Porous-Substrate-Supported MEA

7.4.2.1 Anodic Aluminum Oxide

AAO porous filter substrate has been a popular alternative to free-standing MEA (Wachsman and Lee 2011). The fabrication procedure of porous-substrate-supporting SOFCs does not involve the photolithography patterning process; they can be fabricated by only few sequential depositions of electrodes and electrolytes. However, porous-substrate-supporting SOFCs have inherent pinholes owing to the highly rough surface of the porous substrate. Therefore, completely pinhole-free deposition on AAO substrate is the key to success.

Kwon et al. used patterned AAO as a template to obtain regular gas channels with pore size of 40 nm (Kwon et al. 2011). Anode, electrolyte, and cathode were deposited on the porous structure, and the size of free-standing membrane is $1,000 \times 1,000\,\mu m$. The cell operated at 500°C without significant structural degradation. Ha et al. used a dense 390-nm-thick BYZ electrolyte on an AAO substrate by PLD and tuned the AAO pore size by additional conformal Pt coating by ALD and thereby minimized pinholes as shown in Figure 7.9 (Ha et al. 2013). From BYZ fuel cell supported by 10 mm × 10 mm square AAO substrate with vaporized methanol fuel, 5.6 mW cm^{-2} with OCV of 0.8 V was achieved at 250°C. AAO-supported fuel cell does not require complex MEMS processes to obtain thin-film electrolytes. However, due to the characteristics of porous supporting structure, the challenge to obtain fully dense electrolyte deposition on AAO substrate is still the main concern.

7.4.2.2 Conventional Ni–YSZ Supporting Substrate

Another work that is worth mentioning is a dense sub-micrometer thin-film electrolyte fabricated using sintering processes (Oh et al. 2012). They demonstrated a well-controlled bi-layer thin-film GDC/YSZ with a total electrolyte thickness less than 500 nm. Unlike most of the TF-SOFCs having electrolyte films deposited with vacuum-based method, their electrolyte was coated using chemical solution deposition (CSD) on a typical Ni–YSZ porous substrate. The process requires significant effort to control the sintering schedule and the thermal match between the dense electrolyte film and the porous substrate. The performance was impressive with a maximum power density of 1.3 W cm^{-2} and OCV above 1 V at 650°C.

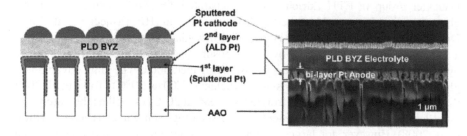

FIGURE 7.9 SEM cross-sectional image of BYZ fuel cell on AAO substrate (100 μm in thickness), Pt bi-layer anode (~300 nm), BYZ electrolyte (~900 nm), and Pt cathode (200 nm). (Reprinted with the permission from Ha et al. (2013) © 2013 Elsevier.)

7.4.2.3 Other Porous Substrates

A porous metal can be a good candidate for TF-SOFC substrate since the metal substrate itself works as a current collector, which resolves the issue of extra design necessary for interconnection when using AAO substrate. As shown in Figure 7.10, Joo and Choi reported a TF-SOFC with thick-film electrolyte of 10–25 μm on porous Ni substrate using screen printing with NiO organic solution with subsequent sintering at 700°C–750°C (Joo and Choi 2008). Similar to the AAO case, the fabrication process is not complicated, and the cell can also be prepared by combining several thin-film deposition processes such as PLD and sputtering. However, judging from the low OCV value which is 0.64 V and power density of 26 mW cm^{-2} at 450°C, inherent problem on pinhole-free deposition was not fully solved.

As another porous substrate, stainless steel has been investigated by Kim et al. (2016). Ni and porous STS dual-layer substrate was introduced using conventional tape-casting and lamination methods as shown in Figure 7.11. The MEA consists of 0.7-μm-thick LSC(La0.7Sr0.3CoO3-δ), 2-μm-thick YSZ (Figure 7.11c), and 0.6-μm-thick Ni–YSZ (Figure 7.11d). (La, Sr)(Ti, Ni)O$_3$ (LSTN)/YSZ layer on dual-layer substrate has pore size of ~500 nm and RMS surface roughness of ~21 nm after polishing. Due to a good surface roughness and robust substrate, the active area can be enlarged to 78 mm^2. They also reported that Cr or Fe diffusion from stainless steel substrate was negligible, and peak power density of 560 mW cm^{-2} and OCV of above 1 V at 550°C were achieved. In Figure 7.11e, thermal stability was also well

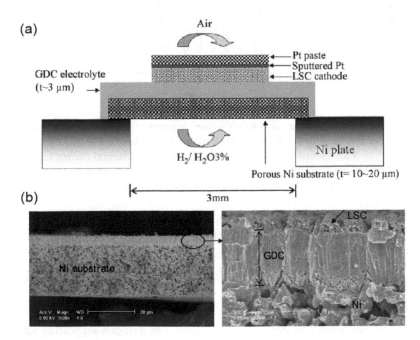

FIGURE 7.10 (a) Schematic of TF-SOFC on porous Ni substrate. (b) Cross-sectional SEM images of TF-SOFC with low and high magnification. (Reprinted with the permission from Joo and Choi (2008) © 2008 Elsevier.)

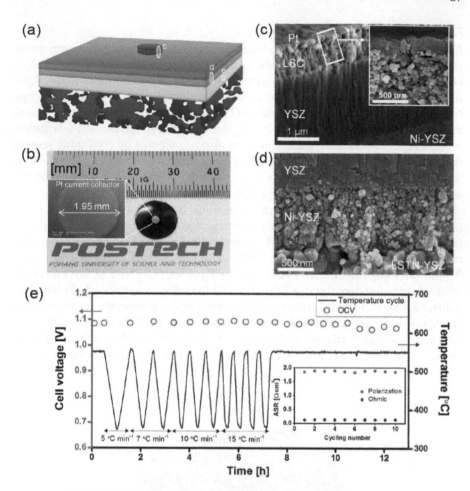

FIGURE 7.11 (a) Schematic of TF-SOFC on a porous stainless steel substrate. (b) Photograph of the fabricated cell. (c) Cross-sectional SEM view of Pt/LSC/YSZ. (d) YSZ/Ni–YSZ/LSTN–YSZ on a porous stainless steel substrate. (e) Thermal cycling test between 350°C and 550°C with heating and cooling rates of 5–15°C min⁻¹. (Reprinted with the permission from Kim et al. (2016) ©2016 Springer Nature Limited.)

maintained under several thermal cycling conditions. This can be attributed to reasonably matched TEC values between cell structures: YSZ (10.5 ppm K⁻¹), Ni-YSZ (12.5 ppm K⁻¹), LSTN (11–12 ppm K⁻¹), and stainless steel (11.2 ppm K⁻¹) (Tietz 1999, Marina, Canfield, and Stevenson 2002).

7.4.3 SCALING-UP OF THE SINGLE CELLS ON MEMBRANE

Several impressive cell performances of TF-SOFCs on a silicon supporting substrate operating at low temperatures below 500°C have been published over the years, showing the potential of utilizing electrolytes in sub-micrometer scale. However, the total current output from the tiny electrolyte membrane is too small to be of practical use. Therefore, continuous effort has been focused on making the cell larger to deliver higher power output.

These scaling methods can be categorized into three directions:

1. Increasing the lateral dimension of a single cell
2. Creating corrugations on the ceramic electrolyte membrane
3. Fabricating a large cell array with the sheer number of cells in a cell chip.

Only a few research groups are actively studying the development of nanoscale TF-SOFCs to a large cell or to a system level because of the fabrication complexity and special thin-film deposition techniques. As listed in Table 7.2, all these microfabricated membranes have thicknesses of 0.1–4 μm and were operated at temperatures

TABLE 7.2
Comparison of Free-Standing TF-SOFC Components and Electrochemical Performance

MEA	Size (μm)	Fuel	Temp. (°C)	OCV (V)	P_{Peak} (mW cm^{-2})	References
Pt/YSZ/Pt	100	H$_2$	350	~1	45	(Jiang et al. 2008)
			400	~1	50	
Pt/YSZ/Pt	240	H$_2$	350	1.10	130	(Huang et al. 2007)
Pt/YSZ/GDC–Pt			350	1.10	200	
Pt/YSZ/GDC–Pt			400	1.10	400	
Pt/YSZ/Pt (corrugated)	600	H$_2$	450	1.09	861	(Su et al. 2008)
Pt/YSZ/Pt	100	H$_2$	500	1.10	1340	(Chao et al. 2011)
Pt/YSZ/Pt	160	H$_2$	500	0.97	1037	(Kerman, Lai, and Ramanathan 2011)
Pt–Ni grid/YSZ/Pt	5,000	H$_2$/Ar	550	0.28	-	(Rey-Mermet and Muralt 2008)
Ru/GDC/YSZ/Pt	160	H$_2$/Ar	520	0.60	1177	(Kerman, Lai, and
		CH$_4$	520	0.85	665	Ramanathan 2012b)
Pt/GDC/YSZ/Pt	160	H$_2$/Ar	510	0.41	1025	(Kerman, Lai, and Ramanathan 2012a)
Pt/YSZ/Pt	100	H$_2$	600	0.57	209	(Tolke et al. 2012)
Pt/YSZ/LSCF/Pt	5,000	H$_2$	510	0.72	155	(Tsuchiya, Lai, and Ramanathan 2011)
Pt/YSZ/Pt	80	H$_2$	500	0.62	92	(Johnson et al. 2009)
Pt/YSZ/LSCF/Pt	80			1.03	60	
Pt/YSZ/LSCF/Pt	350			1.04	20	
LSCF/YSZ/LSCF	160	H$_2$	545	0.18	0.21	(Lai, Kerman, and Ramanathan 2011b)
Pt/YSZ/LSCF	250	H$_2$	560	0.60	120	(Lai, Kerman, and Ramanathan 2010)
Pt–Ni grid/YSZ/LSCF	-	H$_2$	500	0.20	1.0	(Johnson et al. 2010)
						(Continued)

TABLE 7.2 (Continued)

Comparison of Free-Standing TF-SOFC Components and Electrochemical Performance

MEA	Size (μm)	Fuel	Temp. (°C)	OCV (V)	P_{Peak} (mW cm^{-2})	References
Pd/YSZ/Pt	160	CH$_4$	550	0.77	385	(Lai, Kerman, and Ramanathan 2011a)
Pt–YSZ/YSZ–GDC/ Pt-YSZ	180	H$_2$	450	0.68	4.82	(Yan et al. 2012)
Pt/YSZ/Pt	100	H$_2$	350	1.02	270	(Shim et al. 2007)
Pt/YSZ/Pt (AAO)	2,000	H$_2$	450	1.14	180	(Ha, Su, and Cha 2013)
Pt/BYZ/Pt (corrugated)	100	H$_2$	450	0.85	186	(Kim et al. 2011)
Pt/BYZ/Pt (corrugated)	700	H$_2$	400	0.56	7.6	(Su and Prinz 2011)
Ni/YSZ/Pt (nanotubular)	1,500	H$_2$	550	0.66	0.0013	(Motoyama et al. 2014)
Ru/YSZ/Pt	200	CH$_4$	525	0.91	635	(Takagi et al. 2013)
		Natural gas	500	0.96	410	
Pt/BYZ/Pt (AAO)	-	H$_2$	450	1.10	21	(Park et al. 2013)
Pt/YSZ/LSC	390	H$_2$	500	1.05	12	(Evans et al. 2013)
Pt/BYZ/Pt (AAO)	-	H$_2$	450	1.04	44	(Chang, Heo, and Cha 2013)
Pt/YSZ/Pt	250	H$_2$	450	1.10	3.3	(An, Kim, and Prinz 2013)
Pt/YDC/YSZ/Pt (corrugate)	43	H$_2$	450	1.08	1300	(An et al. 2013)
Ru/GDC/YSZ/Pt	200	CH$_4$	485	0.97	275	(Takagi, Adam, and Ramanathan 2012)
Pt/YSZ/Pt (corrugate)	2,000	H$_2$	450	1.13	198	(Su and Prinz 2012)
Pt/ZrO$_2$/Pt	160	H$_2$/Ar	450	0.91	33	(Ko, Kerman, and Ramanathan 2012)
Pt/BCY/Pt	165	H$_2$	400	0.59	30	(Li et al. 2014)
Pt/YDC/YSZ/Pt	57	H$_2$	500	~1.10	1040	(Fan et al. 2012)
Pt/YSZ/Pt	250	H$_2$	400	~1.10	1.9	(Chao, Motoyama, and Prinz 2012)
Pt/YSZ/Pt (AAO)	-	H$_2$	500	1.17	~170	(Ji et al. 2015)
Ru–Pt/GDC/Pt	-	C$_2$H$_5$OH	400	~0.9	~5.5	(Jeong et al. 2015)
Pt/YSZ/LSC	390	H$_2$	425	~1.06	262	(Evans et al. 2015)

of 400°C–550°C with hydrogen fuel gas. However, the membrane lateral dimension varies from a few hundred micrometers to millimeters. The overall cell performance is directly related to the active membrane area at the same operation conditions. The quality of thin films, especially the cathode, is also a limiting factor since pinholes in thin films can easily and detrimentally affect the performance of the entire membrane. Enlarging the membrane laterally would be an intuitive way to increase total power output. However, the fabrication process for enlarging membrane size is extremely challenging because of the fragile nanoscale membranes. Thin-film deposition techniques for fabricating larger pinhole-free membranes without electric short-circuit are also non-trivial.

Until now, "working TF-SOFCs" (meaning those having measurably good cell performance) are limited in lateral dimensions (<0.01 mm²). This is mainly due to the electrolyte membrane that is a brittle and thin (<1 μm) ceramic film and to their high temperature of operation (400°C–550°C) in comparison with other types of polymer electrolyte micro fuel cells. With their small size, the total power output is very limited, even if their peak power density (P_{Peak}) is already some hundreds of mW cm⁻² with some impressive works achieving above 1 W cm⁻² at 500°C.

The cell performances of TF-SOFCs with corresponding MEA dimensions, cell operating temperatures, types of fuels used, and peak power densities reported to date are summarized in Table 7.2. Total power outputs in most cases are below 10 mW over low-temperature region, which is still not enough for application in portable devices. The highest power output is approximately 20 mW due to a larger membrane successfully fabricated (13.5 mm²) with the help of a nickel supporting grid (Tsuchiya, Lai, and Ramanathan 2011). However, the nickel supporting grid took nearly half of the membrane surface area (area utilization percentage: 54.3%), which decreased the total power output. Also, the deposited nickel grid film itself carried undesirable residual stress and introduced more complexity in mechanical stability and integrity of the entire membrane.

7.4.3.1 Enlarging the Lateral Dimension of Free-Standing Membranes

One straightforward way to think of having a more electrochemically active surface area is to fabricate a larger free-standing membrane. Considering the small thickness of electrolyte film in tens of nanometers scale, this is intuitively a challenging approach. The most basic structure of a free-standing membrane on a single crystal silicon substrate can be fabricated by performing either through-wafer etching with wet chemicals to obtain a square membrane (Figure 7.12a) or dry deep reactive ion etching (DRIE) (Figure 7.12b) to obtain an arbitrary shape of interest. However, the former which forms a square membrane always induces severe stress concentration points at the edge of the membrane from membrane buckling, which often causes membrane fracture. The latter DRIE approach requires long etching time (4.5 hours for a 400 μm silicon wafer) and expensive DRIE process (estimated to cost $1 USD/μm of etching) and therefore is not practical for the batch production of TF-SOFCs.

Another architecture with greatly enhanced mechanical stability is a free-standing circular membrane with a special design of membrane-edge support. The fabrication process of a new architecture for a circular membrane combined both anisotropic wet etching and DRIE, so-called combinatorial etching (Figure 7.12c). The shape of the resulting through-hole created by the combinatorial etching is circular, with a thin tapered silicon ring at the edge of the membrane.

To test the capability of the new structure in alleviating membrane stress, the electrolyte thin films were prepared by two deposition methods, ALD and PLD. 100-nm-thick YSZ was deposited by ALD at 250°C of substrate temperature with

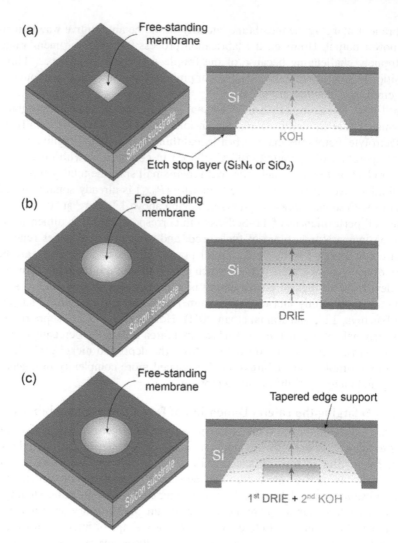

FIGURE 7.12 Fabrication methods for free-standing thin-film electrolyte templates. (a) Square thin-film template using KOH wet etching. (b) Circular thin-film template by DRIE through-wafer etching. (c) Circular thin-film templates with tapered edge support fabricated by a combination of KOH wet etching and DRIE. (Reprinted with the permission from Baek et al. (2015) © 2015 Royal Society of Chemistry.)

a recipe similar to that reported by previous works (Su et al. 2008, Li et al. 2014, Tsuchiya, Lai, and Ramanathan 2011). 100-nm-thick BYZ was deposited by PLD (Coherent 248 nm KrF excimer laser, 2.5 J cm^{-2}, 3 Hz, 1 Pa O$_2$) at a substrate temperature of 700°C. The PLD film was expected to have higher residual stress due to the high operating temperature.

The optical microscope (OM) images of the circular electrolyte membranes architectures are shown in Figure 7.13. 100-nm-thick free-standing electrolyte membranes

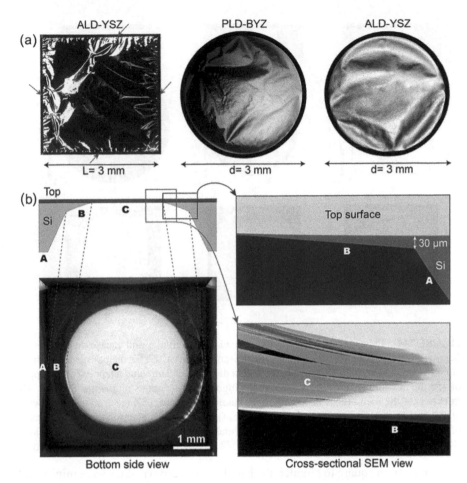

FIGURE 7.13 (a) Top views of square ALD–YSZ (black arrows point to wrinkled regions), circular PLD–BYZ, and circular ALD–YSZ membranes. (b) Optical microscopic image of a circular template with tapered edge support and schematic of cross-section of a circular template with SEM views. (Two-stage supporting structures are illustrated as A and B, and C represents a free-standing membrane.) (Reprinted with the permission from Baek et al. (2015) © 2015 Royal Society of Chemistry.)

with a diameter of 3 mm after fabrication process are shown in Figure 7.13a. Buckling deformation, caused by compressive residual stress, was also observed in the circular membranes for both ALD–YSZ and PLD–BYZ but was much less severe than that with the square ALD–YSZ membrane. In terms of the buckling-induced wrinkles at the clamped edge(s), there were several observed in the square membrane, whereas no apparent wrinkle was observed in the circular ones. From Kerman's calculation (2013), such buckling-induced wrinkles at the clamped edges are stress concentration points where fracture of membranes usually occurs. Here, by changing the

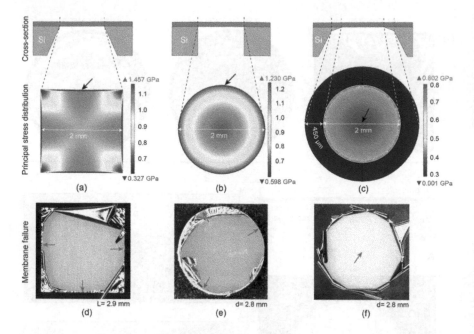

FIGURE 7.14 Principal stress distributions on clamped circular and square membranes. (a) Square YSZ membrane with a lateral length of 2 mm. (b) Circular YSZ membrane with a diameter of 2 mm. (c) Circular YSZ membrane with a diameter of 2 mm and a tapered edge support of 450 μm in width (black arrows point to the highest stress regions on the membranes). The thickness of the membranes is 300 nm in all cases. (d) Square YSZ electrolyte with a width of 2.9 mm broken during fuel cell tests and (e, f) circular YSZ electrolyte with a diameter of 2.8 mm. (Black arrows point to the approximate regions of initiating membrane failure.) (Reprinted with the permission from Baek et al. (2015) © 2015 Royal Society of Chemistry.)

membrane from square to circular, the buckling-induced wrinkles were minimized, and the chances of membrane fracture were expected to decrease significantly.

Figure 7.13b shows the membrane, viewed from the bottom side of the silicon chip, with a schematic of the cross-section. The supporting structure to the free-standing membrane features an additional tapered support (portion B) between the major support (portion A) and the free-standing membrane (portion C). This tapered edge support is a thin and annular single crystal silicon approximately 450 μm in width and 30 μm of height, where the exact dimensions can vary depending on process and design parameters. The addition of this thin taper-shaped support is key to the success of enabling larger lateral dimension of the free-standing because it serves as an effective stress absorber to alleviate the stress concentrated at the clamped edges, pressure difference, as well as vibration and shocks.

To verify the mechanical stability enhancement, principal stress distributions within the square and circular YSZ membranes under fuel cell operating conditions were computed with a finite element method (FEM) simulation. Three different membrane configurations (as shown in Figure 7.14) were constructed to understand how the membrane shape and the addition of tapered edge support have changed the stress distributions on each of the membranes:

1. A 2 mm × 2 mm square membrane clamped at the edge
2. A circular membrane with a diameter of 2 mm clamped at the edge
3. A circular membrane with a diameter of 2 mm and a tapered edge support of 450 μm in width.

The calculation results of principal stress distribution (Figure 7.14a–c) demonstrate a much more uniform stress distribution across the circular membrane with a tapered edge support (Figure 7.14c) than either the square membrane (Figure 7.14a) or the circular membrane without a tapered edge support (Figure 7.14b). Compared with the clamped square and circular membranes, the circular membrane with a support showed a 30%–40% reduction in the maximum principal stress. That is, the maximum principal stress of the circular membrane, located at the clamped edge(s), was reduced significantly by changing the membrane shape from square to circle and reduced further by introducing the tapered edge support.

Figure 7.14d–f shows the corresponding fractured TF-SOFCs after the fuel cell test. For the square membrane and the circular membrane without tapered support, the fracture was initiated at the clamped edge, where the stress is the highest, as confirmed by our simulation results. On the other hand, for the circular membrane with a tapered edge support (Figure 7.14f), the fracture was initiated at the membrane center because the fragments of the fractured membrane were still clamped along the circular boundaries. Therefore, the tapered edge support effectively restrained the edge-fractures typically observed in square membranes, and accordingly, mechanical stability was maintained during fuel cell operation.

7.4.3.2 3D Corrugated Electrolyte Membrane

Electrolyte membrane corrugation using various silicon process techniques was performed to maximize electrochemical reaction area within a confined surface. The idea of having the nanoscale thickness ceramic membrane corrugated and at the same time dense, gas-tight, and pinhole-free was a very challenging one. Su et al. first employed a patterned silicon surface template to create a 3D surface structure with superior surface smoothness, then deposited 70 nm thickness of YSZ electrolyte by ALD to conformally cover the 3D surface. The thin-film electrolyte replicated the 3D surface contour, and after the silicon substrate was etched away, the freestanding corrugated membrane was successfully achieved. The following work also employed such a 3D template method to realize a corrugated membrane electrolyte. Su et al. described the fabrication of SOFCs with corrugated thin-film membrane by patterning the silicon wafer with standard lithography and creating trenches 20 μm in depth and 15 μm in diameter by DRIE (Figure 7.15a) (Su et al. 2008). The electrolyte thin film was deposited onto the patterned silicon template. Etching and sputtering electrodes caused free-standing corrugated membranes with a total thickness of 300 nm and side dimension up to 2 mm which operate from 400°C to 450°C. Depending on the dimension of circular trenches, the active surface can be increased to 2–5 times the projected area occupied by the free-standing membrane alone. Subsequently, Kim et al. fabricated a free-standing membrane electrode assembly area of 100 μm × 100 μm square with surrounding silicon structure supporting the membrane as shown in Figure 7.15b (Kim et al. 2011). It had a closely packed array of crater patterns, providing a corrugated area 1.7–1.8 times larger than the projected area of the planar structure. An et al. demonstrated free-standing nanostructured bilayer electrolytes to obtain high power density at low operating temperature as shown in Figure 7.15c (An et al. 2013). Nanosphere lithography (NSL) was employed, and 50-nm-thick YSZ by ALD and 20-nm-thick YDC by PLD were deposited. From

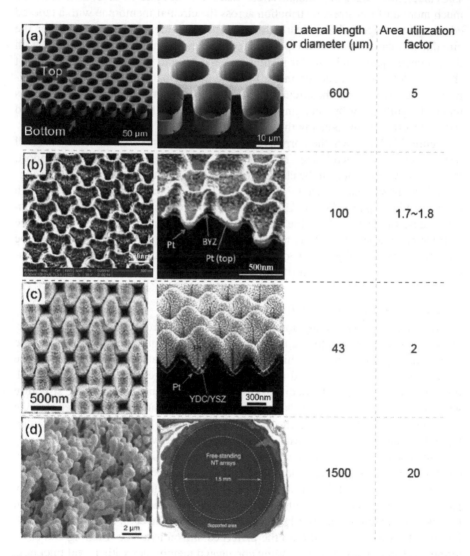

FIGURE 7.15 3D nanostructured SOFC architectures. (a) Cup-shaped array SOFCs. (Reprinted with the permission from Su et al. (2008) © 2013 American Chemical Society.) (b) Crater-shaped array SOFCs. (Reprinted with the permission from Kim et al. (2011) © 2011 Elsevier.) (c) Pyramid-shaped array SOFCs. (Reprinted with the permission from An et al. (2013) © 2013 American Chemical Society.) (d) Nanotubular array SOFCs. (Reprinted with the permission from Motoyama et al. (2014) © 2013 American Chemical Society.)

free-standing Pt/YSZ/YDC/Pt structure with 43 μm in lateral dimension, maximum power density of 1.3 W/cm^2 was obtained at 450°C. Figure 7.15d shows a nanotubular array SOFC structure used by Motoyama et al. to obtain extremely high surface area, and 20 times increase in the effective surface area was attained (Motoyama et al. 2014). The tubular MEAs were 5 μm in length and 500 nm in outer diameter with total MEA thickness of nearly 50 nm. However, cell performance was still miserably low due to chemical shorting by cracks or defective MEA tubes (OCV of 0.6 V and maximum power density of 1.3 μW cm^{-2} at 550°C).

7.4.3.3 Large Membrane Arrays with Mechanical Supporting Layer

In this section, TF-SOFC array architecture with a supporting layer is introduced. The TF-SOFC array incorporates a large number of tiny circular free-standing YSZ electrolyte membranes. In 2012, Su et al. initially fabricated a more robust TF-SOFC array with silicon supporting layer than free-standing corrugated TF-SOFCs (Su and Prinz 2012), and based on this work, Baek et al. further developed a circular TF-SOFC array with more enhanced mechanical and thermal stability (Baek, Yu, and Su 2016).

The complete fabrication sequence of circular TF-SOFC array is shown in Figure 7.16a. On the top side of the wafer, circular trenches were generated with photolithography and DRIE. The diameter of the circular trenches is 50 μm, and the depth is 30 μm as shown in Figure 7.16b. Introducing DRIE circular trench inside the KOH etching window generated a new plane after KOH etching process and formed a tapered edge reinforcement similar to the tapered edge support of circular TF-SOFCs. The newly created structure on the corners of the array is a tapered and rounded silicon support, which reinforces the silicon supporting membrane.

A thin-film YSZ electrolyte with thickness of 80 nm was deposited on the top by ALD with conditions similar to those in previously reported works (Su et al. 2008, Su and Prinz 2011, 2012). The deposited YSZ thin film replicates the surface contour of pre-patterned circular trenches on the top and forms a 3D thin film. By through-wafer etching, silicon was removed until 20 μm of silicon remained, and an edge-reinforced silicon membrane for supporting TF-SOFC array was fabricated. Figure 7.16c shows circular arrayed cells viewed from bottom after KOH etching. Both cathode and anode of the TF-SOFC array were deposited by RF sputtering with 100-nm-thick porous platinum, then each YSZ membrane became an individual fuel cell, and all individual fuel cells in the array were connected in parallel. In the actual arrays, approximately 2,600 individual membranes were embedded in a single circular window with a diameter of 4 mm. Circularly arrayed electrolytes on a circular template are presented in Figure 7.16d.

The thermal and mechanical stability of TF-SOFC array can be verified by repeated thermal cycles as shown in Figure 7.17. Totally seven thermal cycling tests were performed, and the TF-SOFC array was cooled down to 150°C to avoid vapor condensation and heated up to 400°C repetitively. From simple calculation with mechanical properties of YSZ electrolyte (Srikar et al. 2004), a high thermal stress of 700 MPa was applied in the membrane during thermal cycling tests. During the moderate and harsh thermal cycling, no visible membrane deformation or cell

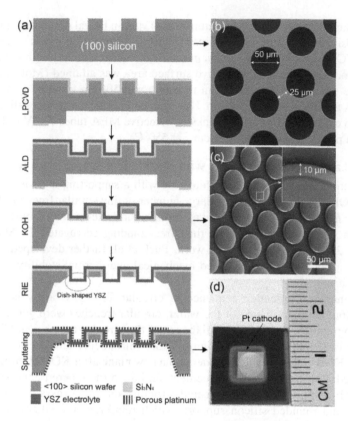

FIGURE 7.16 (a) Fabrication process of circular TF-SOFCs array. (b) Top view of circular trenches. (c) Bottom view of circular array after KOH etching. (d) Circular arrays on 1.5 mm ×1.5 mm silicon chips. (Reprinted with the permission from Baek, Yu, and Su (2016) © 2016 American Chemical Society.)

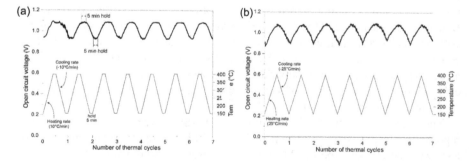

FIGURE 7.17 OCV changes during thermal cycling tests. (a) OCV changes with moderate thermal cycles (10°C min[-1]). (b) OCV changes with harsh thermal cycles with high heating and cooling rates (25°C min[-1]). (Reprinted with the permission from Baek, Yu, and Su (2016) © 2016 American Chemical Society.)

degradation was observed in the TF-SOFC array, which indicates excellent thermo-mechanical integrity of the array architecture.

In order to reinforce free-standing electrolyte membranes, additional mechanical supporting layers were considered with various shapes and materials such as nickel, platinum, and silicon. Ideally, nickel has almost the same TEC as the ceramic membrane (i.e. nickel with 10 ppm K^{-1} differs only by 1 ppm K^{-1} from YSZ). Therefore, many researchers considered it a good choice for a material to support a metal grid. Rey-Mermet et al. fabricated nickel hexagonal or spider-web pattern on the anode side which can mechanically reinforce the 750-nm-thick electrolyte membrane and avoid buckling and cracking (Figure 7.18a) (Rey-Mermet and Muralt 2008). Nickel grid was grown by electroplating with grid spacing in 50–100 μm, and the SOFC membranes having diameter up to 5 mm are mechanically stable at 550°C. However, the low OCV of 0.28 V still needs further investigation. Tsuchiya et al. demonstrated the use of microfabricated metallic grid structures located directly on free-standing 54-nm-thick YSZ membranes to mechanically stabilize the electrochemical active area to over 100 mm², while maintaining high power densities (Tsuchiya, Lai, and

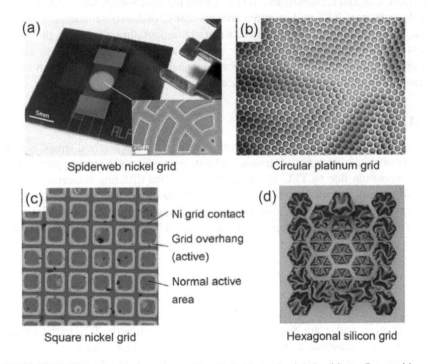

Spiderweb nickel grid   Circular platinum grid

Ni grid contact

Grid overhang
(active)

Normal active
area

Square nickel grid   Hexagonal silicon grid

FIGURE 7.18 Electrolyte supporting grids. (a) Spider-web nickel grids on 5 mm wide circular μ-SOFC. (Reprinted with the permission from Rey-Mermet and Muralt (2008) © 2008 Elsevier.) (b) Circular platinum grid on free-standing LSCF/YSZ membrane and platinum grids after RIE. Each circle is 100 μm in diameter. (Reprinted with the permission from Tsuchiya, Lai, and Ramanathan (2011) © 2011 Springer Nature Limited.) (c) Square-shaped nickel grids on free-standing LSCF/YSZ/Pt membrane. (Reprinted with the permission from Johnson et al. (2010) © 2009 Elsevier.) (d) Hexagonal silicon grids on free-standing YSZ membrane. (Reprinted with the permission from Garbayo et al. (2014) © 2014 Elsevier B.V.)

Ramanathan 2011). Each circle is 100 μm in diameter, and the grid width is 10 μm as shown in Figure 7.18b. The performance of TF-SOFCs with an active area of 13.5 mm² (5 × 5 mm KOH etch hall with 54.3% area utilization) showed higher total power output of ~21 mW.

7.5 ISSUES OF THIN-FILM STRESS

One major reason for the difficulty imposed on scaling of nanothin-film MEA for TF-SOFCs is the inherent thin-film residual stress from the thin-film deposition process. For the vacuum-based processes for electrolyte deposition, the substrates are usually heated in order to promote the crystallization of the electrolyte film and to alleviate possible non-uniformity of stress. However, depending on deposition method, the substrate heating temperature can be very high. For example, PLD deposition for YSZ and BYZ requires the substrate to be maintained at 600°C and 800°C, respectively (Heiroth et al. 2010, Evans et al. 2012), and for reactive or RF sputtering, substrate heating temperature between 500°C and 600°C is preferred (Kerman, Lai, and Ramanathan 2011). Even for the lower-temperature ALD process, the deposition of YSZ still requires around 250°C of substrate temperature (Shim et al. 2007). In general, the higher the substrate temperature, the higher the residual stress in the membrane. Therefore, the study of membrane residual stress and also the design considerations for the supporting substrate to alleviate the high residual stress is critical in membrane mechanical stability.

7.5.1 RESIDUAL STRESS

Almost all thin films grown on a substrate are in a state of internal stress. Residual stresses originate from both intrinsic and extrinsic stresses. Extrinsic stress occurs after deposition due to TEC mismatch between two different materials or phase transformation. Intrinsic stress is closely correlated to film deposition process (Doerner and Nix 1988). Intrinsic stresses vary roughly from 2 GPa compressive to 2 GPa tensile depending on deposition conditions. Therefore, the total residual stress can be expressed as

$$\sigma_{tot} = \sigma_{extrinsic} + \sigma_{intrinsic} = \frac{E_f}{1 - v_f}(\alpha_s - \alpha_f)(T - T_0) + \sigma_{intrinsic} \qquad (7.4)$$

where E_f, v_f, α_f, and α_s are Young's modulus, Poisson's ratio, the TCE of the film, and the TCE of the substrate, respectively. T_0 is the ambient temperature.

Tensile stresses are generated by elastic deformation due to energy minimization under the boundary constraints imposed by substrate, assuming good adhesion between the film and the substrate (Freund and Chason 2001). As the film thickness increases, this energy minimization focuses on reducing surface energy, which induces elastic deformation of grains. A model for tensile stress evolution according to grain growth has been extensively investigated (Windischmann 1992) and experimentally verified with various sputtered materials (Thompson and Carel 1996).

Compressive stresses are generally expected to be caused by high kinetic energy during deposition. Interstitial implantation of sputtered film atoms or gases originating from the evolving surface of the deposited film can cause compressive stresses. The atoms in the developing films are bombarded by the arriving high-energy atoms, which displace the lattice. This is called "atomic peening" and is the most common explanation for compressive stresses.

Another phenomenon to be considered is stress relaxation. Relaxation occurs when lattice distortions are removed by atom diffusions at the surface or grain boundaries with high mobility or thicker films with smaller surfaces (Floro et al. 2002, Floro et al. 2001), and relaxation processes through diffusion can be highly promoted with temperature.

Baersch et al. initially measured residual stresses of YSZ films caused by deposition methods. According to film thickness, residual stress of YSZ film deposited by e-beam evaporator varied between −865 to −155 MPa, whereas residual stress of YSZ deposited by RF sputtering varied between +85 to +235 MPa (Baertsch et al. 2004). Quinn et al. identified as-deposited residual stresses of YSZ film on <100> silicon substrate by varying sputtering conditions. The residual stress deposited at room temperature ranged from −1,400 to 100 MPa and the residual stress deposited at 600°C ranged from −1600 to 400 MPa (Quinn, Wardle, and Spearing 2008). However, these measured stress values include not only residual stress but also thermal stress since the films experience thermal expansion mismatch in cooling process after deposition.

7.5.2 Thermal Stress

Since TF-SOFCs operate at temperatures between 300°C and 600°C, the thermal expansion coefficient mismatch between the substrate and the positive electrode–electrolyte–negative electrode (PEN) is a critical issue. The thermal stress, σ_{th} in the electrolyte membrane can be expressed as the product of the thermal expansion coefficient difference between the film and the substrate, $\Delta\alpha$, the temperature difference during heating, $T_{final} - T_{initial}$, and the biaxial Young's modulus, $Y_{film}/(1 - v_{film})$:

$$\sigma_{th} = \frac{(T_{initial} - T_{final}) \cdot \Delta\alpha \cdot Y_{film}}{1 - v_{film}} \tag{7.5}$$

The simplest way to reduce thermal stress is the choice of a substrate with a thermal expansion coefficient close to that of the electrolyte.

When the ambient temperature is increased to the operation temperature, the thermal mismatch between the supporting structure and the membrane can create buckling or thermal cracks in the membrane. Tang et al. give the allowable maximum temperature difference ΔT_{crit} sustainable by a free-standing membrane as a function of its radius R and its thickness h (Tang et al. 2005):

$$\Delta T_{crit} = \frac{1.22}{1 + v_{film} \cdot \Delta\alpha} \left(\frac{h}{R}\right)^2 \tag{7.6}$$

For a 1 µm thick YSZ membrane supported by silicon substrate at 400°C, the allowable maximum radius is approximately 20 µm. Therefore, the size of free-standing thin SOFCs is limited to hundreds of micrometers due to thermal stress. From the viewpoint of thermal stability, it is very challenging to fabricate free-standing membranes on a large scale.

However, membrane failure has various possible reasons, and then the well-known Weibull statistics can be used to evaluate fracture failure. The probability of failure expanded by Danzer is expressed by (Danzer 1992)

$$P = 1 - \exp\left[-\int_{V(\sigma > 0)} \frac{1}{V_0} \left(\frac{\sigma}{\sigma_0} \right)^m dV \right] \tag{7.7}$$

where m, σ_0, and V_0 are Weibull modulus, characteristic strength, and an arbitrary normalizing volume set as 1 mm³, respectively (Tang et al. 2005). By integration on a flat circular membrane with radius R and thickness h, the failure probability P is given by

$$P = 1 - \exp\left[-\frac{2\pi}{V_0} \left(\frac{\sigma_f}{\sigma_0} \right)^m R^2 h \right] \tag{7.8}$$

From Eq. 7.8, the failure probability of flat film is exponentially increased with $R^2 h$. The mechanical and thermal stability of MEMS thin membranes has been significantly enhanced using corrugated design (van den Boogaart et al. 2006). The corrugated steps as shown in Figure 7.19 help to improve the critical stress of membrane failure without reducing its radius or this thickness (Tang et al. 2005), and it is possible to design large-scale membranes which can modulate a high critical stress by increasing the ratio of H_{step}:H_{mem}. However, the corrugated design requires more complicated microfabrication processes and, flat surfaces are more advantageous in the case of TF-SOFCs.

7.5.3 MEMBRANE BUCKLING

Most TF-SOFCs use a free-standing square membrane as its fabrication using silicon-based micromachining is easy, and severe buckling modes on square membranes due to nonuniform stress distribution are observed in most cases. There have

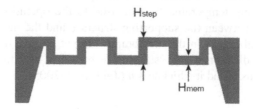

FIGURE 7.19 Schematic of corrugated membrane of thickness H_{mem} and corrugated step size H_{step}.

been extensive researches on TF-SOFCs utilizing nanoscale oxide membranes, mostly without consideration for the buckling responses.

An unstressed membrane is mechanically in stable equilibrium in a flat configuration. As compressive stress increases in the membrane, possibly by raising the temperature, the membrane remains flat before beyond a critical point. At this critical stress, the equilibrium bifurcates into stable deformed equilibrium state and unstable flat equilibrium state.

It is well known that a structure consisting of two or three layers of different materials is susceptible to stress-induced failure by several modes including fracture, buckling, delamination, and spalling (Spearing 1997). As previously mentioned, fuel cells are comprised of two relatively thin porous electrode layers and one dense and brittle electrolyte layer which serves as the primary load-bearing structure. Therefore, it can be reasonable to assume a single-layer membrane for a first approximation (Srikar et al. 2004). This is typically reasonable in thin films of polycrystalline materials, and anisotropic elasticity should be considered for epitaxially grown thin films (Vinci and Vlassak 1996).

Since the electrochemically active area of the fuel cell is free-standing and supported by a relatively large and stiff silicon structure as shown in Figure 7.20, fixed (clamped) boundary conditions can be applied at the edges of the membrane. If the clamped circular membrane expands due to temperature difference between membrane and ambient air, compressive stress occurs, and sufficient compressive stress can induce membrane buckling.

The critical buckling stress on the circular plate σ_{cr} at which buckling occurs is expressed as (Ventsel and Krauthammer 2001)

$$\sigma_{cr} = -1.22 \frac{E}{1-v^2}\left(\frac{h}{b}\right)^2 \tag{7.9}$$

and the critical buckling stress for square membrane is formulated as (Timoshenko and Gere 2009, Kerman et al. 2013)

$$\sigma_{cr} = -4.39 \frac{E}{1-v^2}\left(\frac{h}{L}\right)^2 \tag{7.10}$$

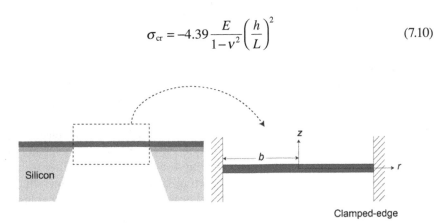

FIGURE 7.20   Modeling scheme of circular plate with clamped edge condition.

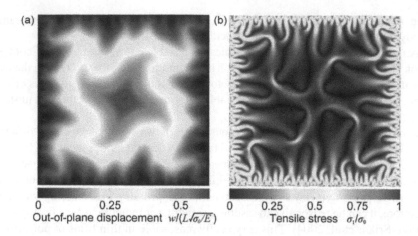

FIGURE 7.21   (a) Out-of-plane displacement. (b) Distribution of tensile stress. (Reprinted with the permission from Kerman et al. (2013) © 2013 Elsevier.)

where h is the thickness of the membrane, b is the radius, L is the lateral length, E is Young's modulus, and v is Poisson's ratio. From Eqs. 7.9 and 7.10, it is shown that thicker and smaller membranes are more robust against buckling and circular membranes show higher critical buckling stresses, which indicates that they are more resistive to buckling compared to square membrane.

Kerman et al. simulated elastic configurations in the post-buckling regime by using the energy method of classical stability analysis and confirmed that edge clamped, free-standing thin membranes show hierarchical wrinkles (Figure 7.21a) and the largest strains are in the vicinity of the clamped boundary, causing membrane failures, as shown in Figure 7.21b (Kerman et al. 2013). Crudely, to avoid membrane failure from buckling, the ultimate tensile strength of the membrane should be greater than residual film stress (σ_{tot}) as expressed in Eq. 7.4.

7.6   CONCLUSIONS AND PERSPECTIVES

The study of TF-SOFCs using nanoscale thin-film electrolyte membranes is an emerging area in providing a potential portable long-lasting power source. The efforts reported have demonstrated the possibility to greatly reduce the operating temperature of SOFCs for the past two decades. MEMS-based micromachining processes used for silicon-supported TF-SOFCs and porous-substrate-supported TF-SOFCs are two prevailing methods to realize the low-temperature device using electrolyte of sub-micrometer thickness. To effectively increase the electrochemically active area, 3D structures of electrolyte membrane were developed, which resulted in performance enhancement. To improve mechanical stability of a large free-standing membrane, additional supporting structures, such as metallic grids and silicon supports, or changing membrane geometry from square to circle is currently implemented. Moreover, various deposition techniques have been developed to tailor residual stress originating from deposition process. Based on these efforts,

TF-SOFCs at low operating temperatures have recently achieved stable and competitive performance compared to conventional micro-SOFCs operating at high temperatures above 700°C. Further development on robust scaling-up of the cell arrays and also on stack configuration design will be the next step toward a working TF-SOFC device for practical applications.

REFERENCES

Adler, S. B. 2004. "Factors governing oxygen reduction in solid oxide fuel cell cathodes." *Chemical Reviews* 104 (10):4791–4843. doi: 10.1021/cr020724o.

An, J., Y. B. Kim, J. Park, T. M. Gur, and F. B. Prinz. 2013. "Three-dimensional nanostructured bilayer solid oxide fuel cell with 1.3 W/cm^{-2} at 450 C." *Nano Letters* 13 (9):4551–4555. doi: 10.1021/Nl402661p.

An, J. W., Y. B. Kim, and F. B. Prinz. 2013. "Ultra-thin platinum catalytic electrodes fabricated by atomic layer deposition." *Physical Chemistry Chemical Physics* 15 (20):7520–7525.

Ashfold, M. N. R., F. Claeyssens, G. M. Fuge, and S. J. Henley. 2004. "Pulsed laser ablation and deposition of thin films." *Chemical Society Reviews* 33 (1):23–31. doi: 10.1039/b207644f.

Bae, K., D. Y. Jang, H. J. Jung, J. W. Kim, J. W. Son, and J. H. Shim. 2014. "Micro ceramic fuel cells with multilayered yttrium-doped barium cerate and zirconate thin film electrolytes." *Journal of Power Sources* 248:1163–1169. doi: 10.1016/j.jpowsour.2013.10.057.

Baek, J. D., K.-Y. Liu, and P.-C. Su. 2017. "A functional micro-solid oxide fuel cell with a 10 nm-thick freestanding electrolyte." *Journal of Materials Chemistry A.* doi: 10.1039/C7TA05245F.

Baek, J. D., Y.-J. Yoon, W. Lee, and P.-C. Su. 2015. "A circular membrane for nano thin film micro solid oxide fuel cells with enhanced mechanical stability." *Energy & Environmental Science* 8 (11):3374–3380. doi: 10.1039/C5EE02328A.

Baek, J. D., C.-C. Yu, and P.-C. Su. 2016. "A silicon-based nanothin film solid oxide fuel cell array with edge reinforced support for enhanced thermal mechanical stability." *Nano Letters* 16 (4):2413–2417. doi: 10.1021/acs.nanolett.5b05221.

Baertsch, C. D., K. F. Jensen, J. L. Hertz, H. L. Tuller, S. T. Vengallatore, S. M. Spearing, and M. A. Schmidt. 2004. "Fabrication and structural characterization of self-supporting electrolyte membranes for a micro solid-oxide fuel cell." *Journal of Materials Research* 19 (9):2604–2615. doi: 10.1557/Jmr.2004.0350.

Baur, E., and H. Preis. 1937. "Über Brennstoff-Ketten mit Festleitern." *Z Elektrochem* 43:727–732.

Beckel, D., A. Bieberle-Hutter, A. Harvey, A. Infortuna, U. P. Muecke, M. Prestat, J. L. M. Rupp, and L. J. Gauckler. 2007. "Thin films for micro solid oxide fuel cells." *Journal of Power Sources* 173 (1):325–345. doi: 10.1016/j.jpowsour.2007.04.070.

Bieberle-Hutter, A., D. Beckel, A. Infortuna, U. P. Muecke, J. L. M. Rupp, L. J. Gauckler, S. Rey-Mermet, P. Muralt, N. R. Bieri, N. Hotz, M. J. Stutz, D. Poulikakos, P. Heeb, P. Muller, A. Bernard, R. Gmur, and T. Hocker. 2008. "A micro-solid oxide fuel cell system as battery replacement." *Journal of Power Sources* 177 (1):123–130. doi: 10.1016/j.jpowsour.2007.10.092.

Bonano, N., B. Ellis, and M. N. Mahmood. 1991. "Construction and operation of fuel cells based on the solid electrolyte BaCeO$_3$:Gd." *Solid State Ionics* 44 (3):305–311. doi: 10.1016/0167-2738(91)90023-5.

Bruschi, P., A. Diligenti, A. Nannini, and M. Piotto. 1999. "Technology of integrable free-standing yttria-stabilized zirconia membranes." *Thin Solid Films* 346 (1–2):251–254. doi: 10.1016/s0040-6090(98)01758-1.

Chang, I., J. Bae, J. Park, S. Lee, M. Ban, T. Park, Y. H. Lee, H. H. Song, Y.-B. Kim, and S. W. Cha. 2016. "A thermally self-sustaining solid oxide fuel cell system at ultra-low operating temperature (319°C)." *Energy* 104:107–113. doi: 10.1016/j.energy.2016.03.099.

Chang, I., P. Heo, and S. W. Cha. 2013. "Thin film solid oxide fuel cell using a pinhole-free and dense Y-doped BaZrO$_3$." *Thin Solid Films* 534:286–290. doi: 10.1016/j.tsf.2013.03.024.

Chang, I., S. Ji, J. Park, M. H. Lee, and S. W. Cha. 2015. "Ultrathin YSZ coating on Pt cathode for high thermal stability and enhanced oxygen reduction reaction activity." *Advanced Energy Materials* 5 (10). doi: 10.1002/aenm.201402251.

Chao, C. C., C. M. Hsu, Y. Cui, and F. B. Prinz. 2011. "Improved solid oxide fuel cell performance with nanostructured electrolytes." *ACS Nano* 5 (7):5692–5696. doi: 10.1021/Nn201354p.

Chao, C. C., M. Motoyama, and F. B. Prinz. 2012. "Nanostructured platinum catalysts by atomic-layer deposition for solid-oxide fuel cells." *Advanced Energy Materials* 2 (6):651–654. doi: 10.1002/aenm.201200002.

Danzer, R. 1992. "A general strength distribution function for brittle materials." *Journal of the European Ceramic Society* 10 (6):461–472. doi: 10.1016/0955-2219(92)90021-5.

Doerner, M. F., and W. D. Nix. 1988. "Stresses and deformation processes in thin-films on substrates." *CRC Critical Reviews in Solid State and Materials Sciences* 14 (3):225–268. doi: 10.1080/10408438808243734.

Evans, A., C. Benel, A. J. Darbandi, H. Hahn, J. Martynczuk, L. J. Gauckler, and M. Prestat. 2013. "Integration of spin-coated nanoparticulate-based La$_{0.6}$Sr$_{0.4}$CoO$_3$-delta cathodes into micro-solid oxide fuel cell membranes." *Fuel Cells* 13 (3):441–444. doi: 10.1002/fuce.201300020.

Evans, A., J. Martynczuk, D. Stender, C. W. Schneider, T. Lippert, and M. Prestat. 2015. "Low-temperature micro-solid oxide fuel cells with partially amorphous La$_{0.6}$Sr$_{0.4}$CoO$_3$-δ cathodes." *Advanced Energy Materials* 5 (1):1400747. doi: 10.1002/aenm.201400747.

Evans, A., M. Prestat, R. Tolke, M. V. F. Schlupp, L. J. Gauckler, Y. Safa, T. Hocker, J. Courbat, D. Briand, N. F. de Rooij, and D. Courty. 2012. "Residual stress and buckling patterns of free-standing yttria-stabilized-zirconia membranes fabricated by pulsed laser deposition." *Fuel Cells* 12 (4):614–623. doi: 10.1002/fuce.201200028.

Fan, Z., J. W. An, A. Iancu, and F. B. Prinz. 2012. "Thickness effects of yttria-doped ceria interlayers on solid oxide fuel cells." *Journal of Power Sources* 218:187–191. doi: 10.1016/j.jpowsour.2012.06.103.

Feduska, W., and A. O. Isenberg. 1983. "High-temperature solid oxide fuel cell—technical status." *Journal of Power Sources* 10 (1):89–102. doi: 10.1016/0378-7753(83)80009-3.

Fergus, J.W. 2006. "Electrolytes for solid oxide fuel cells." *Journal of Power Sources* 162 (1):30–40. doi: 10.1016/j.jpowsour.2006.06.062.

Floro, J. A., E. Chason, R. C. Cammarata, and D. J. Srolovitz. 2002. "Physical origins of intrinsic stresses in Volmer-Weber thin films." *Mrs Bulletin* 27 (1):19–25. doi: 10.1557/mrs2002.15.

Floro, J. A., S. J. Hearne, J. A. Hunter, P. Kotula, E. Chason, S. C. Seel, and C. V. Thompson. 2001. "The dynamic competition between stress generation and relaxation mechanisms during coalescence of Volmer-Weber thin films." *Journal of Applied Physics* 89 (9):4886–4897. doi: 10.1063/1.1352563.

Freund, L. B., and E. Chason. 2001. "Model for stress generated upon contact of neighboring islands on the surface of a substrate." *Journal of Applied Physics* 89 (9):4866–4873. doi: 10.1063/1.1359437.

Garbayo, I., V. Esposito, S. Sanna, A. Morata, D. Pla, L. Fonseca, N. Sabate, and A. Tarancon. 2014. "Porous La$_{0.6}$Sr$_{0.4}$CoO$_3$-delta thin film cathodes for large area micro solid oxide fuel cell power generators." *Journal of Power Sources* 248:1042–1049. doi: 10.1016/j.jpowsour.2013.10.038.

George, S.M. 2010. "Atomic layer deposition: An overview." *Chemical Reviews* 110 (1):111–131. doi: 10.1021/cr900056b.

Ha, S., P. C. Su, and S. W. Cha. 2013. "Combinatorial deposition of a dense nano-thin film YSZ electrolyte for low temperature solid oxide fuel cells." *Journal of Materials Chemistry A* 1 (34):9645–9649. doi: 10.1039/C3ta11758h.

Ha, S. B., P. C. Su, S. H. Ji, and S. W. Cha. 2013. "Low temperature solid oxide fuel cells with proton-conducting Y:BaZrO$_3$ electrolyte on porous anodic aluminum oxide substrate." *Thin Solid Films* 544:125–128. doi: 10.1016/j.tsf.2013.04.058.

Haile, S. M. 2003. "Materials for fuel cells." *Materials Today* 6 (3):24–29. doi: 10.1016/S1369-7021(03)00331-6.

Heiroth, S., T. Lippert, A. Wokaun, M. Döbeli, J. L. M. Rupp, B. Scherrer, and L. J. Gauckler. 2010. "Yttria-stabilized zirconia thin films by pulsed laser deposition: Microstructural and compositional control." *Journal of the European Ceramic Society* 30 (2):489–495. doi: 10.1016/j.jeurceramsoc.2009.06.012.

Huang, H., M. Nakamura, P. C. Su, R. Fasching, Y. Saito, and F. B. Prinz. 2007. "High-performance ultrathin solid oxide fuel cells for low-temperature operation." *Journal of the Electrochemical Society* 154 (1):B20–B24. doi: 10.1149/1.2372592.

Ishihara, T., N. M. Sammes, and O. Yamamoto. 2003. "Chapter 4: Electrolytes." In *High Temperature and Solid Oxide Fuel Cells*, edited by S. C. Singhal and K. Kendall, pp. 83–117. Amsterdam: Elsevier Science.

Iwahara, H., T. Yajima, T. Hibino, K. Ozaki, and H. Suzuki. 1993. "Protonic conduction in calcium, strontium and barium zirconates." *Solid State Ionics* 61 (1–3):65–69. doi: 10.1016/0167-2738(93)90335-Z.

Jeong, H. J., J. W. Kim, K. Bae, H. Jung, and J. H. Shim. 2015. "Platinum-ruthenium heterogeneous catalytic anodes prepared by atomic layer deposition for use in direct methanol solid oxide fuel cells." *ACS Catalysis* 5 (3):1914–1921. doi: 10.1021/cs502041d.

Ji, S., G. Y. Cho, W. Yu, P.-C. Su, M. H. Lee, and S. W. Cha. 2015. "Plasma-enhanced atomic layer deposition of nanoscale yttria-stabilized zirconia electrolyte for solid oxide fuel cells with porous substrate." *ACS Applied Materials & Interfaces* 7 (5):2998–3002. doi: 10.1021/am508710s.

Jiang, X. R., H. Huang, F. B. Prinz, and S. F. Bent. 2008. "Application of atomic layer deposition of platinum to solid oxide fuel cells." *Chemistry of Materials* 20 (12):3897–3905. doi: 10.1021/Cm7033189.

Johnson, A. C., A. Baclig, D. V. Harburg, B. K. Lai, and S. Ramanathan. 2010. "Fabrication and electrochemical performance of thin-film solid oxide fuel cells with large area nanostructured membranes." *Journal of Power Sources* 195 (4):1149–1155. doi: 10.1016/j.jpowsour.2009.08.066.

Johnson, A. C., B. K. Lai, H. Xiong, and S. Ramanathan. 2009. "An experimental investigation into micro-fabricated solid oxide fuel cells with ultra-thin La$_{0.6}$Sr$_{0.4}$Co$_{0.8}$Fe$_{0.2}$O$_3$ cathodes and yttria-doped zirconia electrolyte films." *Journal of Power Sources* 186 (2):252–260. doi: 10.1016/j.jpowsour.2008.10.021.

Joo, J. H., and G. M. Choi. 2008. "Simple fabrication of micro-solid oxide fuel cell supported on metal substrate." *Journal of Power Sources* 182 (2):589–593. doi: 10.1016/j.jpowsour.2008.03.089.

Kang, S., P. Heo, Y. H. Lee, J. Ha, I. Chang, and S.-W. Cha. 2011. "Low intermediate temperature ceramic fuel cell with Y-doped BaZrO$_3$ electrolyte and thin film Pd anode on porous substrate." *Electrochemistry Communications* 13 (4):374–377. doi: 10.1016/j.elecom.2011.01.029.

Kerman, K., B. K. Lai, and S. Ramanathan. 2011. "Pt/Y$_{0.16}$Zr$_{0.84}$O$_{1.92}$/Pt thin film solid oxide fuel cells: Electrode microstructure and stability considerations." *Journal of Power Sources* 196 (5):2608–2614. doi: 10.1016/j.jpowsour.2010.10.068.

Kerman, K., B. K. Lai, and S. Ramanathan. 2012a. "Free standing oxide alloy electrolytes for low temperature thin film solid oxide fuel cells." *Journal of Power Sources* 202:120–125. doi: 10.1016/j.jpowsour.2011.11.062.

Kerman, Kian, B.-K. Lai, and S. Ramanathan. 2012b. "Nanoscale compositionally graded thin-film electrolyte membranes for low-temperature solid oxide fuel cells." *Advanced Energy Materials* 2 (6):656–661. doi: 10.1002/aenm.201100751.

Kerman, K., T. Tallinen, S. Ramanathan, and L. Mahadevan. 2013. "Elastic configurations of self-supported oxide membranes for fuel cells." *Journal of Power Sources* 222:359–366. doi: 10.1016/j.jpowsour.2012.08.092.

Kharton, V. V., F. M. B. Marques, and A. Atkinson. 2004. "Transport properties of solid oxide electrolyte ceramics: A brief review." *Solid State Ionics* 174 (1-4):135–149. doi: 10.1016/j.ssi.2004.06.015.

Kilner, J. A., and C. D. Waters. 1982. "The effects of dopant cation oxygen vacancy complexes on the anion transport-properties of nonstoichiometric fluorite oxides." *Solid State Ionics* 6 (3):253–259. doi: 10.1016/0167-2738(82)90046-7.

Kim, Y. B., T. M. Gur, S. Kang, H. J. Jung, R. Sinclair, and F. B. Prinz. 2011. "Crater patterned 3-D proton conducting ceramic fuel cell architecture with ultra thin $Y:BaZrO_3$ electrolyte." *Electrochemistry Communications* 13 (5):403–406. doi: 10.1016/j.elecom.2011.02.004.

Kim, K. J., B. H. Park, S. J. Kim, Y. Lee, H. Bae, and G. M. Choi. 2016. "Micro solid oxide fuel cell fabricated on porous stainless steel: A new strategy for enhanced thermal cycling ability." *Scientific Reports* 6 (1):1–8.

Ko, C. H., K. Kerman, and S. Ramanathan. 2012. "Ultra-thin film solid oxide fuel cells utilizing un-doped nanostructured zirconia electrolytes." *Journal of Power Sources* 213:343–349. doi: 10.1016/j.jpowsour.2012.04.034.

Krebs, H. U., M. Weisheit, J. Faupel, E. Suske, T. Scharf, C. Fuhse, M. Stormer, K. Sturm, M. Seibt, H. Kijewski, D. Nelke, E. Panchenko, and M. Buback. 2003. "Pulsed laser deposition (PLD)—A versatile thin film technique." *Advances in Solid State Physics* 43:505–517.

Kreuer, K. D. 2003. "Proton-conducting oxides." *Annual Review of Materials Research* 33 (1):333–359. doi: 10.1146/annurev.matsci.33.022802.091825.

Kwon, C. W., J. W. Son, J. H. Lee, H. M. Kim, H. W. Lee, and K. B. Kim. 2011. "High-performance micro-solid oxide fuel cells fabricated on nanoporous anodic aluminum oxide templates." *Advanced Functional Materials* 21 (6):1154–1159. doi: 10.1002/adfm.201002137.

Lai, B. K., K. Kerman, and S. Ramanathan. 2010. "On the role of ultra-thin oxide cathode synthesis on the functionality of micro-solid oxide fuel cells: Structure, stress engineering and in situ observation of fuel cell membranes during operation." *Journal of Power Sources* 195 (16):5185–5196. doi: 10.1016/j.jpowsour.2010.02.079.

Lai, B. K., K. Kerman, and S. Ramanathan. 2011a. "Methane-fueled thin film micro-solid oxide fuel cells with nanoporous palladium anodes." *Journal of Power Sources* 196 (15):6299–6304. doi: 10.1016/j.jpowsour.2011.03.093.

Lai, B.-K., K. Kerman, and S. Ramanathan. 2011b. "Nanostructured $La_{0.6}Sr_{0.4}Co_{0.8}Fe_{0.2}O_3$/ $Y_{0.08}Zr_{0.92}O_{1.96}$/$La_{0.6}Sr_{0.4}Co_{0.8}Fe_{0.2}O_3$ (LSCF/YSZ/LSCF) symmetric thin film solid oxide fuel cells." *Journal of Power Sources* 196 (4):1826–1832. doi: 10.1016/j.jpowsour.2010.09.066.

Lee, Y. H., I. Chang, G. Y. Cho, J. Park, W. Yu, W. H. Tanveer, and S. W. Cha. 2018. "Thin film solid oxide fuel cells operating below 600°C: A review." *International Journal of Precision Engineering and Manufacturing-Green Technology* 5 (3):441–453. doi: 10.1007/s40684-018-0047-0.

Lee, Y. H., G. Y. Cho, I. Chang, S. Ji, Y. B. Kim, and S. W. Cha. 2016. "Platinum-based nano-composite electrodes for low-temperature solid oxide fuel cells with extended lifetime." *Journal of Power Sources* 307:289–296. doi: 10.1016/j.jpowsour.2015.12.089.

Li, Y., P.-C. Su, L. M. Wong, and S. Wang. 2014. "Chemical stability study of nanoscale thin film yttria-doped barium cerate electrolyte for micro solid oxide fuel cells." *Journal of Power Sources* 268:804–809. doi: 10.1016/j.jpowsour.2014.06.128.

Li, Y., L. M. Wong, H. Xie, S. Wang, and P.-C. Su. 2017. "Nanoporous palladium anode for direct ethanol solid oxide fuel cells with nanoscale proton-conducting ceramic electrolyte." *Journal of Power Sources* 340:98–103. doi: 10.1016/j.jpowsour.2016.11.064.

Liu, K. Y., J. D. Baek, C. S. Ng, and P. C. Su. 2018. "Improving thermal stability of nanoporous platinum cathode at platinum/ yttria-stabilized zirconia interface by oxygen plasma treatment." *Journal of Power Sources* 396:73-79. doi: 10.1016/j.jpowsour.2018.06.018.

Liu, K. Y., L. Fan, C.-C. Yu, and P.-C. Su. 2015. "Thermal stability and performance enhancement of nano-porous platinum cathode in solid oxide fuel cells by nanoscale ZrO_2 capping." *Electrochemistry Communications* 56:65–69.

Malavasi, L., C. A. J. Fisher, and M. S. Islam. 2010. "Oxide-ion and proton conducting electrolyte materials for clean energy applications: Structural and mechanistic features." *Chemical Society Reviews* 39 (11):4370–4387. doi: 10.1039/b915141a.

Marina, O. A., N. L. Canfield, and J. W. Stevenson. 2002. "Thermal, electrical, and electro-catalytical properties of lanthanum-doped strontium titanate." *Solid State Ionics* 149 (1):21–28. doi: 10.1016/S0167–2738(02)00140-6.

Mond, L., and C. Langer. 1890. "A new form of gas battery." *Proceedings of the Royal Society of London* 46:296–304.

Motoyama, M., C. C. Chao, J. H. An, H. J. Jung, T. M. Gur, and F. B. Prinz. 2014. "Nanotubular array solid oxide fuel cell." *ACS Nano* 8 (1):340–351. doi: 10.1021/Nn4042305.

Muecke, U. P., D. Beckel, A. Bernard, A. Bieberle-Hutter, S. Graf, A. Infortuna, P. Muller, J. L. M. Rupp, J. Schneider, and L. J. Gauckler. 2008. "Micro solid oxide fuel cells on glass ceramic substrates." *Advanced Functional Materials* 18 (20):3158–3168. doi: 10.1002/adfm.200700505.

Noh, H. S., J. W. Son, H. Lee, H. S. Song, H. W. Lee, and J. H. Lee. 2009. "Low Temperature performance improvement of SOFC with thin film electrolyte and electrodes fabricated by pulsed laser deposition." *Journal of the Electrochemical Society* 156 (12):B1484–B1490. doi: 10.1149/1.3243859.

Oh, E. O., C. M. Whang, Y. R. Lee, S. Y. Park, D. H. Prasad, K. J. Yoon, J. W. Son, J. H. Lee, and H. W. Lee. 2012. "Extremely thin bilayer electrolyte for solid oxide fuel cells (SOFCs) fabricated by chemical solution deposition (CSD)." *Advanced materials* 24 (25):3373-7. doi: 10.1002/adma.201200505.

Park, J., J. Y. Paek, I. Chang, S. Ji, S. W. Cha, and S. I. Oh. 2013. "Pulsed laser deposition of Y-doped $BaZrO_3$ thin film as electrolyte for low temperature solid oxide fuel cells." *Cirp Annals-Manufacturing Technology* 62 (1):563–566. doi: 10.1016/j.cirp.2013.03.025.

Pergolesi, D., E. Fabbri, A. D'Epifanio, E. Di Bartolomeo, A. Tebano, S. Sanna, S. Licoccia, G. Balestrino, and E. Traversa. 2010. "High proton conduction in grain-boundary-free yttrium-doped barium zirconate films grown by pulsed laser deposition." *Nature Materials* 9 (10):846–852. doi: 10.1038/Nmat2837.

Pornprasertsuk, R., P. Ramanarayanan, C. B. Musgrave, and F. B. Prinz. 2005. "Predicting ionic conductivity of solid oxide fuel cell electrolyte from first principles." *Journal of Applied Physics* 98 (10):103513. doi: 10.1063/1.2135889.

Quinn, D. J., B. Wardle, and S. M. Spearing. 2008. "Residual stress and microstructure of as-deposited and annealed, sputtered yttria-stabilized zirconia thin films." *Journal of Materials Research* 23 (3):609–618. doi: 10.1557/Jmr.2008.0077.

Rey-Mermet, S., and P. Muralt. 2008. "Solid oxide fuel cell membranes supported by nickel grid anode." *Solid State Ionics* 179 (27-32):1497–1500. doi: 10.1016/j.ssi.2008.01.007.

Scherrer, B., A. Evans, A. J. Santis-Alvarez, B. Jiang, J. Martynczuk, H. Galinski, M. Nabavi, M. Prestat, R. Tolke, A. Bieberle-Hutter, D. Poulikakos, P. Muralt, P. Niedermann, A. Dommann, T. Maeder, P. Heeb, V. Straessle, C. Muller, and L. J. Gauckler. 2014. "A thermally self-sustained micro-power plant with integrated micro-solid oxide fuel cells, micro-reformer and functional micro-fluidic carrier." *Journal of Power Sources* 258:434–440. doi: 10.1016/j.jpowsour.2014.02.039.

Shim, J. H., C. C. Chao, H. Huang, and F. B. Prinz. 2007. "Atomic layer deposition of yttria-stabilized zirconia for solid oxide fuel cells." *Chemistry of Materials* 19 (15):3850–3854. doi: 10.1021/Cm070913t.

Singhal, S. C. 2000. "Advances in solid oxide fuel cell technology." *Solid State Ionics* 135 (1–4):305–313. doi: 10.1016/S0167-2738(00)00452-5.

Smeacetto, F., M. Salvo, L. C. Ajitdoss, S. Perero, T. Moskalewicz, S. Boldrini, L. Doubova, and M. Ferraris. 2010. "Yttria-stabilized zirconia thin film electrolyte produced by RF sputtering for solid oxide fuel cell applications." *Materials Letters* 64 (22):2450–2453. doi: 10.1016/j.matlet.2010.08.016.

Spearing, S. M. 1997. "Design diagrams for reliable layered materials." *AIAA Journal* 35 (10):1638–1644. doi: 10.2514/2.3.

Srikar, V. T., K. T. Turner, T. Y. A. Ie, and S. M. Spearing. 2004. "Structural design considerations for micromachined solid-oxide fuel cells." *Journal of Power Sources* 125 (1):62–69. doi: 10.1016/j.jpowsour.2003.07.002.

Steele, B. C. H. 1995. "Interfacial reactions associated with ceramic ion-transport membranes." *Solid State Ionics* 75:157–165. doi: 10.1016/0167-2738(94)00182-R.

Su, P. C., C. C. Chao, J. H. Shim, R. Fasching, and F. B. Prinz. 2008. "Solid oxide fuel cell with corrugated thin film electrolyte." *Nano Letters* 8 (8):2289–2292. doi: 10.1021/Nl800977z.

Su, P. C., and F. B. Prinz. 2011. "Cup-shaped yttria-doped barium zirconate membrane fuel cell array." *Microelectronic Engineering* 88 (8):2405–2407. doi: 10.1016/j.mee.2010.12.006.

Su, P. C., and F. B. Prinz. 2012. "Nanoscale membrane electrolyte array for solid oxide fuel cells." *Electrochemistry Communications* 16 (1):77–79. doi: 10.1016/j.elecom.2011.12.002.

Sun, C., R. Hui, and J. Roller. 2009. "Cathode materials for solid oxide fuel cells: A review." *Journal of Solid State Electrochemistry* 14 (7):1125–1144. doi: 10.1007/s10008-009-0932-0.

Takagi, Y., S. Adam, and S. Ramanathan. 2012. "Nanostructured ruthenium - gadolinia-doped ceria composite anodes for thin film solid oxide fuel cells. " *Journal of Power Sources* 217:543–553. doi: 10.1016/j.jpowsour.2012.06.060.

Takagi, Y., K. Kerman, C. Ko, and S. Ramanathan. 2013. "Operational characteristics of thin film solid oxide fuel cells with ruthenium anode in natural gas." *Journal of Power Sources* 243:1–9. doi: 10.1016/j.jpowsour.2013.06.002.

Tang, Y., K. Stanley, J. Wu, D. Ghosh, and J. Zhang. 2005. "Design consideration of micro thin film solid-oxide fuel cells." *Journal of Micromechanics and Microengineering* 15 (9):S185–S192. doi: 10.1088/0960-1317/15/9/s03.

Thiele, E. S., L. S. Wang, T. O. Mason, and S. A. Barnett. 1991. "Deposition and Properties of Yttria-Stabilized Zirconia Thin-Films Using Reactive Direct-Current Magnetron Sputtering." *Journal of Vacuum Science & Technology a-Vacuum Surfaces and Films* 9 (6):3054–3060. doi: 10.1116/1.577172.

Thompson, C. V., and R. Carel. 1996. "Stress and grain growth in thin films." *Journal of the Mechanics and Physics of Solids* 44 (5):657–673. doi: 10.1016/0022-5096(96)00022-1.

Tietz, F. 1999. "Thermal expansion of SOFC materials." *Ionics* 5 (1):129–139. doi: 10.1007/BF02375916.

Timoshenko, S. P., and J. M. Gere. 2009. *Theory of Elastic Stability*, Mineola, NY: Dover Publications.

Tolke, R., A. Bieberle-Hutter, A. Evans, J. L. M. Rupp, and L. J. Gauckler. 2012. "Processing of Foturan (R) glass ceramic substrates for micro-solid oxide fuel cells." *Journal of the European Ceramic Society* 32 (12):3229–3238. doi: 10.1016/j.jeurceramsoc.2012.04.006.

Tsuchiya, M., B. K. Lai, and S. Ramanathan. 2011. "Scalable nanostructured membranes for solid-oxide fuel cells." *Nature nanotechnology* 6 (5):282–286. doi: 10.1038/Nnano.2011.43.

Tuller, H. L. 2000. "Ionic conduction in nanocrystalline materials." *Solid State Ionics* 131 (1–2):143–157. doi: 10.1016/S0167-2738(00)00629-9.

van den Boogaart, M. A. F., M. Lishchynska, L. M. Doeswijk, J. C. Greer, and J. Brugger. 2006. "Corrugated membranes for improved pattern definition with micro/nanostencil lithography." *Sensors and Actuators A: Physical* 130–131:568–574. doi: 10.1016/j.sna.2005.08.037.

Ventsel, E., and T. Krauthammer. 2001. *Thin Plates and Shells: Theory: Analysis, and Applications*, Boca Raton, FL: CRC Press.

Vinci R. P., and J. J. Vlassak. 1996. "Mechanical behavior of thin films." *Annual Review of Materials Science* 26 (1):431–462. doi: 10.1146/annurev.ms.26.080196.002243.

Wachsman, E. D., and K. T. Lee. 2011. "Lowering the temperature of solid oxide fuel cells." *Science* 334 (6058):935–939. doi: -10.1126/science.1204090.

Weissbart, J., and R. Ruka. 1962. "A solid electrolyte fuel cell." *Journal of the Electrochemical Society* 109 (8):723–726.

Windischmann, H. 1992. "Intrinsic stress in sputter-deposited thin films." *Critical Reviews in Solid State and Materials Sciences* 17 (6):547–596. doi: 10.1080/10408439208244586.

Yan, Y., S. C. Sandu, J. Conde, and P. Muralt. 2012. "Experimental study of single triple-phase-boundary and platinum-yttria stabilized zirconia composite as cathodes for micro-solid oxide fuel cells." *Journal of Power Sources* 206:84–90. doi: 10.1016/j.jpowsour.2012.01.113.

8 *In-Situ* Mechanistic Study of Two-Dimensional Energy Materials by Well-Defined Electrochemical On-Chip Approach

Yu Zhou
Central South University

Shuang Yang
East China University of Science and Technology

Fangping Ouyang
Central South University

CONTENTS

8.1 Introduction 285
8.2 Overview of the Development of 2D Materials in Electrochemical
Conversion or Storage Devices...286
8.3 On-Chip Electrochemical Approach 292
    8.3.1 Droplet Confined Electrochemical Reactor....................................292
    8.3.2 On-Chip Polymer Confined Microreactor.......................................295
    8.3.3 Fundamental Studies on the Physicochemical Parameters 302
8.4 Conclusions...306
Acknowledgment 306
References..306

8.1 INTRODUCTION

The structure of materials including crystal structure/symmetry, band structure, and surface structure determines its physicochemical properties and potential applications in electronic, optoelectronic, and electrochemical energy devices.[1] Recently, layered materials with in-plane strong chemical bonding and weak intralayer van der Waals gaps have been exfoliated or synthesized as ultrathin two-dimensional (2D) materials.[2] Differing from the bulk counterpart of layered materials by their unique physical properties, 2D materials have been widely studied by physicists, chemists, and materials scientists.[3] For instance, graphene consisting of a single layer of sp[2]-hybridized carbon atoms shows ultrahigh mobility of ~200,000 cm[2]V s[-1], room temperature quantum hall effect, and also high electrical conductivity for energy applications.[1,3] Other notable 2D materials such as MoS_2 and black phosphorus were also used for electronic, optical, and optoelectronic applications because of their high carrier mobility, tunable bandgap, and unique band structures.[4,5]

Energy consumption of humans is huge and is rising to be the most urgent problem in the 21st century, followed by environmental pollution from traditional power supply systems.[1] Therefore, developing reliable clean energy conversion, storage, and transportation technology is important for science as well as real industry-level application. The conversion and storage of energy from chemical bonds can be achieved by clean, mild, and controllable electrochemical reactions. Ultrathin 2D layered materials have emerged as one of the most promising energy materials due to their large surface area, high intrinsic catalytic activity, controllable chemical coordination bonding states, and high theoretical storage capacity.[6] The anisotropic bonding nature, specific electronic structures, and unique thermodynamic absorptions of graphene-like 2D materials could make them the ideal model systems for electrochemical conversion or storage devices for electro-catalytic/photocatalytic reactions, metal ion batteries, supercapacitors fields, etc.[7–10]

In this chapter, we start with a general overview of the research progress of 2D materials in the field of energy devices. Then, we review the operation mechanism of in-situ observation technology for 2D materials energy devices. Finally, the development history and progress of in-situ on-chip well-defined electrochemical approach are summarized.

8.2   OVERVIEW OF THE DEVELOPMENT OF 2D MATERIALS IN ELECTROCHEMICAL CONVERSION OR STORAGE DEVICES

Since the first exfoliated graphene was awarded the Nobel Prize in 2004, the family of 2D materials has expanded to few-layer or monolayer hexagonal boron nitride (h-BN); silicon; black phosphorus; layered double hydroxide (LDH); transition metal dichalcogenides (TMDs); transition metal carbides, nitrides, and oxides (TMOs); and carbonitrides (MXenes).[3,11–19] The common characteristic structures are those in which crystals are chemically bonded in the x–y plane and weakly boned with van der Waals force along z direction.[20] Based on the nature of crystal structure, tens of synthetic strategies have been invented such as mechanical or chemical exfoliation, ligand confined synthesis, chemical vapor deposition, and molecular beam epitaxy growth. The top-down exfoliation of bulk layered materials benefits from the weak van der Waals interaction between layers, in which external forces could be applied to separate thick layers and thin layers. On the other hand, bottom-up synthesis routes for 2D nanostructures involve formation of chemical bonds with dangling bonds in the in-plane directions, rather than the out-of-plane directions, thus preferring to assemble single layer or few layer atomic crystals on the substrate.[21,22] As the synthesis of 2D materials is rapidly increasing, these 2D materials are being intensively used in energy storage and conversion devices,[23] including photovoltaic solar devices[24,25]; thermoelectric,[26-29] piezoelectric,[30-32] and triboelectric devices[33-35]; and electrochemical devices.[36-38] Here, we focus on the electrochemical process occurring in 2D materials in batteries and supercapacitors.

Water splitting is a basic chemical reaction in which the chemical bonds between hydrogen and oxygen are broken down[39] and during which oxygen and hydrogen gases can be formed as $2H_2O \rightarrow H_2 + O_2$. Two component reactions, i.e., H_2 production and O_2 production, are involved in the process, which can be driven by electrolysis, photosynthesis, and photoelectrochemical and photocatalytic reactions.[40] The electrolysis of water splitting producing hydrogen (H_2) is a hydrogen evolution reaction (HER), converting the chemical energy into hydrogen gas with the reaction formula of $2H^+ + 2e \rightarrow H_2$.[41,42] Producing hydrogen gas from water electrolysis can be an ideal technology for clean hydrogen economy, in which hydrogen can be used as a non-carbon fuel for fuel cell vehicles and transportation of heat and energy over long distances. However, the consumption of electrical power needs to be largely reduced by using highly efficient catalysts.[43] Previous studies have demonstrated that platinum group metals can be used as highly efficient catalysts to drive water electrolysis with small electrochemical overpotentials.[44-48] These catalysts are expensive, and their large-scale fabrication is difficult.[49,50] Due to these problems, several groups have tried to develop earth abundant ultrathin 2D materials as highly efficient hydrogen production catalysts.[51] As shown in Figure 8.1a, Manish Chhowalla group intercalated bulk WS_2 with n-butyl lithium source in the inert environment.[52] Noteworthily, organic n-butyl lithium is spontaneously combustible in air; thus dealing with waste chemicals must be carefully controlled by long-chain alkanes. Also lithium intercalated WS_2 must be dried in the inert gas to avoid oxidation. Afterward, lithium intercalated WS_2 must be transferred into water solution with sonication, in which splitting of hydrogen bubbles from water could expand the interlayers of WS_2 and form monolayer WS_2. Atomic resolution transition electron microscopy image of exfoliated monolayer WS_2 shown in Figure 8.1b clearly shows its distorted 1T structure with a $2a_0 \times a_0$ superlattice, which is similar to the 1T structure of TiS_2. The atom-by-atom observation of zigzag patterns indicates that a high concentration of local strained crystal lattice exists in the exfoliated monolayer $1T–WS_2$.

Standard electrochemical measurements were performed on the bulk WS_2 samples, exfoliated $1T–WS_2$ film, sub-monolayer $1T WS_2$ flakes, and annealed $1T–WS_2$ film (300°C) in which glassy carbon was used as the working electrode and 0.5M H_2SO_4 solution deaerated with Ar was used as the electrolyte. The polarization curves of exfoliated $1T–WS_2$ film in Figure 8.1d show the onset potential in the range of 80–100 mV, the overpotentials around 200 mV at 10 mA cm^{-2} current density, and the Tafel slope at 55~60 mV dec^{-1}. Such an enhanced performance of $1T–WS_2$ films was attributed to local strained structure derived from the neutral thermodynamic hydrogen adsorption. The local strains also have been confirmed by the transmission electron microscopy (TEM) images. However, the transformation from semiconducting $2H–WS_2$ to metallic $1T–WS_2$ not only induces the change of Gibbs energy but also changes the catalysts' conductivity and the interfacial barriers between catalysts and current collectors. At the same time, similar results were observed in the MoS_2

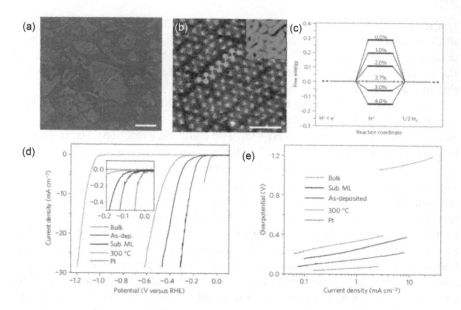

FIGURE 8.1   Chemical exfoliation of bulk WS$_2$ into monolayer 1T WS$_2$ by lithium inter-calation for enhanced HER. (a) Typical dispread exfoliated WS$_2$ on the SiO$_2$/Si substrate characterized by atomic force microscopy. Scale bar, 500 nm. (b) High-resolution STEM image of as-exfoliated WS$_2$ showing the 1T atomic structure. Scale bar, 1 nm. (c) Calculated hydrogen adsorption Gibbs free energy (ΔG_H) as a function of lattice strains. (d) Polarization curves of different treated WS$_2$ samples (1T phase film, sub-monolayer film, 2H phase film, and 2H phase after annealing at 300°C). (e) Extracted Tafel slope of the polarization curves obtained in (d).

family materials also, in which enhanced electrocatalytic performances were either summarized as the increased conductivity or thermo-neutral Gibbs free energy.[53–56] The density functional calculations could not consider the charge transfer and transport process across various interfaces and catalysts, though its results match well with current experimental results. Deep structure–properties studies still need to be carried out for understanding how various controlling factors can influence the overall performance of electrochemical catalytic reaction.[57–59]

Subsequently, ultrathin 2D materials have been widely applied in hydrogen production, oxygen production, and also carbon dioxide reduction. Xie et al. reported the synthesis of atomically thin layered metal cobalt nanosheets, including partially oxidized cobalt nanosheets and defect rich tricobalt oxide nanosheets, whose electrocatalytic activities were much higher than that of their bulk samples in the carbon dioxide reduction reaction.[60,61] Such catalysts that drive the CO$_2$ reduction and convert highly efficient liquid products are very different compared to the traditional catalysts that convert gaseous molecules.[60,61] Zhao Cai et al. have demonstrated that ultrathin Co$_3$O$_4$ nanosheets rich in oxygen defects show a lower oxygen evolution reaction (OER) overpotential of 220 mV with a small Tafel slope of 49.1 mV decade^{-1}, in which the specific O-terminated facets of Co$_3$O$_4$ nanosheets were synthesized by mild solvothermal reduction using ethylene glycol under alkaline condition.[62] The above-mentioned examples indicate that ultrathin 2D materials have the advantages of large specific surface area, high atomic ratios, and unique atomic chemical coordination for achieving efficient conversion.[62-65]

2D materials in high-energy batteries such as Si anodes, Li–S, and Li–air have shown great potential and rapid progress because of their crystal structure, high theoretical capacity, and electrochemical properties.[66-69] Kai Yan et al. tried an advanced electrode of Cu metal with deposited 2D graphene layers and few layers of hexagonal boron nitride (h-BN) acting as confined layers for preventing lithium dendrites. Thus, over 50 cycles with 97% columbic efficiency with current density and areal capacity up to the practical values of 2.0 mA cm^{-2} and 5.0 mAh cm^{-2} were observed without dendritic and mossy Li formation in the lithium deposition process using an organic carbonate electrolyte. This is demonstrated by the improvement of the unprotected electrodes in the same electrolyte by using the 2D materials.[9] Li–S batteries have the potential to surpass the performance of Li-ion batteries, whose energy density is ~2,600 Wh kg^{-1} which is about five times higher than that of Li-ion batteries. The utilization of metallic Li anode is indispensable for the operation of Li–S batteries, in which the prevention of parasitic dendrites' growth, high reactivity with several electrolytes, and the dissolution of polysulfides are not easy controllable. As shown in Figure 8.2a, ultrathin 2D molybdenum disulfide (MoS_2) layer with 10 nm thickness was deposited on the lithium foil acting as a protective barrier. The layered structures of MoS_2 can benefit the intercalation of lithium atoms, thus reducing the interfacial resistance and controlling the Li$^+$ entering and coming out of Li metals. Figure 8.2b shows the galvanostatic charge/discharge (GCD) profiles of Li–MoS_2/CNT–S cells in different cycles of a long-term cycling test at 0.5 C for over 1,200 cycles.[70] The reversible specific capacity of 1,105 mAh g^{-1} at 0.5 C for the first cycle was obtained, and two plateaus in the discharge profile are observed at ~2.1 V and ~1.9 V. Figure 8.2c shows bare Li electrodes was killed around 120th cycle with a sudden increase in polarization, through similar overpotential behavior was initially observed. The combination of a MoS_2 Li-metal anode and a carbon nanotube (CNT)–S composite cathode seems to encapsulate soluble polysulfides, which probably results in the high-performance Li–S batteries.[70] As concluded in Figure 8.2d, a specific energy density of ~589 Wh kg^{-1}, high cycling stability of over 1,200 cycles, and an average Coulombic efficiency of ~98% were obtained for the first time. Through the understanding of a device's operating mechanism, unique atomically layered structure, its phase-change characteristics (semiconductor to metallic transition, 2H phase to 1T phase), and increased conductivity of intercalated MoS_2 the problem of high impedance and/or poor interfacial contacts was solved and the preferential sites for Li dendrite nucleation were suppressed.[70] Such results indicate that it is possible to use 2D materials for components to solve the major challenges of rechargeable batteries.

In addition, controlling the intercalation and deintercalation of ions in the layered materials is the basis of electrochemical energy storage devices. By restacking the exfoliated layered materials, we can control the distances of different layers and assemble other ions, atom and other materials to form composites. With greater tuning ability, more approaches can be discovered for achieving theoretical performance limits.

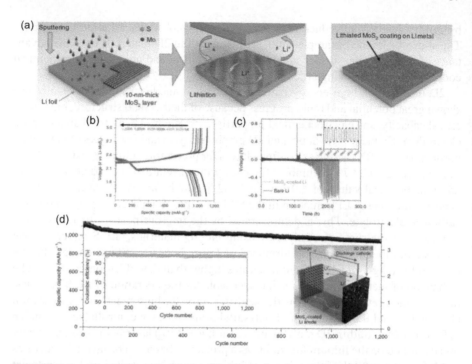

FIGURE 8.2   2D MoS₂ as Stable Interfacial Layer for lithium metal anodes in high-performance Li–S batteries. (a) Schematic illustration of a thin layer MoS₂ coated Li anode via sputtering and lithiation. Left: Sputtering 10 nm MoS₂ layer on the Li foil. Middle: Operating the lithiation of MoS₂ layer through the cell configuration cycling for 15 cycles (30 min charge/discharge per cycle) at 1 mA cm⁻². Right: Disassembling the symmetric cell to obtain the lithiated MoS₂-coated Li electrode. (b) The galvanostatic charge/discharge plots of the corresponding Li–MoS₂/CNT–S cell for up to 1,200 cycles at 0.5 C. (c) Constant current charge/discharge voltage plots for bare Li metal and MoS₂-coated Li metal symmetric cells with a 10 mA cm⁻² cycled current density. Inset: enlarged voltage plots for MoS₂-coated Li electrode. (d) The performance of Li–S battery with the 3D CNT–S cathode (~33 wt% S content) and the MoS₂-coated Li anode with long-term cycling at 0.5 C.

Assembling supercapacitors with restacked graphene nanosheets has shown relatively high volumetric capacitances of ~300 F cm⁻³.[71] Gogotsi, Yury et al. have confirmed MXenes-type 2D materials could be high-performance capacitator materials with volumetric capacitances of over 900 F cm⁻³, which are conductive and hydrophilic intrinsically.[19] As shown in Figure 8.3, exfoliation methods similar to those shown in Figure 8.1 were used for achieving the monolayer 1T MoS₂ from organolithium chemistry assisted by water intercalation.[72] Suspended 1T–MoS₂ nanosheets were fabricated as thick film or "paper" by the traditional filtration methods, as shown in Figure 8.3b and c. The electrical conductivity of 1T phase MoS₂ is about 10⁷ times higher than that of semiconducting 2H phase. The cyclic voltammograms in Figure 8.3e show rectangular curves even at high scan rates (200 mV s⁻¹) with high capacitances of 400–650 F cm⁻³, which were obtained with 0.5 M Na₂SO₄ with the scan rate of 5 mV s⁻¹–1,000 mV s⁻¹. The galvanostatic charge/discharge

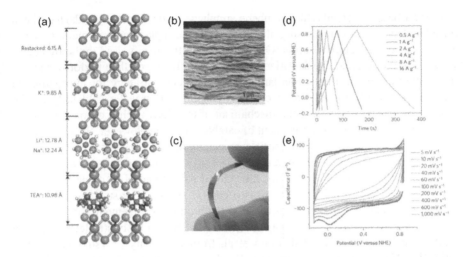

FIGURE 8.3 **Restacked metallic 1T phase MoS$_2$ nanosheets as supercapacitor electrode materials.** (a) Schematics of restacked non-intercalated and intercalated 1T MoS$_2$ nanosheets, in which the spacing is analyzed by the XRD measurement. (b and c) SEM image and photographs of restacked 1T MoS$_2$ electrode. (d) Galvanostatic cycles of the supercapacitors from 0.5 A g^{-1}–16 A g^{-1} in K$_2$SO$_4$. (e) CVs of restacked 1T phase MoS$_2$ electrodes in 0.5 M Na$_2$SO$_4$ with the scan rate of 5 mV s^{-1}–1,000 mV s^{-1}.

characterization of assembled 1T–MoS$_2$ measured at 0.5, 1, 2, 4, 8, and 16 A g^{-1} in Na$_2$SO$_4$ is shown in Figure 8.3d, in which triangular capacitive behaviors are clearly demonstrated. The long-term stability of MoS$_2$ electrodes was long enough to retain 5,000 charge/discharge cycles at a current rate of 2 A g^{-1}. The intercalation of H$^+$, Li$^+$, Na$^+$, and K$^+$ with extraordinary efficiency contributes to high capacitance in various aqueous electrolytes up to 700 F cm^{-3}.

Currently, most of the research labs are focusing on prototype electrochemical devices driven by the performance metric analysis, which is useful for identifying the materials. In reality, the large-scale industry-level application requires a reliable material synthesis method, post-processing method, and also device fabrication. However, the required energy-related technology cannot be achieved without deep understanding of the operation mechanism of electrochemical devices. The structure–property relationships study is highly important for revealing the design principles of materials and devices. Various in-situ or operando methods like synchrotron-radiation-related technology, optical spectroscopy, and in-situ transmission electron microscopy were used to characterize the underlying mechanism under the operation state, including its crystal structure, chemical bonds, and spatial-resolved atomic-level information.[73] The mechanism of energy conversion reaction can be exploited by simultaneous in-situ measurement of the electrical and electrochemical signals of the nanodevice. On-chip electrochemical approaches have not only provided well-defined microreactor systems, but also made it possible to study the electrokinetics, structure–property relationships, and fundamental physicochemical parameters in electrochemical energy conversion field.

Carbon-based materials (like graphene, graphene oxide, and its hybrids) have been considered as the most promising alternatives for the replacement of platinum catalysts in oxygen reduction reaction (ORR), which are frequently used in the fuel cells, metal–air batteries, and dye-sensitized solar cells for their excellent performance.[74–77] Heteroatom doping in the carbon lattices is one efficient strategy for improving the thermodynamic O_2 absorption or charge transport and transfer.[78] Identifying the active sites and understanding the mechanism of the ORR in the carbon catalysts is important for the further improvement of catalysts' design. However, the ORR active sites of nitrogen-doped carbon materials especially are still unclear or under debate. The researchers have discussed that if the active sites are created by graphitic N (grap-N, N bonded to three carbon atoms, also called substituted N or quaternary N) or by pyridinic N (pyri-N, N bonded to two carbon atoms). The structures of doped bulk carbons and nanomaterials, including the atomic arrangement and interfacial bonding, are complex and inhomogeneous and difficult to be controlled during the synthesis procedures. Junji Nakamura et al. from University of Tsukuba (*Science 2016,351,6271*) fabricated edge-patterned carbon catalysts by Ar^+ etching technique with a mask on the highly oriented pyrolytic graphite (HOPG). Thus they smartly fabricated different types of N^+ doped edges on the HOPG: (1) pyridinic N-dominated HOPG (pyri-HOPG), (2) graphitic N-dominated HOPG (grap-HOPG), and (3) edges patterned on the surface without N (edge-HOPG) annealing under NH_3 atmosphere. The nitrogen–carbon chemical bonding states were confirmed by the X-ray photoelectron spectroscopy. Based on the different types of model catalysts, they assembled the ORR measurement systems and compared their performance. Cyclic voltammetry (CV) measurements were used to evaluate their intrinsic catalytic properties in acidic electrolyte (0.1 M H_2SO_4) with oxygen-saturated conditions. They showed that the N-free edge patterned HOPG catalyst exhibits much lower activity at high voltages compared to the pyri-HOPG catalyst. The higher N concentration (N: 0.73 at. %) grap-HOPG sample displays lower activity than the lower N concentration (N: 0.60 at. %) pyri-HOPG sample catalyst. We can infer from the ORR results that pyridinic N sites are better for reducing the ORR overpotentials and can be defined as the most active sites. Finally, a possible mechanism for the ORR on nitrogen-doped carbon materials has been proposed. The carbon atoms next to pyridinic N with Lewis basicity should play as the initial active site to absorb the oxygen molecule. The reaction could happen at a single site with four-electron mechanism or combine two oxygen molecules at two different sites with 2+2-electron mechanism. However, the controlled synthesis methods to expose certain active sites are quite limited, and the ORR performance is also obtained from the statistical estimation rather than the precise estimation, and similar problems were encountered in other energy materials applications.[57,67,79–81]

8.3   ON-CHIP ELECTROCHEMICAL APPROACH

8.3.1   Droplet Confined Electrochemical Reactor

The key question for separating the complex reaction factors is how to miniaturize macroelectrochemical reactors as microreactors or nanoreactors. The earliest well-defined electrochemical reaction is the droplet confined ORR. Wang and Dai et al. demonstrated that the ORR electrocatalytic active sites of highly oriented pyrolytic graphite are located at the edges rather than the basal plane by confining the reaction in a solution droplet with a diameter around 15 μm in Figure 8.4a.[82] The droplet diameter and location can be controlled by a microinjection system, which could precisely define the reaction areas and the types of active sites (Figure 8.4b and c). An ultrafine capillary tube consisting of the electrolyte has been utilized to fill with Pt wire and a Ag/AgCl wire, connecting with the pre-deposited droplet, which could be considered as the well-defined reaction reactor.

Thus, they directly confirmed the overall reaction performance of graphite edges is much higher than that of the graphite basal plane (Figure 8.4d), which helps us understand the relationship between fundamental ORR activities and the carbon-based materials. The fabrication of edge-rich carbon materials could be an alternative way for high-temperature-doped carbon materials or platinum-based materials to perform the ORR.[83-86]

Robert A. W. Dryfe et al. from University of Manchester used a similar droplet reactor to perform the electrochemical and photoelectrochemical reactions on the monolayer, few-layer, and bulk MoS$_2$ with/without incident light illumination[87] (Figure 8.5a). A 200 nm layer of poly (methyl methacrylate) (PMMA) coated Si wafer

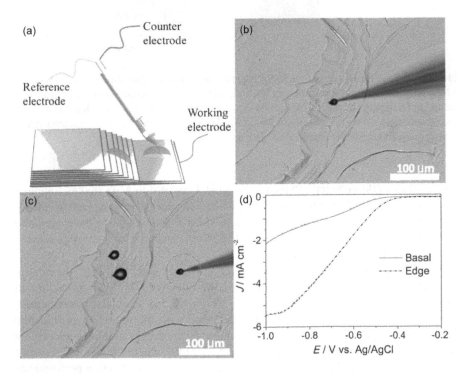

FIGURE 8.4 Droplet confined electrochemical ORR. (a) The schematic of droplet confined reactor on the graphite basal planes or edges connected with counter and reference electrodes. (b) The reaction on the graphite edge. (c) The reaction on the graphite basal plane. (d) The ORR performance for the corresponding reactor in (b) and (c). Reprinted with permission.

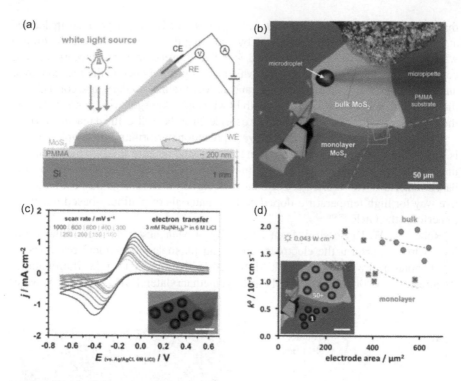

FIGURE 8.5 Photoelectrochemical Property of Pristine Monolayer and Few-Layer MoS₂ with the droplet reactors. (a) The schematic of photoelectrochemical microreactor setup. (b) The droplet reactor on the bulk MoS₂. (c) Cyclic voltammograms of [Ru(NH₃)₆]3+/2+ reduction/oxidation on the monolayer MoS₂. (d) The plots for the relations between electron transfer rate and electrode areas. All scale bars are 50 μm.

was used as a reference for enhancing the optical contrast of MoS₂ flakes compared to that of on the SiO2/Si substrate (Figure 8.5b). The researchers emphasized the use of highly concentrated aqueous electrolyte (6 M LiCl) for preventing the evaporation of electrolyte during the measurements. The silver paste and copper wires serve as the connecting electrical circuits.

From the measurements for the monolayer, few-layer, and bulk MoS₂, the electron transfer kinetics show significant variation in the values of 1.95×10^{-3} cm s⁻¹ for bulk MoS₂ and $\sim10^{-8}$ cm s⁻¹ for monolayer MoS₂ (Figure 8.5d). They also systematically found that the electron transfer kinetics decrease with the reduction in thickness despite the sample variation or the illumination. However, the authors neglected the contact barriers between the monolayer and bulk MoS₂, where the electron transportation across the interface could be a restrictive factor for the reaction performance. They carried out the comparative qualitative correlation for the photoelectrochemical properties of pristine monolayer and few-layer basal plane MoS₂ through the electron transfer and electric double-layer capacitance measurements, which had not been achieved previously.[88–91]

8.3.2 ON-CHIP POLYMER CONFINED MICROREACTOR

In recent years, a universal on-chip polymer confined microreactor[92,93] has been developed as a new tool or technology for in-situ exploration of the mechanism of catalytic reaction, especially the structure–property study of model catalysts.[94,95] Based on the standard or modified micro-/nanodevice fabrication, a metallic micrometer size electrode connected with model catalysts and poly (methyl methacrylate) or photoresistive polymer windows could serve as the electrochemical working electrode and well-defined microreactor. PMMA is EBL resists, photoresistive polymer also can be used as mask materials. Thus, the types and densities of active sites, interface contacts, and multiple interfaces can be preferentially selected for the various 2D materials by recent synthesis strategies, such as liquid exfoliation, chemical vapor deposition, and electrochemical deposition. For the measurements, a regular three-electrode system could be replicated by using a home-made small-size reference electrode/counter electrode and small electrolyte droplets. Thus, this kind of on-chip polymer confined microreactor has new tools with a combination of advanced characterization methods. Precise extraction of fundamental electrochemical parameters[96] can be realized based on certain reaction area, such as the overpotential, exchange current density and Tafel slopes, and turnover frequency.[97,98] On-chip microbatteries also can be assembled to study changes of the 2D materials during charging and discharging.[99–101]

Transition metal dichalcogenides such as MoS_2 and WS_2 were theoretically and experimentally confirmed as the earth-abundant catalysts to replace the platinum group materials for driving hydrogen production in acidic solution.[64,102] Jens K. Nørskov et. al proposed a theoretical model to describe the possibility of proton adsorption with Gibbs free energy on certain active sites.[54] Using density functional theory (DFT) calculations, the calculated hydrogen adsorption Gibbs free energy (ΔG_H) finally suggested the activity of MoS_2 edge is quite close to that of Pt.[39,54] Jaramillo et al. have confirmed their hypothesis with model MoS_2 nanoclusters grown in the vacuum system, in which they demonstrated that metallic edges of MoS_2 are active and the sites on basal plane are inactive.[8] After that, a lot of research work for maximizing the exposed edges has been carried out on the MoS_2 or WS_2 nanostructures for the catalytic performance to reach the limit of thermodynamic predication.[103]

The interaction and influence of catalysts with substrates are also points to be considered for improving their overall activities. Manish Chhowalla et al. have made the microcells by using monolayer MoS_2 deposited on the SiO_2/Si substrate by chemical vapor deposition[104] (Figure 8.6a). The densities and lengths of the edges or the basal plane of monolayer MoS_2 can be controlled as shown in Figure 8.6c and d. The electron transfer through current collectors to catalysts is a decisive step to form final products. The contact resistance was controlled by lithium treatments for getting different concentrations of 1T phase. The band alignment was tuned by the chemically modified 1T phase MoS_2 for reducing the contact resistance, thus speeding the electron injection from the current collectors.[105,106] Figure 8.7a shows different polarization curves under various contact resistances measured in 0.5 M H_2SO_4. The onset potentials and Tafel slopes of 100 mV and 50 mV per decade can be extracted from 2H-phase catalysts with the basal plane exposed. The obvious enhanced catalytic

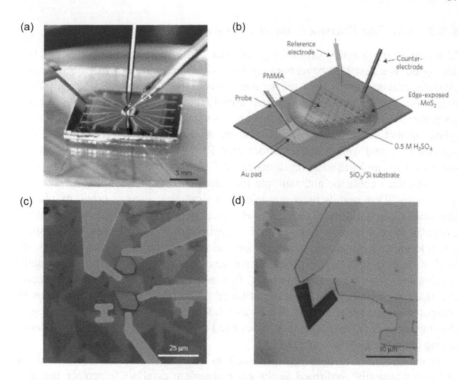

FIGURE 8.6 Microelectrochemical setup for the measurement of HER. (a) Photographs of home-made microcell setup based on probe station. (b) Schematic illustration of the three electrodes' electrochemical measurement setup. (c and d) Optical images of well-defined microreactor on the CVD monolayer MoS$_2$: basal plane exposed (c) and edge exposed (d).

activity of 2H basal plane was finally attributed to natural sulfur vacancies formed during the synthesis process. Scanning transmission electron microscopy (SEM) images of chemically deposited MoS$_2$ basal plane show us the samples have high and different sulfur vacancies up to 9% density. However, it is still under debate that amazingly the turnover frequencies for low-contact-resistance devices were > 100 s^{-1} and > 1,000 s^{-1} at overpotentials of 200 mV and 300 mV, respectively[104] (Figure 8.7b).

Nanostructured MoS$_2$ materials and electrochemical systems are much more complicated, involving defects, coupling effect with other materials, multiple interfaces and inhomogeneous distributions, and thermodynamic or kinetic control.[47,49,102,107–110] For understanding the contribution of single factors to HER, model catalysts of single-crystal MoS$_2$ and WTe$_2$ have been chosen for the HER performance comparison.[111,112] Figure 8.8a shows the thermodynamic hydrogen adsorption of MoS$_2$ is much better than that of WTe$_2$, in which the Mo-related edge is quite close to zero and the W edge γ(W) is far from zero.[113] In thermodynamics, MoS$_2$ is an excellent HER catalyst, and WTe$_2$ is terrible for HER.[112] However, as shown in Figure 8.8b, considering the interfacial barrier between current collectors and catalysts, the Schottky barrier of semiconducting MoS$_2$-current collectors is much higher than that of metallic-current collectors. Graphene flakes and metallic electrodes were used as the current collectors in Figure 8.8c, in which the geometrically constructed devices

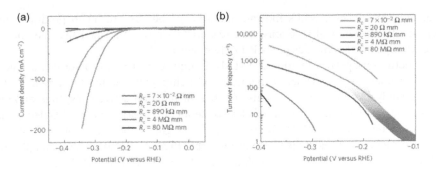

FIGURE 8.7 HER results of the monolayer MoS₂. (a) Polarization curves obtained from MoS₂ microcells under different contact resistances. (b) Calculated turnover frequency as a function of the overpotentials.

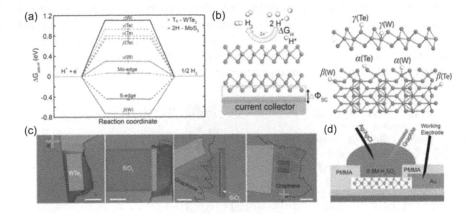

FIGURE 8.8 HER activity analysis based on the model catalyst for revealing the contribution of individual factors. (a) Comparison of thermodynamic hydrogen adsorption Gibbs free energy between T_d–WTe_2 and 2H–MoS_2 at various hydrogen-bonding sites (ΔG_H). (b) Three factors limiting the overall HER performance: thermodynamic hydrogen adsorption free energy (ΔG_H), interfacial barrier ($\Delta \Phi_{sc}$), and crystal structures of T_d–WTe_2 and 2H–MoS_2 (c) The WTe2 and MoS2 HER nanodevices with different current collectors (Au, graphene). (d) Schematic of HER confined nanodevices.

are very clean, and exposed reaction areas are also well-defined. With the polymer confined microreactor (Figure 8.8d), different types of interfacial barriers, different charge transport pathways, and different types of active sites are considered to understand the contribution of individual factors to the HER performance.[112]

Moreover, the electron mobility of WTe_2 ($10^4 cm^2 V^{-1} s^{-1}$) is hundred times higher than that of MoS_2 ($10^2 cm^2 V^{-1} s^{-1}$); hence the electron transport in the WTe_2 is much faster than that in the MoS_2. As a result, it is found that the rational design of current collectors for semiconducting catalysts is much more important than metallic catalysts, which could help eliminate the tunnel barrier to supply enough charge to the catalyst surface driving the electrochemical reaction. In addition, interlayer charge hopping efficiency of semi-metallic WTe_2 is amazingly thousand times higher than that of MoS_2. These results indicate metallic properties are much profitable for certain electrochemical reactions. The researchers also confirmed that the charge effects impeded the kinetics of HER in 2D materials.[114]

Structure–properties relationship study is highly critical for materials science, physical chemistry, and electrochemistry. On-chip polymer reactor provides the in-situ study platform for the precise control of reaction active sites, thus making one-to-one direct determination of electrochemical performance.[112,114] Defects and various boundaries of 2D MoS_2 could be the origin of the active sites[113,115,116] for the electrochemical reactions.[117,118] For instance, Figure 8.9 shows the argon plasma treatment developed by Guangyu Zhang et al. for converting 2H–MoS_2 into 1T–MoS_2, thus forming the domain boundary of 2H–1T phase.[119] Monolayer 2H–MoS_2 exhibits limited HER performance because of low thermodynamic activities and low density of active sites. Inducing different domain boundaries of 2H–2H, 1T–2H, 1T–1T, etc. could improve the activity of monolayer MoS_2.[120]

As shown in Figure 8.9d, monolayer MoS_2 deposited by chemical vapor deposition normally has a high density of 2H–2H domain boundaries that is clear in the dark TEM image. The different 1T–2H and 2H–2H domain boundaries were characterized by scanning tunneling microscopy and scanning tunneling spectroscopy, which provide information on the atomic arrangements and density of electrons (Figure 8.9b and

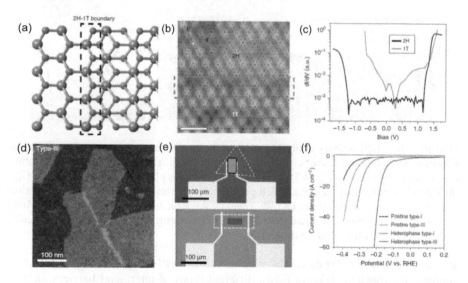

FIGURE 8.9 Specific boundary activated HER revealed by the multiple phase monolayer MoS_2 microreactor. (a) The top views of atomic crystal structure of the zigzag type 2H–1T phase boundaries. (b) Atomically resolved STM zoom-in image of a domain boundary denoted by a dashed square in (a), where 2H and 1T phases are rendered with yellow and purple colors, respectively. (c) Scanning tunneling spectroscopy (STS) spectra taken in the 1T and 2H phases, confirming the metallic 1T phase and 2.5 eV semiconducting 2H phase. (d) Dark-field TEM image of MoS_2 with high-density type-III 2H–2H domain boundaries. (e) Microreactor on the monolayer MoS_2 with 2H–2H boundary and 2H–1Tboundary. (f) Polarization curves of different structures of MoS_2. The pristine type-I MoS_2 has domain boundaries, pristine type-III MoS_2 has 2H–2H domain boundaries, heterophase type-I MoS_2 has 2H–1T domain boundaries, and heterophase type-III MoS_2 has both 2H–2H and 2H–1T domain boundaries.

c).[120] The 2H–2H boundaries show various configurations including arrays of 4–6 rings (4|6), 6–8 rings (6|8), 5–7 rings (5|7), and 4–4 rings (4|4), and the 2H–1T boundary is shown in Figure 8.9a. Therefore, five types of structures of domain boundaries were clearly revealed and constructed (Figure 8.9a). With clear optical contrast of two phases, four types of devices were fabricated (Figure 8.10e) as the pristine type-I MoS₂ (without any domain boundaries), pristine type-III MoS₂ (with 2H–2H domain boundaries), heterophase type-I MoS₂ (with 2H–1T domain boundaries), and heterophase type-III MoS₂ (with both 2H–2H and 2H–1T domain boundaries). Figure 8.9f shows polarization curves of heterophase type-III samples with the lowest overpotential of 200 mV and Tafel slope of 75 mV dec⁻¹. According to the comparison of different samples, creating a high-density domain and phase boundaries (1T–2H > 2H–2H) is most promising for enhanced HER.[120,121]

Monitoring and controlling reactivity and selectivity of electrochemical reactions by applying the external fields are long-term goals and challenges in electrochemistry.[122,123] Electrical and magnetic fields and thermal factors have been considered for accelerating various reactions.[124,125] The carrier type and density of 2D materials can be easily tuned by electrical fields mainly due to their low carrier concentration and the low penetration of electric field.[123,126] It is apparent that external electric fields can provide much more electrons by either increasing the conductivity of catalysts or accelerating the charge transfer process of the catalyst surface.[126] Figure 8.10 shows back-gate few-layer MoS₂ field-effect transistors used to study gate-dependent HER performance.[126] The probe station and electrochemical station were coupled together for measuring the HER under different gate voltages (Figure 8.10a). Inspired by

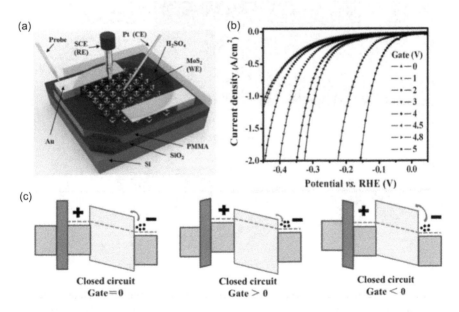

FIGURE 8.10 Gating-dependent HER activity on the MoS₂ microreactor. (a) Schematic of gating-dependent HER measurement setup. (b) The polarization curves of HER under different back-gate voltages. (c) The devices' operating states under different back-gate voltages: positive voltage bias, zero, and negative voltage bias.

the mechanism of field effect transistor,[127] n-type few-layer MoS_2 nanosheets have Fermi levels close to the conduction band, which naturally facilitates electron transfer (Figure 8.10c, **left panel**). The conductance of MoS_2 channels can be tuned to increase by positive gate voltages (Figure 8.10c, **middle panel**) or decrease by negative gate voltages (Figure 8.10c, **right panel**). The gate voltages are applied according to vertical field through the MoS_2–SiO_2–Si capacitors, in which the height of Schottky barriers controls the transport of electrons and thus influences the catalytic activities. Polarization curves in Figure 8.10b confirm the catalytic properties of 2D MoS_2 can be largely improved by the increasing of positive gate voltages.[126] Surprisingly, an ultralow overpotential (38 mV) was demonstrated by the MoS_2 FET-HER devices at 100 mA cm^{-2} current density with gate voltages of 5V. These overpotential values are much lower than those without gate voltages and also any other reported HER performance.[128,129] Therefore, external electric or magnetic field effects can be an effective way to improve electrocatalytic performance in heterogeneous catalysis.[122,124]

Oxygen evolution reaction (OER) of electrochemical water splitting has shown limited efficiency because of slow kinetics, hindered active sites by the binders and additives, and large interfacial barriers.[130,131] Poor kinetics are involved with weak mass end electron transfer processes that are related to the reactants and products at the interface.[131,132] Mai et al. have constructed and built concurrent measurement systems for electrical conductivity of electrode materials for probing the influence of introducing gas (like O_2) at the reaction interface during the OER.[133] The Ni deposited graphene nanosheet was chosen as a model catalyst for OER, as shown in Figure 8.11a.

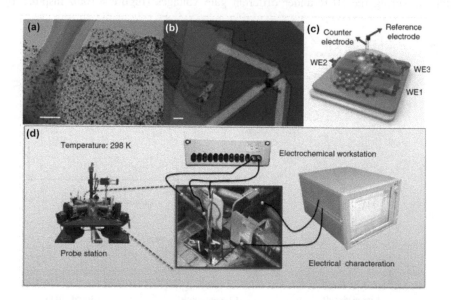

FIGURE 8.11 *In-situ* **measurement setup with the combination of electrical and electrochemical circuits.** (a) TEM image of Ni decorated graphene nanosheets. Scale bar: 200 nm. (b) Optical image of the designed microreactor. (c) Schematic of measurement setup. (d) Photograph of real measurement system by coupling the probe station, electrochemical workstation, and electrical measurement.

The Ni–graphene nanosheets were synthesized through hydrothermal method by heat treatment afterward[133]. Then, standard e-beam lithography was used for defining the working area of electrode on the graphene layers. In this work, the researchers used a negative photoresist called SU8-2002 as the reactor polymer to avoid the leakage current through the aqueous electrolyte (Figure 8.11b), which is checked for electrochemical reaction properties that are not involved with OER.[133]

Figure 8.12a shows the clear difference of the OER performance under the oxygen-absence and oxygen-presence conditions. The onset potentials of two conditions are 1.380 V (saturated O_2) and 1.344 V (without O_2) vs reversible hydrogen electrode (RHE), respectively. The OER performance of nanodevices of same type materials shows similar effects with traditional assembling of nanoparticles by oxygen influence. The enhanced performance is ascribed to the non-hindered active sites and enhanced electron transport. Figure 8.12c shows the on-chip electrochemical impedance spectroscopy (EIS) curves at potentials of 1.2 V vs RHE. According to the fitting with two constant parallel models, it is inferred that the charge transfer resistance is reduced at interface. The time constants are shown with same levels at low frequency, with two orders difference at high frequency (Figure 8.13d). The Tafel curves are shown in Figure 8.12b with smaller Tafel slope under oxygen-absence condition (147.8 mV decade⁻¹). We can see that the OER is a faster kinetic process under the oxygen-rich conditions. The combination

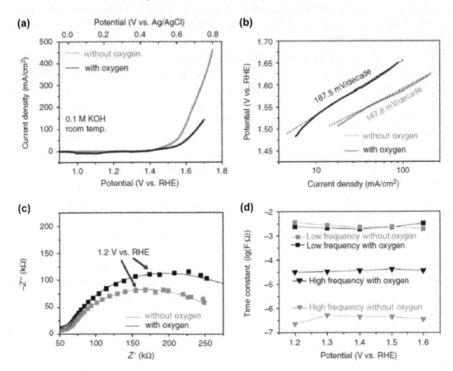

FIGURE 8.12 OER results based on the microreactors. (a) Polarization curves measured from Ni–graphene nanosheets microcells with/without oxygen input. (b) Tafel slope derived from polarization curve. (c) The fitting Nyquist curves with the potential at 1.2 V vs RHE. (d) Plot of the high- and low-frequency time constants vs potentials.

of molecular dynamic calculations, temporary $I–V$ measurement, and impedance spectroscopy measurement helps understand that oxygen is a significant feed to increase the concentration of OH$^-$ ions, thus accelerating the charge transfer of OER.[133]

8.3.3 FUNDAMENTAL STUDIES ON THE PHYSICOCHEMICAL PARAMETERS

Extracting the basic physicochemical parameters precisely from traditional electrochemical reaction is generally difficult and very challenging, because a lot of details are covered by the interface with sophisticated structures and cannot be read by the conventional spectroscopic techniques.[76,97,134–140] Mengning Ding et al. developed electrical transport spectroscopy (ETS) to probe the dynamics of electrochemical reaction.[94,141] They employed the on-chip electrochemical reactor continuously to study the fundamental process of ultrafine Pt nanowires, trying to understand what information the ETS curves reveal, Figure 8.13a. In principle, the field effects applied on metallic Pt nanowires cannot cause a change of electrical conductivity, which indicates the $G_{SD}–V_G$ curves could interpret that the changes of current result from the electrochemical interface: including surface adsorption of selected chemical/ biological species, surface scattering effect and also other dynamic information (Figure 8.13b). Pt surface of polycrystalline and bulk PtNWs membrane has five characteristic redox regions: hydrogen evolution reaction (HER), H adsorption/desorption region, double layer region, surface oxide formation/reduction region, and oxygen evolution reaction (OER).[94]

The molecular adsorption on the metallic nanostructures could induce the surface scattering of conduction electrons causing the resistance to increase. This also happens in the electrochemical reaction.[94,141] Considering the surface chemical process and surface scattering effect, ETS curves can match well with chemical steps in the Figure 8.13d. Based on the cyclic voltammetry (CV) curve ($I_G–V_G$) in Figure 8.13c, the whole area of potentials was identified into three regions. Figure 8.10d shows a double layer (DL) model; a layer of adsorbed water molecules is on the surface of Pt nanowires (PtNWs), and their occupation could correspond to the small change in the DL area (Figure 8.13c). During the HER, Pt surface does not appear with strong H atom adsorption desorption, in which G_{SD} signal shows ultrasmall change. Thus the increase and decrease of G_{SD} signal were ascribed to the H atom adsorption and desorption on the PtNWs (Figure 8.13c). With a similar scheme, the other steps are also discussed in the Figure 8.13d. The relative trend of current (G_{SD}) is more important than its absolute value, which means the differentiated ETS curves should have more instructive information of reaction mechanism. Every peak on the differentiated ETS curves is induced by the surface condition changes on the Pt nanowire, which are recognized as surface hydrogen adsorption/desorption, surface oxide reduction and desorption peak, and surface oxide formation peak. With these powerful ETS techniques, the quantitative correlation between surface conditions, cathodic/anodic currents from redox reactions with G_{SD} signal, and many more new insights of electrochemical surface states can be extracted with/without the electrochemical reaction. This work infers that on-chip measurement can be extended to very broad fields close to electrochemistry, including electrochemical catalysts, biological sensors, and chemical sensors.[94,141–143]

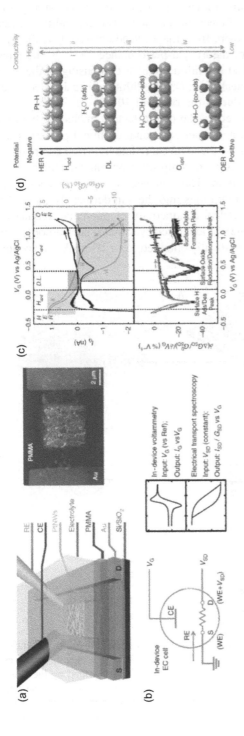

FIGURE 8.13 **Working principle of in-situ ETS and its measurement on the PtNW.** (a) Schematic illustration of the PtNW microreactor and SEM image of a real microreactor showing PMMA covered gold. (b) Schematic diagram of measurement circuits' in-device CV and ETS for in-situ monitoring of the electrochemical interfaces. CE, counter electrode; RE, reference electrode; WE, working electrode; S, source; D, drain. (c) Upper panel: I_G–V_G (black curve) and normalized G_{SD}–V_G (red curve) curves on a typical PtNW device. Bottom panel: Spectral peak characteristics shown on the differentiated ETS curves. (d) Comparison of different Pt surface conditions with the sweeping electrochemical potentials (left black axis) and the corresponding conductivity changes (right red axis, labels shown along the axis correspond to the labels shown on G_{SD} curve in c).

As it is known, the maximum storage capacity of lithium-ion battery is concerned with materials conductivities of electrons and ions. The capability to insert lithium ions insertion and dissociate them from the electrode materials is the first requirement for reaching the limit of theoretical performance.[100,144–149] The physical constant, which describes how fast the lithium ions are diffusing in the electrodes, is the chemical diffusion coefficient of Li, D^δ. The lithium diffusion coefficient of bulk graphite was reported as a value in the range of 10^{-5} to $10^{-12} cm^2 s^{-1}$, in which the extracted parameter is influenced by the mixed factors, including the mixed insertion compound and additives, and other additional chemical processes. Juren Smet et al. selected bilayer graphene as the ideal candidate insertion compound for studying the diffusion coefficient using on-chip electrochemical reactor.[150] Exfoliated bilayer graphene was etched as the Hall bar geometry and deposited on metallic electrodes for intercalating lithium ions into the van der Waals gap and for in-situ measurement of the local electrical conductivity of intercalated bilayer graphene (Figure 8.14a). Bilayer graphene was identified by Raman spectroscopy, in which four Lorentz peaks can be characteristically obtained from 2D peak. The conductivity of

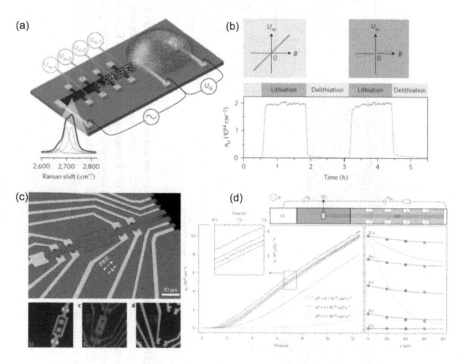

FIGURE 8.14   Extracting the lithium ion diffusion coefficient through intralayer graphene based on on-chip electrochemical cell. (a) Schematic of electrochemical devices of bilayer graphene with Hall-bar geometry. Inset: bilayer graphene confirmed by the Raman spectra. (b) The lithiation or delithiation process can be controlled by the electrode. The conductivity of uncovered graphene can be used to recognize the lithium ion transportation process. (c) Space resolved TOF-SIMS reveals the large Li content of this edge design. (d) The dynamic measurements for probing the Li diffusion process.

bilayer graphene is sensitive to the lithiation and delithation processes, as shown in Figure 8.14b. From the value of Hall voltage $U_{xy(x)}$ in Figure 8.14b, the maximum electron density of Li-intercalated bilayer graphene was confirmed to be about $2 \times 10^{14} cm^{-2}$. Based on the time-dependent density change measurement of uncovered bilayer graphene, the in-plane chemical diffusion coefficient of Lithium can be estimated to be in the range of 1×10^{-5} to $7 \times 10^{-5} cm^2 s^{-1}$ at room temperature by the Fick's second law (Figure 8.14d).[150] Time-of-flight secondary ion mass spectrometry (TOF-SIMS) is used to characterize the Li distribution in the van der Waals channel, graphene edge, and Ti electrodes (Figure 8.14c), which showed highest Li content with stoichiometry of C6LiC6 for bilayer graphene. This work has shown us that the electrochemical on-chip cells and magneto-transport measurements can be coupled to conduct interdisciplinary studies on ion transport, diffusion, local kinetics, etc. for low-dimension materials.[150]

Battery-related intercalation is a unique way to tune the electronic properties of 2D materials. MoS_2-Li nanobattery was used to study the dynamic change of optical and electrical properties under the lithium insertion or removal conditions by Feng Xiong et al. [52,151] It is hard to achieve Li-intercalated MoS_2 and measure its intrinsic properties. Figure 8.15a shows the encapsulated MoS_2-Li was sandwiched by the top glass and bottom Si substrate, in which 1M $LiPF_6$ of 1:1 w/w

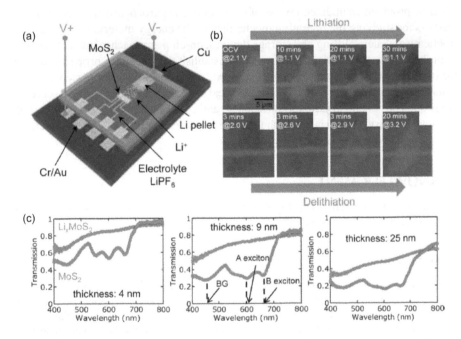

FIGURE 8.15 **Direct observation of dynamic charge and discharge processes with tunable optical and electrical properties on few-layer MoS_2 nanobattery.** (a) Schematic of on-chip MoS_2 nanobattery for lithiation and delithiation processes. (b) In-situ optical images of a thin (~20–35 nm) MoS_2 flake during the lithiation process (upper panel) and delithiation process (bottom panel). (c) Comparison of optical transmission spectra of MoS_2 flakes with different thicknesses (4, 9, and 25 nm) before (red) and after (blue) Li intercalation.

ethylene carbonate/diethyl carbonate was used as the electrolyte and the epoxy was used to cover the whole device to prevent oxidation. The deposited Cr/Au and Cu electrodes served as the working electrodes for measuring the conductivity or driving the lithium diffusion. The semiconducting to metallic phase transformation at ~1.1 V of Li/Li$^+$ was confirmed by observing the Raman spectrum shown in Figure 8.15b. The accumulation of Li in 2H–MoS$_2$ initially started from the edge of layered structure, and it gradually and reversibly changed the optical transmission of MoS$_2$ (Figure 8.15b). The optical transmission spectra of MoS$_2$ nanosheets were measured under quenched Li intercalation at different thicknesses (4, 9, and 25 nm initial thickness). The improvement of optical transmission after Li intercalation of 400–700 nm is obvious both in the spectra in Figure 8.15c and in the optical image in Figure 8.15b. There is slight decrease of optical signal in the range over 700 nm with the dependence on the thickness. In the mechanism, the overlapped band structure of conduction and valance bands at K-point results in a reduction in the absorption of 1T–Li$_x$MoS$_2$, which is consistent with optical transmission enhancement. This is the first in-situ dynamical measurement of both optical and electrical properties of Li$_x$MoS$_2$, thanks to well-defined and protected on-chip electrochemical cells.[36,52,57,151]

8.4 CONCLUSIONS

We have provided an updated review of well-defined electrochemical on-chip micro-reactors for studying the operating mechanism of 2D energy materials. For the electrochemical water splitting and storage devices, on-chip approaches make it possible to probe the electrochemical interfaces and establish the structure–property relation for improving the device performance and extracting certain basic physicochemical constants that contribute to the electrochemical process. This chapter could serve the areas of electrochemistry, materials science, and condensed physics of 2D materials and low-dimension materials, which finally push the real application of the emerging prosperous devices application.[15,152,153]

ACKNOWLEDGMENT

The authors acknowledge the support of National Natural Science Foundation of China (No. 51902346, No. 51902185).

REFERENCES

1. Bonaccorso, F., L. Colombo, G. Yu, M. Stoller, V. Tozzini, A. C. Ferrari, R. S. Ruoff, V. Pellegrini. 2015. "2D materials. Graphene, related two-dimensional crystals, and hybrid systems for energy conversion and storage." *Science* 347 (6217):1246501.
2. Chen, P., Z. Zhang, X. Duan, X. Duan. 2018. "Chemical synthesis of two-dimensional atomic crystals, heterostructures and superlattices." *Chemical Society Reviews* 47 (9):3129.
3. Novoselov, K. S., A. K. Geim, S. V. Morozov, D. Jiang, Y. Zhang, S. V. Dubonos, I. V. Grigorieva, A. A. Firsov. 2004. "Electric field effect in atomically thin carbon films." *Science* 306 (5696):666.

4. Novoselov, K. S., D. Jiang, F. Schedin, T. J. Booth, V. V. Khotkevich, S. V. Morozov, A. K. Geim. 2005. "Two-dimensional atomic crystals." *Proceedings of the National Academy of Sciences of the United States of America* 102 (30):10451.

5. Mouri, S., Y. Miyauchi, K. Matsuda. 2013. "Tunable photoluminescence of monolayer MoS_2 via chemical doping." *Nano Letters* 13 (12):5944.

6. Guo, Y. Q., K. Xu, C. Z. Wu, J. Y. Zhao, Y. Xie. 2015. "Surface chemical-modification for engineering the intrinsic physical properties of inorganic two-dimensional nanomaterials." *Chemical Society Reviews* 44 (3):637.

7. Guo, D., R. Shibuya, C. Akiba, S. Saji, T. Kondo, J. Nakamura. 2016. "Active sites of nitrogen-doped carbon materials for oxygen reduction reaction clarified using model catalysts." *Science* 351 (6271):361.

8. Jaramillo, T. F., K. P. Jorgensen, J. Bonde, J. H. Nielsen, S. Horch, I. Chorkendorff. 2007. "Identification of active edge sites for electrochemical H_2 evolution from MoS_2 nanocatalysts." *Science* 317 (5834):100.

9. Yan, K., H.-W. Lee, T. Gao, G. Zheng, H. Yao, H. Wang, Z. Lu, Y. Zhou, Z. Liang, Z. Liu, S. Chu, Y. Cui. 2014. "Ultrathin two-dimensional atomic crystals as stable interfacial layer for improvement of lithium metal anode." *Nano Letters* 14 (10):6016.

10. Ding, Q., F. Meng, C. R. English, M. Caban-Acevedo, M. J. Shearer, D. Liang, A. S. Daniel, R. J. Hamers, S. Jin. 2014. "Efficient photoelectrochemical hydrogen generation using heterostructures of Si and chemically exfoliated metallic MoS_2." *Journal of the American Chemical Society* 136 (24):8504.

11. Caldwell, J. D., I. Aharonovich, G. Cassabois, J. H. Edgar, B. Gil, D. N. Basov. 2019. "Photonics with hexagonal boron nitride." *Nature Reviews Materials* 4 (8):552.

12. Tao, L., E. Cinquanta, D. Chiappe, C. Grazianetti, M. Fanciulli, M. Dubey, A. Molle, D. Akinwande. 2015. "Silicene field-effect transistors operating at room temperature." *Nature Nanotechnology* 10 (3):227.

13. Xia, F., H. Wang, J. C. M. Hwang, A. H. C. Neto, L. Yang. 2019. "Black phosphorus and its isoelectronic materials." *Nature Reviews Physics* 1 (5):306.

14. Cai, Z., D. Zhou, M. Wang, S.-M. Bak, Y. Wu, Z. Wu, Y. Tian, X. Xiong, Y. Li, W. Liu, S. Siahrostami, Y. Kuang, X.-Q. Yang, H. Duan, Z. Feng, H. Wang, X. Sun. 2018. "Introducing Fe^{2+} into nickel–iron layered double hydroxide: Local structure modulated water oxidation activity." *Angewandte Chemie* 130 (30):9536.

15. Zhang, X., Y. Xie. 2013. "Recent advances in free-standing two-dimensional crystals with atomic thickness: design, assembly and transfer strategies." *Chemical Society Reviews* 42 (21):8187.

16. Xu, C., L. Wang, Z. Liu, L. Chen, J. Guo, N. Kang, X.-L. Ma, H.-M. Cheng, W. Ren. 2015. "Large-area high-quality 2D ultrathin Mo_2C superconducting crystals." *Nature Materials* 14 (11):1135.

17. Anasori, B., M. R. Lukatskaya, Y. Gogotsi. 2017. "2D metal carbides and nitrides (MXenes) for energy storage." *Nature Reviews Materials* 2 (2):16098.

18. Yang, J., Z. Zeng, J. Kang, S. Betzler, C. Czarnik, X. Zhang, C. Ophus, C. Yu, K. Bustillo, M. Pan, J. Qiu, L.-W. Wang, H. Zheng. 2019. "Formation of two-dimensional transition metal oxide nanosheets with nanoparticles as intermediates." *Nature Materials* 18 (9):970.

19. Ghidiu, M., M. R. Lukatskaya, M.-Q. Zhao, Y. Gogotsi, M. W. Barsoum. 2014. "Conductive two-dimensional titanium carbide 'clay' with high volumetric capacitance." *Nature* 516 (7529):78.

20. Zheng, W. S., T. Xie, Y. Zhou, Y. L. Chen, W. Jiang, S. L. Zhao, J. X. Wu, Y. M. Jing, Y. Wu, G. C. Chen, Y. F. Guo, J. B. Yin, S. Y. Huang, H. Q. Xu, Z. F. Liu, H. L. Peng. 2015. "Patterning two-dimensional chalcogenide crystals of Bi_2Se_3 and In_2Se_3 and efficient photodetectors." *Nature Communications* 6:6972.

21. Ji, Q. Q., Y. F. Zhang, T. Gao, Y. Zhang, D. L. Ma, M. X. Liu, Y. B. Chen, X. F. Qiao, P. H. Tan, M. Kan, J. Feng, Q. Sun, Z. F. Liu. 2013. "Epitaxial monolayer MoS$_2$ on mica with novel photoluminescence." *Nano Letters* 13 (8):3870.

22. Liu, K. K., W. J. Zhang, Y. H. Lee, Y. C. Lin, M. T. Chang, C. Su, C. S. Chang, H. Li, Y. M. Shi, H. Zhang, C. S. Lai, L. J. Li. 2012. "Growth of large-area and highly crystalline MoS$_2$ thin layers on insulating substrates." *Nano Letters* 12 (3):1538.

23. Mendoza-Sanchez, B., Y. Gogotsi. 2016. "Synthesis of two-dimensional materials for capacitive energy storage." *Advanced Materials* 28 (29):6104.

24. Tsai, M.-L., S.-H. Su, J.-K. Chang, D.-S. Tsai, C.-H. Chen, C.-I. Wu, L.-J. Li, L.-J. Chen, J.-H. He. 2014. "Monolayer MoS$_2$ heterojunction solar cells." *ACS Nano* 8 (8):8317.

25. Lin, S., X. Li, P. Wang, Z. Xu, S. Zhang, H. Zhong, Z. Wu, W. Xu, H. Chen. 2015. "Interface designed MoS$_2$/GaAs heterostructure solar cell with sandwich stacked hexagonal boron nitride." *Scientific Reports* 5 (1):15103.

26. Zhao, L.-D., S.-H. Lo, Y. Zhang, H. Sun, G. Tan, C. Uher, C. Wolverton, V. P. Dravid, M. G. Kanatzidis. 2014. "Ultralow thermal conductivity and high thermoelectric figure of merit in SnSe crystals." *Nature* 508 (7496):373.

27. Lee, M.-J., J.-H. Ahn, J. H. Sung, H. Heo, S. G. Jeon, W. Lee, J. Y. Song, K.-H. Hong, B. Choi, S.-H. Lee, M.-H. Jo. 2016. "Thermoelectric materials by using two-dimensional materials with negative correlation between electrical and thermal conductivity." *Nature Communications* 7 (1):12011.

28. Mleczko, M. J., R. L. Xu, K. Okabe, H. H. Kuo, I. R. Fisher, H. S. P. Wong, Y. Nishi, E. Pop. 2016. "High current density and low thermal conductivity of atomically thin semimetallic WTe$_2$." *ACS Nano* 10 (8):7507.

29. Zhou, Y., H. Jang, J. M. Woods, Y. Xie, P. Kumaravadivel, G. A. Pan, J. Liu, Y. Liu, D. G. Cahill, J. J. Cha. 2017. "Direct synthesis of large-scale WTe$_2$ thin films with low thermal conductivity." *Advanced Functional Materials* 27 (8):1605928.

30. Cui, C., F. Xue, W.-J. Hu, L.-J. Li. 2018. "Two-dimensional materials with piezoelectric and ferroelectric functionalities." *NPJ 2D Materials and Applications* 2 (1):18.

31. Wu, W., L. Wang, Y. Li, F. Zhang, L. Lin, S. Niu, D. Chenet, X. Zhang, Y. Hao, T. F. Heinz, J. Hone, Z. L. Wang. 2014. "Piezoelectricity of single-atomic-layer MoS$_2$ for energy conversion and piezotronics." *Nature* 514 (7523):470.

32. Dai, M., W. Zheng, X. Zhang, S. Wang, J. Lin, K. Li, Y. Hu, E. Sun, J. Zhang, Y. Qiu, Y. Fu, W. Cao, P. Hu. 2020. "Enhanced piezoelectric effect derived from grain boundary in MoS$_2$ monolayers." *Nano Letters* 20 (1):201.

33. Seol, M., S. Kim, Y. Cho, K.-E. Byun, H. Kim, J. Kim, S. K. Kim, S.-W. Kim, H.-J. Shin, S. Park. 2018. "Triboelectric series of 2D layered materials." *Advanced Materials* 30 (39):1801210.

34. Kim, D. W., J. H. Lee, J. K. Kim, U. Jeong. 2020. "Material aspects of triboelectric energy generation and sensors." *NPG Asia Materials* 12 (1):6.

35. Morales-Guio, C. G., X. L. Hu. 2014. "Amorphous molybdenum sulfides as hydrogen evolution catalysts." *Accounts of Chemical Research* 47 (8):2671.

36. Wang, H. T., Z. Y. Lu, S. C. Xu, D. S. Kong, J. J. Cha, G. Y. Zheng, P. C. Hsu, K. Yan, D. Bradshaw, F. B. Prinz, Y. Cui. 2013. "Electrochemical tuning of vertically aligned MoS$_2$ nanofilms and its application in improving hydrogen evolution reaction." *Proceedings of the National Academy of Sciences of the United States of America* 110 (49):19701.

37. Yan, H. J., C. G. Tian, L. Wang, A. P. Wu, M. C. Meng, L. Zhao, H. G. Fu. 2015. "Phosphorus-modified tungsten nitride/reduced graphene oxide as a high-performance, non-noble-metal electrocatalyst for the hydrogen evolution reaction." *Angewandte Chemie-International Edition* 54 (21):6325.

38. Lukowski, M. A., A. S. Daniel, C. R. English, F. Meng, A. Forticaux, R. J. Hamers, S. Jin. 2014. "Highly active hydrogen evolution catalysis from metallic WS_2 nanosheets." *Energy & Environmental Science* 7 (8):2608.

39. Greeley, J., T. F. Jaramillo, J. Bonde, I. B. Chorkendorff, J. K. Norskov. 2006. "Computational high-throughput screening of electrocatalytic materials for hydrogen evolution." *Nature Materials* 5 (11):909.

40. McEvoy, J. P., G. W. Brudvig. 2006. "Water-splitting chemistry of photosystem II." *Chemical Reviews* 106 (11):4455.

41. Wang, D. Y., M. Gong, H. L. Chou, C. J. Pan, H. A. Chen, Y. P. Wu, M. C. Lin, M. Y. Guan, J. Yang, C. W. Chen, Y. L. Wang, B. J. Hwang, C. C. Chen, H. J. Dai. 2015. "Highly active and stable hybrid catalyst of cobalt-doped FeS_2 nanosheets-carbon nanotubes for hydrogen evolution reaction." *Journal of the American Chemical Society* 137 (4):1587.

42. Li, X. L., W. Liu, M. Y. Zhang, Y. R. Zhong, Z. Weng, Y. Y. Mi, Y. Zhou, M. Li, J. J. Cha, Z. Y. Tang, H. Jiang, X. M. Li, H. L. Wang. 2017. "Strong metal-phosphide interactions in core-shell geometry for enhanced electrocatalysis." *Nano Letters* 17 (3):2057.

43. Wang, Q., K. Domen. 2020. "Particulate photocatalysts for light-driven water splitting: mechanisms, challenges, and design strategies." *Chemical Reviews* 120 (2):919.

44. Jiao, Y., Y. Zheng, M. T. Jaroniec, S. Z. Qiao. 2015. "Design of electrocatalysts for oxygen- and hydrogen-involving energy conversion reactions." *Chemical Society Reviews* 44 (8):2060.

45. Morales-Guio, C. G., L. A. Stern, X. L. Hu. 2014. "Nanostructured hydrotreating catalysts for electrochemical hydrogen evolution." *Chemical Society Reviews* 43 (18):6555.

46. Zou, X. X., Y. Zhang. 2015. "Noble metal-free hydrogen evolution catalysts for water splitting." *Chemical Society Reviews* 44 (15):5148.

47. Deng, J., P. J. Ren, D. H. Deng, L. Yu, F. Yang, X. H. Bao. 2014. "Highly active and durable non-precious-metal catalysts encapsulated in carbon nanotubes for hydrogen evolution reaction." *Energy & Environmental Science* 7 (6):1919.

48. Zheng, Y., Y. Jiao, M. Jaroniec, S. Z. Qiao. 2015. "Advancing the electrochemistry of the hydrogen-evolution reaction through combining experiment and theory." *Angewandte Chemie-International Edition* 54 (1):52.

49. Kong, D. S., J. J. Cha, H. T. Wang, H. R. Lee, Y. Cui. 2013. "First-row transition metal dichalcogenide catalysts for hydrogen evolution reaction." *Energy & Environmental Science* 6 (12):3553.

50. Vesborg, P. C. K., B. Seger, I. Chorkendorff. 2015. "Recent development in hydrogen evolution reaction catalysts and their practical implementation." *Journal of Physical Chemistry Letters* 6 (6):951.

51. Chia, X. Y., A. Y. S. Eng, A. Ambrosi, S. M. Tan, M. Pumera. 2015. "Electrochemistry of nanostructured layered transition-metal dichalcogenides." *Chemical Reviews* 115 (21):11941.

52. Voiry, D., H. Yamaguchi, J. W. Li, R. Silva, D. C. B. Alves, T. Fujita, M. W. Chen, T. Asefa, V. B. Shenoy, G. Eda, M. Chhowalla. 2013. "Enhanced catalytic activity in strained chemically exfoliated WS_2 nanosheets for hydrogen evolution." *Nature Materials* 12 (9):850.

53. Voiry, D., M. Salehi, R. Silva, T. Fujita, M. W. Chen, T. Asefa, V. B. Shenoy, G. Eda, M. Chhowalla. 2013. "Conducting MoS_2 nanosheets as catalysts for hydrogen evolution reaction." *Nano Letters* 13 (12):6222.

54. Hinnemann, B., P. G. Moses, J. Bonde, K. P. Jørgensen, J. H. Nielsen, S. Horch, I. Chorkendorff, J. K. Nørskov. 2005. "Biomimetic hydrogen evolution: MoS_2 nanoparticles as catalyst for hydrogen evolution." *Journal of the American Chemical Society* 127 (15):5308.

55. Li, Y. G., H. L. Wang, L. M. Xie, Y. Y. Liang, G. S. Hong, H. J. Dai. 2011. "MoS₂ nanoparticles grown on graphene: An advanced catalyst for the hydrogen evolution reaction." *Journal of the American Chemical Society* 133 (19):7296.

56. Lukowski, M. A., A. S. Daniel, F. Meng, A. Forticaux, L. S. Li, S. Jin. 2013. "Enhanced hydrogen evolution catalysis from chemically exfoliated metallic MoS₂ nanosheets." *Journal of the American Chemical Society* 135 (28):10274.

57. Wang, H. T., Q. F. Zhang, H. B. Yao, Z. Liang, H. W. Lee, P. C. Hsu, G. Y. Zheng, Y. Cui. 2014. "High electrochemical selectivity of edge versus terrace sites in two-dimensional layered MoS₂ materials." *Nano Letters* 14 (12):7138.

58. Yu, Y. F., S. Y. Huang, Y. P. Li, S. N. Steinmann, W. T. Yang, L. Y. Cao. 2014. "Layer-dependent electrocatalysis of MoS₂ for hydrogen evolution." *Nano Letters* 14 (2):553.

59. Pan, J., Z. L. Wang, Q. Chen, J. G. Hu, J. L. Wang. 2014. "Band structure engineering of monolayer MoS₂ by surface ligand functionalization for enhanced photoelectrochemical hydrogen production activity." *Nanoscale* 6 (22):13565.

60. Gao, S., Y. Lin, X. Jiao, Y. Sun, Q. Luo, W. Zhang, D. Li, J. Yang, Y. Xie. 2016. "Partially oxidized atomic cobalt layers for carbon dioxide electroreduction to liquid fuel." *Nature* 529 (7584):68.

61. Gao, S., Z. Sun, W. Liu, X. Jiao, X. Zu, Q. Hu, Y. Sun, T. Yao, W. Zhang, S. Wei, Y. Xie. 2017. "Atomic layer confined vacancies for atomic-level insights into carbon dioxide electroreduction." *Nature Communications* 8 (1):14503.

62. Cai, Z., Y. M. Bi, E. Y. Hu, W. Liu, N. Dwarica, Y. Tian, X. L. Li, Y. Kuang, Y. P. Li, X. Q. Yang, H. L. Wang, X. M. Sun. 2018. "Single-crystalline ultrathin Co₃O₄ nanosheets with massive vacancy defects for enhanced electrocatalysis." *Advanced Energy Materials* 8 (3):1701694.

63. Voiry, D., J. Yang, M. Chhowalla. 2016. "Recent strategies for improving the catalytic activity of 2D TMD nanosheets toward the hydrogen evolution reaction." *Advanced Materials* 28 (29): 6197–6206.

64. Hansen, M. H., L. A. Stern, L. G. Feng, J. Rossmeisl, X. L. Hu. 2015. "Widely available active sites on Ni₂P for electrochemical hydrogen evolution - insights from first principles calculations." *Physical Chemistry Chemical Physics* 17 (16):10823.

65. Voiry, D., H. S. Shin, K. P. Loh, M. Chhowalla. 2018. "Low-dimensional catalysts for hydrogen evolution and CO₂ reduction." *Nature Reviews Chemistry* 2 (1):0105.

66. Shao, Q., Z.-S. Wu, J. Chen. 2019. "Two-dimensional materials for advanced Li-S batteries." *Energy Storage Materials* 22:284.

67. Chen, K.-S., I. Balla, N. S. Luu, M. C. Hersam. 2017. "Emerging opportunities for two-dimensional materials in lithium-ion batteries." *ACS Energy Letters* 2 (9):2026.

68. Wang, X., Q. Weng, Y. Yang, Y. Bando, D. Golberg. 2016. "Hybrid two-dimensional materials in rechargeable battery applications and their microscopic mechanisms." *Chemical Society Reviews* 45 (15):4042.

69. Peng, L., Y. Zhu, D. Chen, R. S. Ruoff, G. Yu. 2016. "Two-dimensional materials for beyond-lithium-ion batteries." *Advanced Energy Materials* 6 (11):1600025.

70. Cha, E., M. D. Patel, J. Park, J. Hwang, V. Prasad, K. Cho, W. Choi. 2018. "2D MoS₂ as an efficient protective layer for lithium metal anodes in high-performance Li–S batteries." *Nature Nanotechnology* 13 (4):337.

71. Tao, Y., X. Xie, W. Lv, D.-M. Tang, D. Kong, Z. Huang, H. Nishihara, T. Ishii, B. Li, D. Golberg, F. Kang, T. Kyotani, Q.-H. Yang. 2013. "Towards ultrahigh volumetric capacitance: Graphene derived highly dense but porous carbons for supercapacitors." *Scientific Reports* 3 (1):2975.

72. Acerce, M., D. Voiry, M. Chhowalla. 2015. "Metallic 1T phase MoS₂ nanosheets as supercapacitor electrode materials." *Nature Nanotechnology* 10 (4):313.

73. Hynek, D. J., J. V. Pondick, J. J. Cha. 2019. "The development of 2D materials for electrochemical energy applications: A mechanistic approach." *APL Materials* 7 (3):030902.

74. Dumont, J. H., U. Martinez, K. Artyushkova, G. M. Purdy, A. M. Dattelbaum, P. Zelenay, A. Mohite, P. Atanassov, G. Gupta. 2019. "Nitrogen-doped graphene oxide electrocatalysts for the oxygen reduction reaction." *ACS Applied Nano Materials* 2 (3):1675.

75. Ly, Q., B. V. Merinov, H. Xiao, W. A. Goddard, T. H. Yu. 2017. "The oxygen reduction reaction on graphene from quantum mechanics: Comparing armchair and zigzag carbon edges." *The Journal of Physical Chemistry C* 121 (44):24408.

76. Yang, L., J. Shui, L. Du, Y. Shao, J. Liu, L. Dai, Z. Hu. 2019. "Carbon-based metal-free ORR electrocatalysts for fuel cells: Past, present, and future." *Advanced Materials* 31 (13):1804799.

77. Lv, Q., W. Si, J. He, L. Sun, C. Zhang, N. Wang, Z. Yang, X. Li, X. Wang, W. Deng, Y. Long, C. Huang, Y. Li. 2018. "Selectively nitrogen-doped carbon materials as superior metal-free catalysts for oxygen reduction." *Nature Communications* 9 (1):3376.

78. Li, J.-C., M. Cheng, T. Li, L. Ma, X. Ruan, D. Liu, H.-M. Cheng, C. Liu, D. Du, Z. Wei, Y. Lin, M. Shao. 2019. "Carbon nanotube-linked hollow carbon nanospheres doped with iron and nitrogen as single-atom catalysts for the oxygen reduction reaction in acidic solutions." *Journal of Materials Chemistry A* 7 (24):14478.

79. Wang, H. T., Z. Y. Lu, D. S. Kong, J. Sun, T. M. Hymel, Y. Cui. 2014. "Electrochemical tuning of MoS_2 nanoparticles on three-dimensional substrate for efficient hydrogen evolution." *ACS Nano* 8 (5):4940.

80. Zeng, M., Y. G. Li. 2015. "Recent advances in heterogeneous electrocatalysts for the hydrogen evolution reaction." *Journal of Materials Chemistry A* 3 (29):14942.

81. Cummins, D. R., U. Martinez, R. Kappera, D. Voiry, A. Martinez-Garcia, J. Jasinski, D. Kelly, M. Chhowalla, A. D. Mohite, M. K. Sunkara, G. Gupta. 2015. "Catalytic activity in lithium-treated core-shell MoO_x/MoS_2 nanowires." *Journal of Physical Chemistry C* 119 (40):22908.

82. Shen, A., Y. Zou, Q. Wang, R. A. W. Dryfe, X. Huang, S. Dou, L. Dai, S. Wang. 2014. "Oxygen reduction reaction in a droplet on graphite: Direct evidence that the edge is more active than the basal plane." *Angewandte Chemie International Edition* 53 (40):10804.

83. San Roman, D., D. Krishnamurthy, R. Garg, H. Hafiz, M. Lamparski, N. T. Nuhfer, V. Meunier, V. Viswanathan, T. Cohen-Karni. 2020. "Engineering three-dimensional (3D) out-of-plane graphene edge sites for highly selective two-electron oxygen reduction electrocatalysis." *ACS Catalysis* 10 (3):1993.

84. Yang, S., Y. Yu, M. Dou, Z. Zhang, L. Dai, F. Wang. 2019. "Two-dimensional conjugated aromatic networks as high-site-density and single-atom electrocatalysts for the oxygen reduction reaction." *Angewandte Chemie International Edition* 58 (41):14724.

85. Ye, Y., H. Li, F. Cai, C. Yan, R. Si, S. Miao, Y. Li, G. Wang, X. Bao. 2017. "Two-dimensional mesoporous carbon doped with Fe–N active sites for efficient oxygen reduction." *ACS Catalysis* 7 (11):7638.

86. Chen, D., Y. Zou, S. Wang. 2019. "Surface chemical-functionalization of ultrathin two-dimensional nanomaterials for electrocatalysis." *Materials Today Energy* 12:250.

87. Velický, M., M. A. Bissett, C. R. Woods, P. S. Toth, T. Georgiou, I. A. Kinloch, K. S. Novoselov, R. A. W. Dryfe. 2016. "Photoelectrochemistry of pristine mono- and few-layer MoS_2." *Nano Letters* 16 (3):2023.

88. Ritzert, N. L., V. A. Szalai, T. P. Moffat. 2018. "Mapping electron transfer at MoS_2 using scanning electrochemical microscopy." *Langmuir* 34 (46):13864.

89. Tributsch, H., J. C. Bennett. 1977. "Electrochemistry and photochemistry of MoS₂ layer crystals. I." *Journal of Electroanalytical Chemistry and Interfacial Electrochemistry* 81 (1):97.

90. Zhang, L., X. Ji, X. Ren, Y. Ma, X. Shi, Z. Tian, A. M. Asiri, L. Chen, B. Tang, X. Sun. 2018. "Electrochemical ammonia synthesis via nitrogen reduction reaction on a MoS₂ catalyst: Theoretical and experimental studies." *Advanced Materials* 30 (28):1800191.

91. Biroju, R. K., D. Das, R. Sharma, S. Pal, L. P. L. Mawlong, K. Bhorkar, P. K. Giri, A. K. Singh, T. N. Narayanan. 2017. "Hydrogen evolution reaction activity of graphene–MoS₂ van der Waals heterostructures." *ACS Energy Letters* 2 (6):1355.

92. Mai, L., Y. Dong, L. Xu, C. Han. 2010. "Single nanowire electrochemical devices." *Nano Letters* 10 (10):4273.

93. Su, Y., C. Liu, S. Brittman, J. Tang, A. Fu, N. Kornienko, Q. Kong, P. Yang. 2016. "Single-nanowire photoelectrochemistry." *Nature Nanotechnology* 11 (7):609.

94. Ding, M., Q. He, G. Wang, H.-C. Cheng, Y. Huang, X. Duan. 2015. "An on-chip electrical transport spectroscopy approach for in situ monitoring electrochemical interfaces." *Nature Communications* 6 (1):7867.

95. Ding, M., H.-Y. Shiu, S.-L. Li, C. K. Lee, G. Wang, H. Wu, N. O. Weiss, T. D. Young, P. S. Weiss, G. C. L. Wong, K. H. Nealson, Y. Huang, X. Duan. 2016. "Nanoelectronic investigation reveals the electrochemical basis of electrical conductivity in shewanella and geobacter." *ACS Nano* 10 (11):9919.

96. Feng, J., K. Liu, M. Graf, M. Lihter, R. D. Bulushev, D. Dumcenco, D. T. L. Alexander, D. Krasnozhon, T. Vuletic, A. Kis, A. Radenovic. 2015. "Electrochemical reaction in single layer MoS₂: Nanopores opened atom by atom." *Nano Letters* 15 (5):3431.

97. Zhang, G., H. Liu, J. Qu, J. Li. 2016. "Two-dimensional layered MoS₂: Rational design, properties and electrochemical applications." *Energy & Environmental Science* 9 (4):1190.

98. Acerce, M., E. K. Akdoğan, M. Chhowalla. 2017. "Metallic molybdenum disulfide nanosheet-based electrochemical actuators." *Nature* 549 (7672):370.

99. David, L., R. Bhandavat, G. Singh. 2014. "MoS₂/graphene composite paper for sodium-ion battery electrodes." *ACS Nano* 8 (2):1759.

100. Wang, Y., B. Chen, D. H. Seo, Z. J. Han, J. I. Wong, K. Ostrikov, H. Zhang, H. Y. Yang. 2016. "MoS₂-coated vertical graphene nanosheet for high-performance rechargeable lithium-ion batteries and hydrogen production." *NPG Asia Materials* 8 (5):e268.

101. Ghazi, Z. A., X. He, A. M. Khattak, N. A. Khan, B. Liang, A. Iqbal, J. Wang, H. Sin, L. Li, Z. Tang. 2017. "MoS₂/celgard separator as efficient polysulfide barrier for long-life lithium–sulfur batteries." *Advanced Materials* 29 (21):1606817.

102. Wang, H. T., C. Tsai, D. S. Kong, K. R. Chan, F. Abild-Pedersen, J. Norskov, Y. Cui. 2015. "Transition-metal doped edge sites in vertically aligned MoS₂ catalysts for enhanced hydrogen evolution." *Nano Research* 8 (2):566.

103. Kiriya, D., P. Lobaccaro, H. Y. Y. Nyein, P. Taheri, M. Hettick, H. Shiraki, C. M. Sutter-Fella, P. Zhao, W. Gao, R. Maboudian, J. W. Ager, A. Javey. 2016. "General thermal texturization process of MoS₂ for efficient electrocatalytic hydrogen evolution reaction." *Nano Letters* 16 (7):4047.

104. Voiry, D., R. Fullon, J. E. Yang, C. D. C. E. Silva, R. Kappera, I. Bozkurt, D. Kaplan, M. J. Lagos, P. E. Batson, G. Gupta, A. D. Mohite, L. Dong, D. Q. Er, V. B. Shenoy, T. Asefa, M. Chhowalla. 2016. "The role of electronic coupling between substrate and 2D MoS₂ nanosheets in electrocatalytic production of hydrogen." *Nature Materials* 15 (9):1003.

105. Lee, S. Y., U. J. Kim, J. Chung, H. Nam, H. Y. Jeong, G. H. Han, H. Kim, H. M. Oh, H. Lee, H. Kim, Y. G. Roh, J. Kim, S. W. Hwang, Y. Park, Y. H. Lee. 2016. "Large work function modulation of monolayer MoS_2 by ambient gases." *ACS Nano* 10 (6):6100.

106. Pierucci, D., H. Henck, J. Avila, A. Balan, C. H. Naylor, G. Patriarche, Y. J. Dappe, M. G. Silly, F. Sirotti, A. T. C. Johnson, M. C. Asensio, A. Ouerghi. 2016. "Band alignment and minigaps in monolayer MoS_2-graphene van der Waals heterostructures." *Nano Letters* 16 (7):4054.

107. Liu, W., E. Y. Hu, H. Jiang, Y. J. Xiang, Z. Weng, M. Li, Q. Fan, X. Q. Yu, E. I. Altman, H. L. Wang. 2016. "A highly active and stable hydrogen evolution catalyst based on pyrite-structured cobalt phosphosulfide." *Nature Communications* 7:10771.

108. Gao, M. R., J. X. Liang, Y. R. Zheng, Y. F. Xu, J. Jiang, Q. Gao, J. Li, S. H. Yu. 2015. "An efficient molybdenum disulfide/cobalt diselenide hybrid catalyst for electrochemical hydrogen generation." *Nature Communications* 6: 5982.

109. Wang, H. T., D. S. Kong, P. Johanes, J. J. Cha, G. Y. Zheng, K. Yan, N. A. Liu, Y. Cui. 2013. "$MoSe_2$ and WSe_2 nanofilms with vertically aligned molecular layers on curved and rough surfaces." *Nano Letters* 13 (7):3426.

110. Kong, D. S., H. T. Wang, Z. Y. Lu, Y. Cui. 2014. "$CoSe_2$ nanoparticles grown on carbon fiber paper: An efficient and stable electrocatalyst for hydrogen evolution reaction." *Journal of the American Chemical Society* 136 (13):4897.

111. Ali, M. N., J. Xiong, S. Flynn, J. Tao, Q. D. Gibson, L. M. Schoop, T. Liang, N. Haldolaarachchige, M. Hirschberger, N. P. Ong, R. J. Cava. 2014. "Large, non-saturating magnetoresistance in WTe_2." *Nature* 514 (7521):205.

112. Zhou, Y., J. L. Silva, J. M. Woods, J. V. Pondick, Q. L. Feng, Z. X. Liang, W. Liu, L. Lin, B. C. Deng, B. Brena, F. N. Xia, H. L. Peng, Z. F. Liu, H. L. Wang, C. M. Araujo, J. J. Cha. 2018. "Revealing the contribution of individual factors to hydrogen evolution reaction catalytic activity." *Advanced Materials* 30 (18):1706076.

113. Zhang, J., J. J. Wu, H. Guo, W. B. Chen, J. T. Yuan, U. Martinez, G. Gupta, A. Mohite, P. M. Ajayan, J. Lou. 2017. "Unveiling active sites for the hydrogen evolution reaction on monolayer MoS_2." *Advanced Materials* 29 (42):1701955.

114. Zhou, Y., J. V. Pondick, J. L. Silva, J. M. Woods, D. J. Hynek, G. Matthews, X. Shen, Q. Feng, W. Liu, Z. Lu, Z. Liang, B. Brena, Z. Cai, M. Wu, L. Jiao, S. Hu, H. Wang, C. M. Araujo, J. J. Cha. 2019. "Unveiling the interfacial effects for enhanced hydrogen evolution reaction on MoS_2/WTe_2 hybrid structures." *Small* 15 (19):1900078.

115. Ye, G. L., Y. J. Gong, J. H. Lin, B. Li, Y. M. He, S. T. Pantelides, W. Zhou, R. Vajtai, P. M. Ajayan. 2016. "Defects engineered monolayer MoS_2 for improved hydrogen evolution reaction." *Nano Letters* 16 (2):1097.

116. Tran, P. D., T. V. Tran, M. Orio, S. Torelli, Q. D. Truong, K. Nayuki, Y. Sasaki, S. Y. Chiam, R. Yi, I. Honma, J. Barber, V. Artero. 2016. "Coordination polymer structure and revisited hydrogen evolution catalytic mechanism for amorphous molybdenum sulfide." *Nature Materials* 15 (6):640–646.

117. Azizi, A., X. L. Zou, P. Ercius, Z. H. Zhang, A. L. Elias, N. Perea-Lopez, G. Stone, M. Terrones, B. I. Yakobson, N. Alem. 2014. "Dislocation motion and grain boundary migration in two-dimensional tungsten disulphide." *Nature Communications* 5:1–7.

118. Zhou, W., X. L. Zou, S. Najmaei, Z. Liu, Y. M. Shi, J. Kong, J. Lou, P. M. Ajayan, B. I. Yakobson, J. C. Idrobo. 2013. "Intrinsic structural defects in monolayer molybdenum disulfide." *Nano Letters* 13 (6):2615.

119. Zhu, J., Z. Wang, H. Yu, N. Li, J. Zhang, J. Meng, M. Liao, J. Zhao, X. Lu, L. Du, R. Yang, D. Shi, Y. Jiang, G. Zhang. 2017. "Argon plasma induced phase transition in monolayer MoS_2." *Journal of the American Chemical Society* 139 (30):10216.

120. Zhu, J., Z.-C. Wang, H. Dai, Q. Wang, R. Yang, H. Yu, M. Liao, J. Zhang, W. Chen, Z. Wei, N. Li, L. Du, D. Shi, W. Wang, L. Zhang, Y. Jiang, G. Zhang. 2019. "Boundary activated hydrogen evolution reaction on monolayer MoS$_2$." *Nature Communications* 10 (1):1348.

121. Tsai, C., H. Li, S. Park, J. Park, H. S. Han, J. K. Norskov, X. L. Zheng, F. Abild-Pedersen. 2017. "Electrochemical generation of sulfur vacancies in the basal plane of MoS$_2$ for hydrogen evolution." *Nature Communications* 8:15113.

122. Shaik, S., S. P. de Visser, D. Kumar. 2004. "External electric field will control the selectivity of enzymatic-like bond activations." *Journal of the American Chemical Society* 126 (37):11746.

123. Wang, Z., H.-H. Wu, Q. Li, F. Besenbacher, Y. Li, X. C. Zeng, M. Dong. 2020. "Reversing interfacial catalysis of ambipolar wse$_2$ single crystal." *Advanced Science* 7 (3):1901382.

124. Pacchioni, G., J. R. Lomas, F. Illas. 1997. "Electric field effects in heterogeneous catalysis." *Journal of Molecular Catalysis A: Chemical* 119 (1):263.

125. Wang, L., H. Yang, J. Yang, Y. Yang, R. Wang, S. Li, H. Wang, S. Ji. 2016. "The effect of the internal magnetism of ferromagnetic catalysts on their catalytic activity toward oxygen reduction reaction under an external magnetic field." *Ionics* 22 (11):2195.

126. Wang, J., M. Yan, K. Zhao, X. Liao, P. Wang, X. Pan, W. Yang, L. Mai. 2017. "Field effect enhanced hydrogen evolution reaction of MoS$_2$ nanosheets." *Advanced Materials* 29 (7):1604464.

127. Berthod, C., N. Binggeli, A. Baldereschi. 2003. "Schottky barrier heights at polar metal/semiconductor interfaces." *Physical Review B* 68 (8).

128. Wang, F. M., J. S. Li, F. Wang, T. A. Shifa, Z. Z. Cheng, Z. X. Wang, K. Xu, X. Y. Zhan, Q. S. Wang, Y. Huang, C. Jiang, J. He. 2015. "Enhanced electrochemical H$_2$ evolution by few-layered metallic WS$_{2(1-x)}$Se$_{2x}$ nanoribbons." *Advanced Functional Materials* 25 (38):6077.

129. Woods, J. M., Y. Jung, Y. J. Xie, W. Liu, Y. H. Liu, H. H. Wang, J. J. Cha. 2016. "One-step synthesis of MoS$_2$/WS$_2$ layered heterostructures and catalytic activity of defective transition metal dichalcogenide films." *ACS Nano* 10 (2):2004.

130. Xu, Y., B. Li, S. Zheng, P. Wu, J. Zhan, H. Xue, Q. Xu, H. Pang. 2018. "Ultrathin two-dimensional cobalt–organic framework nanosheets for high-performance electrocatalytic oxygen evolution." *Journal of Materials Chemistry A* 6 (44):22070.

131. Zaffran, J., M. C. Toroker. 2017. "Understanding the oxygen evolution reaction on a two-dimensional NiO$_2$ catalyst." *ChemElectroChem* 4 (11):2764.

132. Rupp, C. J., S. Chakraborty, J. Anversa, R. J. Baierle, R. Ahuja. 2016. "Rationalizing the hydrogen and oxygen evolution reaction activity of two-dimensional hydrogenated silicene and germanene." *ACS Applied Materials & Interfaces* 8 (2):1536.

133. Wang, P., M. Yan, J. Meng, G. Jiang, L. Qu, X. Pan, J. Z. Liu, L. Mai. 2017. "Oxygen evolution reaction dynamics monitored by an individual nanosheet-based electronic circuit." *Nature Communications* 8 (1):645.

134. Yin, Y., J. C. Han, Y. M. Zhang, X. H. Zhang, P. Xu, Q. Yuan, L. Samad, X. J. Wang, Y. Wang, Z. H. Zhang, P. Zhang, X. Z. Cao, B. Song, S. Jin. 2016. "Contributions of phase, sulfur vacancies, and edges to the hydrogen evolution reaction catalytic activity of porous molybdenum disulfide nanosheets." *Journal of the American Chemical Society* 138 (25):7965.

135. Esposito, D. V., S. T. Hunt, A. L. Stottlemyer, K. D. Dobson, B. E. McCandless, R. W. Birkmire, J. G. G. Chen. 2010. "Low-cost hydrogen-evolution catalysts based on monolayer platinum on tungsten monocarbide substrates." *Angewandte Chemie-International Edition* 49 (51):9859.

136. Wang, T. Y., L. Liu, Z. W. Zhu, P. Papakonstantinou, J. B. Hu, H. Y. Liu, M. X. Li. 2013. "Enhanced electrocatalytic activity for hydrogen evolution reaction from self-assembled monodispersed molybdenum sulfide nanoparticles on an Au electrode." *Energy & Environmental Science* 6 (2):625.
137. Xie, J. F., H. Zhang, S. Li, R. X. Wang, X. Sun, M. Zhou, J. F. Zhou, X. W. Lou, Y. Xie. 2013. "Defect-rich MoS$_2$ ultrathin nanosheets with additional active edge sites for enhanced electrocatalytic hydrogen evolution." *Advanced Materials* 25 (40):5807.
138. Yang, J., D. Voiry, S. J. Ahn, D. Kang, A. Y. Kim, M. Chhowalla, H. S. Shin. 2013. "Two-dimensional hybrid nanosheets of tungsten disulfide and reduced graphene oxide as catalysts for enhanced hydrogen evolution." *Angewandte Chemie-International Edition* 52 (51):13751.
139. Benck, J. D., T. R. Hellstern, J. Kibsgaard, P. Chakthranont, T. F. Jaramillo. 2014. "Catalyzing the hydrogen evolution reaction (HER) with molybdenum sulfide nanomaterials." *ACS Catalysis* 4 (11):3957.
140. Tsai, C., F. Abild-Pedersen, J. K. Norskov. 2014. "Tuning the MoS$_2$ edge-site activity for hydrogen evolution via support interactions." *Nano Letters* 14 (3):1381.
141. Ding, M., G. Zhong, Z. Zhao, Z. Huang, M. Li, H.-Y. Shiu, Y. Liu, I. Shakir, Y. Huang, X. Duan. 2018. "On-chip in situ monitoring of competitive interfacial anionic chemisorption as a descriptor for oxygen reduction kinetics." *ACS Central Science* 4 (5):590.
142. Valera, A. E., N. T. Nesbitt, M. M. Archibald, M. J. Naughton, T. C. Chiles. 2019. "On-chip electrochemical detection of cholera using a polypyrrole-functionalized dendritic gold sensor." *ACS Sensors* 4 (3):654.
143. Tokuda, T., K. Tanaka, M. Matsuo, K. Kagawa, M. Nunoshita, J. Ohta. 2007. "Optical and electrochemical dual-image CMOS sensor for on-chip biomolecular sensing applications." *Sensors and Actuators A: Physical* 135 (2):315.
144. Choi, J. W., D. Aurbach. 2016. "Promise and reality of post-lithium-ion batteries with high energy densities." *Nature Reviews Materials* 1 (4):16013.
145. Rodrigues, M.-T. F., G. Babu, H. Gullapalli, K. Kalaga, F. N. Sayed, K. Kato, J. Joyner, P. M. Ajayan. 2017. "A materials perspective on Li-ion batteries at extreme temperatures." *Nature Energy* 2 (8):17108.
146. Stoddart, A. 2017. "Lithium-ion batteries: Stress relief for silicon." *Nature Reviews Materials* 2 (8):17057.
147. Assat, G., J.-M. Tarascon. 2018. "Fundamental understanding and practical challenges of anionic redox activity in Li-ion batteries." *Nature Energy* 3 (5):373.
148. Mauger, A., C. M. Julien. 2017. "Critical review on lithium-ion batteries: Are they safe? Sustainable?" *Ionics* 23 (8):1933.
149. Goodenough, J. B., K.-S. Park. 2013. "The Li-ion rechargeable battery: A perspective." *Journal of the American Chemical Society* 135 (4):1167.
150. Kühne, M., F. Paolucci, J. Popovic, P. M. Ostrovsky, J. Maier, J. H. Smet. 2017. "Ultrafast lithium diffusion in bilayer graphene." *Nature Nanotechnology* 12 (9):895.
151. Xiong, F., H. Wang, X. Liu, J. Sun, M. Brongersma, E. Pop, Y. Cui. 2015. "Li intercalation in MoS$_2$: In situ observation of its dynamics and tuning optical and electrical properties." *Nano Letters* 15 (10):6777.
152. Akinwande, D., C. Huyghebaert, C.-H. Wang, M. I. Serna, S. Goossens, L.-J. Li, H. S. P. Wong, F. H. L. Koppens. 2019. "Graphene and two-dimensional materials for silicon technology." *Nature* 573 (7775):507.
153. Butler, S. Z., S. M. Hollen, L. Cao, Y. Cui, J. A. Gupta, H. R. Gutiérrez, T. F. Heinz, S. S. Hong, J. Huang, A. F. Ismach, E. Johnston-Halperin, M. Kuno, V. V. Plashnitsa, R. D. Robinson, R. S. Ruoff, S. Salahuddin, J. Shan, L. Shi, M. G. Spencer, M. Terrones, W. Windl, J. E. Goldberger. 2013. "Progress, challenges, and opportunities in two-dimensional materials beyond graphene." *ACS Nano* 7 (4):2898.

9 Phase Change Materials for Thermal Energy Storage

Baris Burak Kanbur, Zhen Qin,
Chenzhen Ji, and Fei Duan
Nanyang Technological University

CONTENTS

9.1   Introduction 318
9.2   Research Directions for PCM-Based TES 320
     9.2.1   Thermal Conductivity Enhancement of PCMs 320
     9.2.2   Numerical Simulation Study Related to Latent Heat Storage 321
9.3   Thermal Analysis of PCMs 322
     9.3.1   Theoretical Model 322
     9.3.2   Differential Scanning Calorimetry Test 325
           9.3.2.1   Experimental Setup and Procedures of DSC 326
           9.3.2.2   Results and Discussion 326
     9.3.3   T-history Method Experiment 328
           9.3.3.1   Experimental Setup and Procedure of T-history Method ...328
           9.3.3.2   T-history Analysis for Reference Material: KNO_3 328
9.4   On-Site Design and Operation Criteria of the PCMs 330
     9.4.1   Design of PCM-Based TES Unit 331
     9.4.2   PCM Selection Criteria 331
     9.4.3   Performance of PCM-Based TES Tank 333
9.5   Numerical Studies of the PCM-Based TES Applications 341
     9.5.1   Design of PCM-Based TES Tank in 2D and 3D Domains 341
     9.5.2   Mesh Generation and Independency Study 345
     9.5.3   Analysis Procedure and Results of Numerical Simulations 346
9.6   Performance Enhancement Studies for the PCM-Based TES Tank 353
     9.6.1   Effect of Fin Location on the PCM Performance 353
     9.6.2   Effect of Double-Fin Integration with Various Arrangements 356
     9.6.3   Effect of Angled Double Fins on the PCM Performance 359
9.7   Feasibility Assessments of PCMs in Thermal Energy Storage 364
     9.7.1   Economic Analysis of PCM-Integrated Combined
           Energy Systems 365

    9.7.2   Thermoeconomic Analysis of PCM-Integrated
            Combined Energy Systems..366
    9.7.3   Sustainability Analysis of PCM-Integrated Combined
            Energy Systems 367
9.8   Summary 370
References...370

9.1 INTRODUCTION

Energy storage systems have always been significant since the first day of electricity use in daily life. They have been achieving cost savings and providing additional advantages for the resilient energy infrastructure which are important factors for the utilities and consumers. Energy storage systems can be classified into five different groups which are mechanical, chemical, biological, magnetic, and thermal energy storage systems (Dincer and Rosen 2010). Thermal energy storage (TES) systems have been being widely used to increase the thermal component effectiveness as well as large-scale switching. In many applications, the mismatch is seen between the energy demand and energy supply, so the TES is a generally beneficial way for correcting the mismatch. TES systems are categorized as sensible heat storage, latent heat storage, and thermochemical storage, but the sensible and latent heat storage are the two methods most preferred for applications. Latent heat systems store the thermal energy during the phase change period of material. The materials used for latent heat TES are called as the phase change materials (PCMs) which provide isothermal thermal energy storage (Gil et al. 2010). PCMs provide the latent heat storage, thanks to their endothermic and exothermic reactions in heat absorption and desorption steps, respectively, in order to achieve reliable temperature control (Delgado et al. 2019). The most known advantage of the PCM-based TES systems is their relatively small component size when compared to the sensible heat storage systems; therefore, they have been becoming one of the emerging methods to reduce waste heat in some thermal engineering applications. However, there are some drawbacks for PCM applications such as the selection of heat transfer medium and system design (Farid et al. 2004). The TES with PCMs has been a hot research topic in energy field for around 40 years (Zalba et al. 2003). The first interest in the PCM-based TES started with ice-based cold storage. Nowadays, the TES with PCMs is being widely applied in various sectors such as power generation, building energy efficiency systems, electronic package cooling, peak load demands for the renewables, energy security, etc. Although there have been many technological developments in the PCM-based TES systems, there are still some research gaps. In this chapter, the theoretical, numerical, and experimental procedures of the PCM analysis are explained with real-scale unit setup and feasibility assessments.

PCMs can be operated in a wide temperature range from −100°C to 800°C (Delgado et al. 2019). Aqueous salts are the most feasible PCMs for the operations below 0°C. Above 0°C, paraffin, fatty acids, water, salt hydrates, and eutectic mixtures are generally preferred up to 100°C. The use of paraffin can reach even nearly 120°C–130°C. From 100°C to 400°C; sugar alcohols, nitrates, and hydroxides can be used. For 400°C to 800°C and above, chlorides, carbonates, and fluorides are the well-known PCMs. It must be noted that the melting enthalpy of PCMs increases with increase of the operating temperature. Besides the temperature-based classification, PCMs can be categorized according to material types, which are organic, inorganic, and eutectic, as shown in Figure 9.1. The details of PCM types are explained in the subsection below.

Organic PCMs are divided into two sub-groups as the paraffins and fatty acids (non-paraffins). Paraffins have a wide operating temperature range, and they are especially preferred in building energy efficiency systems since they do not perform phase separation, but they have some disadvantages such as low thermal conductivity and moderate flammability (Sharma et al. 1999). Non-paraffins are more preferable than the paraffins in building applications due to the fact that they provide satisfying properties like reliable stability (both chemical and thermal), small volume change, low vapor pressure, low sub-cooling, and non-toxicity. Fatty acids, alcohols, and glycols are some of the well-known non-paraffins. Similar to paraffins (Yuan et al. 2014, Sharma et al. 2015), low thermal conductivity and moderate flammability are the main drawbacks of the non-paraffins. Hydrated salts and metallics are two well-known inorganic PCMs that have higher thermal conductivity values than the organic PCMs. Hydrated salts have effective storage capacity, so they are more popular than the metallics that have a critical disadvantage from the point of weight. Also, strict controls may be required for the reliable long-term performance of the inorganic PCMs. Super-cooling and corrosion are the other drawbacks of the inorganic PCMs. The eutectic mixtures, which are simply compositions obtained by melting of multiple components, are different kinds of PCMs. The mixtures can be either inorganic–inorganic, organic–inorganic, or organic–organic. They are more preferable for the high storage capacity and a specific melting temperature range. However, their biggest drawback is the lack of thermophysical investigations on them for commercial applications. The comparative overview of the organic, inorganic, and eutectic PCMs is presented in Table 9.1.

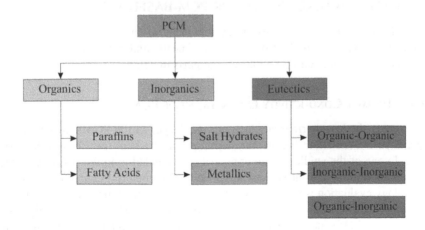

FIGURE 9.1   Classification of PCMs.

TABLE 9.1

Pros and Cons of the Organic, Inorganic, and Eutectic PCMs

PCM Type	Pros	Cons
Organic PCMs	Low vapor pressure during phase change period Available in large temperature range Large heat of fusion Non-reactive and safe Recyclable Congruent melting phase Directly Incorporable Freeze without significant sub- or supercooling	Low thermal conductivity Insoluble in water (some of them) Flammable Low phase change enthalpy, density, and heat capacity Large surface area requirement due to the low volumetric latent heat storage capacity Costly Large volumetric expansion
Inorganic PCMs	Large heat storage capacity and high thermal conductivity High heat of fusion Non-flammable Large phase change enthalpy and low volumetric expansion Cheap	Thermal and chemical instability External support requirement (container, etc.) High weight and tendency to degrade Incongruent melting Dehydration during thermal cycling and phase segregation High supercooling degree
Eutectic PCMs	Large volumetric thermal storage density Sharp melting temperature Incongruent melting without phase segregation	Lack of experimental thermophysical database Low latent heat capacity and strong odor Supercooling effect (some of them) Costly

Source: Redrawn from Nazir et al. (2019).

9.2 RESEARCH DIRECTIONS FOR PCM-BASED TES

The PCM-based TES studies focus on the material-based thermal conductivity enhancement, numeric-method based improvements, and component scale experiments. Hence, this section investigates these parts in detail.

9.2.1 THERMAL CONDUCTIVITY ENHANCEMENT OF PCMs

To overcome the low thermal conductivity challenges of the PCMs, different studies have been conducted; herein, the increment in the heat transfer rate has been aimed. In one of the studies (Xiao et al. 2014), expanded graphite (EG) was added to the PCM (nitrate salt) at various mass rates. The material characterization and morphology evaluation were done by different methods including scanning electron microscopy (SEM), transmission electron microscopy (TEM), energy dispersive spectrometry (EDS), and X-ray diffraction (XRD). It was seen that the EG inclusion increased the thermal conductivity while it decreased the latent heat nearly by 11%.

Also, Zhao et al. (2014) used additive and natural EG which was highly conductive and treated with sulfuric acid (ENG-TSA) for the binary salt PCM. Thermal stability, conductivity, phase change transition properties, and microstructures were investigated. The results revealed that the thermal conductivity increased around 110 times compared to the salt powder, while the latent heat dropped slightly, and the phase change temperature was nearly unchanged. Apart from the EG inclusions, chloride was added to the PCM by Peng et al. (2010). The chlorine inclusion rate of 5% achieved higher thermal stability and conductivity as well as 50°C higher operating temperature value, but the freezing point was observed at a lower value.

9.2.2 Numerical Simulation Study Related to Latent Heat Storage

Numerical simulations based on commercial software like ANSYS Fluent or COMSOL Multiphysics are the fast and low-cost methods to analyze and investigate the thermal performance of the TES system with PCMs. Instead of using a costly experimental setup, the initial parametric studies can be performed via numerical techniques. In numerical studies, the enthalpy and the modified heat capacity methods are two well-known techniques to simulate PCM phase change behaviors. The mathematical models and details about the numerical studies can be found in Liu et al. (2014) in detail. When a solid–liquid PCM is considered in operation, the melting occurs during the heat absorption (charging), while the solidification is observed during the heat desorption (discharging). In both melting and solidification steps, two heat transfer mechanisms, which are convection and conduction, dominate the process. Therefore, both mechanisms should be considered in the numerical method. The comparison of both enthalpy and heat capacity methods infers that the enthalpy method may encounter some difficulties about simulating the super-cooling and the temperature oscillation issues, while the modified heat capacity method has some difficulties if the simulated PCM has a narrow phase change temperature range (Liu et al. 2014). The enthalpy method is a convenient technique for both two-dimensional (2D) (Das et al. 2016) and three-dimensional (3D) (Shamsundar and Sparrow 1975) PCM simulations. 2D studies employ simpler approaches, but they can still provide sufficient information on the melting or solidification process. Guo and Zhang (2008) proved the positive impact of aluminum foil addition on the heat transfer performance enhancement, whereas Yimer and Adami (1997) performed nonlinear parametric simulations in radial and axial dimensions both with 2D numerical models. The study of Shamsundar and Sparrow (1975) is one of the pioneer 3D PCM simulation studies. Although several geometric shapes have been investigated for PCM-based TES systems, the cylinder-tube shape is the widely used one amongst the others. He and Zhang (2000) designed a numerical model to solve the coupled unsteady freezing-forced convection problem with finite difference approach that successfully operated the center-difference and fully implicit schemes. In finite element methods, semi-discrete equations can represent the melting and freezing simulations in good agreement for 3D PCM studies (Gong and Mujumdar 1996). By using the finite volume method, a simple explicit scheme can also solve the 3D PCM simulations. A working fluid tube with a cavity radiation model was solved with this scheme for a cylindrical solar heat receiver including PCM (Cui et al. 2003).

Besides the aforementioned approaches, different functions such as if, else-if, and else clauses; tangent functions; and error functions are able to analyze the 3D PCM simulations (Qin 2016).

Though the numerical studies give crucial insights on the design concept, the experimental studies are also required to confirm the reliability of latent heat storage performance. To overcome the low thermal conductivity drawback of PCMs, the PCM containers must be designed as effectively as possible. Many researchers experimentally investigated different kinds of heat exchanger designs for the best PCM performance. Kabbara and Abdallah (2013) studied the cylindrical PCM tanks, which use air as heat transfer fluid during the discharging and charging periods, in order to better understand their performance in different scales in residential solar heating operations. Zhang et al. (2014) validated the transient heat transfer models with phase transitions by using experimental temperature change hot chamber technique. Both enthalpy and heat capacity methods were successfully validated. Thapa et al. (2014) analyzed the low-cost thermal conductivity enhancements such as copper foams and metallic inserts via millimeter-scale-constructed experimental setup to better observe the specific impacts.

9.3 THERMAL ANALYSIS OF PCMs

9.3.1 THEORETICAL MODEL

Two-dimensional (2D) or three-dimensional (3D) heat transfer phenomena are complex to be defined via analytical equations. However, one-dimensional (1D) analytical models can be used to visualize some critical heat transfer mechanisms before the numerical and/or experimental studies. The analytical model is also known as the theoretical model, and it helps to analyze the charging and discharging of PCMs under steady conditions. The 1D heat transfer model schematic is seen in Figure 9.2.

For 1D illustrations, the PCM tank is considered to be operated with waste heat of a microturbine plant. The exhaust gas transfers the heat to the oil flow through the middle PCM. At first, the heat transfer rate, $\dot{Q}$, is defined by the gas properties as shown in Eq. 9.1, which is equivalent to the convective heat transfer rate between the gas and the PCM surface, as shown in Eq. (9.2),

$$\dot{Q} = C_{p,\,\text{gas}} \cdot \dot{m}_{\text{gas}} \cdot \left(T_{\text{gas,in}} - T_{\text{gas,out}} \right) \tag{9.1}$$

$$\dot{Q} = h_{\text{gas}} \cdot A_{\text{gas}} \cdot \left(\frac{T_{\text{gas,in}} - T_{\text{gas,out}}}{2} - T_5 \right) \tag{9.2}$$

where $C_{p,\text{gas}}$ is the specific heat of exhaust gas, $\dot{m}_{\text{gas}}$ is the mass flow rate of exhaust gas, h_{gas} is the convective heat transfer coefficient of the exhaust gas, A_{gas} is the heat transfer area of exhaust gas on the PCM surface, T_5 is the PCM temperature at the gas side, and $T_{\text{gas,in}}$ and $T_{\text{gas,out}}$ are the exhaust gas inlet and outlet temperatures, respectively. The heat transfer mechanism is conductive heat transfer until the melting process starts; therefore, the heat transfer rate can be defined in Eq. (9.3),

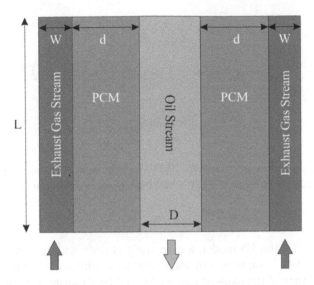

FIGURE 9.2 Simplified schematic of the 1D heat transfer model for PCM-based TES tank. (Modified and redrawn from Qin (2016).)

which also equals to the heat transfer from PCM to the oil as shown in Eqs. (9.4) and (9.5), respectively,

$$\dot{Q} = k_{PCM} \cdot A_{gas} \cdot \left(\frac{T_5 - T_6}{d_{PCM}} \right) \tag{9.3}$$

$$\dot{Q} = C_{p,oil} \cdot \dot{m}_{oil} \cdot \left(T_{oil,in} - T_{oil,out} \right) \tag{9.4}$$

$$\dot{Q} = h_{oil} \cdot A_{oil} \cdot \left(T_6 - \frac{T_{oil,in} + T_{oil,out}}{2} \right) \tag{9.5}$$

where k_{PCM} is the thermal conductivity of the PCM, T_6 is the PCM surface temperature at the heat transfer oil side, d_{PCM} is the thickness of PCM in the designed container, $C_{p,oil}$ is the specific heat of the heat transfer oil, $\dot{m}_{oil}$ is the mass flow rate, h_{oil} is the convective heat transfer coefficient, A_{oil} is the heat transfer area of the heat transfer oil on the PCM surface, and $T_{oil,in}$ and $T_{oil,out}$ are the heat transfer oil inlet and outlet temperature, respectively. According to the 1D heat transfer model, the exhaust gas outlet temperature, $T_{gas,out}$; the oil outlet temperature, $T_{oil,out}$; and PCM surface temperatures at both gas and oil sides, T_5 and T_6, are calculated as shown in Eqs. (9.6)–(9.9), respectively.

$$T_{gas,out} = \frac{T_{gas,in} \left(\dfrac{2C_{p,gas} \cdot \dot{m}_{gas}}{h_{oil} \cdot A_{oil}} + \dfrac{C_{p,gas} \cdot \dot{m}_{gas}}{C_{p,oil} \cdot \dot{m}_{oil}} + \dfrac{C_{p,gas} \cdot \dot{m}_{gas}}{h_{gas} \cdot A_{gas}} + \dfrac{C_{p,gas} \cdot \dot{m}_{gas} \cdot d_{PCM}}{k \cdot A_{gas}} - \dfrac{1}{2} \right) + T_{oil,in}}{\dfrac{2C_{p,gas} \cdot \dot{m}_{gas}}{h_{oil} \cdot A_{oil}} + \dfrac{C_{p,gas} \cdot \dot{m}_{gas}}{C_{p,oil} \cdot \dot{m}_{oil}} + \dfrac{C_{p,gas} \cdot \dot{m}_{gas}}{h_{gas} \cdot A_{gas}} + \dfrac{C_{p,gas} \cdot \dot{m}_{gas} \cdot d_{PCM}}{k \cdot A_{gas}} + \dfrac{1}{2}}$$

$$\tag{9.6}$$

$$T_{\text{oil,out}} = T_{\text{oil,in}} + \frac{2C_{p,\text{gas}} \cdot \dot{m}_{\text{gas}}}{C_{p,\text{oil}} \cdot \dot{m}_{\text{oil}}} \left(T_{\text{gas,in}} - T_{\text{gas,out}} \right) \tag{9.7}$$

$$T_5 = \frac{T_{\text{gas,in}} + T_{\text{gas,out}}}{2} - \frac{C_{p,\text{gas}} \cdot \dot{m}_{\text{gas}} \left(T_{\text{gas,in}} - T_{\text{gas,out}} \right)}{h_{\text{gas}} \cdot A_{\text{gas}}} \tag{9.8}$$

$$T_6 = \frac{C_{p,\text{gas}} \cdot \dot{m}_{\text{gas}} \cdot d_{\text{PCM}}}{k \cdot A_{\text{gas}}} \left(T_{\text{gas,in}} - T_{\text{gas,out}} \right) \tag{9.9}$$

Following the procedure given in Eqs. (9.1)–(9.9), the total heat transfer rate of a TES tank that includes N number of PCM chambers, $\dot{Q}_{\text{TES}}$, is defined in Eq. (9.10).

$$\dot{Q}_{\text{TES}} = N2 \cdot C_{p,\text{gas}} \cdot \dot{m}_{\text{gas}} \cdot \left(T_{\text{gas,in}} - T_{\text{gas,out}} \right) \tag{9.10}$$

To better illustrate the 1D model, a case study is done with the parameters given in Table 9.2, which is supported by the operating conditions in Table 9.3. Also, the thermal properties of the exhaust gas, PCM, and heat transfer oil are presented in Table 9.4.

The results are projected in Figure 9.3a–d according to the various exhaust gas inlet temperatures (from 265°C to 290°C), exhaust gas mass flow rates (0.15–0.4 kg s⁻¹), oil inlet temperatures (from 227°C to 250°C), and oil mass flow rates (0.2–0.9 kg s⁻¹), respectively. It is seen that oil inlet temperature has more dominant effects than the exhaust gas inlet temperature on the PCM boundary temperature and $\dot{Q}_{\text{TES}}$.

TABLE 9.2
Baseline Geometric Parameters for the TES System

Gas Side		PCM Side		Oil Side	
Channel height	0.6 m	Chamber number	8	Tube outer diameter	0.008 m
Channel width	0.02 m	Chamber height	0.6 m	Tube thickness	1 mm
Channel length	1.53 m	PCM width	0.015 m	Tube inner diameter	0.006 m
Height to width ratio	30	Chamber length	0.015 m	Tube length	8 m

TABLE 9.3
Baseline Working Conditions for the TES System

Gas Side		Oil Side	
Mass flow rate	0.295 kg s⁻¹	Mass flow rate	0.25
Gas average velocity[a]	4.047 m s⁻¹	Mass flow rate per tube	0.03125
Inlet temperature	285°C	Inlet temperature	227°C

[a] The average velocity of exhaust gas in each air channel.

TABLE 9.4
Thermal Properties of Exhaust Gas, PCM, Heat Transfer Oil

	Thermal Conductivity (W m^{-1}·K)	Heat Capacity (J kg^{-1}·K)	Density (kg m^{-3})	Kinetic Viscosity (m^{2}s^{-1})	Prandtl's Number	Dynamic Viscosity (Pa·s)
Exhaust gas (250°C)	0.0421	1,034	0.675	4.1×10^{-5}	0.68	2.779×10^{-5}
PCM (221°C)	0.45	1,481	1948	-	-	4.343×10^{-3}
Oil (210°C)	0.11235	2,134	904.3			3.685×10^{-4}

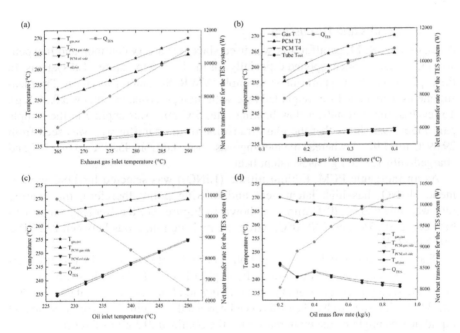

FIGURE 9.3 Effect of the (a) exhaust gas inlet temperature, (b) exhaust gas mass flow rate, (c) oil inlet temperature, and (d) oil mass flow rate on the output temperatures and the net heat transfer rate of the TES system. (Modified and redrawn from Qin (2016).)

With respect to these results, the suggested oil temperature is 240°C since the PCM oil side temperature must be lower than the melting point.

9.3.2 DIFFERENTIAL SCANNING CALORIMETRY TEST

Differential Scanning Calorimetry (DSC) technique is a well-known thermal analysis method that focuses on the difference between the amount of heat required to increase the sample temperature and the measured reference. During the DSC test, sample material and reference material are maintained at the same temperature. Also, the holder temperature increases linearly (as a function of time) during the

DSC test. It is a reliable technique to measure the latent heat of fusion, melting point, and heat capacity for PCMs.

9.3.2.1 Experimental Setup and Procedures of DSC

For DSC tests, the experimental instrument TA Q200 was used for the sample materials with weight 0–10 mg. To hold the samples, an aluminum pan (Perkin Elmer) was used, while an empty pan was considered as the reference. Q200 chamber was used for heating and cooling of samples at various rates. Data collection was done with a program, TA Universal Analysis; then, the convenient cycles were chosen for the analysis. Q200 chamber and pans (both reference and sample) are shown in Figure 9.4. By using the experimental setups, different commercial PCMs were tested.

9.3.2.2 Results and Discussion

As an organic PCM, paraffin wax was used since it is a very common and inexpensive PCM that is widely applied especially for the low-temperature TES systems. In the experiments, 6.9 mg paraffin wax sample (VWR, Singapore) was used. The measurements were done by a weight balance with high precision (accuracy: ±0.0001 g). To reach accurate results, a slow heating rate, 5°C/min, was applied to the sample pan. Figure 9.5 shows that the glass transition of the paraffin wax was seen from 29.09°C to 33.92°C, whereas the melting was observed from 45.01°C to 61.73°C and charged with 128.20 J g^{-1} as its latent heat.

As an inorganic PCM, lithium nitrate ($LiNO_3$) was selected for DSC experiments. $LiNO_3$ has high latent heat and melting point; therefore, the heating rate was selected as 20°C/min. As shown in Figure 9.6, the freezing range was between 250.25°C and 220.70°C, while the melting range was between 232.21°C and 275.13°C.

Since the melting and freezing periods showed different behaviors, that difference must be considered as well to really apply $LiNO_3$. The latent heat values were observed to be different as well. For the melting and freezing periods, they were 304 and 325 J g^{-1}, respectively. The difference between the specific heats of the solid and liquid phases might be the main reason for the obtained 21 J g^{-1} difference.

(a)   (b)

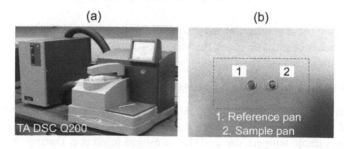

FIGURE 9.4 DSC test equipment: (a) TA Q200 and (b) aluminum pans. (Modified and redrawn from Qin (2016).)

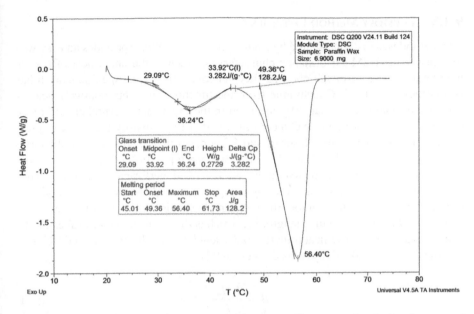

FIGURE 9.5 The DSC melting curve of 6.9 mg paraffin wax under the heating rate of 5°C/min. (Modified and redrawn from Qin (2016).)

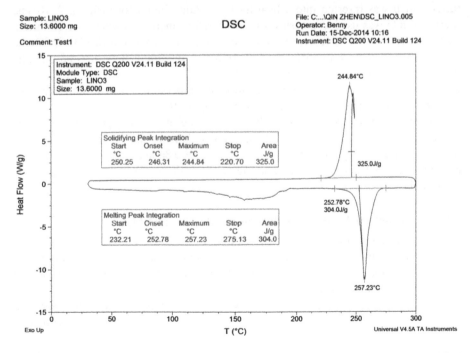

FIGURE 9.6 The DSC curve for 13.6 mg LiNO₃ under heating rate 20°C/min. (Modified and redrawn from Qin (2016).)

9.3.3 T-HISTORY METHOD EXPERIMENT

The method was firstly presented by Zhang and Jiang (1999). It provides an easy way to measure the PCM performance characteristics such as the latent heat of fusion, melting point, degree of sub-cooling, thermal conductivity, and specific heat. It has similarities with the DSC technique from the function angle, but relatively heavier samples, normally more than 10 g, are used in the experiments. Another difference between the T-history and DSC techniques is that the T-history method can present thermal property values closer to real PCM performance since real applications use a large amount of PCM, unlike the DSC technique. In the T-history method, the cooling curve is plotted first; then, all the PCM performance characteristics are obtained from the cooling curve by using the lumped capacitance method. For uniform temperature distribution, the test sample should have the Biot number (Bi) value <0.1. Bi depends on tube length, L_c, which is the radius of the test tubes in this experiment, r; convective heat transfer coefficient between the ambient and PCM, h; and the thermal conductivity, k, as defined in 9.11.

$$Bi = \frac{h \cdot L_c}{k} \tag{9.11}$$

9.3.3.1 Experimental Setup and Procedure of T-history Method

A mortar was used to grind the samples; then, they were used in the test unit (a test tube), which was inserted into a heating block (aluminum) for melting. A K-type thermocouple was used to measure the temperature increment after its calibration. The calibration provides less error for more accurate results. From 30°C to 90°C, the calibration was done with 15°C intervals in a bath calibrator with thermal oil. The reverse cooling was conducted following the same intervals. After the melting process, the PCM was cooled down to the ambient temperature. The calibration curves of three selected thermocouples are shown in Figure 9.7.

The observed difference was ±0.96°C; therefore, it was accepted that the thermocouples were calibrated in good agreement. During the heating and cooling processes, the cooling curve was observed. Figure 9.8a and b presents the experimental setup for T-history method and the heating/cooling steps of the experiments, respectively.

9.3.3.2 T-history Analysis for Reference Material: KNO$_3$

As a reference material for T-history experiments, KNO$_3$ was selected due to its high melting point at 334°C, and it does not experience any phase transition in the T-history method temperature range. The obtained cooling curve of KNO$_3$ is shown in Figure 9.9, and the h value can be calculated through the curve.

For the calculation of h, the heat transfer balance between the convective heat loss and the heat stored by the tube and material were built first, as defined in Eq. (9.12),

$$h \cdot A_c \cdot A = \left(m_t \cdot c_{p,t} + m_r \cdot c_{p,r} \right) \cdot \left(T_0 - T_f \right) \tag{9.12}$$

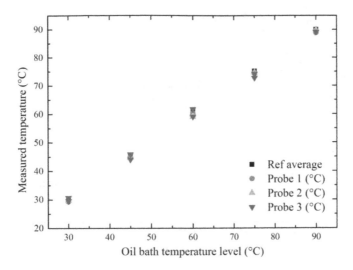

FIGURE 9.7 The thermocouples calibration results for T-history experiments. (Modified and redrawn from Qin (2016).)

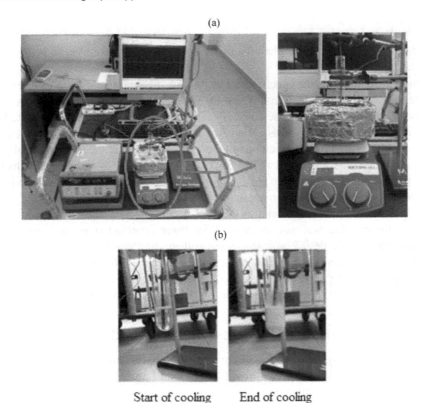

FIGURE 9.8 T-history method: (a) the experimental setup and (b) the heating/cooling steps of the experiments. (Modified and redrawn from Qin (2016).)

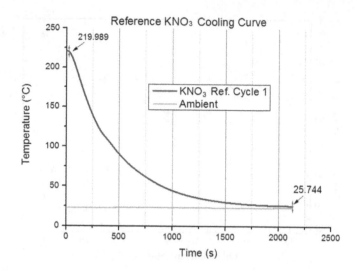

FIGURE 9.9 Cooling curve for KNO_3 as reference material. (Modified and redrawn from Qin (2016).)

where $A = \int_{t_0}^{t_1}(T_m - T_{amb})dt$ is the area under temperature–time curve, m_t is the mass of the test tube, $c_{p,t}$ is the specific heat of the test tube, m_r is the mass of reference material KNO_3, $c_{p,r}$ is the specific heat of KNO_3, T_0 is the initial temperature, and T_f is the final temperature. The convective heat transfer area, A_c, was calculated by assuming the PCM test tube as cylindrical, which allowed assuming A_c as the area by multiplication of tube diameter and tube height. Following the mathematical procedure, the h value was found as 26.05 W m^{-2}.°C. The experiments were conducted three times to present accurate results, and the repeated measurements showed that the results were in good agreement with one another, as shown in Figure 9.10.

The mean h value was found as 26.4 W m^{-2}.°C. With respect to the obtained curves, the h value was also calculated for three repeated runs, and the results are presented in Table 9.5 with the calculated thermal conductivity value, ≥0.934 W m^{-1}.K, which was obtained by considering the maximum limit of Bi number as 0.1 (see Eq. (9.11)). The experimental procedure presented for KNO_3 is able to give the main procedure for the eutectic mixtures (e.g. $LiNO_3$–KNO_3) that are a type of PCM.

9.4   ON-SITE DESIGN AND OPERATION CRITERIA OF THE PCMs

PCM-based TES systems have wide application areas, and the power generation sector is one of them. PCMs can provide significant improvement for the waste heat utilization from the exhaust gas of power generation systems. However, the design and operation of the PCMs are vital for reliability. Hence, in this section, the design and operation criteria of the PCM-based TES systems are presented with the on-site experimental data.

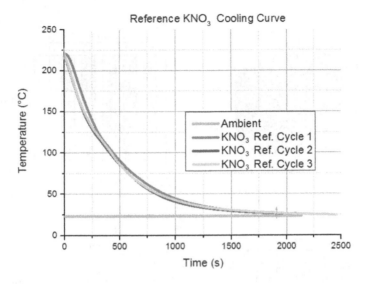

FIGURE 9.10 Three repeated runs for KNO_3 cooling curves. (Modified and redrawn from Qin (2016).)

TABLE 9.5
Heat Transfer Coefficients for Multiple Tests for KNO_3

	h_1 (W m$^{-2.}$°C)	h_2 (W m$^{-2.}$°C)	h_3 (W m$^{-2.}$°C)	h_{avg} (W m$^{-2.}$°C)	L_c (m)	k (W m^{-2}·k)
Value	26.05	28.05	25.13	26.41	0.0036	≥0.94

9.4.1 DESIGN OF PCM-BASED TES UNIT

The cuboidal PCM encapsulation structure was chosen as the experimental proto-type since it has some advantages such as easy fabrication, large contact area, and small pressure drop (provides small pressure drops less than the back pressure of turbine in the power generation system) when compared to the other designs. The schematic of the PCM-based TES unit is shown in Figure 9.11a. PCMs were filled in a stainless-steel shell and aluminum chambers (five chambers). Four thermocouples were integrated into the chambers labeled B (label numbers: 101, 102, 301, and 302) and D (label numbers: 111, 112, 311, and 312) in order to collect interior PCM tem-perature data. The positions of thermocouples are projected in Figure 9.11b–d.

9.4.2 PCM SELECTION CRITERIA

There are some critical points to determine the most convenient PCM in the rel-evant application. The thermal capacity is always a key issue, but the environmental concerns, cost-effective performance, and other technical details must be taken into

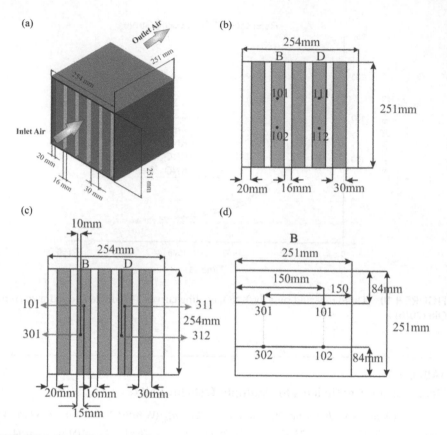

FIGURE 9.11 PCM-based TES tank: (a) simplified schematic, (b) front view, (c) top view, and (d) side view. (Modified and redrawn from Qin (2016).)

account as well. For the thermal capacity concerns, some of the important details are listed below (Agyenim et al. 2010, Johra and Heiselberg 2017):

- PCMs should be selected by considering the melting point value, which can vary within a small range, according to the operating conditions of the thermal process.
- High latent heat is a desired property for the PCMs to be applied in a small amount.
- Besides the latent heat storage, a high specific heat value is also desired for satisfying sensible heat storage performance.
- To minimize the temperature gradient during the charging and discharging process, high thermal conductivity is a key factor.
- The phase transition can cause volume changes, which negatively affect the PCMs if they are located in a closed tank; therefore, small volume changes are preferable.
- Safety criteria such as being non-corrosive, chemically stable, non-toxic with fire resistance, and non-explosive should also be taken into account.

In the PCM unit design, paraffin wax with the melting point of 50°C (approx. value) was selected as the storage media. The thermophysical properties of the paraffin wax and the relevant DSC test result are shown in Table 9.6 and Figure 9.12, respectively.

9.4.3 PERFORMANCE OF PCM-BASED TES TANK

Performance investigation of the TES tank can give a general opinion about the impact of operating parameters and system limitations. To investigate the performance of the PCM-based TES unit, a test platform was built modifying the test unit for air conditioning purpose. In the conventional air conditioning test unit, which has a rectangular profile with plastic duct material (PMMA duct with 245×259 mm dimensions), an evaporator is located in the middle of the unit. In the modified design, a PCM unit was integrated instead of the evaporator. In addition, two adjustable coil heaters (1 kW at 220 V each, extended fin electric heating element), one

TABLE 9.6

Physical Properties of Paraffin Wax (Received from the Manufacturing Company)

Physical Properties	Melting Point (°C)	Freezing Point (°C)	Latent Heat (kJ kg⁻¹)	Density (kg m⁻³)
Value	55.8	53	210	860 (solid) 780 (liquid)
Physical Properties	Specific Heat (kJ kg⁻¹·K)	Thermal Conductivity (W m⁻¹·K)		Dynamic Viscosity (N·s m⁻²)
Value	2.9 (solid) 2.1 (liquid)	0.24 (solid)	0.15 (liquid)	0.205

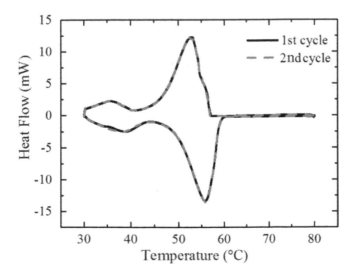

FIGURE 9.12 DSC curve of 6.9 mg paraffin wax under the heating rate of 5°C/min.

velocity-adjusted axial fan, and several mixers (for uniform air distribution) were used for the test platform. A voltmeter and ammeter were used for the measurements. To control the recirculation, a damper at the duct exit orifice was adjustable for different positions. It must be noted that the temperature measurement part was well-insulated (with 13 mm thickness of class 0 Armaflex) to block the heat loss through the surface. For measurements, K type thermocouples were located in the unit. The calibration of thermocouples was done, and the accuracy of ±0.5°C was observed. Figure 9.13a and b shows the simplified schematic and real scale setup, respectively.

Three different air flow velocities by adjusting fan speeds (25, 35, 45 rpm) and two heating rates (1 and 2 kW) were used to better characterize the PCM performance. The charging process was 4 hours; then, the heaters were turned off. Similarly, the discharging process was 4 hours so that the entire process was completed. Before the experiments, the insulation performance test was also done since the heat loss could occur through surfaces due to the high operating conditions (melting point) of PCM although all significant surfaces were well insulated. There were two main heat loss regions in the PCM unit, which were (1) around the heaters and (2) in the duct. Both heat losses were determined by the heat loss coefficients α_1 and α_2, respectively. The insulation performance test was done without the PCM unit. Nine different thermocouples were located at inlet and outlet sections of the test unit in order to monitor the temperature values. During the test, the heater power and fan speed were adjusted in several steps. At first, 1 kW heater power and 55 rpm fan speed were applied until a stable temperature was observed. After that, the heater power was increased to 2 kW. Once the temperature was stabilized again, the fan speed was dropped to 45 rpm. This adjustment was done until the fan speed was decreased to 25 rpm (at 10 per reduction) at the heating power of 2 kW. Figure 9.14 shows the simplified schematic of the insulation test setup.

The heat loss coefficients α_1 and α_2 were calculated separately. For α_1, the generated heat via heaters is defined in Eq. (9.13),

$$Q_{heater} = Q_{eff} + Q_{loss} \qquad (9.13)$$

where Q_{heater} is total heater power rate, Q_{loss} is the heat loss from the heater to the ambient environment, and Q_{eff} is the effective heating power rate of the heater. The definitions of Q_{eff} and Q_{loss} are given in Eqs. (9.14) and (9.15), respectively,

$$Q_{eff} = \dot{m}_a \cdot C_{p,a} \cdot \left(T_{ave,inlet} - T_{amb}\right) \qquad (9.14)$$

$$Q_{loss} = \alpha_1 \cdot \left(T_{ave,1} - T_{amb}\right) \qquad (9.15)$$

where $\dot{m}_a$ is the mass flow rate of air, $C_{p,a}$ is the specific heat of air, $T_{ave,inlet}$ is the average air inlet temperature (after heaters), T_{amb} is the ambient air temperature, and $T_{ave,1}$ is the average value of $T_{ave,inlet}$ and T_{amb}. The accuracy and sensitivity of the voltmeter and ammeter were validated by parametric studies, and their accuracies were found as ±1 V and ±0.1 A, respectively. Also, the mass flow meter was validated by parametric studies at different operating conditions. The heat transfer coefficient α_2 was calculated similarly as shown in Eq. (9.16),

(a)

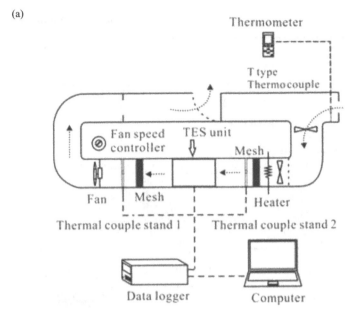

(b)

1) Preheaters 2)Thermoanemometer 3)Thermocouple for inlet air temperature
4) Thermal energy storage unit 5) Thermocouple for outlet air temperature
6) Data acquisition system

FIGURE 9.13 PCM test unit: (a) the simplified schematic and (b) the actual setup. (Modified and redrawn from Qin (2016).)

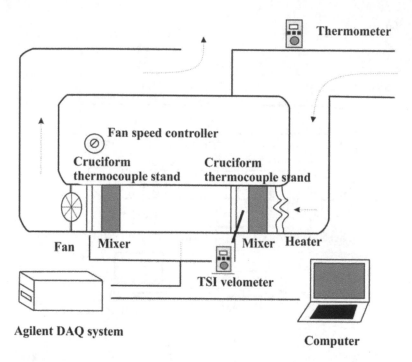

FIGURE 9.14 Schematic diagram of the insulation test setup. (Modified and redrawn from Qin (2016).)

$$Q_{\text{loss}} = \alpha_2 \cdot \left(T_{\text{avg},2} - T_{\text{amb}} \right) \tag{9.16}$$

where $T_{\text{ave},2}$ is the average temperature value calculated by the average inlet and outlet temperatures that are shown in Figure 9.15.

Also, the calculated heat loss coefficients and their percentages are presented in Tables 9.7 and 9.8 for α_1 and α_2, respectively. In summary, the insulation tests showed that the heat loss was maintained minimum at low heating power rate and high fan speed, whereas it could be assumed low even at high heating power rate and low fan speed.

The results show the satisfactory insulation performance of the PCM-based TES tank. Also, the calculated heat transfer coefficients can allow us to calculate TES performance more accurately. After investigating the impact of insulation on the test setup, different parametric studies were performed. Three different storage media, which were the water, air (empty chamber), and paraffin wax (VWR Singapore), were tested at three different fan speeds and two different heater power rates as 25/35/45 rpm and 1/2 kW, respectively. The objectives of the tests were to see the impact of these parameters on the temperature inside PCM, charging and discharging power, charging and discharging efficiency, and stored heat in the PCM unit.

Figure 9.16 shows the comparative results of the PCM curves at two different heater power rates for the same position (chamber D, position 311) at constant fan speed of 25 rpm. It is known that a higher inlet air temperature is obtained by

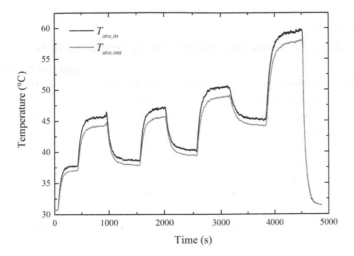

FIGURE 9.15 Temperature history of average inlet temperature and outlet temperature to determine α_2. (Modified and redrawn from Qin (2016).)

TABLE 9.7

Calculated Heat Loss Coefficients α_1 in the Insulation Tests

Working Conditions	Heat Loss Power (%)	Heat Loss Coefficient α_1 (W/°C)
Fan speed: 45; heater power: 1 kW	21.7	20.7
Fan speed: 45; heater power: 2 kW	23.7	23.0
Fan speed: 54; heater power: 1 kw	24.5	50.9
Fan speed: 54; heater power: 2 kW	24.3	51.2
Fan speed: 65; heater power: 1 kW	21.8	55.8
Fan speed: 65; heater power: 2 kW	25.5	67.7
Fan speed: 75; heater power: 1 kW	20.8	57.2
Average value	23.2	46.7

increasing the heater power. Thus, it can be seen that the increase of heater power provides a higher PCM melting rate. The highest temperature values were 52.4°C ± 0.5°C and 62.6°C ± 0.5°C for the 1 and 2 kW heating rates, respectively. When 2 kW heater power was applied, a sudden jump was realized at around 11,625 seconds. The reason could be related to the natural convection that made the temperature distribution non-uniform in the liquid phase of PCM (paraffin wax) due to the significant temperature gradient between the solid and liquid phases. A comprehensive study was done between two different positions in chamber D, positions 311 and 312, at 1 and 2 kW heater power rates while the fan speed was constant at 25 rpm.

Figure 9.17 shows that the PCM melting curve of 311 does not have a remarkable difference compared to the trends of 312, except for the end of heating of 311 that has a sudden jump because of the complete melting of paraffin wax at this specific

TABLE 9.8

Calculated Heat Loss Coefficient α_2 in the Insulation Tests

Working Conditions	Heat Loss Power (%)	Heat Loss Coefficient α_2 (W/°C)
Fan speed: 45; heater power: 1 kW	53.1	2.4
Fan speed: 45; heater power: 2 kW	35.5	0.8
Fan speed: 54; heater power: 1 kW	52.8	5.2
Fan speed: 54; heater power: 2 kW	94.0	4.7
Fan speed: 65; heater power: 1 kW	61.5	7.5
Fan speed: 65; heater power: 2 kW	126.8	8.0
Fan speed: 75; heater power: 1 kW	61.9	8.1
Average value	63.4	5.3

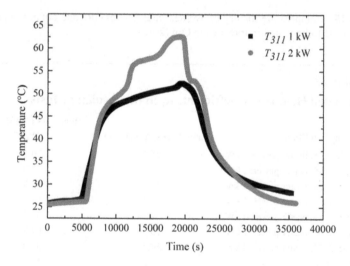

FIGURE 9.16 Temperature history of the PCM at position 311 of chamber D under 1 and 2 kW heater power at fan speed 25 rpm. (Modified and redrawn from Qin (2016).)

position. Therefore, it is inferred that the upper PCM melted first by means of the natural convection conditions. A similar temperature jump was seen at 1 kW heating power rate for position 311. Like the heater power rate, the fan speed also had an impact on the PCM performance since it crucially affected the mass flow rate of the air.

Figure 9.18a and b illustrates the PCM temperatures at the middle and one-third sections of chamber D with respect to different fan speeds as well as the constant heater power rate of 2 kW. It was observed that the PCM melting performance was faster at low speeds, while faster freezing was seen at high speeds. Also, the upper position of 112 presented faster freezing than the lower position of 112. Similar sensible heating regions were observed for the PCM at the speeds of 25 and 35 rpm,

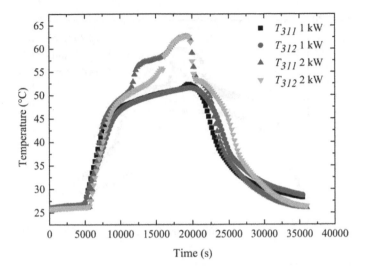

FIGURE 9.17 Temperature history of the PCM at positions 311 and 312 of chamber D under 1 and 2 kW heater power at fan speed 25 rpm. (Modified and redrawn from Qin (2016).)

while the melting curves were different. Besides, complete PCM melting was not observed at 45 rpm.

Instantaneous heat transfer rate, Q_{ins}, is a significant parameter to monitor the PCM performance; therefore, the energy balance equation was used to calculate it with a good agreement under the assumption of the quasi-steady conditions as presented in Eq. (9.17),

$$Q_{ins} = \dot{m}_a \cdot C_{p,a} \cdot \left(\overline{T}_{inlet} - \overline{T}_{outlet} \right) \qquad (9.17)$$

where $\overline{T}_{inlet}$ and $\overline{T}_{outlet}$ are the average inlet and outlet temperatures of air. Figure 9.19 projects that the highest Q_{ins} was obtained at 25 rpm and 2 kW (lowest speed and highest heater power rate), while the lowest one was seen at 45 rpm and 1 kW (highest speed and lowest heater power rate) during the charging period. The obtained result was expected since the stored energy decreased over time with decrease of the temperature difference between PCM and inlet air, which caused a lower heat transfer rate. In the case of the discharging period, Q_{ins} became negative due to the release of stored energy. As in the charging period, lower Q_{ins} was observed at the heater power rate of 1 kW. Also, Q_{ins} gradually came close to zero. The parametric results showed that the heater power rate had a more dominant effect than the fan speed for transient PCM performance.

Besides the fan speed and heater power rate, the effectiveness of paraffin wax was investigated with comparative studies. The optimal operating condition, 25 rpm and 2 kW, was selected for the comparison of paraffin wax with the water and air (empty chamber). Figure 9.20a and b shows the comparative results in the middle and one-third section of chamber D, respectively.

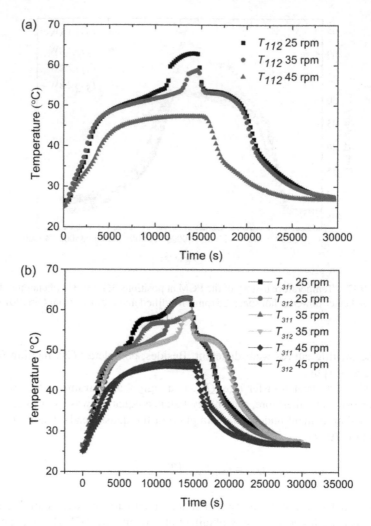

FIGURE 9.18 Comparison of the PCM temperatures at the (a) middle and (b) one-third sections in chamber D. (Modified and redrawn from Qin (2016).)

The peak temperatures of air and water seen were as 65°C and 62°C, respectively; whereas the heating curve gradient of air was found close to 1 (steeper than that of water). Therefore, it can be said that air provided faster heating and cooling than water. Moreover, paraffin wax reached the peak temperature of 62°C, which was very close to that of water. However, phase change was observed between 46°C and 53°C, while water and air did not have any phase change period. The temperature curve gradient of paraffin wax was found to be the lowest compared to that of water and air. The charging and discharging periods are shown in Figure 9.21 in detail. It can be seen that the PCM was found to be the most feasible choice followed by water and air.

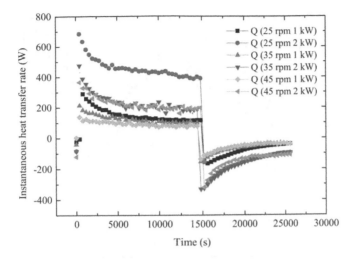

FIGURE 9.19 Instantaneous heat transfer rate of the TES unit at different working conditions. (Modified and redrawn from Qin (2016).)

9.5 NUMERICAL STUDIES OF THE PCM-BASED TES APPLICATIONS

PCM-based TES systems are two-phase models, and they can be simulated by using different numerical techniques. Numerical simulations of the PCMs can be performed in two ways which are 2D and 3D studies. In this section, the design of PCM geometry will be discussed for both 2D and 3D cases.

9.5.1 Design of PCM-Based TES Tank in 2D and 3D Domains

To better explain the 2D and 3D design criteria, the commercial simulation software COMSOL Multiphysics (finite element method) was used in the simulations. In both 2D and 3D models, the turbulent flow was modeled via κ–ε model due to the fact that the Reynolds number was calculated as 4,237 based on the boundary conditions, which are defined in the following sentences. Equations (9.18) and (9.19) are the general governing equations for the numerical studies,

$$\rho\frac{\partial u}{\partial t} + \rho(u \cdot \nabla)u = \nabla \cdot \left[-\rho I + (\mu + \mu_T)(\nabla u + \nabla u^T)\right] + F \quad (9.18)$$

$$\rho\nabla \cdot (u) = 0 \quad (9.19)$$

where ρ, T, and u are the density, absolute temperature, and the velocity vector, respectively. Dynamic and turbulent viscosity terms are denoted by μ and μ_T, respectively. The turbulence intensity is denoted by I. F is the volume force, equal to $F = g\rho\alpha(T - T_0)$ where α is thermal diffusivity, and it is related to the natural convection calculations. In the κ–ε model, the turbulent viscosity, μ_T, is modeled as shown in Eq. (9.20),

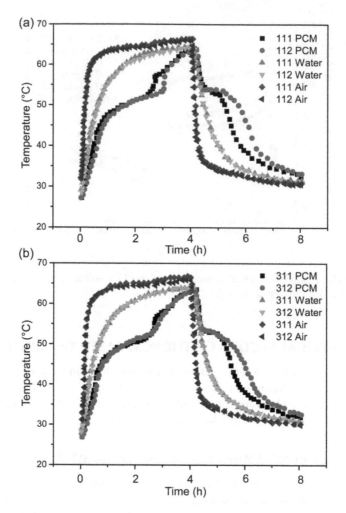

FIGURE 9.20 Comparison of PCM temperatures in the (a) middle and (b) one-third sections of chamber D. (Modified and redrawn from Qin (2016).)

$$\mu_T = \rho C_\mu + \frac{k^2}{\varepsilon} \qquad (9.20)$$

where C_μ is a constant, k is the turbulent energy, and ε is the turbulent dissipation rate. By using the general definitions and turbulent viscosity, transport equations for k and ε values are given in Eqs. (9.21), and (9.22), respectively,

$$\rho \frac{\partial k}{\partial t} + \rho(u \cdot \nabla)\varepsilon = \nabla \cdot \left[\left(\mu + \frac{\mu_T}{\sigma_k} \right) \nabla k \right] + P_k - \rho \varepsilon \qquad (9.21)$$

$$\rho \frac{\partial \varepsilon}{\partial t} + \rho(u \cdot \nabla)k = \nabla \cdot \left[\left(\mu + \frac{\mu_T}{\sigma_\varepsilon} \right) \nabla \varepsilon \right] + C_{e1} \frac{\varepsilon}{k} P_k - C_{e2} \frac{\varepsilon^2}{k} \rho \qquad (9.22)$$

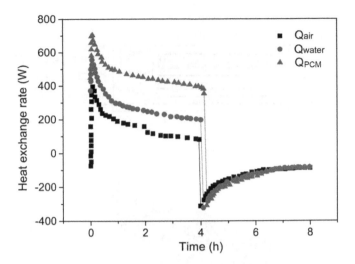

FIGURE 9.21 Energy stored per unit time of PCM, water, and air. (Modified and redrawn from Qin (2016).)

where C_mu is 0.09, C_e1 is 1.44, C_e2 is 1.92, Sigma_k is 1.00, and Sigma_epsilon is 1.30. The constant values are the assumed values in the COMSOL Multiphysics environment. P_k is the production term of the κ–ε model, and it is defined in Eq. 9.23.

$$P_k = \mu_T \left[\nabla u : \left(\nabla u + \nabla u^T \right) \right] \qquad (9.23)$$

Both 2D and 3D models have three different material regions which are the PCM, aluminum, and air. PCMs are located in aluminum chambers, and air flows through the aluminum chamber surfaces by means of the air ducts. The 2D and 3D domains are shown in Figure 9.22a and b, respectively.

Following the experimental setup details, 2D design domain was constituted with the symmetric characteristic to save the computational costs. For the numerical solution of the TES tank, the governing equation of the heat transfer mechanism is presented in Eq. (9.24),

$$d_z \rho C_p u \cdot \nabla T + q \cdot \nabla = d_z Q + q_0 + d_z Q_p + d_z Q_{vd} \qquad (9.24)$$

where Q denotes the heat sources other than the viscous dissipation that is given as Q_{vd}. Q_p denotes the heat sources related to the pressure work. The conductive heat flux, q, can be better understood from Eq. (9.25).

$$q = -d_z k \cdot \nabla T \qquad (9.25)$$

The properties of the PCM are determined by the heat capacity formulation. The material phase change temperature is defined as T_{pc}; therefore, it is assumed that the transformation occurs in a temperature interval which is between $T_{pc} - \nabla T/2$ and $T_{pc} + \nabla T/2$. In the defined range, a smooth function (θ), which represents the phase

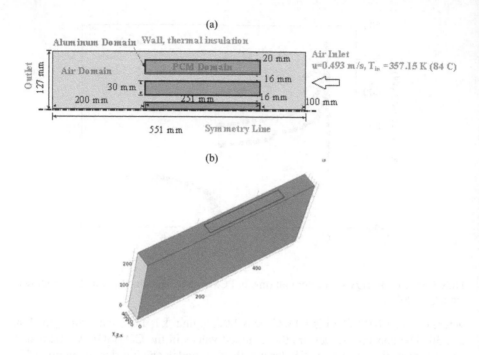

FIGURE 9.22 (a) 2D and (b) 3D design domains for the PCM-based TES tank. (Modified and redrawn from Qin (2016).)

fraction before the transition, is used for modeling of the material phase. The value of θ equals to 0 and 1 at $T_{pc} + \frac{\nabla T}{2}$ and $T_{pc} - \frac{\nabla T}{2}$, respectively. To define the density term, the material properties in two phases (liquid–solid, liquid–gas, etc.) must be defined well. Equation (9.26) expresses the density as a function of θ and the densities at different phases (ρ_{phase1} and ρ_{phase2}),

$$\rho = \theta \rho_{\text{phase1}} + (1 - \theta) \rho_{\text{phase2}} \qquad (9.26)$$

Similar to the density, the specific heat, C_p, the mass fraction, α_m, and the thermal conductivity, λ, can be defined with θ and densities at different phases, as shown in Eqs. (9.27)–(9.29), respectively,

$$C_p = \frac{1}{\rho} \left(\theta \rho_{\text{phase1}} C_{p,\text{phase1}} + (1 - \theta) \rho_{\text{phase2}} C_{p,\text{phase 2}} \right) + L \frac{\partial \alpha_m}{\partial T} \qquad (9.27)$$

$$\alpha_m = \frac{1}{2} \frac{(1 - \theta) \rho_{\text{phase2}} - \theta \rho_{\text{phase1}}}{\theta \rho_{\text{phase1}} + (1 - \theta) \rho_{\text{phase2}}} \qquad (9.28)$$

$$\lambda = \theta \lambda_{\text{phase1}} + (1 - \theta) \lambda_{\text{phase2}} \qquad (9.29)$$

where L is the latent heat. In the 2D model, natural convection was not considered; i.e., the model was considered according to the heat conduction mechanism. As the boundary conditions, the air inlet velocity was set to 0.493 m s^{-1}, whereas the air inlet temperature was considered as 357.15 K. These values were determined from the experiments. The upper wall of the 2D model was well insulated; therefore, a strong thermal insulation was assumed. The outlet of the air duct was defined as the outlet boundary condition. Inlet temperature was equal to 303.15 K which was the room temperature. The COMSOL Multiphysics software already had the thermal properties of the aluminum and air, and the properties of PCM (paraffin wax) were inputted to the software manually. The thermal properties of the PCM are shown in Table 9.9. Unlike the 2D model, the 3D model considered the natural convection, but only one chamber was modeled to save computational time. The height of chamber geometry was designed as 250 mm (z-direction).

9.5.2 MESH GENERATION AND INDEPENDENCY STUDY

For the air and aluminum regions of the PCM-based TES domain, triangular meshing was used in COMSOL Multiphysics. However, the quadrilateral mesh was applied in the PCM region for better observation of the melting. In the 2D model, the number of total elements was 30140 (numbers of triangular and quadrilateral elements were 18237 and 11903, respectively). Figure 9.23a and b illustrates the meshing of the whole 2D domain and the 2D PCM–aluminum region, respectively. In the 3D domain, the number of meshes dramatically increases compared to the 2D domain; therefore, the computational time and costs increase as well. To minimize the computational cost and time, the mesh sweep method was applied with free triangular elements through the z-direction. Denser mesh structures were applied to the critical regions (e.g., boundaries and interfaces) for more accurate simulations. The total number of elements was calculated as 147648. Figure 9.23c shows 3D meshing.

Apart from the mesh structures and the number of elements, the effect of mesh elements on the simulation performance was investigated via mesh independence study. For this purpose, four different mesh structures were built with 12094, 16663, 30140, and 56037 elements, respectively. Figure 9.24 shows the different mesh structures.

The fan speed and heater power rate were decided as 25 rpm and 2 kW, respectively; then the comparative simulations were plotted as shown in Figures 9.25 and 9.26 for the air outlet temperature and PCM temperature, respectively. The results showed that the mesh independency was achieved for the study, and the mesh

TABLE 9.9
Thermal Properties of Paraffin Wax as PCM

PCM	Paraffin Wax
Density	775 kg m^{-3}
Conductivity	0.25 W m^{-1}·K
Heat capacity at constant pressure	2,384 J kg^{-1}·K
Phase change temperature range	322.615–328.765 K
Latent heat	126.2 kJ kg^{-1}

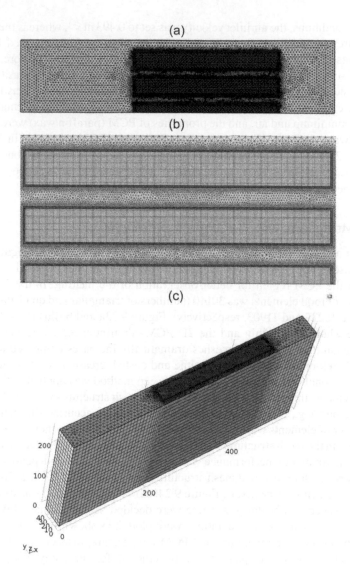

FIGURE 9.23 The mesh structure for (a) 2D domain, (b) PCM region in 2D domain, and (c) 3D domain. (Modified and redrawn from (Qin 2016).)

structure which had the minimum number of elements was preferred for less computational time and cost.

9.5.3 ANALYSIS PROCEDURE AND RESULTS OF NUMERICAL SIMULATIONS

Both 2D and 3D models were simulated under transient conditions with the time period of 14,400 seconds. The default COMSOL Multiphysics settings were applied in the simulations; and the temperature, pressure, and velocity data were saved

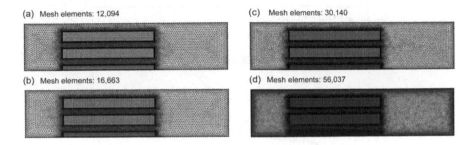

(a) Mesh elements: 12,094

(c) Mesh elements: 30,140

(b) Mesh elements: 16,663

(d) Mesh elements: 56,037

FIGURE 9.24 Meshes generated for 2D grid independence study. (Modified and redrawn from Qin (2016).)

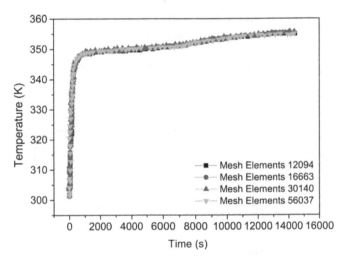

FIGURE 9.25 Air outlet temperature histories in cases with three different mesh elements. (Modified and redrawn from Qin (2016).)

automatically. At first, the trends of the steady air flow velocity and streamlines were observed. It was seen that the air velocity became steady after a short time in the transient simulation. The air flow reached a higher velocity while the air was flowing into the gaps between PCM chambers; also a decrease in velocity was observed while air was leaving the gaps. Figure 9.27 shows the simulation results for the velocity and streamline trends. At the inlet region of the PCM model, a boundary layer developing region was observed which is shown with blue color in Figure 9.27.

Several vortexes were observed at the exit region of the PCM model because of the non-uniform air-flow distribution and momentary pressure changes. It is seen that the current case can be improved by applying the following: (1) integration of a return guide vane into the front part of the chamber to smooth the momentary pressure drop and (2) the redesign of PCM chamber (in a hydraulic form) to get better air flow uniformity. Temperature contours are shown in Figure 9.28 at 1,000, 5,000, and 10,000 seconds respectively. It was seen that the uniformity increased with time

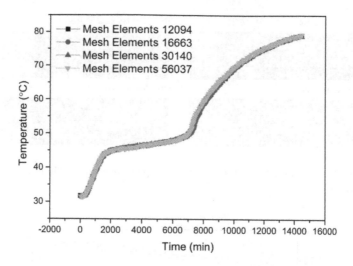

FIGURE 9.26 PCM temperature at position (200, 46). (Modified and redrawn from Qin (2016).)

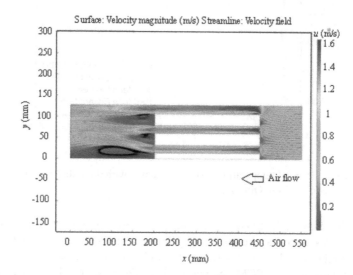

FIGURE 9.27 Velocity contour and streamline achieved from the simulation results. (Modified and redrawn from Qin (2016).)

due to the temperature differences between the PCMs, aluminum, and hot air flow, and the heat transfer rate decreased. Also, it was observed that the heat transfer at the middle was slower than the heat transfer at the boundary. Based on these conclusions, a better PCM encapsulation geometry can be designed.

Similar to the temperature contours, the phase indicator contours are plotted in Figure 9.29 at 1,000, 5,000, and 10,000 seconds, respectively. At 1,000 seconds, the PCM was in the solid state, while it was in the phase transition and the completely

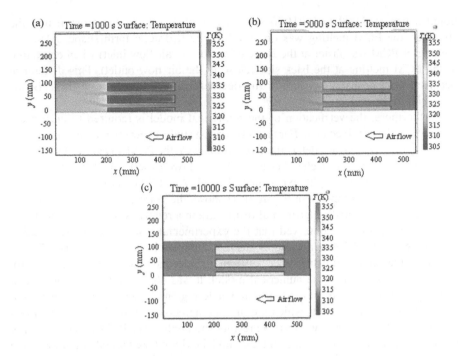

FIGURE 9.28 Temperature contours of the whole domain at (a) 1,000 seconds, (b) 5,000 seconds, and (c) 10,000 seconds respectively. (Modified and redrawn from Qin (2016).)

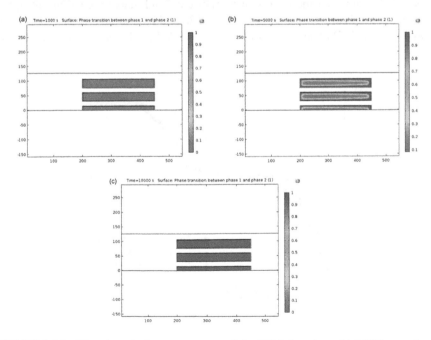

FIGURE 9.29 The phase indicator contours of the PCM domain at (a) 1,000 seconds, (b) 5,000 seconds, and (c) 10,000 seconds, respectively. (Modified and redrawn from Qin (2016).)

liquid states at 5,000 and 10,000 seconds, respectively. At the front inside the chamber, the PCM melting was observed as the hydraulic form shape. Also, the melting of PCM was faster at the front side (close to air flow inlet) when compared to the PCM melting at the back side (close to the air flow outlet). This difference occurred most probably because of the decrease in heat transfer rate from the air flow inlet to the outlet.

Furthermore, the verification of the numerical model is required to ensure the accuracy of the obtained data. For the verification study, experimental and numerical results were compared at a determined position inside the PCM and air outlet regions. A location was selected in chamber B (PCM), and two experimental results (experimental result 1 at 2/3 height, experimental result 2 at 1/3 height in the z-direction) were used for the comparison. Figure 9.30 shows the comparative results, and it is seen that numerical and experimental results are in a relatively good agreement.

However, it was also observed that the experimental results had shorter melting period. Besides the PCM temperatures, the instantaneous heat transfer rate (see Figure 9.31a) and the temperature difference between the inlet and outlet air (see Figure 9.31b) were compared to verify the numerical model. It was observed that both numerical and experimental results showed similar stable regions which were caused by the PCM latent heat. The results which are in good agreement show that the constituted 2D numerical model provides accurate results and predicts the PCM-based TES tank performance well. However, since the 2D model did not take natural convection into account, the PCM melting period was longer than that in the experimental studies.

To create a more accurate numerical model, natural convection was considered in the 3D domain, and the results of temperature and velocity contours at 6,000 and 7,200 seconds are presented in Figure 9.32a and b, respectively. Both contours showed that the PCM was in the melting period. When the upper and lower parts

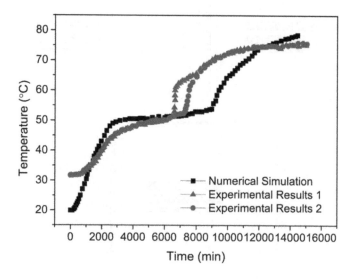

FIGURE 9.30 Simulation temperature profiles compared with experimental results inside PCM. (Modified and redrawn from Qin (2016).)

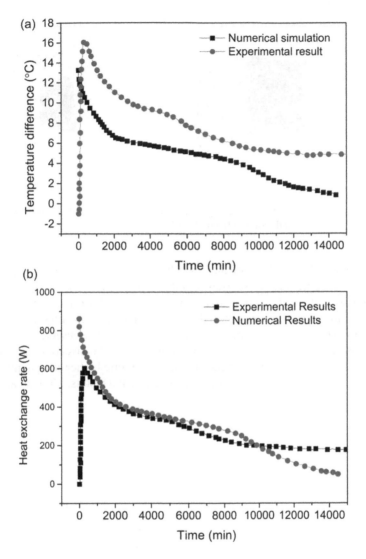

FIGURE 9.31 Verification of 2D domain for (a) temperature difference between inlet and outlet during charging process and (b) instantaneous heat transfer rate during the charging process. (Modified and redrawn from Qin (2016).)

were compared, it was seen that faster melting occurred at the upper part, which was the result of the enhancement of heat transfer via natural convection. Moreover, it was seen that the PCM melting proceeded from the chamber sidewalls to the center, excepting the bottom and top wall. Another important observation is that the heat conduction mechanism in the solid PCM was less efficient than the convective heat transfer in the liquid PCM. This result shows the reason why many studies have focused on the enhancement of PCM thermal conductivity. In Section 9.6, some of the enhancement suggestions will be explained in detail.

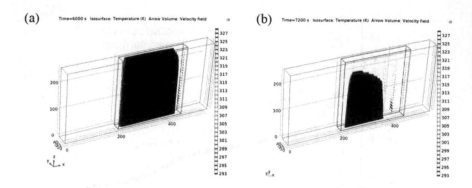

FIGURE 9.32 Field contours of (a) temperature and (b) velocity at 6,000 and 7,200 seconds, respectively. (Modified and redrawn from Qin (2016).)

Compared to 2D domain verification results for PCM melting period (Figure 9.30), numerical results provided closer trends to the experimental results since the natural convection was taken into account as shown in Figure 9.33. Besides the closer trends, a sudden jump, which was not observed for the numerical simulations in Figure 9.30, was seen in 3D simulations. Thus, it is possible to say that the observation of the sudden jump is directly related to the consideration of natural convection in the numerical model. That is to say, although the 3D models require higher computational time and costs, they achieve trends closer to the experimental results. More details can be found in Ji et al. (2017).

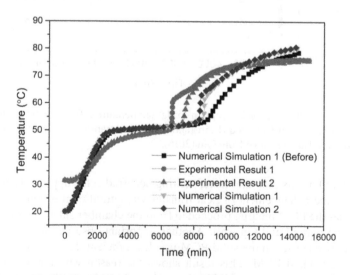

FIGURE 9.33 The 3D simulation temperature profiles compared with experimental results inside PCM. (Modified and redrawn from Qin (2016).)

9.6   PERFORMANCE ENHANCEMENT STUDIES
FOR THE PCM-BASED TES TANK

As mentioned in Section 9.5 in detail, PCM's low thermal conductivity is one of the main drawbacks for the PCM-based TES operations. To overcome this challenge, different heat transfer enhancement and material development studies have been conducted and are available in the scientific literature. This section highlights the heat-transfer-based performance enhancement techniques for the PCM-based TES operations.

9.6.1   EFFECT OF FIN LOCATION ON THE PCM PERFORMANCE

Structured fins are efficient inserts to increase the heat transfer surface area that provides a better heat transfer rate. Therefore, the structured fins were inserted in the PCM tank, and the effect of fin locations was simulated by Qin et al. (2017). The PCM-based TES tank had a copper surface, and a piece of stainless steel fin was located at different heights, which were 1/4H, 1/2H, and 3/4H (H is the height of the tank), in the TES tank as can be seen in Figure 9.34.

The labels of P.1, P.2, and P.3 corresponded to the different heights in a sequence. The fin had 4 mm thickness. The constant heat flux condition was defined on the left side (copper side), while the adiabatic boundary conditions were defined for the other sides of the tank. The PCM applied was RT_{42} (Rubitherm) with the thermal properties given in Table 9.10.

The numerical solution of the phase change problem was obtained by the method of Voller and Prakash (1987). At each iteration step, the liquid fraction was calculated. As mentioned in Section 9.5 in detail, the natural convection has a crucial impact on the PCM performance; therefore, the Boussinesq equation was used to model the natural convection in the simulation. The numerical simulations were done in the ANSYS CFD environment. Also, the mesh independency study was done for three different number of mesh elements, which were 6862, 14900, and 26554.

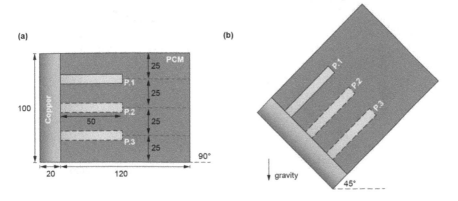

FIGURE 9.34   Geometric configuration of the designed PCM-based TES tank; P.1: Plate 1, P.2: Plate 2, and P.3: Plate 3. (Modified and redrawn from Qin et al. (2017) with permission.)

TABLE 9.10

Thermal Properties of Rubitherm 42 and Copper in the Single-Fin Location Study

Properties	RT$_{42}$	Copper
Density (kg m^{-3})	760	8,978
Specific heat (J kg^{-1}·K)	2,000	381
Thermal conductivity (W m^{-1} K)	0.2	387.6
Dynamic viscosity (kg m^{-1}·s)	0.02351	-
Thermal expansion rate (1/K)	0.0005	-
Heat of fusion (J kg^{-1})	165,000	-
Solidification/melting temperature (K)	311.5–315.5	-

Source: Qin et al. (2017).

The mesh structure with 14900 elements presented a good agreement with the experimental results and the higher number of mesh elements (26554); therefore, it was selected as the mesh structure in the numerical simulation. More details can be found in Qin et al. (2017). The verification of the numerical model was achieved by using the experiments previously conducted by Kamkari et al. (2014). Figure 9.35 shows that the numerical results are in very good agreement with the experimental results.

The simulation runs were applied for 22,000 seconds, but the comparative assessments were done at the time steps of 6,000, 10,000, and 14,000 seconds since the major part of the PCM was melted after 14,000 s. The heat flux input applied was 1,200 W m². Figure 9.36 presents the transient melting process of PCM at different fin locations.

It is seen that the location of fin significantly affected the initial behavior of the PCM. The PCM region that was close to the fin surface always melted first.

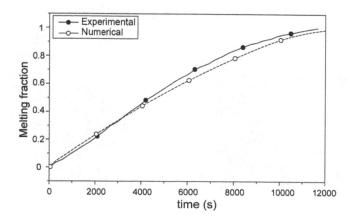

FIGURE 9.35 Comparison of the current numerical results with the experimental outputs of Kamkari et al. (2014). (Modified and redrawn from Qin et al. (2017) with permission.)

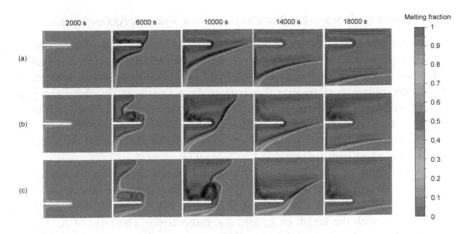

FIGURE 9.36 PCM melting fraction contours at different time steps for the labels (a) P.1, (b) P.2, and (c) P.3.

In addition, even if the fin was located at the bottom part, the top surface near the heat source surface melted first. Compared to the bottom surface, top surface melted faster in all cases, but the fin location at the bottom part (P.3) provided more uniform melting period. Besides the transient melting period, the melting time was investigated by using the melting fraction evolution. The fraction value of 1 represents the completion of melting, while the 0 (zero) value represents the beginning of the melting. Figure 9.37 illustrates the melting fraction evolution for three cases (P.1, P.2, and P.3). The results show that the fin at a lower location provided the fastest melting period, which agreed with the uniform melting during the transient period as shown in Figure 9.36. The total melting time was found as 21,620, 21,600, and 20,820 seconds for the locations P.1, P.2, and P.3, respectively.

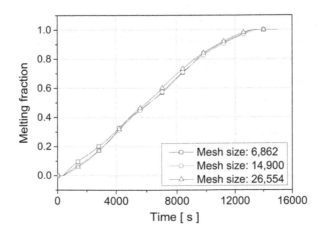

FIGURE 9.37 Transient melting fraction of the PCM at different fin locations. (Modified and redrawn from Qin et al. (2017) with permission.)

9.6.2 Effect of Double-Fin Integration with Various Arrangements

Section 9.6.1 proposes that the fins improve the PCM melting performance, and the single-fin location has an impact on the heat transfer enhancement. In this section, the double-fin structure was located on the PCM tank with different arrangements to see the enhancement performance. Apart from the single-fin study, aluminum plate fins were used in the numerical study, and the heat source was defined as the temperature input at the constant value of 70°C from the aluminum plate surface (Ji et al. 2018a). In the PCM tank, RT_{42} (Rubitherm) was considered. Thermal properties of both RT_{42} and aluminum are presented in Table 9.11.

The total length of fins was 50 mm, but nine different double-fin configurations were designed with different lengths. Figure 9.38 illustrates the schematic of the PCM-based TES tank. The double-fin structure had two fins which were the upper and lower fins with the length of L_u and L_d, respectively. The ratio of L_u/L_d was defined as the length ratio, and the nine different double-fin structures had different length ratios in the range of 0.04–4, as can be seen in Table 9.12 in detail.

The numerical model was 2D and in steady state with the laminar and incompressible Newtonian flow of PCM motions in the liquid phase. Thermophysical properties were assumed constant, whereas the volume expansion of PCM was not taken into account. The natural convection was modeled by using the Boussinesq equation. Mesh size independency was provided by comparing three different mesh structures with the number of cells of 6304, 12933, and 25258, respectively. To avoid the errors in numerical accuracy, the mesh with 12933 cells was selected in the numerical calculations. Before the analysis of numerical results, the verification of the numerical model was done by the experiments of Kamkari et al. (2014). Figure 9.39 shows the verification plots which show the errors were less than ±5%. Therefore, it can be said that the numerical results were in good agreement with those of the previously performed experiments.

TABLE 9.11
Thermal Properties of Rubitherm 42 and Aluminum in the Double-Fin Arrangement Study

Properties	RT_{42}	Aluminum
Density (solid/liquid) (kg m^{-3})	880/760	2,179
Specific heat (J kg^{-1} K)	2,000	871
Thermal conductivity (W m^{-1}·K)	0.2	202.4
Dynamic viscosity (kg m^{-1}·s)	0.0235	-
Thermal expansion rate (1/K)	0.00001	-
Heat of fusion (J kg^{-1})	165,000	-
Solidification/melting temperature (K)	311.5–315.5	-
Prandtl number	235.07	-

Source: Ji et al. (2018a).

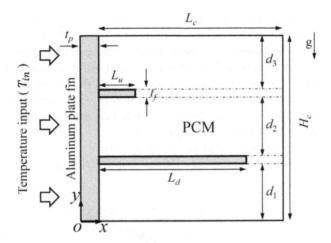

FIGURE 9.38 PCM-based TES tank configuration with different double-fin arrangements. (Modified and redrawn from Ji et al. (2018a) with permission.)

TABLE 9.12
Nine Different Double-Fin Arrangements

Case Number	L_u (mm)	L_d (mm)	$L_u + L_d$ (mm)	L_u/L_d
I	25	25	50	1
II	30	20	50	1.5
III	40	10	50	4
IV	20	30	50	0.67
V	10	40	50	0.25
VI	8	42	50	0.19
VII	5	45	50	0.11
VIII	3.5	46.5	50	0.08
IX	2	48	50	0.04

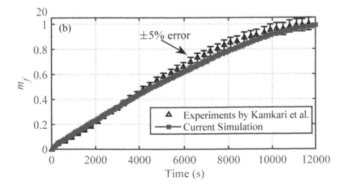

FIGURE 9.39 Verification of numerical model with the experimental results of Kamkari et al. (2014). (Modified and redrawn from Ji et al. (2018a) with permission.)

Figure 9.40a shows the evolution of melting fraction to understand the melting time and PCM melting performances. The results infer that the fastest melting was provided at the lowest length ratio. The length ratio above 1, the melting trends were very close to one another. In addition, Figure 9.40b illustrates the absorbed sensible and latent heat rates for different configurations. Since the lowest length ratio provided the fastest melting, the fastest latent heat storage was observed for the lowest length ratio, inherently. However, for the sensible heat storage rates, there was no significant difference between the configurations.

Figure 9.41 supports the outcomes of Figure 9.40a and b by projecting the liquid fraction and velocity contours for different length ratios. It can be seen that the

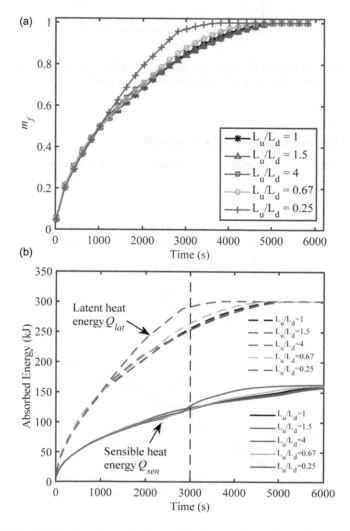

FIGURE 9.40   (a) Melt fraction evolution and (b) absorbed energy rates for different length ratios. (Modified and redrawn from Ji et al. (2018a) with permission.)

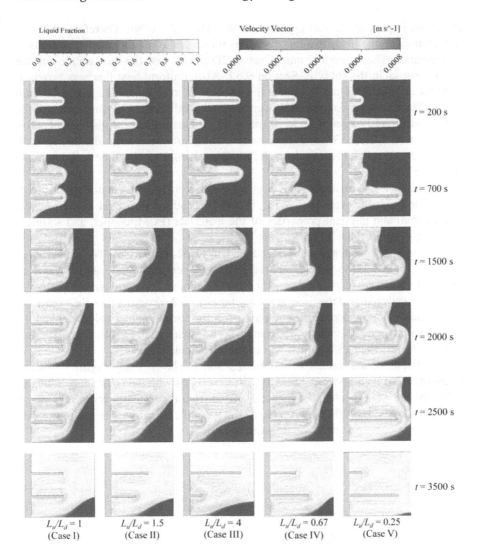

FIGURE 9.41 Instantaneous velocity vector and liquid fraction contours for different length ratios between 200 and 3,500 seconds. (Modified and redrawn from Ji et al. (2018a) with permission.)

lowest length ratio provided the fastest melting between 200 and 3,500 seconds. These results deduce the impact of natural convection on the melting. The longer bottom fin length makes the melting more uniform and preferable.

9.6.3 EFFECT OF ANGLED DOUBLE FINS ON THE PCM PERFORMANCE

Double-fin structure provided crucial heat transfer enhancement for the PCM melting period. The lowest length ratio presented the fastest melting time. In addition to the length ratio between the upper and lower fins, the fin angles can have also an important impact on the melting performance (Ji et al. 2018b). Therefore, this section focuses on the investigation of various fin angles to obtain better heat transfer performance. The numerical model was a 2D model, and the heat source surface was copper as in the study in Section 9.6.1. Also, uniform heat flux was provided from the copper heat plate that had a thickness of 20 mm (L_p) and height of 100 mm (W_{wall}). The rest of the domain walls were defined as adiabatic. The dimensions of the 2D domain were 120 mm (L_{wall})×100 mm (W_{wall}). The insulated walls were of stainless steel with 1.5 and 33 mm thicknesses on the right/bottom walls and the top wall, respectively. The PCM material was RT$_{42}$ (Rubitherm). The 2D numerical domain is shown in Figure 9.42.

The comparative fin angle study was done with a no-fin and five double-fin structures. The fin angles were −30°, −15°, 0°, 15°, and 30° as projected in Figure 9.43. The thermal and physical properties of the materials (PCM, copper, and stainless steel) are given in Table 9.13.

The mesh independency study was done before the simulations. The grid sizes of 8296, 17215, and 29186 were compared in the independency study, and it was seen that the results were in good agreement with one another. The grid size of 17215 elements was selected, and the time step was chosen as 0.2 seconds. Also, the numerical verification was provided by the comparison of the current numerical model and the experimental study previously performed by Kamkari et al. (2014). The error was found to be less than ±5%, which was in the acceptable range.

The first numerical results were obtained from the no-fin study. Figure 9.44 illustrates the PCM melting from 2,000 to 12,000 seconds with the constant heat flux of 2,500 W m^{-2}. As expected, the melting started from the top-left corner, and then it ended at the bottom-right corner. Due to the natural convection, the top region had faster melting performance than the bottom region.

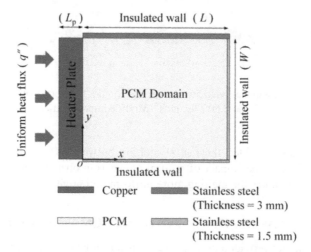

FIGURE 9.42 2D numerical domain of the study. (Modified and redrawn from Ji et al. (2018b) with permission.)

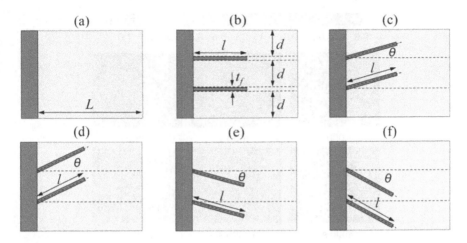

FIGURE 9.43 Angled fins for the numerical study: (a) no-fin case and parallel fins at angles of (b) 0°, (c) 15°, (d) 30°, (e) –15°, (f) –30°. (Modified and redrawn from Ji et al. (2018b) with permission.)

TABLE 9.13
Thermal Properties of Materials Used in Angled Double-Fin Study

Properties	RT$_{42}$	Copper	Stainless Steel
Density (solid/liquid) (kg m^{-3})	880/760	8,978	8,030
Specific heat (J kg^{-1} K)	2,000	381	502
Thermal conductivity (W m^{-1} K)	0.2	387.6	16.27
Dynamic viscosity (kg m^{-1}s)	0.0235	-	
Thermal expansion rate (1/K)	0.0001	-	
Heat of fusion (J kg^{-1})	165,000	-	
Solidification melting temperature (K)	311.5–315.5	-	
Prandtl number	235.07	-	

Source: Ji et al. (2018b).

The parallel double-fin structure ($\theta = 0°$) performance is presented with the fin length ratio of 0.5 (*l/L* see Figure 9.43) in Figure 9.45a. Up to 4,000 seconds, the heat conduction was the dominant mechanism, but the natural convection became more dominant after 4,000–12,000 seconds. Although the double-fin insert increased the heat transfer rate in both top and bottom parts, the bottom part had a still remarkable solid PCM region. Figure 9.45b and c shows the double-fin structures with the angles of 15° and 30°, respectively. The upward angled fins provided more uniform heat transfer for the top region, but they could not solve the non-uniform distribution problem for the bottom part. Therefore, it can be said that the parallel and upward angled fins showed similar melting trends for PCMs at the bottom region. However, the

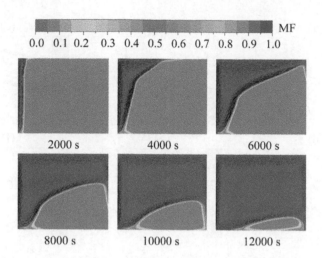

FIGURE 9.44 Melt fraction contours with natural convection-driven flow vectors for the no-fin case. (Modified and redrawn from Ji et al. (2018b) with permission.)

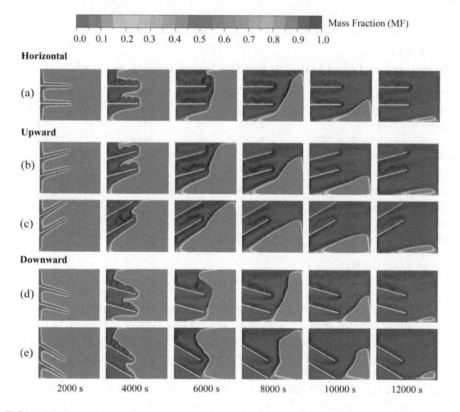

FIGURE 9.45 Melt fraction contours with natural convection-driven flow vectors for the (a) parallel-fin case (0°), the upward angled fins with (b) 15° and (c) 30°, and the downward angled fins with (d) −15° and (e) −30°. (Modified and redrawn from Ji et al. (2018b) with permission.)

downward angled fins with the angles of −15° and −30° (see Figure 9.45d and e) achieved more uniform heat distribution since they increased the melting performance at the bottom region.

The melting fraction trends for the parallel, upward, and downward fins are shown in Figure 9.46 with the enhancement ratio, which shows how the PCM melting is positively affected by means of the angled fins. As discussed for Figure 9.45, the downward angled fins showed the fastest melting performance. That is, the angle of −15° had the fastest melting, and it was followed by the fin with the downward angle of −30°. The parallel fin (no-angle fins, 0°) came after the downward angled fins. The fast PCM melting brought extra advantage from the point of saving time

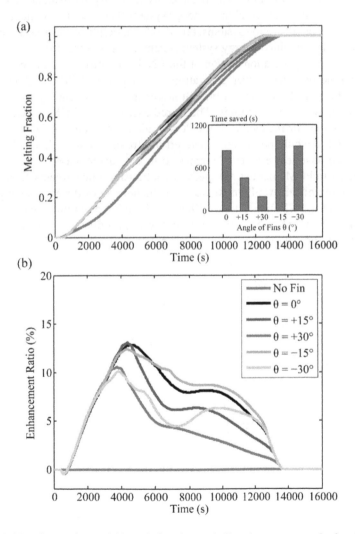

FIGURE 9.46 Comparison of (a) melt fraction and (b) enhancement ratio for the no-fin, parallel angled fin, upward angled fins, and downward angled fins. (Modified and redrawn from Ji et al. (2018b) with permission.)

during the melting period. The downward angled fins saved more than 900 seconds, while the minimum time saving belonged to the upward angled fin with 30°.

9.7 FEASIBILITY ASSESSMENTs OF PCMs IN THERMAL ENERGY STORAGE

PCMs can be applied in different thermal energy storage applications from building energy efficiency systems to the power generation. They increase the thermal performance, energy savings, and system stability while their investment and operation costs can be a drawback from the point of economics. Therefore, this section presents the simple economic, thermoeconomic, and sustainability assessment techniques for the PCMs in a small-scale combined energy system. The calculation methods are not only for the power generation systems, but also for all application areas of the PCMs. The small-scale combined energy system schematic is presented in Figure 9.47. It is a combined power generation system of the LNG cold utilization and microturbine/ Stirling-engine-operated power generation systems. The PCM-based TES tank is located at the exit of the microturbine. It stores the thermal energy from the waste heat of the microturbine (stream 9). The stored energy in the PCM is used to operate the hot end of the Stirling engine. If the PCM tank is not located in the combined system, the hot end would be directly operated by stream 9, but the hot end of Stirling engine would have less efficiency compared to the current case. The reason is the stored energy in PCM operates the special heat transfer fluid (streams 12 and 13) that provides more uniform wall temperature for the hot end of the Stirling engine. The exit stream of the PCM tank (stream 10) has still high temperature, and it is utilized in the heat exchanger for thermal energy (hot water) production. More details on the system design and details can be found in Kanbur et al. (2017).

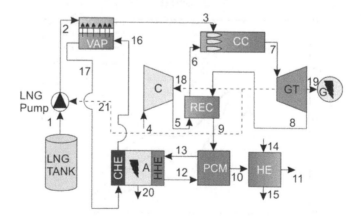

HE: Heat Exchanger **REC:** Recuperator **GT:** Gas Turbine **PCM:** PCM Tank
CC: Combustion Chamber **G:** Generator **C:** Compressor **A:** Alternator
HHE: Hot Heat Exchanger **CHE:** Cold Heat Exchanger **VAP:** LNG Vaporizer

FIGURE 9.47 Simplified schematic of the combined energy system including PCM-based TES tank. (Modified and redrawn from Kanbur et al. (2017) with permission.)

9.7.1 ECONOMIC ANALYSIS OF PCM-INTEGRATED COMBINED ENERGY SYSTEMS

Economic feasibility assessments are generally called as economic analyses. They can be done with many different methods and approaches, but the payback period (PP) is one of the simplest and clearest definitions of the economic analysis. PP is simply defined as the return on investment time for the system. For the combined energy generation (including power generation) systems, the payback period uses the investment, operation, and maintenance costs, and they are evaluated by using the total generated electricity rate and the unit cost of electricity. Equation 9.30 defines the payback period for the combined energy generation system in Figure 9.47,

$$PP = \left[(PEC + OM) \cdot 4.3 \right] / \left[\dot{W}_{net} \cdot C_{el} \cdot 3,600 \cdot \tau \right] \qquad (9.30)$$

where PEC and OM are the purchased equipment cost and operation & maintenance cost, respectively, $\dot{W}_{net}$ is the net generated power rate of the combined energy system, c_{el} is the unit electricity price, and τ is the annual operation hours. Equation 9.30 was proposed by Bejan et al. (1996) to make the general economic analysis simple. The PEC value of the system is the summation of the purchased equipment costs of all components. OM is simply defined as the $1.092 \times$ PEC value. The annual operation hours, τ, is 8,000 hours. The PEC value of PCM tank is considered to be 3,600 SGD that is a crucial amount in the combined system design. $\dot{W}_{net}$ is the net generated power rate of the combined energy system. The details of the PEC, OM, τ, and $\dot{W}_{net}$ values can be found in Kanbur et al. (2017). The electricity price, c_{el}, is assumed 4.24×10^{-5} kJ^{-1} (Kanbur et al. 2018). For the system design presented, the economic assessment was done in the range of 288.15–313.15 K and 50%–90% for the ambient air temperature and the relative humidity, respectively. Figure 9.48 shows the payback period of the combined design.

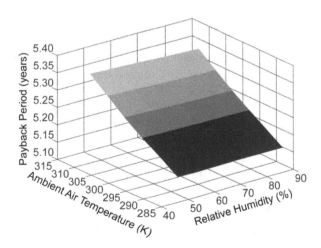

FIGURE 9.48 The payback period of the PCM-based TES-tank-integrated combined energy system.

The results show that the PP value varies between 5.15 and 5.33 years in the defined range. The maximum PP value is seen at the maximum ambient air temperature (313.15 K) and the minimum relative humidity (50%), while the lowest PP belongs to the minimum ambient air temperature (288.15 K) and the maximum relative humidity (90%). Also, it is seen that the impact of ambient air temperature on the PP trends is significant, while the impact of relative humidity can be neglected. If the PCM-based TES tank was not integrated into the combined energy system, the system could not operate the Stirling engine efficiently; therefore, the microturbine would be the only power generation device. In that case, both $\dot{W}_{net}$ and PEC would decrease, which means the single power generation system would have less power generation rate while it has cheaper investment costs. Therefore, the decision-making process and more detailed analyses must be performed to better understand and evaluate the feasibility of PCM-based TES applications. Thermoeconomic assessment is one of the ways for the detailed assessments, and it is explained in the next section.

9.7.2 THERMOECONOMIC ANALYSIS OF PCM-INTEGRATED COMBINED ENERGY SYSTEMS

Thermoeconomics combines the objective functions of thermodynamics and economics to present an interdisciplinary approach for the feasibility assessments of the energy conversion systems. Thermodynamics part can be built by the energetic or exergetic models, but the exergetic models have been the most preferable in the past and the present (Bejan et al. 1996). In thermoeconomics, each stream in the combined system has its own unit cost, c. The multiplication of unit cost and the exergy rate, $\dot{E}$, gives the levelized cost of the stream, $\dot{C}$. After that, the thermoeconomic balance equation, which is similar to the general exergy balance equation, is written for each system component, individually. When the combined system presented and the PCM-based TES tank are considered, the thermoeconomic balance equation for the PCM-based TES tank is given in Eq. (9.31) (see Figure 9.47),

$$\dot{C}_9 + \dot{C}_{12} = \dot{Z}_{PCM} + \dot{C}_{10} + \dot{C}_{13} \qquad (9.31)$$

where $\dot{Z}_{PCM}$ is the levelized component cost of the PCM-based tank. It is calculated by using the capital recovery factor (CRF), interest time (i), system lifetime (n), and PEC values as shown in Eqs. 9.32 and 9.33.

$$\dot{Z}_{PCM} = \frac{(CRF + OM) \cdot PEC}{\tau} \qquad (9.32)$$

$$CRF = \frac{(1+i)^n \cdot i}{(1+i)^n - 1} \qquad (9.33)$$

Similar to that of the PCM-based TES tank, the thermoeconomic balance equation is written for all the components; then, the general thermoeconomic balance equation of the system is calculated as shown in Eq. (9.34),

$$\dot{C}_P = \dot{Z}_{system} + \dot{C}_F \qquad (9.34)$$

where $\dot{C}_F$ and $\dot{Z}_{system}$ are the levelized fuel cost and the total levelized component cost for the combined system, respectively. The levelized product cost, $\dot{C}_P$, is one of the well-known assessment parameters in thermoeconomics. The minimum $\dot{C}_P$ value means that the system is more feasible from the viewpoint of thermoeconomics. Therefore, the general aim for both system-based and component-based thermoeconomics is to minimize the $\dot{C}_P$ value. Besides $\dot{C}_P$ values, the relative product cost difference, r, and the exergoeconomic factor, f, are two preferred thermoeconomic assessment parameters. Equations (9.35) and (9.36) present the definitions of r and f, respectively,

$$r = \frac{C_{P,K} - C_{F,K}}{C_{F,K}} \tag{9.35}$$

$$f = \frac{\dot{Z}_k}{\dot{Z}_k + \dot{C}_{D \cdot k}} \tag{9.36}$$

where $C_{P,k}$ and $C_{F,k}$ denote the unit product and unit fuel costs of the *component k*, whereas the $\dot{Z}_k$ and $\dot{C}_{D.k}$ are the levelized component cost and the levelized destruction cost, respectively. Instead of *component k*, the r and f can also be calculated for the whole system. More details on the exergy and thermoeconomics can be found in Kanbur et al. (2017). According to the given data, the thermoeconomic performance trend of the overall system is shown in Figure 9.49.

The results show that the levelized product cost decreases with increase of the ambient air temperature; hence, high-temperature environments (closed operation room or open environment) are more feasible for the operation of the combined energy system. From the minimum temperature value (288.15 K) to maximum (313.15 K), the levelized product cost decreases by around 12%, which is an important decrement rate. The relative product cost also decreases with the increase of the ambient air temperature. Similar to the levelized product cost, the minimum value of the relative product cost difference is a desired case since a minimum difference between the unit product and fuel costs simply means that the unit cost of the stream is less affected by the component/or system costing. The rate of decrease between the ambient air temperature of 288.15 and 313.15 K is calculated as 25.43%. Unlike the relative product cost difference, the exergoeconomic factor trends are stable. It must be noted that the PCM-based TES tank can significantly affect the overall thermoeconomic performance. For example, the PCM with higher melting temperature and cost increases the power generation rate and the levelized product cost at the same time; therefore, the multiobjective optimization between the thermodynamic efficiency and levelized product cost is required for different PCM types. The details of multiobjective optimization approach can be found in Kanbur et al. (2019).

9.7.3 SUSTAINABILITY ANALYSIS OF PCM-INTEGRATED COMBINED ENERGY SYSTEMS

The global concerns on the greenhouse gas (GHG) emissions by the energy systems have been rising for a long time since the fossil fuels are still the major energy sources

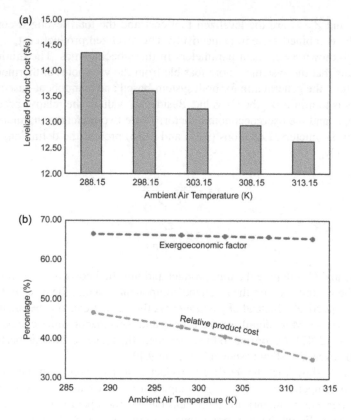

FIGURE 9.49 Thermoeconomic assessment of the combined energy system: (a) levelized product cost and (b) relative product cost difference and exergoeconomic factor.

for the energy generation and they produce high amount of GHG emissions which negatively affect the climate change and global warming. Hence, besides the economic and thermodynamic perspectives, the sustainability perspective is becoming important for the combined energy systems. PCM-based TES applications increase the energy efficiency of the energy systems; therefore, it has a positive contribution on the sustainability. However, the PCM type, its life cycle assessment, and PCM cost are the other factors which can negatively affect the sustainability performance. To better understand the impact of PCM applications on the system sustainability, the thermodynamic efficiency (both energetic and exergetic), economic costs, environment-friendly properties (studies related to life cycle assessment), and the exergy destruction facts must be evaluated well. The literature has different kinds of sustainability methods for both PCM applications and energy system analysis; however, Rosen et al. (2008) defined a very simple and effective sustainability indicator that is known as the sustainability index (SI), shown in Eq. (9.37).

$$SI = \frac{\dot{E}_F}{\dot{E}_D} \tag{9.37}$$

SI simply means the ratio of the fuel exergy rate to the exergy destruction rate. That is, the fuel types with higher exergy efficiency increase the system sustainability, while the exergy destruction due to the thermodynamic operations of the components/system cause the decrease in sustainability. SI can be applied to both components and the overall system, and the different fuel types and various component/system configurations can be compared. Figure 9.50 shows the SI trends of the PCM-based TES tank and overall system.

Results show that the PCM-based TES tank and the overall system configuration have the similar trends at the minimum and maximum ambient air temperatures. The ambient air temperature has more dominant effects than the relative humidity. The minimum ambient air temperature and relative humidity presented the highest SI values. To increase the SI performance of the PCM-based TES tank, the PCMs with higher melting temperature can be considered if their costs are affordable.

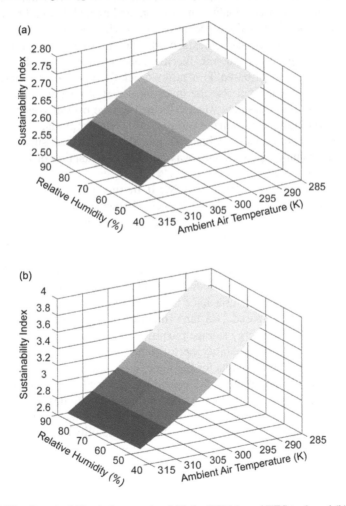

FIGURE 9.50 Sustainability index trends of (a) the PCM-based TES tank and (b) the whole combined system.

Also, the system enhancement studies decrease the exergy destruction and provide higher SI trends as mentioned in Section 9.6. It must be noted that different PCM types can increase or decrease the SI trends for both TES tank and overall system.

9.8 SUMMARY

This chapter focused on the PCMs for their properties, selection criteria, modeling, analysis, and application for thermal energy storage and relevant engineering applications. The classification was done according to their structures. The performance evaluation and enhancement techniques were mentioned in detail by using various experimental procedures. After that, the thermal model and testing were explained with respect to the designed TES tank. PCM performance and its impact on the unit design were discussed with experimental results, and further enhancement studies such as single-fin insert, double-fin location, and angled double-fin design were simulated with computational methods; then, the verification steps were performed. In the end, the feasibility assessments of the PCM-based TES application were explained according to the economic, thermoeconomic, and sustainability concepts. It was seen that the PCM-based TES tanks crucially increased not only the thermodynamic performance, but also the investment cost. To better evaluate this fact, the thermoeconomic assessment was performed for the overall system, and it was evaluated according to the levelized product cost, relative product cost difference, and the exergoeconomic factor, which are globally known assessment parameters. Then, the sustainability performance was defined as a function of fuel exergy and exergy destruction rates.

The comprehensive assessment of PCMs in thermal energy storage showed that the thermal conductivity was still the biggest challenge for better melting performance; thus, the heat transfer enhancement studies have significant roles in thermal engineering applications. Since the natural convection has a big impact on the melting, it was seen that the melting period is always fast at the top part of a PCM tank. To make the melting more uniform, different inserts with convenient angles were required to be inserted. Besides the technical details, the cost of PCM was seen as another important criteria since the cost can dramatically increase the payback period of the PCM-based TES-tank-integrated energy system. If a good PCM can be used for thermal energy storage purpose in an energy system, the economic, thermoeconomic, and sustainability performances increase. The obtained results and assessment gave new insights for the near future studies. For the PCM melting cases, various fin configurations are designed for higher PCM performance. For the system-based approach, the multiobjective optimization studies are considered to better define the best trade-off point of the PCM-based tank according to the economic, thermodynamic, thermoeconomic, and sustainability aspects.

REFERENCES

Agyenim, F., N. Hewitt, P. Eames, and M. Smyth. 2010. A review of materials, heat transfer and phase change problem formulation for latent heat thermal energy storage system (LHTESS). *Renewable and Sustainable Energy Reviews* 14:615–628.

Bejan, A., G. Tsatsaronis, and M. Moran. 1996. *Thermal Design and Optimization*. New York: Wiley.

Cui, H., X. Hou, and X. Yuan. 2003. Energy analysis of space solar dynamic heat receivers. *Solar Energy* 74:303–308.

Das, N., Y. Takata, M. Kohno, and S. Harish. 2016. Melting of graphene based phase change nanocomposites in vertical latent heat thermal energy storage unit. *Applied Thermal Engineering* 107:101–113.

Delgado, J.M.P.Q., J.C. Martinho, A.V. Sa, A.S. Guimaraes, and V. Abrantes. 2019. *Thermal Energy Storage with Phase Change Materials: A Literature Review of Applications for Building Materials, Springer Briefs in Applied Science and Technology*. Switzerland: Springer Nature.

Dincer, I. and M.A. Rosen 2010. *Thermal Energy Storage*. Chichester, UK: John Wiley & Sons, Ltd.

Farid, M.M., A.M. Khudhair, S.A.K. Razack, and S. Al-Hallaj. 2004. A review on phase change energy storage: Materials and applications. *Energy Conversion and. Management* 45:1597–1615.

Gil, A., M. Medrano, I. Martorell, A. Lazaro, P. Dolado, B. Zalba, et al. 2010. State of the art on high temperature thermal energy storage for power generation. Part 1-concepts, materials and modellization. *Renewable and Sustainable Energy Reviews* 14:31–55.

Gong, Z.X. and A.S. Mujumdar 1996. Cyclic heat transfer in a novel storage unit of multiple phase change materials. *Applied Thermal Engineering* 16:807–815.

Guo, C. and W. Zhang. 2008. Numerical simulation and parametric study on new type of high temperature latent heat thermal energy storage system. *Energy Conversion and Management* 49:919–927.

He, Q. and W.N. Zhang 2000. A study on latent heat storage exchangers with the high temperature PCM. *World Renewable Energy Congress* VI:1044–1047.

Ji, C., Z. Qin, S. Dubey, F.H. Choo, and F. Duan. 2018a. Simulation on PCM melting enhancement with double-fin length arrangements in a rectangular enclosure induced by natural convection. *International Journal of Heat and Mass Transfer* 127:255–265.

Ji, C., Z. Qin, Z. Low, S. Dubey, F.H. Choo, and F. Duan. 2018b. Non-uniform heat transfer suppression to enhance PCM melting by angled fins. *Applied Thermal Engineering* 129:269–279.

Ji, C., Z. Qin, S. Dubey, F.H. Choo, and F. Duan. 2017. Three-dimensional transient numerical study on latent heat thermal storage for waste heat recovery from a low temperature gas flow. *Applied Energy* 205:1–12.

Johra, H. and P. Heiselberg. 2017. Influence of internal thermal mass on the indoor thermal dynamics and integration of phase change materials in furniture for building energy storage: A review. *Renewable and Sustainable Energy Reviews* 69:19–32.

Kabbara, M.J. and N.B. Abdallah. 2013. Experimental investigation on phase change material based thermal energy storage unit. *Procedia Computer Science* 19:694–701.

Kanbur, B.B., L. Xiang, S. Dubey, F.H. Choo, and F. Duan. 2017. Thermoeconomic and environmental assessments of a combined cycle for the small scale LNG cold utilization. *Applied Energy* 204:1148–1162.

Kanbur, B.B., L. Xiang, S. Dubey, F.H. Choo, and F. Duan. 2018. Mitigation of carbon dioxide emission using the liquefied natural gas cold energy in small scale power generation systems. *Journal of Cleaner Production* 200:982–995.

Kanbur, B.B., L. Xiang, S. Dubey, F.H. Choo, and F. Duan. 2019. Sustainability and thermoenvironmental indicators on the multiobjective optimization of the liquefied natural gas fired micro-cogeneration systems. *Chemical Engineering Science* 202:429–426.

Kamkari, B., H. Shokouhmand, and F. Bruno. 2014. Experimental investigation of the effect of inclination angle on convection-driven melting of phase change material in a rectangular enclosure. *International Journal of Heat and Mass Transfer* 72:186–200.

Liu, S., Y. Li, and Y. Zhang 2014. Mathematical solutions and numerical models employed for the investigations of PCMs' phase transformations. *Renewable and Sustainable Energy Reviews* 33:659–674.

Nazir, H., M. Batool, F.J.B. Osorio, M. Isaza-Ruiz, X. Xu, K. Vignarooban, et al. 2019. Recent developments in phase change materials for energy storage applications. *International Journal of Heat and Mass Transfer* 129:491–523.

Peng, Q., J. Ding, X. Wei, J. Yang, and X. Yang. 2010. The preparation and properties of multi-component molten salts. *Applied Energy* 87:2812–2817.

Rosen, M.A., I. Dincer, M. Kanoglu. 2008. Role of exergy in increasing efficiency and sustainability and reducing impact. *Energy Policy* 36:128–137.

Qin, Z. 2016. Thermal Energy Storage by Phase Change Materials in Power Generation. M.Sc. Thesis in Nanyang Technological University.

Qin, Z., C. Ji, Z. Low, S. Dubey, F.H. Choo, and F. Duan. 2017. Effect of fin location on the latent heat storage: A numerical study. *Energy Procedia* 143:320–326.

Shamsundar, N. and E.M. Sparrow 1975. Analysis of multidimensional conduction phase change via the enthalpy model. *Journal of Heat Transfer* 97:333–340.

Sharma, S.D., D. Buddhi, and R.L. Sawhney. 1999. Accelerated thermal cycle test of latent heat storage materials. *Solar Energy* 66:483–490.

Sharma, R.K., P. Ganesan, V.V. Tyagi, H.S.C. Metselaar, and S.C. Sandaran. 2015. Developments inorganic solid–liquid phase change materials and their applications in thermal energy storage. *Energy Conversion and Management* 95:193–228.

Thapa, S., S. Chukwu, A. Khaliq, and L. Weiss. 2014. Fabrication and analysis of small-scale thermal energy storage with conductivity enhancement. *Energy Conversion and Management* 79:161–170.

Voller, V.R. and C. Prakash. 1987. A fixed grid numerical modelling methodology for convection-diffusion mushy region phase-change problems. *International Journal of Heat and Mass Transfer* 30:1709–1719.

Xiao, J., J. Huang, P. Zhu, C. Wang, and X. Li. 2014. Preparation, characterization and thermal properties of binary nitrate salts/expanded graphite as composite phase change material. *Thermochimica Acta* 587:52–58.

Yimes, B. and M. Adami. 1997. Parametric study of phase change thermal energy storage systems for space application. *Energy Conversion and Management* 38:253–262.

Yuan, Y., N. Zhang, W. Tao, X. Cao, and Y. He. 2014. Fatty acids as phase change materials: a review. *Renewable and Sustainable Energy Reviews* 29:482–498.

Zalba, B., J.M. Marin, L.F. Cabeza, and H. Mehling. 2003. Review on thermal energy storage with phase change: Materials, heat transfer analysis and applications. *Applied Thermal Engineering* 23:251–283.

Zhao, Y.J., R.Z. Wang, L.W. Wang, and N. Yu. 2014. Development of highly conductive $KNO_3/NaNO_3$ composite for TES (thermal energy storage). *Energy* 70:272–277.

Zhang, Y. and Y. Jiang. 1999. A simple method, the -history method, of determining the heat of fusion, specific heat and thermal conductivity of phase-change materials. *Measurement Science and Technology* 10:201–205.

Zhang, Y., K. Du, M.A. Medina, and J. He. 2014. An experimental method for validating transient heat transfermathematical models used for phase change materials (PCMs) calculations. *Phase Transitions* 87:541–558.

10 Strategies for Performance Improvement of Organic Solar Cells

Wei-Long Xu
Shandong University
Changzhou Institute of Technology

Xiao-Tao Hao
Shandong University

CONTENTS

10.1 Introduction 374
    10.1.1 Prologue 374
    10.1.2 Working Mechanism of OSCs 375
        10.1.2.1 Photon Absorption 375
        10.1.2.2 Exciton Diffusion 375
        10.1.2.3 Charge Separation 375
        10.1.2.4 Charge Collection 376
    10.1.3 Device Structure of OSCs 376
        10.1.3.1 Conventional Structure 376
        10.1.3.2 Inverted Structure 376
    10.1.4 OSC Preparation Process 377
    10.1.5 The Parameters of OSCs 378
10.2 Localized Surface Plasmon Resonance 379
    10.2.1 Plasmonic Effect 379
    10.2.2 Plasmonic in Active Layer 381
    10.2.3 Plasmonic in Charge Transport Layer 383
    10.2.4 Plasmonic as Electrodes 384
10.3 Interface Engineering 385
    10.3.1 Electron Transport Layer 386
    10.3.2 Hole Transport Layer 394
10.4 Ternary Solar Cells 396
    10.4.1 Charge Transfer 396
    10.4.2 Energy Transfer 398

10.4.3 Parallel Working Mechanism 399
10.4.4 Alloy Model 402
10.5 Opportunities and Challenges 405
References 406

10.1 INTRODUCTION

10.1.1 Prologue

Energy issues have always plagued the rapid economy development since the 21st century. First of all, people are now using non-renewable energy sources such as coal, oil, and natural gas. As the energy consumption increases, the demand for traditional energy sources will become more severe. Secondly, the burning of fossil fuels can cause environmental pollution. The Great Smog of London in 1952 was a severe air-pollution event, and 4,000 people died as a direct result of the smog. In December 2015, the severe smog weather in Beijing and North China also made people realize the seriousness of environmental pollution. Solar energy is a kind of clean energy, which is inexhaustible. Solar cell is a device that converts solar energy into electrical energy. The mainstream of solar cells is silicon-based inorganic solar cells because of their stable performance, high photoelectric conversion efficiency, and long service life.

New organic semiconductor materials have been continuously synthesized with the development of organic electronics in recent decades. The mobility, spectral absorption, and solubility of organic materials have been improved gradually. These materials have been applied in the preparation of organic solar cells (OSCs). OSCs have many advantages over inorganic solar cells, such as light weight, solution-based treatment, without high temperature and high vacuum, and flexibility.[1-5] The rich source and tunable optical and electronic properties of organic materials have attracted much attention. The potential of the OSCs' market is huge, not only for solving the current energy crisis and environmental pollution problems, but also for obtaining huge social and economic benefits. Tang[6] prepared a donor–acceptor double-layer planar heterojunction solar cell with a photoelectric conversion efficiency of about 1% in 1986. However, the device performance of planar solar cells is mainly limited by two aspects: a limited contact area between donor and acceptor and short diffusion length of carriers. These problems can be solved by introducing a bulk heterojunction (BHJ) of the active layer. The first BHJ was achieved by Hiramoto et al.[7] who co-evaporated donor and acceptor molecules at high vacuum condition. The first effective BHJ solar cell was achieved by Heeger group and Friend group in 1995.[8,9] Heeger group mainly adopted conjugated polymers and small molecule fullerenes to form a heterojunction, while Friend et al. emphasized the BHJ between polymer and polymer. The BHJ based on polymer and fullerene dominated in the next 20 years. An efficiency of 10% was achieved in single-junction heterojunction OSCs in 2015.[2,10,11] With the development of new narrow-bandgap donor materials, non-fullerene materials, interface modification, morphology control, and ternary OSCs in recent years, the efficiency of OSCs has now exceeded 18%.[12-14] In addition, the products of OSCs can be flexible and semi-transparent. They are getting closer to practical application and commercialization.

10.1.2 Working Mechanism of OSCs

The typical working process of OSCs mainly includes the following four steps: (1) photon absorption, (2) exciton diffusion, (3) charge separation, and (4) charge collection.

10.1.2.1 Photon Absorption

The active layer of the BHJ is composed of conjugated polymers and fullerenes. When the light is incident on the active layer, the electrons will be excited from the highest occupied molecular orbital (HOMO) to the lowest unoccupied molecular orbital (LUMO) of the polymers. Light absorption efficiency depends on the optical characteristics of organic materials. In general, a large absorption coefficient of organic materials is in favor of light absorption. The absorption spectra of organic materials are narrow and always located in the visible regions. Therefore, the absorption efficiency of organic materials is less than 40%, which is one reason why the photoelectric conversion efficiency (PCE) of OSCs is less than inorganic solar cells. Organic semiconductors have low carrier mobility, so the thickness of the active layer is in the range of tens to hundreds of nanometers, which also reduces the light absorption capacity.

10.1.2.2 Exciton Diffusion

The electron–hole pairs are tightly bound together to form a Frenkel exciton due to the low dielectric constant ($\varepsilon_r \approx 2$–4) of the polymer, and the exciton binding energy is in the range of 0.3–$1.0\,eV$.[15,16] According to the spin state, excitons can be divided into singlet and triplet states. Since excitons are neutral, exciton motion is also described as a diffusion process. After exciton formation, it will rapidly diffuse to the donor–acceptor interface. Due to the low charge mobility of the polymer, the effective diffusion distance of the excitons is within $10\,nm$.[15] When the size of the phase separation between the donor and acceptor is too large, the exciton will recombine, leading to the reduced photocurrent generation.

10.1.2.3 Charge Separation

Exciton separation requires an additional driving force owing to the strong Coulomb interaction of electron–hole pairs and low dielectric constant of organic materials.[17–19] The conventional view is that the energy-level difference between the donor and the acceptor provides the driving force for exciton separation. The larger the LUMO difference between the donor and the acceptor, the easier the excitons are separated. However, recent studies have shown that the exciton separation has no absolute relationship with the energy-level difference between donor and acceptor. Friend et al.[20,21] used ultrafast spectroscopy to confirm that the driving force of charge separation is to make the excited state delocalized. This process has a short lifetime and can replace Coulomb interaction. Charge separation does not require too much energy to overcome the Coulomb interaction. If the electron energy state generated on the donor overlaps with the fullerene eigenstate, the electrons can be transferred by resonance. Charge separation could be achieved in a long range. Charge transfer efficiencies are different during several polymers even though the similar polymer energy levels.[22] Charge transfer efficiencies are caused by the charge transfer state in the molecules. This further confirms that charge separation is not completely determined by the energy difference between the donor and acceptor.

10.1.2.4   Charge Collection

When the excitons are separated to free charges, the electrons are transported in the acceptor, and the holes are transported in the donor. Under the action of the external electric field, the electrons move toward the cathode, and the holes move toward the anode. The continuous transport channel is critical to the charge collection. The donor and acceptor have large contact areas, which facilitate exciton separation. At the same time, the interconnected channels can improve the charge collection efficiency.

10.1.3   DEVICE STRUCTURE OF OSCs

10.1.3.1   Conventional Structure

The device is mainly composed of the following parts: transparent conductive layer, hole transport layer, active layer, electron transport layer, and cathode. Figure 10.1 shows the conventional structure of OSCs. Tin-doped indium oxide (ITO) is the commonly used transparent conductive layer. The mixture of polyethylene dioxythiophene with polystyrene sulfonate (PEDOT:PSS) serves as the hole transport layer, which makes the ITO surface more flat. It could also enhance the contact between ITO and the active layer and enhance the work function of ITO. The active layer is composed of conjugated polymer and fullerene or non-fullerene BHJ which absorbs sunlight. The electron transport layer is mainly composed of materials such as LiF and Ca, which can effectively reduce the work function of the cathode electrode and can also protect the active layer from the damage of evaporated metal atom layer. A metal with a low work function, such as Al, always plays the role of the cathode.

10.1.3.2   Inverted Structure

The acidic property of PEDOT:PSS has a corrosive effect on ITO. Meanwhile, the cathode with low work function metal material is easily oxidized. Therefore, the stability of

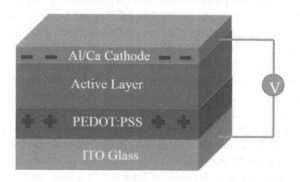

FIGURE 10.1   Conventional structure of OSCs.

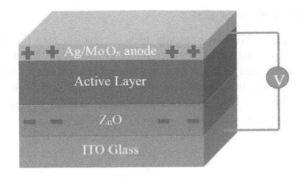

FIGURE 10.2 Inverted structure of OSCs.

the device with conventional structure is poor. In order to solve this problem, the new device structure, that is, inverted structure is proposed, as shown in Figure 10.2.

The main feature of the inverted structure is that ITO is used as the cathode of the device and the electron transport layer is usually made of inorganic materials such as ZnO, TiO_2, and Cs_2CO_3. High work function oxides are used as the hole transport layer. The metals with high work function are used as the anode, which can effectively reduce the air oxidation damage.

10.1.4 OSC PREPARATION PROCESS

The preparation process of OSCs will be described briefly. The classic P3HT:PCBM conventional structure OSC was selected.[23] The preparation process of other OSCs may be different from this preparation process, but it can be used as a reference. In the first step, the pre-patterned ITO glass needs to be cleaned. Acetone, isopropanol, and deionized water are required for the cleaning process. It is usually necessary to filter PEDOT:PSS solution with a needle tube to remove large particles before spin-coating process. It is not suitable to use rubber syringes because the acidity of PEDOT:PSS can cause corrosion to the rubber. PEDOT:PSS films can be spin-coated after filtering. The spin-coating speed is usually 5,000–6,000 rpm, and the time is 30 seconds, which can form a film with a thickness of 30–40 nm. The film is usually annealed on a hotplate in order to achieve better performance. The active layer can then be spin-coated on the hole transport layer (PEDOT:PSS). The speed and time of spin coating vary depending on the solvent and concentration of the composition of donor and acceptor to achieve the desired film thickness and surface coverage. The edge of the active layer needs to be wiped clean and then placed under the mask for vacuum evaporation. The electrode is usually fabricated by vacuum evaporation instead of the sputtering technique owing to the high energy of metal atoms during the sputtering process, which can lead to metal atoms' penetration into the active layer, causing a short circuit of the device. Annealing treatment is required to improve the photoelectric conversion efficiency of OSCs based on P3HT:PCBM.[24] Encapsulation with epoxy can improve the long-term stability of the device.

10.1.5　THE PARAMETERS OF OSCs

The current–voltage curve is commonly used to characterize the device performance. The main parameters are open circuit voltage (V_{OC}), short circuit current (J_{SC}), fill factor (FF), and external quantum conversion efficiency (EQE). Its illumination conditions are an incident power of $100\,mW/cm^2$ and angle of 48.2°. This condition is called AM1.5. The photoelectric conversion efficiency can be expressed by the following formula:

$$\eta_e = \frac{V_{oc} \times J_{sc} \times FF}{P_{in}}$$

where V_{oc} is the open circuit voltage, J_{sc} is the short circuit current, FF is the fill factor, and P_{in} is the incident light intensity. Scharber concluded that V_{oc} was determined by the difference between the HOMO of donor and LUMO of acceptor based on 26 kinds of OSCs.[25]

$$V_{oc} = (1/e)\left(\left|E^D_{HOMO}\right| - \left|E^A_{LUMO}\right|\right) - 0.3\ V$$

It is easy to find that V_{oc} is proportional to the energy difference between HOMO (donor) and LUMO (acceptor). In addition, V_{oc} is influenced by the interface materials and the morphology of the active layer.

The short-circuit current can be expressed by the following formula $J_{sc} = ne\mu E$, where n is the charge density, e is the electron charge, μ is the mobility, and E is the electric field strength. The short-circuit current is mainly determined by the light-induced charge density and the mobility of the carriers. BHJ structure enhances the contact area of the donor/acceptor and also provides channels for efficient charge transport. The FF can be expressed as the ratio of the maximum power to the input power, as shown in Figure 10.3.

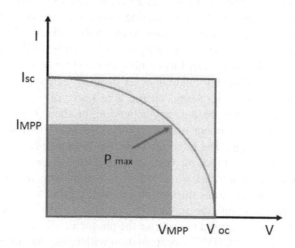

FIGURE 10.3　Schematic illustration of FF calculation.

FF = $I_{MPP}*V_{MPP}/I_{sc}*V_{oc}$, where I_{MPP} and V_{MPP} represent the current and voltage at the maximum power, respectively. The optimal condition is that the photogenerated carriers are quickly extracted without large recombination loss. Another important parameter is the EQE, which is defined as the ratio of photogenerated charge carriers to the number of incident photons at a specific incident wavelength. A high EQE value is a necessary condition to achieve high conversion efficiency for the OSCs.

10.2 LOCALIZED SURFACE PLASMON RESONANCE

10.2.1 PLASMONIC EFFECT

Plasma is a kind of electromagnetic wave, which has the largest field strength at the surface and exponentially decays in the vertical direction. Plasmonic effect has been applied in the field of OSCs. Plasmonic effect can be divided into far-field effect and near-field effect. The far-field effect is often called the scattering effect.[26–28]

Usually, a normal incidence mode is required to maximize the light intensity. However, in this case, a part of the light is transmitted through the film, resulting in insufficient absorption.[29–31] The movement path of the photons can be increased by the scattering effect of the nanoparticles under normal incidence. Thus, the absorption efficiency can be improved, as shown in Figure 10.4. For spherical nanoparticles, small particles will produce more scattering in the forward direction, while large particles will produce more scattering in the back direction. Therefore, the size of the nanoparticles can be designed according to the scattering direction.[32] For example, small-size nanoparticles are placed on the top of the active layer to increase the incident light transmission path, which increases the probability of photon absorption. On the other hand, plasma particles can be placed at the bottom of

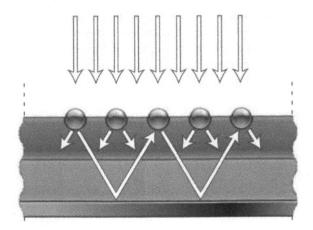

FIGURE 10.4 Capturing more photons by surface particle scattering. (Adapted from H. A. Atwater and A. Polman, *Nat. Mater.*, 2010, **9**, 205. Reprinted with permission from Springer Nature.)

the active layer to reflect the photons that are not absorbed by the active layer. In special cases, the photons which are not absorbed will be reflected back and forth at the two sides of active layer until they are totally absorbed by the active layer. Since absorption enhancement is achieved by scattering effect, scattering is also referred to as "light trapping" or "far-field effect."[33,34]

The near-field enhancement effect means that the excited plasma confined to the surface has a stronger electromagnetic field than the incident light.[35,36] The excitation mode may be electronic excitation or photon excitation. Near-field enhancement can be divided into two modes: plasmon and plasmon resonance. The collective oscillation of surface electrons generates a strong oscillating electric field and propagates along the surface. The electromagnetic waves propagating on these surfaces are called surface plasmons. Unlike surface plasmons, plasmon resonance typically occurs on surface-distorted or curved noble metal surfaces. Metal colloidal particles are the most popular systems.[37,38] The excited plasma is confined to the surface of the particle and cannot propagate along the plane. Therefore, a high intensity of local electric field force is generated during the electronic vibration, which increases the number of absorbed photons. The strength of plasmon resonance depends mainly on the size, shape, and geometry of the particles.[31,39] The absorption intensity is the largest at the plasmon resonance frequency. For noble metal nanoparticles, resonance usually occurs in the visible and near-infrared wavelength regions. The maximum value of resonance is red-shifted as the size of the nanoparticles increases. A higher order mode will appear as the structure of particle deviates from the sphere, as shown in Figure 10.5.[40]

In OSCs, the plasma-enhanced materials are usually Au and Ag because the electrons in valence band can be delocalized. They are expensive as they are noble metals, so copper or aluminum is also used as a plasma element in the literature. The localized surface plasmon resonance (LSPR) peak position can be adjusted by the shape, size, and dielectric environment of the metal nanoparticles. The nanostructure of the metal may be spherical, ellipsoidal, hemispherical, prismatic, cylindrical, and so on.[41–49] Metal nanocomposite structures or metals doped by organic molecules may be employed to achieve broad spectrum absorption of the plasma. For example, the mixture of gold nanoparticles and silver nanoparticles can achieve broad absorption spectrum enhancement.

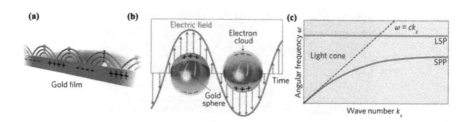

FIGURE 10.5 (a) Plasma surface excimer; (b) plasmon resonance; (c) the relationship between plasma intensity and angular frequency. (Adapted from M. L. Juan, M. Righini and R. Quidant, *Nat. Photonics*, 2011, **5**, 349. Reprinted with permission from Springer Nature.)

10.2.2 PLASMONIC IN ACTIVE LAYER

Plasma nanoparticles can be mixed with organic semiconductors as an active layer in BHJ OSCs. The LSPR effect is obvious when the size of the nanoparticles is less than 20 nm, which increases the absorption efficiency of the active layer. At the same time, the incident light is scattered by the nanoparticles, which increases the transmission path and further enhances the absorption. When the active layer is very thick, the nanoparticles are redistributed in the active layer, possibly near the anode/cathode or in the middle of the active layer,[50,51] as shown in Figure 10.6.

Some specific examples will be given to illustrate the effect of metal nanomaterials on device performance. Lee et al. connected carbon nanotubes to gold nanoparticles to form a hybrid structure.[52] A high-resolution transmission electron micrograph is shown in Figure 10.7. These hybrid gold nanoparticles were incorporated into the active layer. The result demonstrated that gold nanoparticles can promote the LSPR effect to induce charge generation and separation. Doped carbon nanotubes can promote charge selective transport and improve the regularity of organic semiconductor in local area. The photoelectric conversion efficiency of OSCs can be increased to 9.98% under the effect of LSPR.

Gollu et al.[42] compared the effects of gold nanoparticles and gold nanorods on device performance. The time domain finite difference (TDFD) calculations show that gold nanoparticles/gold nanorods can enhance the LSPR effect and light scattering intensity. The performance of devices based on gold nanorods is superior to that based on gold nanoparticles, because the synergistic effect of horizontal and vertical resonance of gold nanorods makes the LSPR spectrum wider. Heeger's group used

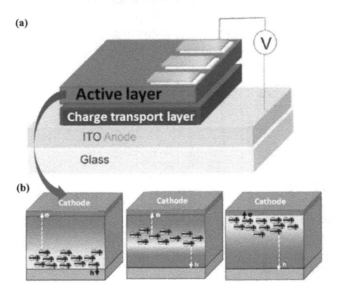

FIGURE 10.6 (a) Schematic illustration of the sandwich-type BHJ solar cell; (b) Ag NPs located in sublayers. (Adapted from Y.-S. Hsiao et al., *J. Phys. Chem. C*, 2012, **116**, 20731–20737; E. Wei et al., *Sci. Rep.*, 2015, **5**, 8525. Reprinted with permission from American Chemical Society (ACS); Nature Publishing.)

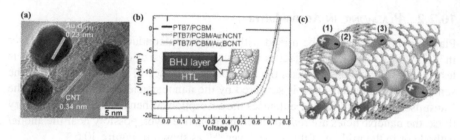

FIGURE 10.7 (a) TEM image of gold particles decorated by carbon nanotubes; (b) I–V curves of OSCs with different active layer; (c) the working mechanism of the gold nanoparticles and carbon nanotube composites in OSCs. (Adapted from J. M. Lee et al., *Adv. Mater.*, 2015, **27**, 1519–1525.) Reprinted with permission from John Wiley & Sons.)

ultraviolet photoelectron spectroscopy (UPS) to analyze the electron transfer mechanism in OSCs. The result shows that silver nanostructures can reduce the vacuum level of PCDTBT/PCBM mixture, which will reduce the electron transfer barrier in the device and facilitate electron extraction from the electrodes. As a comparison, octahedral gold nanoparticles can act as a hole transport medium, reducing the probability of electron–hole recombination, as shown in Figure 10.8.[53,54]

The plasma enhancement effect also promotes Förster resonance energy transfer (FRET). For example, a near-field plasma can be generated when gold nanoparticles serve as the core. The gold nanoparticles could be introduced into the vicinity of the FRET pair. In this case, the charge generation efficiency of the donor material is improved by near-field enhancement. When the absorption spectrum of the donor material overlaps with the LSPR band of the plasmonic nanostructure, additional energy is transferred to the donor, further enhancing charge generation in the donor material. The metal nanoparticles may form a recombination center of electrons/

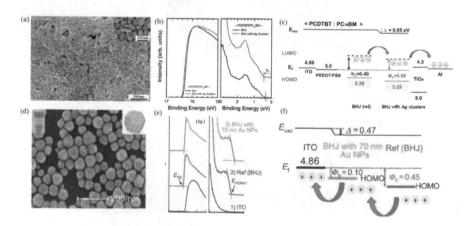

FIGURE 10.8 (a, d) SEM images of Ag and Au nanoparticles, respectively; (b, e) UPS spectra of the BHJ with different components; (c, f) Energy-level diagrams of the BHJ with different components. (Adapted from D. H. Wang, et al., *Adv. Energy Mater.*, 2011, **1**, 766–770. Reprinted with permission from John Wiley & Sons.)

holes in the active layer, which is disadvantageous for device performance improvement. One solution is to encapsulate the metal core with TiO_2 or SiO_2 as the shell structure. However, plasma strength may be reduced after encapsulation. A balance between current and plasma effects is required. Another solution is to transfer the plasma to the charge transport layer. The following section focuses on the role of the plasma in the charge transport layer.

10.2.3 Plasmonic in Charge Transport Layer

When metal nanoparticles are incorporated into the active layer, the metal nanoparticles may form carrier recombination centers, reducing the free charge collection efficiency at the electrodes. In addition, continuous transport channels are not formed in the nanostructures, which is not in favor of charge transfer. Metal nanoparticles can be placed in the charge transport layer to avoid these defects. For example, gold nanoparticles are placed in the electron transport layer, as shown in Figure 10.9. The light transmission path can be increased by the scattering effect, leading to the enhanced absorption efficiency.[55,56] The nanoparticles are placed near the aluminum electrode, which can increase the absorption efficiency by light backscattering. At the same time, the absorption of metal electrode can be suppressed.

Some specific examples[57] are given to illustrate the role of metal nanomaterials in the charge transport layer. In the P3HT:PCBM BHJ solar cell, when silver is used as the plasmon, plasma far-field scattering can promote light trapping. It is found that the electric field of the active layer can be enhanced by further fitting the light absorption, although the metal nanomaterials are located in the charge transport layer, as shown in Figure 10.10.

Park et al.[58] used gold nanoparticle clusters as plasma. It has been found that the interaction of gold nanoparticles can achieve plasma coupling. The near-field coupling between particles can greatly enhance the local electromagnetic field strength,

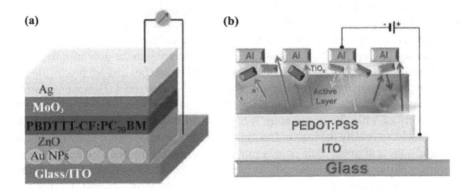

FIGURE 10.9 (a) Forward and (b) backward scattering when Au nanostructure are imbedded in electron transport layer. (Adapted from G. Kakavelakis et al., *Adv. Energy Mater.*, 2016, **6**, 1501640; D. Chi et al., *Nanoscale*, 2015, **7**, 15251–15257. Reprinted with permission from John Wiley & Sons; Royal Society of Chemistry (RSC).)

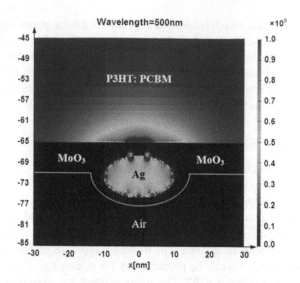

FIGURE 10.10 The simulated absorption profile of the device with Ag NPs by TDFD method. (Adapted from P. Xu et al., *Appl. Phys. Lett.*, 2013, **102**, 53. Reprinted with permission from AIP Publishing.)

which is beneficial for exciton generation and separation. By placing the gold nanoparticle clusters at the bottom of the active layer, the photoelectric conversion efficiency of the device is significantly improved to 9.48%, as shown in Figure 10.11.

10.2.4 PLASMONIC AS ELECTRODES

The main feature of ITO is its combination of electrical conduction and optical transparency, which can be fabricated over a large area by sputtering method. It has been widely used in transparent conductive electrodes. However, ITO also has some drawbacks, such as the high price of indium, brittleness and lack of flexibility, and the expensive vacuum equipment in the deposition process. Therefore, people are trying to find alternatives for ITO. In OSCs, metal nanostructures have the potential to replace ITO because of their flexibility and absorption enhancement in the near infrared.

Silver nanowires have good optical and electrical properties and are promising for future application in the field of transparent conductive electrodes. However, pristine silver nanowires have a low melting point. The silver nanowires tend to accumulate at the interface when the temperature is higher than 100°C. The performance of silver nanowires is easily damaged. The multilayer structure can inhibit the migration of metal ions and avoid their aggregation. Therefore, their optical, electrical, and thermal stability will be improved. At the same time, the multilayer structure can avoid the interference of the electromagnetic field in the structure and improve the plasma effect in the device. For example, Chalh et al.[59] prepared an electrode with the structure of ZnO NP/Ag NW/ZnO NP by solution method, and its structure is shown in the Figure 10.12. The photoelectric conversion efficiency of P3HT:PCBM solar cell based on this electrode is 3.53%, which is better than 3.16% based on ITO.

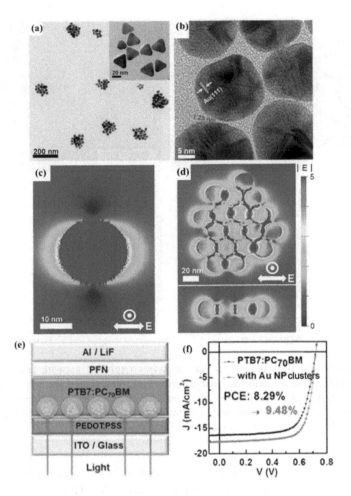

FIGURE 10.11 (a, b) TEM images of Au nanoparticle clusters; (c, d) FDTD-simulated electric field distribution of single Au nanoparticle and Au nanoparticle clusters; (e) device structure. (f) J–V curves of the device with/without Au nanoparticle clusters. (Adapted from H. I. Park et al., *ACS Nano*, 2014, **8**, 10305–10312. Reprinted with permission from American Chemical Society (ACS).)

The reason is that high optical transmission and plasma effect increase the electric field strength in the active layer.

Nanoimprint method can be used to prepare the metal electrode with the nano-structure array. The process is simple and can also be compatible with the flexible processing technology.[60–62] Figure 10.13 shows the nanopatterned metal electrodes.

10.3 INTERFACE ENGINEERING

The open circuit voltage is typically determined by the difference between the HOMO of the donor and the LUMO of the acceptor in BHJ OSCs, provided that the active layer forms Ohmic contact with the cathode/anode.[19,63] If a Shockley contact is

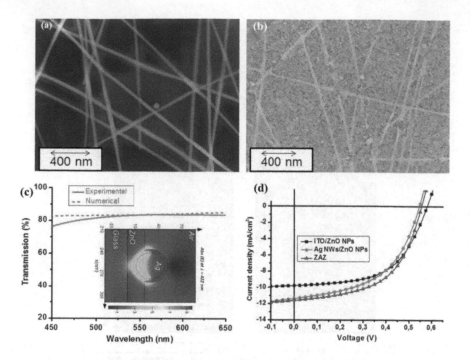

FIGURE 10.12 SEM images of (a) Ag nanowires (AgNWs) network and (b) ZnO/Ag/ ZnO (ZAZ) electrode; (c) numerical and experimental transmission spectra of a glass/ZAZ structure; inset is electric field distribution in the ZAZ electrode at $\lambda = 422$ nm. (d) I–V curves of OSCs based on different electrodes. (Adapted from M. Chalh et al., *Sol. Energy Mat. Sol. Cells.*, 2016, **152**, 34–41. Reprinted with permission from Elsevier.)

formed, the open circuit voltage will be determined by the work function difference in electrode according to the metal–insulator–metal (MIM) model. In the conventional device configuration, holes are extracted from the bottom electrode, while electrons are extracted from the bottom electrode in an inverted device structure. The device performance of OSCs can be improved by interface engineering.

10.3.1 ELECTRON TRANSPORT LAYER

Among the electron transport layers, n-type metal oxides with a high energy level occupy the dominant position because of their stability in air, processability in solution, high light transmittance, and good electron extraction ability. To date, effective electron transport layers include ZnO, TiO_x, Nb_2O_5, SnO_x and the emerging Al-doped ZnO, Mg-doped ZnO, and Cs-doped metal oxides.

ZnO has many superior properties, such as low price, simple synthesis, low toxicity, high stability, and special optoelectronic properties, which make it to be one of the best options for the electron transport layer.[64] The work function of ZnO is about 4.3 eV, which can reduce the work function of ITO or metal electrode and match the LUMO energy level of the acceptor material well. The optical transmittance,

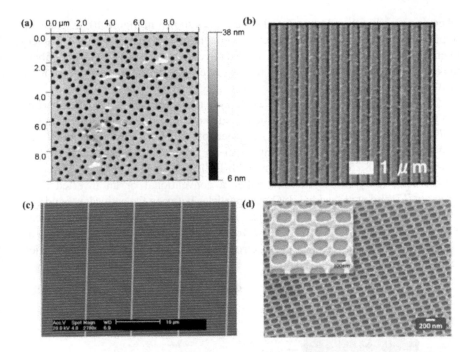

FIGURE 10.13 (a) Atomic force micrograph of 30 nm silver nanohole film; (b) SEM image of patterned Ag electrode; (c) SEM image of an as-fabricated metal electrode on glass; (d) SEM image of Au array. (Adapted from M. G. Kang et al., *Adv. Mater.*, 2008, **20**, 4408–4413; M. G. Kang et al., *Adv. Mater.*, 2010, **22**, 4378–4383; E. C. Garnett et al., *Nat. Mater.*, 2012, **11**, 241. Reprinted with permission from John Wiley & Sons; John Wiley & Sons; Royal Society of Chemistry (RSC.)

mobility, and interfacial properties of ZnO can be controlled by lattice structure, film morphology, surface energy, composition, and film thickness. Many methods are used to synthesize ZnO layer, including electrochemical deposition, sol-gel, hydrothermal method, nanoparticles method, and atomic layer deposition.[65-69] Although the sol-gel method requires annealing at high temperature to achieve a hydrolysis reaction, it is still a very promising method. The ZnO film prepared by sol-gel method was mainly used in P3HT system in the early stage, and the efficiency of the device was always less than 5%.[64] The efficiency of the device based on narrow bandgap semiconductor system could exceed 6%.[70] In the recent work, the efficiency of single junction OSCs has exceeded 10% when ZnO serves as the electron transport layer.[71] The quality of ZnO thin films prepared by sol-gel method is affected by the precursor and annealing treatment. Yin et al. obtained ZnO thin films with different light transmittance and electron mobility by controlling the concentration of the ZnO precursor.[72] The solar cells based on ZnO film exhibited long-term stability and high open circuit voltage (1.00–1.06 V). The introduction of ZnO reduces energy loss at the interface and improves electron collection efficiency. Tang et al.[10] used nanoimprinting method to fabricate patterned ZnO/ITO glass. The transmission in the visible light range of 350–800 nm was enhanced in patterned ZnO/ITO glass when compared with ordinary ZnO/ITO glass, as shown in Figure 10.14. The nano-imprinted structure has a great influence on the light field distribution. High-energy intensity light flux will be confined in the active layer, which will further enhance the absorption efficiency. Therefore, the device efficiency will increase from 8.46% to 10.1% accordingly.

ZnO nanoparticles can also serve as electron transport layers. ZnO nanoparticles can be processed by solution spin-coating method. The efficiency of OSC is close to 7% in poly[[4,8-bis[(2-ethylhexyl)oxy] benzo[1,2-b:4,5-b'] dithiophene-2,6-diyl] [3-fluoro-2- [(2-ethylhexyl) carbonyl] thieno [3,4-b] thiophenediyl]] (PTB7): 3'H-Cyclopropa[8,25] [5,6] fullerene-C70-D5h(6)-3' - butanoicacid, 3'-phenyl-, methyl ester (PC$_{70}$BM) system in either a conventional or an inverted device structure.[73] Dkhil et al.[74] added ZnO nanoparticles between the active layer and Al as the optical spacers. Although the optical absorption of the device did not increase, ZnO nanoparticles promoted electron extraction and reduced contact resistance and charge recombination between the electrodes and active layers. ZnO nanoparticles

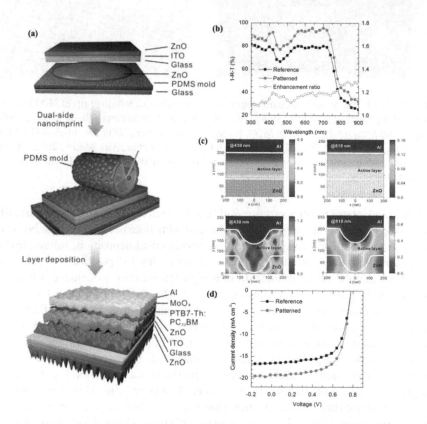

FIGURE 10.14 (a) Schematic illustration of the fabrication process; (b) total transmittance and haze values of ITO glass substrates without and with deterministic aperiodic nanostructure (DAN) patterns; (c) optical simulation of photon flux and field intensity distributions in patterned OSCs; (d) I–V characteristics of OSCs. (Adapted from J. D. Chen et al., *Adv. Mater.*, 2015, **27**, 1035–1041. Reprinted with permission from John Wiley & Sons.)

need to be treated with ultraviolet light and ozone, which can reduce the defect state and increase the carrier lifetime. ZnO needs to be further improved by some chemical/physical doping to overcome its own drawbacks.

Titanium oxide is another n-type semiconductor material with good optical transparency, high electron mobility, and environmental stability.[75] As the electron transport layer, titanium oxide nanoparticles can improve the photoelectric conversion efficiency and stability of the device. Similarly, the electron transport layer prepared by the sol-gel method can prevent oxygen/humidity from intruding into the active layer.[76–79] Therefore, the lifetime of the device can be increased by two orders of magnitude in a humidity environment. The precursor of titanium and the annealing temperature affect the film structure and photoelectric properties of the final TiO_x film in sol-gel method. Short-time illumination can be used to fill shallow electronic defects which is advantageous to device performance.[77] TiO_x treated at low temperature can enhance the efficiency and stability of the device. TiO_2 nanoparticles can reduce the work function of ITO to facilitate electron collection in an inverted device structure. It should be noted that TiO_x has a large number of defects and vacancies, which will form a Shockley barrier at the high work function electrode and metal oxide interface. In addition, the energy-level mismatch between the metal oxide and active layers hinders the ability of TiO_x to transport electrons.[80–82] TiO_2 can be exposed to ultraviolet radiation to reduce oxygen defects, reduce electrical resistance, increase carrier density, and finally eliminate the S-type J–V curve.

Tin oxide is also an n-type wide bandgap oxide semiconductor. The device based on tin oxide exhibits good stability under humidity and high temperature conditions even in the case of no package. This property is not available in other oxide semiconductors. The work function of zinc oxide or titanium oxide is lowered under the illumination. The work function of tin oxide is about 4.2 eV which is stable with or without illumination treatment.[83] The solvothermal method can be used to prepare uniformly dispersed SnO_2 nanoparticles with a size of 2–4 nm, which is important for forming a uniform SnO_2 film. However, SnO_2 bulk structure as electron transport layer leads to current shunts and leakage due to the presence of large particles and defects.[84,85] In general, SnO_2 is a relatively new class of electron transport material.

In addition to oxide semiconductors, Al_2O_3 or ZrO_2 insulating nanolayers' low work function can promote charge collection by reducing the work function of the cathode.[86,87] The aluminum oxide film can be prepared by atomic layer deposition method or ultraviolet-ozone-treated aluminum-thin-film method. It is necessary to carefully control the thickness of Al_2O_3 and use ultraviolet light to activate its conductivity. The Al_2O_3 film can be prepared by the low-temperature-solution method. The work function of Al_2O_3 film can be controlled by the annealing temperature. The work function of Al_2O_3 can be reduced to 3.89 eV at 150°C.[88] The light transmittance of Al_2O_3 film is high. The band gap of Al_2O_3 is large, which can prevent hole transport and reduce the carrier recombination at the interface. Al_2O_3 and ZrO_2 nanolayers prepared by atomic layer deposition method can also be used in combination with TiO_x films.

Metal oxides have been extensively studied as electron transport layers. However, the metal oxide is limited by its own fixed characteristics such as bandgap, energy level, light transmittance, and conductivity. It is necessary to design the components and properties of ternary metal oxides to improve this situation. For example, the strong absorption in the ultraviolet region of ZnO and TiO_x causes the weakened absorption of the active layer. Yin et al. used ternary $Zn_{1-x}Mg_xO$ metal oxide as the electron transport layer to modify the bandgap, work function, energy level, and optical light transmittance by the amount of Mg doping, as shown in Figure 10.15.[89]

The conductivity of ZnO and TiO_x is poor, so the thickness of the film is greatly limited. Element doping can be used to overcome this problem. The representative materials are aluminum-doped zinc oxide (AZO), gallium-doped zinc oxide (GZO), and indium-doped zinc oxide (IZO).[90–95] The conductivity of AZO is three orders of magnitude higher than that of ZnO. Therefore, the thickness of AZO can be 100 nm or even 680 nm as the electron transport layer, which has no obvious effect on the device performance. The AZO layer can be prepared by low-temperature-solution method. Since GZO and IZO have good conductivity, the thickness has little effect on device performance. The effects of doping concentration, precursor, annealing temperature, surface topography, and roughness need to be considered before doping. The light transmittance and work function of MoO_3 electron transport layer can be improved by changing the doping content of aluminum. The monovalent metal elements Li and Cs are also used for doping to prepare ternary metal oxides, such as Li-doped ZnO, Cs-doped ZnO, Cs-doped TiO_2, Cs-doped MoO_x, etc.[96–99] Cs doping can increase the conductivity and electron transport properties of the metal oxide. The Cs doping

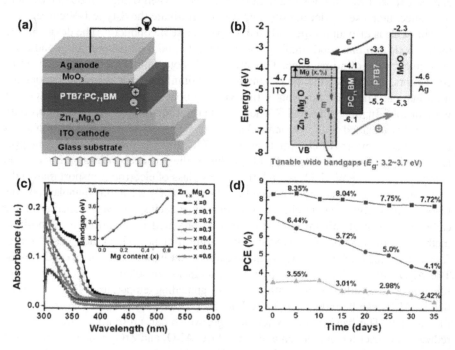

FIGURE 10.15 (a) A schematic device structure based on $Zn_{1-x}Mg_xO$ (ZMO); (b) energy levels of the components in the OSCs with various ZMO; (c) optical absorption spectra of ZMO films; the inset shows an increase in the bandgap of ZMO films; (d) device stability of three types of devices. (Adapted from Z. Yin et al., *Adv. Energy Mater.*, 2014, **4**, 1301404. Reprinted with permission from John Wiley & Sons.)

method can also convert the hole transport layer to an electron transport layer by controlling the work function, energy level, and electrical properties in a wide range. Doping can be introduced during the sol gel process or co-sputtering process.

Polymers and small molecules are employed as interface materials, which is compatible with the preparation process of the active layer. A single layer can be formed due to the intermolecular dipole moment and self-assembly effect. The electron transport layer can induce the formation of an interface dipole from the cathode to the active layer, thereby effectively reducing the cathode work function and increasing built-in potential.[96] Water-soluble/alcohol-soluble conjugated polymers are used effectively in OSCs.[100–104] These materials possess the conjugated backbone with delocalized electronic structure and the polar groups of the side chain with good solubility in water or polar solvents. In PTB7-based devices, the photoelectric conversion efficiency can be increased from 5.0% to 8.37% when Poly[(9,9-bis(3'- (N,N-dimethylamino)propyl)-2,7-fluorene) (PFN) serves as the electron transport layer. The PFN can enhance the built-in potential in addition to function as the ordinary electron transport layer. The efficiency of the device was increased from 9.21% to 10.61% in the PTB7-Th based device using PFN as the electron transport layer. PCCn6 can induce higher interface dipoles to increase the open circuit voltage. The optical electric field can be redistributed through optical interference effects to improve the absorption efficiency. Therefore, the photoelectric conversion efficiency of the solar cells is also significantly improved. Non-conjugated polymers, such as PEI, PEIE, and polyallylamine (PAA), can also be used as interfacial layers.[105–107] A strong electrostatic force dipole can be formed by self-assembly at the ITO–PEI interface, reducing the work function of ITO from 4.8 to 4.0 eV. Small molecules have their own characteristics as electron transport layers, such as clear chemical structure, easy synthesis and high purity. Ge et al. designed a small molecular material MSAPBS as the electron transport layer with a LUMO level of −2.31 eV and a vertical electron mobility of $1.18 \times 10^{-4} cm^2 V^{-1} s^{-1}$. The short-circuit current based on PTB7:PCBM devices can be as high as $19.37 mA\, cm^{-2}$; the efficiency is also higher than 10%.

The low work function metal as the electron transport layer can reduce the work function of the cathode, form an Ohmic contact with the acceptor material, and improve the carrier extraction efficiency.[108–110] The metal films with low work function are usually prepared by vacuum method, which limits their application in inexpensive large-area preparation techniques. The use of metal salts is a better solution and expands the selective range of materials. The addition of metal salts can increase the wettability of the electrode and the surface of the active layer, enhance interface contact, and reduce electrical resistance, contributing to the improvement of device performance. However, such an electron transport layer is prone to decomposition. The decomposed elements would diffuse into the active layer and the electrode material causing device degradation. The second type of metal complex, including zinc chelate, titanium chelate, copper chelate, and zirconium chelate, can be used as an electron transport layer.[111–113] The preparation method is usually thermal deposition or solution processing method. They have good vertical conductivity and high carrier mobility and a suitable energy level which can reduce the work function of the cathode. For example, zirconium acetylacetonate was spin-coated on the active layer as an electron transport layer. The photoelectric conversion efficiency of the device can be as high as 9.23% without post-treatment, which is higher than that of Al (5.72%) or Ca/Al (7.34%).[114] At the same time, the stability of the device is greatly improved.

The electron transport layer with organic/inorganic composite has better compatibility between the active layer and the electrode layer than the organic electron transport layer or the inorganic electron transport layer. There are many organic/inorganic composites including ZnO–polyvinylpyrrolidone, ZnO–PFN, PEO-modified ZnO, ZnO–diethanolamine.[115–118] A composite structure of polyethylene glycol (PEG) and TiO$_x$ was synthesized by Yin group,[119] which can be used as an electron transport layer and applied to a variety of polymer systems with good universality. The oxygen atoms in the PEG backbone fill the surface defect state of the TiO$_x$. The surface dipole momentum of PEG-TiO$_2$ directs to ITO. The work function of ITO is reduced from 4.78 to 4.69 eV with PEG–TiO$_2$ adjustment, as shown in Figure 10.16. The composite electron transport layer can reduce the energy-level difference between the active layer and the ITO layer, facilitating the electron transport. In addition, it has advantages such as low annealing temperature, low cost, and good photoelectric characteristics at the interface.

Dye molecules can also be used to modify ZnO. Nian et al.[120] reported a highly conductive electron transport layer material that incorporates a small amount of light absorbing material PBI-H into the ZnO sol-gel. The PBI-H molecule can be bound to ZnO through a Zn–N bond during annealing, and its structure is shown in Figure 10.17. The electron mobility can be increased

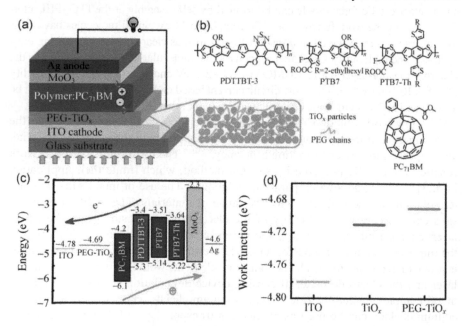

FIGURE 10.16 (a) Schematic illustration of the inverted OSCs with hybrid PEG–TiO$_x$ electron transport layer; (b) molecular structures of PDTTBT-3, PTB7, PTB7-Th, and PC71BM; (c) energy levels of each component used in the OSCs; (d) work functions of ITO, TiO$_x$, and PEG–TiO$_x$ films. (Adapted from Z. Yin et al., *Nano Res.*, 2015, **8**, 456–468. Reprinted with permission from Springer Nature.)

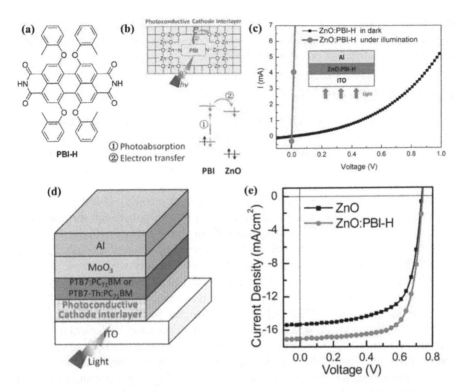

FIGURE 10.17 (a) Chemical structure of PBI-H; (b) PBI-H molecules are chemically bound to ZnO, which accelerates electron transfer between them upon illumination, resulting in high photoconductivity; (c) I–V curves for the device with ITO/ZnO:PBI-H (90 nm)/Al in the dark and under AM 1.5G illumination; (d) schematic device structure based on the electron transport layer of ZnO:PBI-H; (e) I–V curves of OSCs based on ZnO or ZnO:PBI-H electron transport layer. (Adapted from L. Nian et al., *J. Am. Chem. Soc.*, 2015, **137**, 6995–6998. Reprinted with permission from American Chemical Society (ACS).)

from 5.10×10^{-4} cm^2 V^{-1}S^{-1} (in pristine ZnO) to 2.02×10^{-3} cm^2 V^{-1}S^{-1} (in the composite). The light-induced electrons in PBI-H will be transferred to ZnO under the light, which will greatly increase the electron concentration in ZnO conduction band. The thickness of the electron transport layer can be used in a wide range. The efficiency of the device based on the ZnO:PBI-H electron transport layer was 10.59%, which was much higher than that of the device based on the intrinsic ZnO electron transport layer (8.45%).

A single layer of small molecules can be formed to modify the metal oxide. Woo et al.[121] combined PEI and ZnO to form a two-layer structure, which serves as an electron transport layer. The PTB7:PC$_{71}$BM device based on this two-layer electron transport layer has a photoelectric conversion efficiency of 8.88%, which is higher than that of ZnO (6.99%) or PEI (7.49%). The reason is that the dipole formed at the ZnO–PEI interface reduces the conduction band level of ZnO and the series resistance of the device. The modification of ZnO with small organic molecules can play the role of passivation on the surface of the metal oxide, which can significantly improve the stability of the device.[122] The surface energy of ZnO can also be controlled by using different polar groups.[123,124]

10.3.2 Hole Transport Layer

The electron transport layer typically requires a low work function to collect electrons, while the hole transport layer needs a high work function to match the HOMO energy level of the donor material to promote hole collection. At the same time, the hole transport layer also needs to have an effective hole transporting ability to reduce the series resistance.

PEDOT: PSS is a widely used hole transport layer material with the work function of 5.1 eV.[125] It can form a good energy-level matching between the conductive electrode and the donor material. It can also form ohmic contact with the electrodes. It has high light transmittance and can be processed by solution method. However, PEDOT:PSS has some defects; for example, its own acidity will corrode ITO, and indium ions in the electrode may diffuse into the PEDOT:PSS layer, which is fatal to the stability of the device. Therefore, some researches are focused on other hole transport layers to replace PEDOT:PSS to improve the performance of OSCs.

P-type transition metal oxide semiconductor materials are the typical hole transport layers. Firstly, they can form ohmic contact with the active layer to reduce the contact resistance due to their work function. Secondly, the high conduction band position can hinder the electron recombination. The high light transmittance from the visible to infrared range can realize the photon reaching the active layer. In the early studies, metal oxides were obtained by a vacuum deposition process. However, this was not compatible with high-volume roll-to-roll processes. In order to solve this problem, the solution processing method has been developed. NiO_x, V_2O_5, CuO, WO_3, RuO_2, and CrO_x have been successfully prepared.[126–129] MoO_x is one of the most used hole transport layers in OSCs. There are many preparation methods for MoO_x.[130–132] MoO_x can be prepared by preheating the Mo precursor and then annealing it in air. With the development of technology, a smooth MoO_x film can be realized under low-temperature annealing (70°C). The device performance of P3HT:PCBM depends on the density of pentavalent molybdenum in MoO_x. When the pentavalent molybdenum content is reduced, the device efficiency based on MoO_x as the hole transport layer can be comparable to that of PEDOT:PSS. The nanoparticles of MoO_x could be prepared by oxidizing metal Mo powder. When MoO_x serves as the hole transport layer, the process is very simple no matter whether it is in the conventional device structure or the inverted structure, as shown in Figure 10.18.[73] It is not necessary to consider the influence of external conditions such as high temperature on the device performance.

Graphene oxide (GO) has a suitable work function and good film formability as the hole transport layer. Compared with other charge transport materials, GO has some other characteristics, such as a two-dimensional structure, easy functionalization, adjustable bandgap, solution-based treatment, and low cost. In 2010, Li et al. reported the use of GO as a hole transport layer for the first time.[133] GO can be uniformly deposited on the ITO anode and have a good energy match with P3HT donor material.

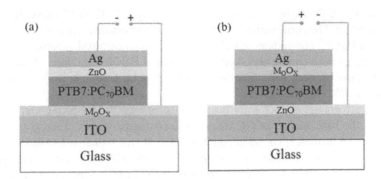

FIGURE 10.18 Conventional and inverted device structures. (Adapted from W. L. Xu, et al., *Org. Electronics*, 2015, **25**, 266–274. Reprinted with permission from Elsevier.)

The energy-level relationship is shown in Figure 10.19. The photoelectric conversion efficiency of the device based on GO as the hole transport layer can reach 3.5%, which is equivalent to the device performance based on PEDOT:PSS. The authors also studied the thickness effect of GO layer on device performance.[134] The result shows that the photoelectric conversion efficiency decreases as the thickness of GO increases from 2 to 10 nm, which is related to the increase in series resistance of the insulating properties of GO. In the inverted structure, GO can also function as a hole transport layer.[135]

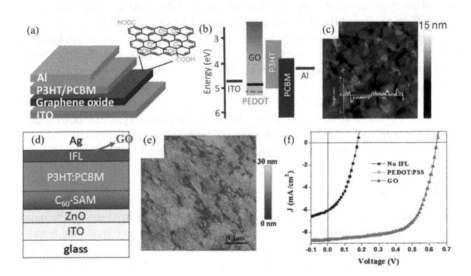

FIGURE 10.19 (a) Schematic illustration of the device structure with GO as the hole transport layer; (b) energy-level diagrams of the bottom electrode ITO; interlayer materials PEDOT:PSS, GO, P3HT, and PCBM; and the top electrode Al; (c) an AFM height image of a GO thin film with a thickness of approximately 2 nm; (d) device configuration of the inverted device using the GO interfacial layer as hole transport layer; (e) AFM image of interfacial layer; (f) J–V characteristics of the inverted OSCs. (Adapted from J. M. Yun, et al., *Adv. Mater.*, 2011, **23**, 4923–4928. Reprinted with permission from John Wiley & Sons.)

A thin layer of GO can be uniformly deposited on the P3HT:PCBM layer by a spin-coating method. The performance of the device is comparable to that of a device based on PEDOT:PSS.[136] The authors found that GO contains carboxyl groups, phenols and alcohols, etc., which can be doped on the surface of P3HT. Therefore, ohmic contact can be formed between the active layer and the metal electrode to improve device performance.[137]

The disadvantage of GO is its insulating property.[138] The epoxy and hydroxyl groups on the base surface of GO disturb the conjugate length of the graphene lattice to form an insulator. Therefore, the FF of the device based on GO hole transport layer is typically 0.65, which is less than 0.7 of PEDOT:PSS. The device performance is greatly affected by the thickness of the GO, so it is necessary to improve the conductivity of the GO layer. Removal of the oxygen-containing groups by the reduction process will enhance the electrical conductivity of the GO. The reduction of GO can be achieved by thermal annealing, microwave irradiation, laser irradiation, and chemical reduction.[138,139]

10.4 TERNARY SOLAR CELLS

In OSCs, the absorption spectra of the active layer in the near-infrared region can be effectively enhanced by designing novel narrow bandgap polymer donors and non-fullerene acceptors. However, it is difficult to cover the spectral range of sunlight by a single combination of donor and acceptor, which is mainly limited by the absorption spectrum of the organic material itself.[140] By using a ternary component solar cell, the range of the absorption spectra can be expanded while preserving the simple processing technique of the binary solar cell. In general, the active layer of a ternary solar cell is usually composed of a main donor–acceptor combination with a third component, such as a polymer, a small molecule, a dye, or a nanoparticle.[141–147] The third material usually needs to have complementary absorption characteristics with the active layer, so that the absorption spectrum can be covered from the short-wavelength to the near-infrared region. The third component typically plays roles in the active layer, such as charge transfer, energy transfer, parallel connection, or alloy mode, which can be performed individually or synergistically.[148–150]

10.4.1 CHARGE TRANSFER

The charge transfer process in the ternary system is different from that in the binary device, which is affected by the third component content, energy level, and position. The addition of the third element depends on the bulk composition of the active layer and requires a suitable energy level to avoid exciton recombination and carrier trapping. The LUMO and HOMO energy levels of the added donor or acceptor should be located between the LUMO and HOMO energy levels of the host donor/acceptor, which facilitates formation of energy-level cascades and charge transport and collection.[151–153] When the donor 1 is excited to generate excitons, the donor 2 can play the energy cascade and transfer carriers to the acceptor efficiently, and finally current is collected by the electrodes. The position of the third element material is preferably located at the interface between the donor and the acceptor to form a cross-interconnection structure, which facilitates charge transport and collection. The position of the third element is affected by the surface energy and crystallinity of the active layer.[154–156] For example, when the dye is added into the P3HT:PCBM system, the dye will be located at the interface between P3HT and PCBM due to the different surface energies of the three materials and the influence of P3HT crystallization phase.

Photoluminescence (PL) spectroscopy is a very practical technique that can be used to distinguish charge transfer or energy transfer in different materials.[157] If charge transfer occurs, it will result in a fluorescence quenching effect on one donor material, while the fluorescence intensity of the other material will not increase. If energy transfer occurs between the two donors, the fluorescence intensity of the narrow bandgap semiconductor material will increase, while the intensity of the other donor material will decrease. For example, consider that SMPV1 is incorporated in the P3HT:PCBM system as the third component. The PL spectra indicate that the luminescence intensity of P3HT or SMPV1 decreases with the increase of SMPV1 doping amount, which indicates that charge transfer occurs between P3HT and SMPV1, as shown in Figure 10.20a.[157]

There are some drawbacks in using the PL characterization method. For example, the luminescence intensity of PL is affected by the thickness, uniformity, and concentration of the fluorophore, so the time-resolved fluorescence spectroscopy may be more convincing to provide optical characteristic information.[73,158,159,160] Xu et al.[159] incorporated PTB7 into P3HT and found that the lifetime of P3HT will be reduced from 226 to 13 ps, as shown in Figure 10.20b. The fluorescence intensity of narrow-bandgap semiconductor is always weak, and the fluorescence lifetime is short, so it is difficult to detect changes in fluorescence intensity or fluorescence lifetime. The absorption coefficient of narrow-bandgap semiconductors is usually high. Thus, the absorption spectrum is more suitable for studying the charge dynamics process in the ternary system.

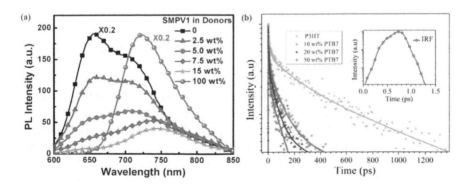

FIGURE 10.20 (a) PL spectra of P3HT:SMPV1 films with different SMPV1 doping ratios; (b) fluorescence decay profiles of pure P3HT and P3HT/PTB7 blend films. (Adapted from Q. An, et al., *ACS Appl. Mater. Interfaces*, 2015, **7**, 3691; W. L. Xu, et al., *J. Phys. Chem. C*, 2015, **119**, 21913–21920. Reprinted with permission from American Chemical Society (ACS); American Chemical Society (ACS).)

10.4.2 ENERGY TRANSFER

In ternary mixed films, energy transfer and charge transfer have a competitive relationship. Forster Resonance Energy Transfer (FRET) is a non-radiative energy transfer process that transfers photoexcitation energy into the acceptor material.[161–163] Energy transfer generally involves the inductive Coulombic interaction between the materials. The efficiency of energy transfer is related to the size of the phase separation between the donor and acceptor. One of the main parameters is the Forster radius which can be expressed as

$$R_0^6 = \frac{9000(\ln 10)Q_D k^2}{128\pi^5 n^4 N_A} J$$

where k^2 is the relative orientation of the donor and acceptor dipoles, Q_D is the emission quantum efficiency of donor in the absence of the acceptor, N_A is the Avogadro's number, and n is the refractive index. J is the spectral overlap integral between the donor emission and acceptor absorption, defined as

$$J = \frac{\int F_D(\lambda)\varepsilon_A(\lambda)\lambda^4 d\lambda}{\int F_D(\lambda)d\lambda}$$

where $F_D(\lambda)$ is the normalized emission spectrum of the donor, ε_A is the acceptor molar extinction coefficient. The energy transfer efficiency can be calculated by the following formula:

$$E = 1 - \frac{\tau_{DA}}{\tau_D}$$

where τ_{DA} and τ_D are the fluorescence lifetimes of donor in the presence and absence of the acceptor, respectively.

Hao group has done a lot of research on FRET in the ternary system.[159,164–166] For example, the energy transfer efficiency between P3HT and PTB7 (50% mixing) was found to be as high as 94.2%. It is calculated that the energy transfer radius is 5.9 nm, which is about half of the phase separation between the two donors. The cross-interconnection between the two donor materials is also the reason for the high efficiency FRET. In the P3HT:PCBM system, the energy transfer efficiency between donors was as high as 87.9% with 15% amount of PffBT4T-2OD incorporation, as shown in Figure 10.21.[164]

In the DRCN5T/PBDB-T/PCBM ternary system, when the addition amount of PBDB-T is 20%, the photoelectric conversion efficiency is the highest about 9.32%.[165] Time-resolved fluorescence spectroscopy revealed that both charge transfer and FRET occur simultaneously. When the two donors are mixed, there is no obvious

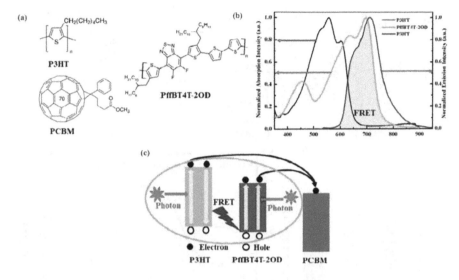

FIGURE 10.21 (a) Chemical geometries of P3HT, PC71BM, and PffBT4T-2OD; (b) normalized UV–Vis absorption spectra of P3HT and PffBT4T-2OD and the PL spectrum of P3HT film; (c) the schematic diagram of the energy transfer from P3HT to PffBT4T-2OD. (Adapted from F. Wang, et al., *Org. Electronics*, 2019, **67**, 146–152. Reprinted with permission from Elsevier.)

aggregation state. The fluorescence lifetime is relatively uniform, which shows that the two donor materials have good compatibility, as shown in Figure 10.22.

In addition to the dual donor system, Hao group also investigated the energy transfer and charge transfer processes by time-resolved fluorescence spectroscopy in the ternary system with two acceptors.[166]

10.4.3 PARALLEL WORKING MECHANISM

The parallel working mechanism differs from the charge transfer/energy transfer mechanism. It does not require precise control of material position, energy level, and bandwidth. For dual-donor and one-acceptor systems, excitons generated by each donor can migrate to the acceptor interface and then separate into free electrons and holes. In parallel working mode, there is no energy transfer or charge transfer between the two donors, which is equivalent to the two binary cells. The parallel working mechanism can be confirmed by the PL spectrum. Different ratios of PBDT-TS1 were incorporated into PTB7, and the results showed that the luminescence intensity increased in the region from 750 to 850 nm with the increase of PBDT-TS1 content in the mixed film.[167,168] The luminescence peak is located between PTB7 and PBDT-TS1, indicating that there is no energy transfer or charge transfer between the two donor materials. In the current–voltage curve, the short-circuit current of the mixed film is in the middle of the individual film position, which also indicates that the charge transfer between the two donors is negligible. At the same time, the overlap between the emission spectrum of PTB7 and the absorption

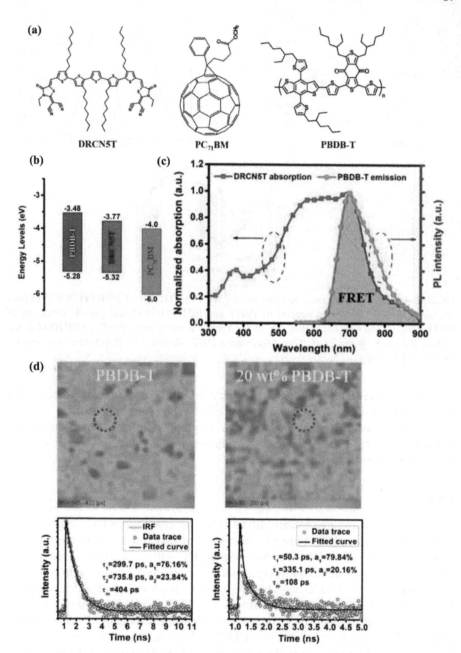

FIGURE 10.22 (a) Chemical structures of DRCN5T, PBDB-T, and PC71BM; (b) energy-level diagram of DRCN5T, PBDB-T, and PC71BM; (c) normalized absorption spectra of neat DRCN5T film and the PL spectra of PBDB-T film under 500 nm excitation; (d) time-resolved fluorescence images and corresponding decay curves in the selected spot. (Adapted from K.-N. Zhang, et al., *Org. Electronics*, 2018, **62**, 643–652. Reprinted with permission from Elsevier.)

spectrum of PTDT-TS1 is small, which also shows that the energy transfer between the two donors is small, as shown in Figure 10.23.

You et al.[167] were the first to report the parallel working mechanism in the ternary system. They selected two polymer donors and mixed them at a mass ratio of 0.5:0.5:1 (poly(benzodithiophenedithienylbenzotriaole)(TAZ):DTBT:PCBM). The thickness of the ternary blend films was 100 nm. The ratio of polymer:fullerene was 1:1 in binary blend film with the thickness of 50 nm. The authors found that the absorption spectrum of the ternary film is basically equal to the linear superposition of two binary films. The EQE of ternary solar cell is almost the sum of that in the two binary films. This indicates that both donors contribute to the formation of excitons. Carriers generated by exciton separation can be efficiently collected by the electrodes. It can be seen from Figure 10.24 that the photocurrent is significantly increased. The open circuit voltage in the ternary system is close to that of the smaller one of the binary cells because the open circuit voltage in the parallel working mechanism is determined by the higher HOMO level in the donor material. This further indicates that the two donor materials work in parallel.

Yang et al.[140] reported series of polymers with different absorption spectral ranges. It was found that compatibility between donors (similar molecular

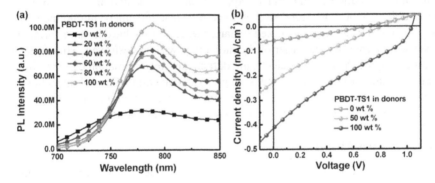

FIGURE 10.23   (a) PL spectra of PTB7:PBDT-TS1 films with different PBDT-TS1 doping ratios; (b) I–V curves of solar cells with PTB7, PBDT-TS1, and PTB7:PBDT-TS1 (1:1) as active layers. (Adapted from M. Zhang et al., *J. Mater. Chem. C*, 2015, **3**, 11930–11936. Reprinted with permission from Royal Society of Chemistry (RSC).)

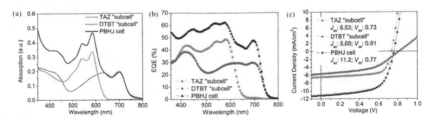

FIGURE 10.24   (a–c) Absorption spectra, EQEs, and J–V curves of the parallel-linkage ternary solar cells based on TAZ/DTBT and their "sub-cells." (Adapted from L. Yang, et al., *J. Am. Chem. Soc.*, 2012, **134**, 5432–5435. Reprinted with permission from American Chemical Society (ACS).)

orientation, crystallization, and particle size) is critical to device performance. For example, in the P3HT:PBDTT-DPP:PCBM ternary system, although the photocurrent response range is wider, the current and FF are not increased, which means that additional spectral absorption does not contribute to device performance improvement. This may be related to the physical incompatibility between the donor materials. It is well known that P3HT tends to form an edge-on layered arrangement, while the narrow bandgap polymer PBDTT-DPP forms a face-on molecular orientation. When two donor materials with different orientations are mixed, the non-conductive side chain of one donor will be close to the conjugated backbone of the other material. In this case, the coherence length will be reduced with defect formation, which decreases the charge transport. Therefore, structure compatibility is very important.

Wide-angle X-ray scattering is very sensitive to crystalline structures and regions and is ideal for studying organic photovoltaic systems. In transmission mode, the collected data is relatively simple, and the signal-to-noise ratio is good. In the grazing incidence mode, the contact area between the X-ray and the sample is large with high surface sensitivity. Therefore, the application range is wider. Figure 10.25 shows the grazing incidence wide-angle X-ray scattering (GIWAXS) images of several typical polymers. PBDTTT-C, PBDTT-DPP, PTB7, and PBDTT-SeDPP are face-on molecular orientations, and P3HT is a typical edge-on orientation.[140,169,170] When two compatible polymers are blended, the molecular structure can be maintained. While the two materials are incompatible no matter whether they have in-plane orientation or vertical orientation, the order of the molecules will be destroyed. This molecular disorder may be the origin of electronic defects and recombination centers, limiting the performance of the ternary OSCs. In general, two compatible polymers have less effect on molecular stack orientation, crystal size, and domain morphology; meanwhile the electronic structure of the polymer can be maintained.

10.4.4 ALLOY MODEL

In the parallel working mechanism, photoexcited carriers have their own transport channels. The alloy model means that two donors (acceptors) with similar electronic structures will form the same quasi-HOMO and quasi-LUMO levels.[149,171] This result is similar to the changes in valence bands and conduction bands in inorganic semiconductor alloys. The position of quasi-HOMO level and the quasi-LUMO level depends on the doping amount of the third element. Unlike the inorganic semiconductor alloy, the optical absorption of the exciton state maintains its own characteristics, not the alloy absorption, mainly because of the high localized exciton characteristics and the high delocalized electron and hole characteristics. Therefore, the change in the open circuit voltage depends on the composition of the donor, and the short circuit current is increased by the ternary complementary absorption. The prerequisite for alloy formation is the good blending and compatibility between the two donor materials. Zhang et al. selected two non-fullerene acceptors (MeIC and MeIC2) with similar molecular structure, as shown in Figure 10.26, due to the good compatibility between these two materials.[149,172,173] The efficiency of the device based on three components can be as high as 12.55%, which is much higher than that of a binary device (11.4%). The improvement in performance is mainly due to the formation of the acceptor alloy, which promotes the transport of electrons.

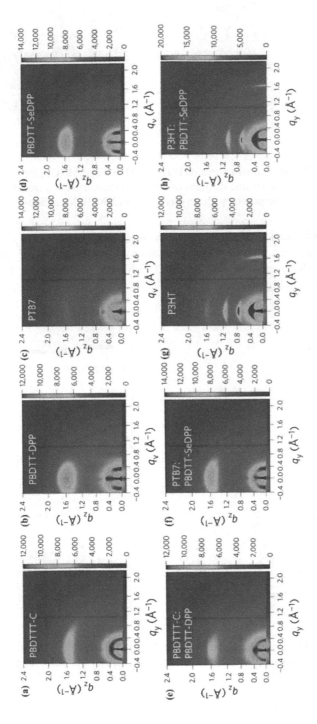

FIGURE 10.25 GIWAXS patterns of PBDTTT-C, PBDTT-DPP, PTB7, PBDTT-SeDPP, PBDTTT-C/PBDTT-DPP, PTB7:PBDTT-SeDPP, P3HT, and P3HT:PBDTT-SeDPP. (Adapted from Y. Yang et al., *Nat. Photonics*, 2015, **9**, 190–198. Reprinted with permission from Springer Nature.)

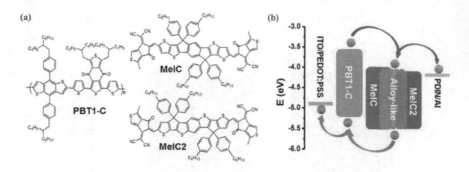

FIGURE 10.26 (a) Chemical structures of PBT1-C, MeIC, and MeIC2; (b) energy levels of the used materials and schematic diagram of alloy-like state. (Adapted from Q. An et al., *Small*, 2018, **14**, 1802983. Reprinted with permission from John Wiley & Sons.)

In addition to formation of alloys between small molecules, Wei et al.[150] found that alloy structures can also be formed between polymers and small molecules. Such high-crystallinity and face-on-oriented polymers are favorable for charge transport, as shown in Figure 10.27.

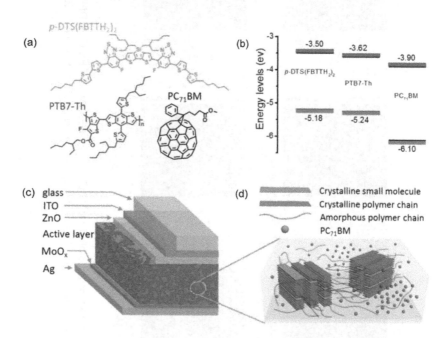

FIGURE 10.27 (a) Chemical structures of p-DTS(FBTTH2)2, PTB7-Th, and PC71BM; (b) energy-level diagrams for p-DTS(FBTTH2)2, PTB7-Th, and PC71BM; (c) device structures of the OSC (glass/ITO/ZnO/active layer/MoO$_x$/Ag); (d) illustration of the active layer of ternary solar cells, in which the small molecules induce the face-on π–π stacking. (Adapted from J. Zhang, et al., *J. Am. Chem. Soc.*, 2015, **137**, 8176–8183. Reprinted with permission from American Chemical Society (ACS).)

10.5 OPPORTUNITIES AND CHALLENGES

In the economic evaluation of OSCs, when the photoelectric conversion efficiency of OSCs is more than 7% with large area and the OSC maintains 5-year lifetime, it can produce a competitive power output with silicon solar cells.[174] The efficiency of OSCs in large scale can be 7% with a module area of $4\,cm^2$.[175] However, the poor stability limits its industrial application. In recent years, many factors have been found to limit the stability of devices, including metastable morphology, diffusion of electrodes and charge transfer layers, oxygen and water, and light radiation. The main approaches to solve the device stability problem are new organic materials, the inverted structure, interface engineering, and packaging technology. Donor and acceptor materials determine the device performance of OSCs. Fullerene with the spherical structure is easy to aggregate and crystallize, which affects the long-term stability of the device seriously. Nowadays, the device based on the combination of wide-bandgap polymer donor and low-bandgap no-fullerene acceptor shows the highest PCE and good stability. This may be a trend in the synthesis of new organic semiconductors. The main reason for the good performance of the device with inverted structure is that the top electrode reacts less with oxygen and water. Interface engineering has been introduced in detail in the previous section; it is only briefly introduced here. PEDOT: PSS is usually used as the hole transport layer material which has a corrosive effect on ITO electrode. In order to improve the stability of the device, metal oxide is also a good choice. For example, MoO_x has better long-term air stability because of its hydrophobicity and better oxidation resistance. Encapsulation is a convenient, effective, and universal method to improve the device stability. Encapsulation materials include organic materials, inorganic materials, and organic–inorganic composite materials. Epoxy resin is usually used as the encapsulation material, which has high sealing property, high transparency, and low cost. The efficiency of the device encapsulated with epoxy resin can be maintained at 90% for 1,000 hours.[176] The stability of the device is improved after encapsulation, which is mainly related to the similar thermal expansion between organic materials.

The solvents used in the preparation of OSCs are usually dichlorobenzene, chlorobenzene, or chloroform, which will cause damage to human health and the environment. Chlorine-based solvents are used because they can promote the homogeneity of the films with smaller particles, which are favorable of the sunlight absorption and the exciton separation.[177] With the continuous improvement of the device efficiency, how to remove toxic solutions and use environmentally friendly solvents is still a great challenge in the practical application of OSCs. In order to solve the problem of solvent toxicity, we can design donor materials which are soluble in environmentally friendly solvents. For example, some functional groups were introduced into the conjugated main chain of the donor to make it dissolve in environmentally friendly solvents.[178–180] A lot of attempts have been made to replace chlorine-based solvents in the classical P3HT:PCBM system. For high-performance donor materials with narrow bandgap, the device efficiency can exceed 9% when non-chlorine solvents are used as solvents.[181–186] The mixed solvent will cause the metastable morphology, current–voltage S-shape curve, reduced filling factor, and poor repeatability of the preparation process. Therefore, the selection of a single green solvent will be more attractive and economical. It should be pointed out that most of the device efficiency has not been certified. It is not universal for the solvents in different material systems when compared with chlorine-based solvents. Another problem is that the real green solvents in OSCs have not yet been realized from the perspective of sustainable development.

REFERENCES

1. X. T. Hao, T. Hosokai, N. Mitsuo, S. Kera, K. Mase, K. K. Okudaira and N. Ueno, *Appl. Phys. Lett.*, **89** 182113, (2006).
2. Z. He, B. Xiao, F. Liu, H. Wu, Y. Yang, S. Xiao, C. Wang, T. P. Russell and Y. Cao, *Nat. Photonics*, **9** 174–179, (2015).
3. G. Li, V. Shrotriya, J. Huang, Y. Yao, T. Moriarty, K. Emery and Y. Yang, *Nat. Mater.*, **4** 864–868, (2005).
4. G. Li, R. Zhu and Y. Yang, *Nat. Photonics*, **6** 153–161, (2012).
5. J. You, L. Dou, K. Yoshimura, T. Kato, K. Ohya, T. Moriarty, K. Emery, C. C. Chen, J. Gao, G. Li and Y. Yang, *Nat Commun*, **4** 1446, (2013).
6. C. W. Tang, *Appl. Phys. Lett.*, **48** 183, (1986).
7. M. Hiramoto, H. Fujiwara and M. Yokoyama, *J. Appl. Phys.*, **72** 3781–3787, (1992).
8. J. J. M. Halls, C. A. Walsh, N. C. Greenham, E. A. Marseglia, R. H. Friend, S. C. Moratti and A. B. Holmes, *Nature*, **376** 498–500, (1995).
9. G. Yu, J. Gao, J. C. Hummelen, F. Wudl and A. J. Heeger, *Science*, **270** 1789–1791, (1995).
10. J. D. Chen, C. Cui, Y. Q. Li, L. Zhou, Q. D. Ou, C. Li, Y. Li and J. X. Tang, *Adv Mater*, **27** 1035–1041, (2015).
11. X. Ouyang, R. Peng, L. Ai, X. Zhang and Z. Ge, *Nat. Photonics*, **9** 520–524, (2015).
12. Q. Liu, Y. Jiang, K. Jin et al., *Sci. Bull.*, doi:10.1016/j.scib.2020.01.001.
13. C. Sun, S. Qin, R. Wang, S. Chen, F. Pan, B. Qiu, Z. Shang, L. Meng, C. Zhang, M. Xiao, C. Yang, and Y. Li, *J. Am. Chem. Soc.*, **142** 1465–1474, (2020).
14. K. Jiang, Q. Wei, J. Y. L. Lai, Z. Peng, H. K. Kim, J. Yuan, L. Ye, H. Ade, Y. Zou, and H. Yan, *Joule*, **3** 18, (2019).
15. M. Hallermann, I. Kriegel, E. Da Como, J. M. Berger, E. von Hauff and J. Feldmann, *Adv. Funct. Mater.*, **19** 3662–3668, (2009).
16. S. Westenhoff, W. J. D. Beenken, R. H. Friend, N. C. Greenham, A. Yartsev and V. Sundström, *Phys. Rev. Lett.*, **97** 166804, (2006).
17. B. M. Savoie, S. Dunaisky, T. J. Marks and M. A. Ratner, *Adv. Energy Mater.*, **5** 1400891, (2015).
18. J. M. Szarko, B. S. Rolczynski, S. J. Lou, T. Xu, J. Strzalka, T. J. Marks, L. Yu and L. X. Chen, *Adv. Funct. Mater.*, **24** 10–26, (2014).
19. B. C. Thompson and J. M. J. Frechet, *Angew. Chem. Int. Ed.*, **47** 58–77, (2008).
20. A. A. Bakulin, A. Rao, V. G. Pavelyev, P. H. M. van Loosdrecht, M. S. Pshenichnikov, D. Niedzialek, J. Cornil, D. Beljonne and R. H. Friend, *Science*, **335** 1340–1344, (2012).
21. S. Gelinas, A. Rao, A. Kumar, S. L. Smith, A. W. Chin, J. Clark, T. S. van der Poll, G. C. Bazan and R. H. Friend, *Science*, **343** 512–516, (2014).
22. B. S. Rolczynski, J. M. Szarko, H. J. Son, Y. Liang, L. Yu and L. X. Chen, *J. Am. Chem. Soc.*, **134** 4142–4152, (2012).
23. S. D. Stranks, G. E. Eperon, G. Grancini, C. Menelaou, M. J. P. Alcocer, T. Leijtens, L. M. Herz, A. Petrozza and H. J. Snaith, *Science*, **342** 341–344, (2013).
24. M. T. Dang, L. Hirsch and G. Wantz, *Adv. Mater.*, **23** 3597–3602, (2011).
25. M. C. Scharber, D. Mühlbacher, M. Koppe, P. Denk, C. Waldauf, A. J. Heeger and C. J. Brabec, *Adv. Mater.*, **18** 789–794, (2006).

26. H. A. Atwater and A. Polman, *Nat. Mater.*, **9** 205, (2010).
27. H. Deckman, C. Roxlo and E. Yablonovitch, *Opt. Lett.*, **8** 491–493, (1983).
28. E. Yablonovitch and G. D. Cody, *IEEE T. Electron Dev.*, **29** 300–305, (1982).
29. Y. H. Jang, Y. J. Jang, S. Kim, L. N. Quan, K. Chung and D. H. Kim, *Chem. Rev.*, **116** 14982–15034, (2016).
30. S. Pillai, K. Catchpole, T. Trupke and M. Green, *J. Appl. Phys.*, **101** 093105, (2007).
31. T. Temple, G. Mahanama, H. Reehal and D. Bagnall, *Sol. Energy Mat. Sol. Cells.*, **93** 1978–1985, (2009).
32. H. M. Chen, C. K. Chen, C.-J. Chen, L.-C. Cheng, P. C. Wu, B. H. Cheng, Y. Z. Ho, M. L. Tseng, Y.-Y. Hsu and T.-S. Chan, *ACS Nano.*, **6** 7362–7372, (2012).
33. K. Catchpole and A. Polman, *Appl. Phys. Lett.*, **93** 191113, (2008).
34. K. A. Catchpole and A. Polman, *Opt. Express.*, **16** 21793–21800, (2008).
35. F. Beck, A. Polman and K. Catchpole, *J. Appl. Phys.*, **105** 114310, (2009).
36. S. Lim, W. Mar, P. Matheu, D. Derkacs and E. Yu, *J. Appl. Phys.*, **101** 104309, (2007).
37. P. Matheu, S. Lim, D. Derkacs, C. McPheeters and E. Yu, *Appl. Phys. Lett.*, **93** 113108, (2008).
38. P. Spinelli, M. Verschuuren and A. Polman, *Nat. Commun.*, **3** 692, (2012).
39. H. F. Zarick, O. Hurd, J. A. Webb, C. Hungerford, W. R. Erwin and R. Bardhan, *ACS Photonics*, **1** 806–811, (2014).
40. M. L. Juan, M. Righini and R. Quidant, *Nat. Photonics*, **5** 349, (2011).
41. L. Feng, M. Niu, Z. Wen and X. Hao, *Polymers*, **10** 123, (2018).
42. S. R. Gollu, R. Sharma, G. Srinivas, S. Kundu and D. Gupta, *Org. Electronics*, **29** 79–87, (2016).
43. I. Khan, H. Keshmiri, F. Kolb, T. Dimopoulos, E. J. List-Kratochvil and J. Dostalek, *Adv. Opt. Mater.*, **4** 435–443, (2016).
44. A. C. Liapis, M. Y. Sfeir and C. T. Black, *Appl. Phys. Lett.*, **109** 201101, (2016).
45. A. M. Nardes, S. Ahn, D. Rourke, C. Mao, J. van de Lagemaat, A. J. Ferguson, W. Park and N. Kopidakis, *Org. Electronics*, **39** 59–63, (2016).
46. X. Ren, J. Cheng, S. Zhang, X. Li, T. Rao, L. Huo, J. Hou and W. C. Choy, *Small*, **12** 5200–5207, (2016).
47. C. Stelling, C. R. Singh, M. Karg, T. A. König, M. Thelakkat and M. Retsch, *Sci. Rep.*, **7** 42530, (2017).
48. K. Yao, H. Jiao, Y.-X. Xu, Q. He, F. Li and X. Wang, *J. Mater. Chem. A*, **4** 13400–13406, (2016).
49. M. Yao, P. Shen, Y. Liu, B. Chen, W. Guo, S. Ruan and L. Shen, *ACS Appl. Mater. Inter.*, **8** 6183–6189, (2016).
50. Y.-S. Hsiao, S. Charan, F.-Y. Wu, F.-C. Chien, C.-W. Chu, P. Chen and F.-C. Chen, *J. Phys. Chem. C*, **116** 20731–20737, (2012).
51. E. Wei, H. L. Zhu, L. Chen, W. C. Chew and W. C. Choy, *Sci. Rep.*, **5** 8525, (2015).
52. J. M. Lee, J. Lim, N. Lee, H. I. Park, K. E. Lee, T. Jeon, S. A. Nam, J. Kim, J. Shin and S. O. Kim, *Adv. Mater.*, **27** 1519–1525, (2015).
53. D. H. Wang, D. Y. Kim, K. W. Choi, J. H. Seo, S. H. Im, J. H. Park, O. O. Park and A. J. Heeger, *Angew. Chem. Int. Ed.*, **50** 5519–5523, (2011).
54. D. H. Wang, K. H. Park, J. H. Seo, J. Seifter, J. H. Jeon, J. K. Kim, J. H. Park, O. O. Park and A. J. Heeger, *Adv. Energy Mater.*, **1** 766–770, (2011).
55. D. Chi, S. Lu, R. Xu, K. Liu, D. Cao, L. Wen, Y. Mi, Z. Wang, Y. Lei and S. Qu, *Nanoscale*, **7** 15251–15257, (2015).
56. G. Kakavelakis, I. Vangelidis, A. Heuer-Jungemann, A. G. Kanaras, E. Lidorikis, E. Stratakis and E. Kymakis, *Adv. Energy Mater.*, **6** 1501640, (2016).
57. P. Xu, L. Shen, F. Meng, J. Zhang, W. Xie, W. Yu, W. Guo, X. Jia and S. Ruan, *Appl. Phys. Lett.*, **102** 53, (2013).

58. H. I. Park, S. Lee, J. M. Lee, S. A. Nam, T. Jeon, S. W. Han and S. O. Kim, *ACS Nano*, **8** 10305–10312, (2014).
59. M. Chalh, S. Vedraine, B. Lucas and B. Ratier, *Sol. Energy Mat. Sol. Cells.*, **152** 34–41, (2016).
60. E. C. Garnett, W. Cai, J. J. Cha, F. Mahmood, S. T. Connor, M. G. Christoforo, Y. Cui, M. D. McGehee and M. L. Brongersma, *Nat. Mater.*, **11** 241, (2012).
61. M. G. Kang, M. S. Kim, J. Kim and L. J. Guo, *Adv. Mater.*, **20** 4408–4413, (2008).
62. M. G. Kang, T. Xu, H. J. Park, X. Luo and L. J. Guo, *Adv. Mater.*, **22** 4378–4383, (2010).
63. M. C. Scharber, C. Lungenschmied, H.-J. Egelhaaf, G. Matt, M. Bednorz, T. Fromherz, J. Gao, D. Jarzab and M. A. Loi, *Energy Environ. Sci.*, **4** 5077, (2011).
64. J. Huang, Z. Yin and Q. Zheng, *Energy Environ. Sci.*, **4** 3861–3877, (2011).
65. S. Kim, J. H. Koh, X. Yang, W. S. Chi, C. Park, J. W. Leem, B. Kim, S. Seo, Y. Kim and J. S. Yu, *Adv. Energy Mater.*, **4** 1301338, (2014).
66. S. Kundu, S. R. Gollu, R. Sharma, G. Srinivas, A. Ashok, A. Kulkarni and D. Gupta, *Org. Electronics*, **14** 3083–3088, (2013).
67. S. Sanchez, S. Berson, S. Guillerez, C. Lévy-Clément and V. Ivanova, *Adv. Energy Mater.*, **2** 541–545, (2012).
68. J.-C. Wang, W.-T. Weng, M.-Y. Tsai, M.-K. Lee, S.-F. Horng, T.-P. Perng, C.-C. Kei, C.-C. Yu and H.-F. Meng, *J. Mater. Chem.*, **20** 862–866, (2010).
69. M.-S. White, D. Olson, S. Shaheen, N. Kopidakis and D. S. Ginley, *Appl. Phys. Lett.*, **89** 143517, (2006).
70. Y. Sun, J. H. Seo, C. J. Takacs, J. Seifter and A. J. Heeger, *Adv. Mater.*, **23** 1679–1683, (2011).
71. Y. Liu, J. Zhao, Z. Li, C. Mu, W. Ma, H. Hu, K. Jiang, H. Lin, H. Ade and H. Yan, *Nat. Commun.*, **5** 5293, (2014).
72. Z. Yin, Q. Zheng, S. C. Chen and D. Cai, *ACS Appl. Mater. Inter.*, **5** 9015–9025, (2013).
73. W.-L. Xu, B. Wu, F. Zheng, H.-B. Wang, Y.-Z. Wang, F.-G. Bian, X.-T. Hao and F. Zhu, *Org. Electronics*, **25** 266–274, (2015).
74. S. B. Dkhil, D. Duché, M. Gaceur, A. K. Thakur, F. B. Aboura, L. Escoubas, J. J. Simon, A. Guerrero, J. Bisquert and G. Garcia-Belmonte, *Adv. Energy Mater.*, **4** 1400805, (2015).
75. K. Lee, J. Y. Kim, S. H. Park, S. H. Kim, S. Cho and A. J. Heeger, *Adv. Mater.*, **19** 2445–2449, (2007).
76. X. Bao, L. Sun, W. Shen, C. Yang, W. Chen and R. Yang, *J. Mater. Chem. A*, **2** 1732–1737, (2014).
77. S. H. Park, A. Roy, S. Beaupré, S. Cho, N. Coates, J. S. Moon, D. Moses, M. Leclerc, K. Lee and A. J. Heeger, *Nat. Photonics*, **3** 297, (2009).
78. C. Waldauf, M. Morana, P. Denk, P. Schilinsky, K. Coakley, S. Choulis and C. Brabec, *Appl. Phys. Lett.*, **89** 233517, (2006).
79. J. Xiong, B. Yang, C. Zhou, J. Yang, H. Duan, W. Huang, X. Zhang, X. Xia, L. Zhang and H. Huang, *Org. Electronics*, **15** 835–843, (2014).
80. A. R. bin Mohd Yusoff, H. P. Kim and J. Jang, *Sol. Energy Mat. Sol. Cells.*, **109** 63–69, (2013).
81. J. Bok Kim, S. Ahn, S. Ju Kang, C. Nuckolls and Y.-L. Loo, *Appl. Phys. Lett.*, **102** 45, (2013).
82. J. Liu, S. Shao, B. Meng, G. Fang, Z. Xie, L. Wang and X. Li, *Appl. Phys. Lett.*, **100** 213906, (2012).
83. S. Trost, K. Zilberberg, A. Behrendt and T. Riedl, *J. Mater. Chem.*, **22** 16224–16229, (2012).
84. B. Bob, T.-B. Song, C.-C. Chen, Z. Xu and Y. Yang, *Chem. Mater.*, **25** 4725–4730, (2013).

85. S. Trost, A. Behrendt, T. Becker, A. Polywka, P. Görrn and T. Riedl, *Adv. Energy Mater.*, **5** 1500277, (2015).
86. H. Zhang and J. Ouyang, *Appl. Phys. Lett.*, **97** 063509, (2010).
87. Y. Zhou, H. Cheun, W. J. Potscavage Jr, C. Fuentes-Hernandez, S.-J. Kim and B. Kippelen, *J. Mater. Chem.*, **20** 6189–6194, (2010).
88. J. Peng, Q. Sun, Z. Zhai, J. Yuan, X. Huang, Z. Jin, K. Li, S. Wang, H. Wang and W. Ma, *Nanotechnology*, **24** 484010, (2013).
89. Z. Yin, Q. Zheng, S. C. Chen, D. Cai, L. Zhou and J. Zhang, *Adv. Energy Mater.*, **4** 1301404, (2014).
90. T. Z. Oo, R. D. Chandra, N. Yantara, R. R. Prabhakar, L. H. Wong, N. Mathews and S. G. Mhaisalkar, *Org. Electronics*, **13** 870–874, (2012).
91. A. Puetz, T. Stubhan, M. Reinhard, O. Loesch, E. Hammarberg, S. Wolf, C. Feldmann, H. Kalt, A. Colsmann and U. Lemmer, *Sol. Energy Mat. Sol. Cells.*, **95** 579–585, (2011).
92. K.-S. Shin, K.-H. Lee, H. H. Lee, D. Choi and S.-W. Kim, *J. Phys. Chem. C*, **114** 15782–15785, (2010).
93. T. Stubhan, I. Litzov, N. Li, M. Salinas, M. Steidl, G. Sauer, K. Forberich, G. J. Matt, M. Halik and C. J. Brabec, *J. Mater. Chem. A*, **1** 6004–6009, (2013).
94. T. Stubhan, H. Oh, L. Pinna, J. Krantz, I. Litzov and C. J. Brabec, *Org. Electronics*, **12** 1539–1543, (2011).
95. M. Thambidurai, J. Y. Kim, C.-M. Kang, N. Muthukumarasamy, H.-J. Song, J. Song, Y. Ko, D. Velauthapillai and C. Lee, *Renewable Energy*, **66** 433–442, (2014).
96. X. Li, F. Xie, S. Zhang, J. Hou and W. C. Choy, *Adv. Funct. Mater.*, **24** 7348–7356, (2014).
97. X. Li, F. Xie, S. Zhang, J. Hou and W. C. Choy, *Light Sci. Appl.*, **4** e273, (2015).
98. J. You, C. C. Chen, L. Dou, S. Murase, H. S. Duan, S. A. Hawks, T. Xu, H. J. Son, L. Yu and G. Li, *Adv. Mater.*, **24** 5267–5272, (2012).
99. S. Chen, J. R. Manders, S.-W. Tsang and F. So, *J. Mater. Chem.*, **22** 24202–24212, (2012).
100. W. Cai, P. Liu, Y. Jin, Q. Xue, F. Liu, T. P. Russell, F. Huang, H. L. Yip and Y. Cao, *Adv. Sci.*, **2** 1500095, (2015).
101. Z. He, C. Zhong, X. Huang, W. Y. Wong, H. Wu, L. Chen, S. Su and Y. Cao, *Adv Mater*, **23** 4636–4643, (2011).
102. X. Hu, C. Yi, M. Wang, C. H. Hsu, S. Liu, K. Zhang, C. Zhong, F. Huang, X. Gong and Y. Cao, *Adv. Energy Mater.*, **4** 1400378, (2014).
103. Y. L. Li, Y. S. Cheng, P. N. Yeh, S. H. Liao and S. A. Chen, *Adv. Funct. Mater.*, **24** 6811–6817, (2014).
104. Z. Tang, W. Tress, Q. Bao, M. J. Jafari, J. Bergqvist, T. Ederth, M. R. Andersson and O. Inganäs, *Adv. Energy Mater.*, **4** 1400643, (2014).
105. H. Kang, S. Hong, J. Lee and K. Lee, *Adv. Mater.*, **24** 3005–3009, (2012).
106. E. Saracco, B. Bouthinon, J. M. Verilhac, C. Celle, N. Chevalier, D. Mariolle, O. Dhez and J. P. Simonato, *Adv. Mater.*, **25** 6534–6538, (2013).
107. Y. Zhou, C. Fuentes-Hernandez, J. Shim, J. Meyer, A. J. Giordano, H. Li, P. Winget, T. Papadopoulos, H. Cheun and J. Kim, *Science*, **336** 327–332, (2012).
108. V. Gupta, A. K. K. Kyaw, D. H. Wang, S. Chand, G. C. Bazan and A. J. Heeger, *Sci. Rep.*, **3** 1965, (2013).
109. C. Y. Jiang, X. W. Sun, D. W. Zhao, A. K. K. Kyaw and Y. N. Li, *Sol. Energy Mat. Sol. Cells.*, **94** 1618–1621, (2010).
110. D. Zhao, P. Liu, X. Sun, S. Tan, L. Ke and A. Kyaw, *Appl. Phys. Lett.*, **95** 275, (2009).
111. J. Huang, G. Li and Y. Yang, *Adv. Mater.*, **20** 415–419, (2008).
112. G. Wang, T. Jiu, C. Sun, J. Li, P. Li, F. Lu and J. Fang, *ACS Appl. Mater. Inter.*, **6** 833–838, (2014).

113. T. Xiao, W. Cui, M. Cai, W. Leung, J. W. Anderegg, J. Shinar and R. Shinar, *Org. Electronics*, **14** 267–272, (2013).

114. Z. A. Tan, S. Li, F. Wang, D. Qian, J. Lin, J. Hou and Y. Li, *Sci. Rep.*, **4** 4691, (2014).

115. S. Shao, K. Zheng, T. Pullerits and F. Zhang, *ACS Appl. Mater. Inter.*, **5** 380–385, (2013).

116. C. E. Small, S. Chen, J. Subbiah, C. M. Amb, S.-W. Tsang, T.-H. Lai, J. R. Reynolds and F. So, *Nat. Photonics*, **6** 115, (2012).

117. N. Wu, Q. Luo, Z. Bao, J. Lin, Y.-Q. Li and C.-Q. Ma, *Sol. Energy Mat. Sol. Cells.*, **141** 248–259, (2015).

118. Z.-Q. Xu, J.-P. Yang, F.-Z. Sun, S.-T. Lee, Y.-Q. Li and J.-X. Tang, *Org. Electronics*, **13** 697–704, (2012).

119. Z. Yin, Q. Zheng, S.-C. Chen, J. Li, D. Cai, Y. Ma and J. Wei, *Nano Research*, **8** 456–468, (2015).

120. L. Nian, W. Zhang, N. Zhu, L. Liu, Z. Xie, H. Wu, F. Würthner and Y. Ma, *J. Am. Chem. Soc.*, **137** 6995–6998, (2015).

121. S. Woo, W. Hyun Kim, H. Kim, Y. Yi, H. K. Lyu and Y. Kim, *Adv. Energy Mater.*, **4** 1301692, (2014).

122. S. R. Cowan, P. Schulz, A. J. Giordano, A. Garcia, B. A. MacLeod, S. R. Marder, A. Kahn, D. S. Ginley, E. L. Ratcliff and D. C. Olson, *Adv. Funct. Mater.*, **24** 4671–4680, (2014).

123. H. W. Lee, Y. O. Jin, T. I. Lee, W. S. Jang, Y. B. Yoo, S. S. Chae, J. H. Park, J. M. Myoung, K. M. Song and K. B. Hong, *Appl. Phys. Lett.*, **102** 1474, (2013).

124. Y. Zhang, S. Yuan, Y. Li and W. Zhang, *Electrochimica Acta*, **117** 438–442, (2014).

125. H.-L. Yip and A. K.-Y. Jen, *Energy Environ. Sci.*, **5** 5994–6011, (2012).

126. F. Jiang, W. C. Choy, X. Li, D. Zhang and J. Cheng, *Adv. Mater.*, **27** 2930–2937, (2015).

127. T. Stubhan, N. Li, N. A. Luechinger, S. C. Halim, G. J. Matt and C. J. Brabec, *Adv. Energy Mater.*, **2** 1433–1438, (2012).

128. X. Tu, F. Wang, C. Li, Z. A. Tan and Y. Li, *J. Phys. Chem. C*, **118** 9309–9317, (2014).

129. F. Wang, Q. Xu, Z. A. Tan, L. Li, S. Li, X. Hou, G. Sun, X. Tu, J. Hou and Y. Li, *J. Mater. Chem. A*, **2** 1318–1324, (2014).

130. K. S. Tan, M. K. Chuang, F. C. Chen and C. S. Hsu, *ACS Appl. Mater. Inter.*, **5** 12419–12424, (2013).

131. Z. A. Tan, D. Qian, W. Zhang, L. Li, Y. Ding, Q. Xu, F. Wang and Y. Li, *J. Mater. Chem. A*, **1** 657–664, (2013).

132. J. Wang, J. Zhang, B. Meng, B. Zhang, Z. Xie and L. Wang, *ACS Appl. Mater. Inter.*, **7** 13590–13596, (2015).

133. S.-S. Li, K.-H. Tu, C.-C. Lin, C.-W. Chen and M. Chhowalla, *ACS Nano*, **4** 3169–3174, (2010).

134. Y. Gao, H.-L. Yip, S. K. Hau, K. M. O'Malley, N. C. Cho, H. Chen and A. K.-Y. Jen, *Appl. Phys. Lett.*, **97** 251, (2010).

135. Y. Gao, H. L. Yip, K. S. Chen, K. M. O'Malley, O. Acton, Y. Sun, G. Ting, H. Chen and A. K. Y. Jen, *Adv. Mater.*, **23** 1903–1908, (2011).

136. J. M. Yun, J. S. Yeo, J. Kim, H. G. Jeong, D. Y. Kim, Y. J. Noh, S. S. Kim, B. C. Ku and S. I. Na, *Adv. Mater.*, **23** 4923–4928, (2011).

137. S. Mao, H. Pu and J. Chen, *RSC Adv.*, **2** 2643–2662, (2012).

138. D. R. Dreyer, S. Park, C. W. Bielawski and R. S. Ruoff, *Chem. Soc. Rev.*, **39** 228–240, (2010).

139. S. Pei and H.-M. Cheng, *Carbon*, **50** 3210–3228, (2012).

140. Y. Yang, W. Chen, L. Dou, W.-H. Chang, H.-S. Duan, B. Bob, G. Li and Y. Yang, *Nat. Photonics*, **9** 190–198, (2015).

141. S.-T. Chuang, S.-C. Chien and F.-C. Chen, *Appl. Phys. Lett.*, **100** 013309, (2012).

142. F. Guo, B. Yang, Y. Yuan, Z. Xiao, Q. Dong, Y. Bi and J. Huang, *Nat. Nanotech.*, **7** 798–802, (2012).

143. Y. Han, C. Fan, G. Wu, H.-Z. Chen and M. Wang, *J. Phys. Chem. C*, **115** 13438–13445, (2011).
144. J. M. Melancon and S. R. Živanović, *Appl. Phys. Lett.*, **105** 163301, (2014).
145. H. Wang, Z. Li, C. Fu, D. Yang, L. Zhang, S. Yang and B. Zou, *IEEE Photonics Tech. Lett.*, **27** 612–615, (2015).
146. Y. Xie, M. Gong, T. A. Shastry, J. Lohrman, M. C. Hersam and S. Ren, *Adv Mater*, **25** 3433–3437, (2013).
147. D. J. Xue, J. J. Wang, Y. Q. Wang, S. Xin, Y. G. Guo and L. J. Wan, *Adv Mater*, **23** 3704–3707, (2011).
148. Q. An, F. Zhang, J. Zhang, W. Tang, Z. Deng and B. Hu, *Energy Environ. Sci.*, **9** 281–322, (2016).
149. Q. An, J. Zhang, W. Gao, F. Qi, M. Zhang, X. Ma, C. Yang, L. Huo and F. Zhang, *Small*, **14** 1802983, (2018).
150. J. Zhang, Y. Zhang, J. Fang, K. Lu, Z. Wang, W. Ma and Z. Wei, *J. Am. Chem. Soc.*, **137** 8176–8183, (2015).
151. Q. An, F. Zhang, J. Zhang, W. Tang, Z. Wang, L. Li, Z. Xu, F. Teng and Y. Wang, *Sol. Energy Mat. Sol. Cells.*, **118** 30–35, (2013).
152. J.-H. Huang, M. Velusamy, K.-C. Ho, J.-T. Lin and C.-W. Chu, *J. Mater. Chem.*, **20** 2820–2825, (2010).
153. M. Koppe, H. J. Egelhaaf, E. Clodic, M. Morana, L. Lüer, A. Troeger, V. Sgobba, D. M. Guldi, T. Ameri and C. J. Brabec, *Adv. Energy Mater.*, **3** 949–958, (2013).
154. J.-S. Huang, T. Goh, X. Li, M. Y. Sfeir, E. A. Bielinski, S. Tomasulo, M. L. Lee, N. Hazari and A. D. Taylor, *Nat. Photonics*, **7** 479, (2013).
155. L. Lu, T. Xu, W. Chen, E. S. Landry and L. Yu, *Nat. Photonics*, **8** 716, (2014).
156. L. Ye, H.-H. Xu, H. Yu, W.-Y. Xu, H. Li, H. Wang, N. Zhao and J.-B. Xu, *J. Phys. Chem. C*, **118** 20094–20099, (2014).
157. Q. An, F. Zhang, L. Li, J. Wang, Q. Sun, J. Zhang, W. Tang and Z. Deng, *ACS Appl. Mater. Inter.*, **7** 3691, (2015).
158. X.-T. Hao, L. J. McKimmie and T. A. Smith, *J. Phys. Chem. Lett.*, **2** 1520–1525, (2011).
159. W.-L. Xu, B. Wu, F. Zheng, X.-Y. Yang, H.-D. Jin, F. Zhu and X.-T. Hao, *J. Phys. Chem. C*, **119** 21913–21920, (2015).
160. W. L. Xu, P. Zeng, B. Wu, F. Zheng, F. Zhu, T. A. Smith, K. P. Ghiggino and X. T. Hao, *J. Phys. Chem. Lett.*, **7** 1872, (2016).
161. N. Y. Chan, M. Chen, X.-T. Hao, T. A. Smith and D. E. Dunstan, *J. Phys. Chem. Lett.*, **1** 1912–1916, (2010).
162. A. R. Clapp, I. L. Medintz, J. M. Mauro, B. R. Fisher, M. G. Bawendi and H. Mattoussi, *J. Am. Chem. Soc.*, **126** 301–310, (2004).
163. G. D. Scholes, *Annu. Rev. Phys. Chem.*, **54** 57–87, (2003).
164. F. Wang, X.-Y. Yang, M.-S. Niu, L. Feng and X.-T. Hao, *Org. Electronics*, **67** 146–152, (2019).
165. K.-N. Zhang, P.-Q. Bi, Z.-C. Wen, M.-S. Niu, Z.-H. Chen, T. Wang, L. Feng, J.-L. Yang and X.-T. Hao, *Org. Electronics*, **62** 643–652, (2018).
166. K.-N. Zhang, X.-Y. Yang, M.-S. Niu, Z.-C. Wen, Z.-H. Chen, L. Feng, X.-J. Feng and X.-T. Hao, *Org. Electronics*, **66** 13–23, (2019).
167. L. Yang, H. Zhou, S. C. Price and W. You, *J. Am. Chem. Soc.*, **134** 5432–5435, (2012).
168. M. Zhang, F. Zhang, J. Wang, Q. An and Q. Sun, *J. Mater. Chem. C*, **3** 11930–11936, (2015).
169. W. Chen, T. Xu, F. He, W. Wang, C. Wang, J. Strzalka, Y. Liu, J. Wen, D. J. Miller and J. Chen, *Nano Lett.*, **11** 3707–3713, (2011).
170. J. Rivnay, R. Noriega, R. J. Kline, A. Salleo and M. F. Toney, *Phys. Rev. B*, **84** 045203, (2011).

171. J. Zhang, Y. Zhang, J. Fang, K. Lu, Z. Wang, W. Ma and Z. Wei, *J. Am. Chem. Soc.*, **137** 8176–8183, (2015).
172. T. Liu, L. Huo, S. Chandrabose, K. Chen, G. Han, F. Qi, X. Meng, D. Xie, W. Ma and Y. Yi, *Adv. Mater.*, **30** 1707353, (2018).
173. Z. Luo, H. Bin, T. Liu, Z. G. Zhang, Y. Yang, C. Zhong, B. Qiu, G. Li, W. Gao and D. Xie, *Adv. Mater.*, **30** 1706124, (2018).
174. B. Azzopardi, C. J. Emmott, A. Urbina, F. C. Krebs, J. Mutale and J. Nelson, *Energy Environ. Sci.*, **4** 3741–3753, (2011).
175. L. Zuo, S. Zhang, H. Li and H. Chen, *Adv. Mater.*, **27** 6983–6989, (2015).
176. S. B. Sapkota, A. Spies, B. Zimmermann, I. Dürr and U. Würfel, *Sol. Energy Mat. Sol. Cells.*, **130** 144–150, (2014).
177. L. Ye, X. Jiao, M. Zhou, S. Zhang, H. Yao, W. Zhao, A. Xia, H. Ade and J. Hou, *Adv. Mater.*, **27** 6046–6054, (2015).
178. Y. Chen, S. Zhang, Y. Wu and J. Hou, *Adv. Mater.*, **26** 2744–2749, (2014).
179. C. Duan, W. Cai, B. B. Y. Hsu, C. Zhong, K. Zhang, C. Liu, Z. Hu, F. Huang, G. C. Bazan, A. J. Heeger and Y. Cao, *Energy Environ. Sci.*, **6** 3022–3034, (2013).
180. B. Meng, Y. Fu, Z. Xie, J. Liu and L. Wang, *Polym. Chem.*, **6** 805–812, (2015).
181. Y. Deng, W. Li, L. Liu, H. Tian, Z. Xie, Y. Geng and F. Wang, *Energy Environ. Sci.*, **8** 585–591, (2015).
182. X. Dong, Y. Deng, H. Tian, Z. Xie, Y. Geng and F. Wang, *J. Mater. Chem. A*, **3** 19928–19935, (2015).
183. J. Griffin, A. J. Pearson, N. W. Scarratt, T. Wang, A. D. F. Dunbar, H. Yi, A. Iraqi, A. R. Buckley and D. G. Lidzey, *Org. Electronics*, **21** 216–222, (2015).
184. G. Susanna, L. Salamandra, C. Ciceroni, F. Mura, T. M. Brown, A. Reale, M. Rossi, A. Di Carlo and F. Brunetti, *Sol. Energy Mat. Sol. Cells.*, **134** 194–198, (2015).
185. W. Zhao, L. Ye, S. Zhang, M. Sun and J. Hou, *J. Mater. Chem. A*, **3** 12723–12729, (2015).
186. C. Sprau, F. Buss, M. Wagner, D. Landerer, M. Koppitz, A. Schulz, D. Bahro, W. Schabel, P. Scharfer and A. Colsmann, *Energy Environ. Sci.*, **8** 2744–2752, (2015).

11 Surface Passivation Materials for High-Efficiency Silicon Solar Cells

Shui-Yang Lien, Chia-Hsun Hsu, and Xiao-Ying Zhang
Xiamen University of Technology

Pao-Hsun Huang
Jimei University

CONTENTS

11.1 Introduction 414
    11.1.1 Prologue 414
    11.1.2 Device Structure 414
        11.1.2.1 Conventional Structure 414
        11.1.2.2 Passivated Emitter and Rear Contact Structure 416
    11.1.3 Atomic Layer Deposition 416
        11.1.3.1 Spatial Atomic Layer Deposition 417
        11.1.3.2 Thermal Atomic Layer Deposition 417
        11.1.3.3 Plasma-Enhanced Atomic Layer Deposition 418
11.2 Passivation Thin Films 419
    11.2.1 Mechanism and Advancement 420
    11.2.2 Aluminum Oxide Thin Films 422
    11.2.3 Hafnium Oxide Thin Films 423
11.3 Fabrication of Passivation Thin Films 424
    11.3.1 Effect of Surface Morphology 424
    11.3.2 Effect of Post-Annealing Ambient Temperature 426
        11.3.2.1 Aluminum Oxide 426
        11.3.2.2 Hafnium Oxide 430
11.4 Solar Cell Applications 434
    11.4.1 Parameters 434
    11.4.2 Device Structure 434
    11.4.3 Performance 434
11.5 Summary and Outlook 436
References 438

11.1 INTRODUCTION

11.1.1 PROLOGUE

Since the first silicon p–n junction solar cell was successfully fabricated in Bell Labs, silicon solar cells have always been the dominant product in photovoltaics (PV). In the early stage, the conversion efficiency of the silicon solar cells was mainly improved by classical semiconductor technologies like diffusion. Afterward, in order to improve the cell efficiency and reduce the fabrication cost, some techniques such as surface texturing, screen printing, passivated emitter and rear contact (PERC) solar cells, and firing process were introduced. These techniques played a key role and promoted the industrialization of Si solar cells. Nowadays, two types of silicon solar cells, namely, monocrystalline silicon (c-Si) and polycrystalline silicon solar cells, are the most important PV technology with a global market share of more than 80%. The efficiencies of the monocrystalline and polycrystalline silicon solar cells with conventional aluminum (Al) back surface field (BSF) are currently 18.5% and 19.8%, respectively [1–3]. Due to the demand for further enhancing the cell efficiency and reducing the fabrication cost, the major PV manufacturers such as SolarWorld, Trina Solar, Jinko Solar, Canadian Solar Inc., Hebei JA SOLAR, GCL Solar Energy, Wuxi Suntech, and Zhejiang Astronergy have been working in the research and development of various high-efficiency, low-cost c-Si solar cells such as heterojunction-intrinsic-thin film, integrated back contact, and Topcon c-Si solar cells [4–6].

Considering the development history of c-Si solar cells in recent decades, the key technical factors to improve the c-Si cell efficiency can be summarized as follows: photolithographically defined metallization, surface texturization, shallow junction, improvements in antireflection coatings, selective emitter, front-and-rear surface passivation, and elimination of optical shading losses. Among them, the surface passivation is the most crucial as it effectively suppresses the recombination of photogenerated carriers. This is also important for future thinner substrates. Extensive research and development have been made to reduce the surface recombination of c-Si, and some valuable device structures have been proposed. In the past few years, the recorded efficiency of c-Si solar cells is refreshed over and over again. In particular, PERC solar cells are the most promising for industrialization and commercialization owing to their compatibilities with the industrial p-type c-Si solar cell processes. It is estimated that PERC solar cells will gradually replace the traditional Al-BSF c-Si solar cells in the next few years. The evolution of the commercial conversion efficiencies of PERC silicon solar cells using each technology is shown in Figure 11.1. At present, LONGi's PERC solar cells reach the conversion efficiency of 24.02%.

11.1.2 DEVICE STRUCTURE

11.1.2.1 Conventional Structure

The conventional c-Si solar cells are fabricated using the full Al-BSF and screen-printing technologies. Figure 11.2 shows the structure and fabrication flow of the conventional c-Si solar cell, which is mostly fabricated in the 2010s [7].

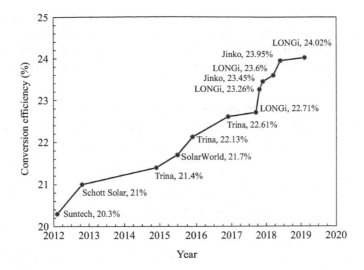

FIGURE 11.1 The evolution of the conversion efficiencies of PERC c-Si solar cells with the years (2012–2019).

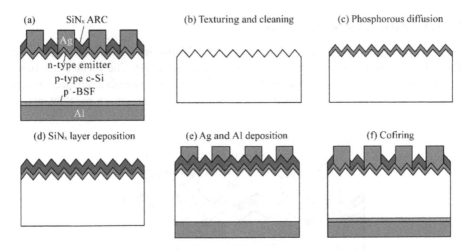

FIGURE 11.2 Structure and fabrication flow for a conventional Al-BSF c-Si solar cell. (a) Schematic diagram of the cross section of the solar cell. (b–f) Five core steps that are used for the fabrication of the basic Al-BSF c-Si cells.

The absorber material is based on p-type c-Si wafer. In general, the processes in the industrial screen-printing technology include five core steps, namely, texturing and surface cleaning, phosphorus dopant diffusion to form p–n junction, coating antireflective layer on the front side, printing Al paste on the rear to form BSF and silver (Ag) pastes on the front, and co-firing the printed pastes to form ohmic contacts to rear base and front emitter. A detailed description of the Al-BSF and screen-printing technologies can be found in several studies [8,9].

The surface of the front emitter is passivated by silicon nitride (SiN$_x$), which also acts as an antireflective layer. During the co-firing process, Al atoms are doped into c-Si to form a BSF that prevents minority carriers from recombination at the rear surface. No additional dielectric thin films are used for passivating the rear surface of c-Si cells.

11.1.2.2 Passivated Emitter and Rear Contact Structure

For a PERC solar cell as shown in Figure 11.3, the front and rear surfaces of the c-Si are passivated by dielectrics [10]. The rear dielectric layer is partly opened by laser, and then metal can be contacted to the rear surface of c-Si. Compared to a conventional Al-BSF c-Si solar cell, PERC solar cell has a higher conversion efficiency mainly due to the additional passivating dielectric layer on the rear side that avoids the recombination of minority carriers at the rear surface. In addition, the rear dielectric layer can reflect the long-wavelength light from the rear surface back to the device to increase light absorption.

A square image with zigzag on top presents the structure of PERC solar cells. The structure from top to bottom includes Ag front contact, silicon nitride anti-reflection coating (ARC) layer, n-type c-Si emitter, p-type c-Si base, dielectric rear passivation layer, laser openings, and Al rear contact.

11.1.3 ATOMIC LAYER DEPOSITION

The atomic layer deposition (ALD) consists of self-limiting surface reactions, and the substrate surface is exposed to gas-phase precursors [11]. The surface reactions are self-limiting as they automatically stop when all the available surface groups have reacted. Every surface reaction occurs between a gas-phase reactant and a

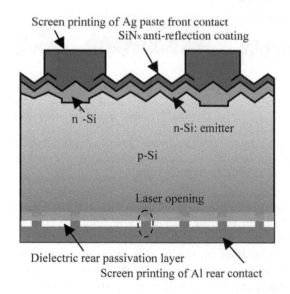

FIGURE 11.3   Diagram of the structure of a PERC solar cell.

surface functional group [11,12]. A typical ALD process consists of several steps and utilizes two precursors: precursor A reacts with the substrate and the remaining precursor and by-products are pumped away, and precursor B reacts with the new surface group; at the end, the initial surface groups are restored [12].

11.1.3.1 Spatial Atomic Layer Deposition

Spatial ALD is also referred to as "zone-separated ALD," as shown in Figure 11.4. Gases are supplied via the pipes to the input side of the injector under choked flow condition at atmospheric pressure. The two precursor injection zones are separated by a relatively large physical distance, and the zones of an inert gas act as a diffusion barrier. The substrate is moving back and forth beneath the injector so that the substrate is exposed to the precursors one by one. Between and around the reaction zones, inert gas curtains are again used to separate the precursor flows. When the gas curtain is used, the distance between the reactor and the substrate can be minimized. The small gap heights combined with the resulting high flow rates in the gaps result in great diffusion barriers with relatively narrow widths. An additional advantage of using gas curtains is that they completely seal off the reaction zones, so that the reactor is independent of the environment, enabling operation under atmospheric pressure conditions. Some research groups or companies—Eastman Kodak (Rochester, USA) [13], ASM International (Almere, The Netherlands) [14], and TNO (Eindhoven, The Netherlands) [15]—have made efforts on gas bearing-based spatial ALD concepts.

11.1.3.2 Thermal Atomic Layer Deposition

The chemical mechanism in ALD involves two gas-phase chemical species, namely, a metal-organic precursor and a co-reactant like an oxygen source [16,17]. The precursor and co-reactant are sequentially introduced into a heated surface, resulting in two time-separated half-reaction steps. An inert gas purge process is required between the reactant exposure steps to ensure the time-separated half-reactions. Figure 11.5 shows a typical thermal ALD cycle for the deposition of aluminum oxide (Al_2O_3). The metal

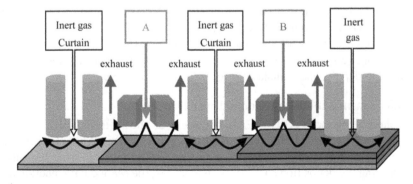

FIGURE 11.4 Schematic diagram of spatial ALD reactor concept, where the precursor reacting zones are separated by inert gas curtains. The substrate moves horizontally underneath the injector, so that the reactions of the two precursors with the substrate surface will take place sequentially to form an ALD monolayer.

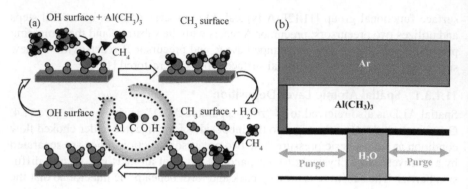

FIGURE 11.5 (a) Diagram of one cycle of thermal ALD of Al_2O_3 using sequential satura-tion exposures of $Al(CH_3)_3$ and H_2O. After the full cycle, the starting hydroxylated surface is reproduced, allowing the cycle to be repeated to build up a coating with near monolayer precision. (b) Schematic representation of the typical thermal ALD cycles for Al_2O_3.

precursor source and oxidant are usually trimethylaluminum (TMA), $Al(CH_3)_3$, and water (H_2O), respectively. As shown in the top-left of the figure, the first precursor exposure contributes to the first ALD half-reaction. In this step, the precursor chemi-cally reacts and binds to the surface without its complete decomposition. The remain-ing precursors and by-products are pumped using an inert purge gas. For the second ALD half-reaction, the co-reactant is transferred to the reactor and reacts on the growth surface. The vapor products are purged, and one "ALD cycle" is completed. After the complete cycle, the starting hydroxylated surface is restored, thus allowing the cycle to be repeated. It is noted that enough precursor must be delivered to reach full saturation; otherwise, the deposition will be non-ideal and nonuniform.

One advantage of the self-saturated half-reactions is that with long exposure times, the precursor and co-reactant can find all available reaction sites with high aspect ratio or tortuous substrates, resulting in very conformal growth without excess growth on the top of the sample. The self-limiting nature is achieved by care-fully matching the precursor and co-reactant, and by optimizing the deposition tem-perature and purging steps. To achieve ALD growth, the temperature should be low enough to avoid the decomposition of precursors during surface adsorption, but must be high enough to thermally activate the chemical reaction. There is thus a tem-perature range, commonly referred to as "ALD window," where the temperature is optimized to deposit one monolayer of growth per ALD cycle.

11.1.3.3 Plasma-Enhanced Atomic Layer Deposition

In the plasma-enhanced ALD (PEALD) process, as shown in Figure 11.6, the first half-reaction is identical to the thermal ALD process, where TMA molecules react with surface hydroxyl (OH) groups to form the CH_3 groups covering the surface. After purging the deposition chamber with inert gas, the second half-reaction of the ALD cycle starts. An oxygen plasma is ignited above the substrate, and the gener-ated oxygen radicals react with the $Al-CH_3$ groups on the surface [18–20]. Usually, an inductively coupled plasma source is used, and the oxygen plasma is not in direct contact with the silicon wafer during Al_2O_3 deposition. This type of remote plasma

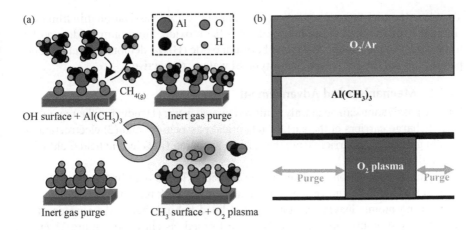

FIGURE 11.6 (a) Schematic of one cycle of a PEALD process. (b) Schematic representation of typical PEALD cycles for Al$_2$O$_3$.

deposition technique leads to almost no plasma damage on the growth surface and thus is especially suited for the surface passivation of c-Si wafers. Table 11.1 compares the performance of spatial, thermal, and PEALD systems.

11.2 Passivation Thin Films

Reduction in surface recombination rate of the c-Si cells is essential for improving the efficiency of c-Si solar cells. Surface recombination is a special case of Shockley–Read–Hall (SRH) recombination, in which the localized states are presented on the surface. Unlike bulk SRH centers occupying a single energy level, these localized states form a set of states distributed across the band gap. The surface recombination is evaluated in terms of its rate. In a simple case, we can consider a sample with a thickness W, constant bulk lifetime τ_b, and a surface recombination velocity S. The effective lifetime (τ_{eff}) is given by:

$$\tau_{eff} = \frac{1}{\tau_b} + \frac{W}{2S} \quad (11.1)$$

TABLE 11.1

Comparison between Spatial, Plasma, and Thermal ALD Systems

	Spatial ALD	Plasma ALD	Thermal ALD
Cycle time	<5 seconds	~1 minute	~1 minute
Deposition rate	1.6 Å/cycle	~1.05 Å/cycle	~0.8 Å/cycle
Film quality	good	excellent	good
Precursor	TMA, H$_2$O	TMA, O$_2$	TMA, H$_2$O
Temperature	125°C–200°C	25°C–400°C	125°C–400°C
Clean gas	N$_2$	N$_2$ or Ar	N$_2$ or Ar
Equipment cost	low	very expensive	expensive

To reduce the surface recombination, the commonly used passivation thin film materials are hydrogenated amorphous silicon, silicon oxide, aluminum oxide, and hafnium oxide. In particular, Al_2O_3 and hafnium oxide (HfO_2) demonstrate an excellent passivation quality for p-type and n-type Si wafers, respectively.

11.2.1 Mechanism and Advancement

Surface passivation can be mainly achieved by two ways: (1) reducing the recombination of charge carriers at the surface (chemical passivation) and (2) electrostatically repelling the charge carriers from the surface by an internal electric field (field-effect passivation).

Chemical passivation is related to the reduction in the dangling bonds (DBs) on the c-Si surface. It is known that silicon atom in the bulk region bonds with its four neighboring atoms, leaving no unsaturated bond behind, but at the surface of the silicon crystal atoms, missing and DBs are formed, as shown in Figure 11.7a. An electrically neutral DB having an unpaired electron creates a deep defect state in the band gap. In the SRH model, a recombination of electrons and holes occurs through deep-level defects in the semiconductor band gap. An electron is trapped at a deep defect state, and then a hole is trapped at the same deep-level defect. The sequence can obviously be reversed, with a hole being captured first at a deep-level defect, followed by an electron. This process is illustrated in Figure 11.7b. The charge carriers are hardly released once they are trapped by these deep defect states.

The density of these deep defect states is approximately $10^{14}cm^{-2}eV^{-1}$, resulting in severe surface recombination. The surface recombination rate is expressed by the following equation [21]:

$$U_s = \int_{E_v}^{E_c} \frac{V_{th}D_{it}(E_t)dE_t}{\left[n_s + n_1(E_t)\sigma_p^{-1}(E_t) + \left[p_s + p_1(E_t)\right]\right]\sigma_n^{-1}(E_t)} \tag{11.2}$$

where U_s is the surface recombination rate per unit area, E_v is the valence band energy level, E_c is the conduction band level, V_{th} is the thermal voltage, D_{it} is the interface defect density, E_t is the defect energy level, n_s and n_p are the electron and hole density per unit volume at the surface, and n_1 and p_1 are the SRH electron and

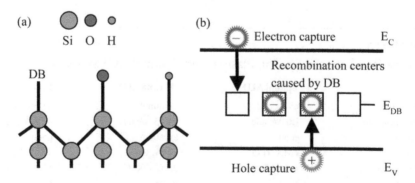

FIGURE 11.7 Diagrams of (a) silicon surface DBs and passivation of the DBs by oxygen and hydrogen atoms. (b) Defect level introduced by silicon surface DBs.

hole density, respectively. From Eq. (11.2), it is observed that the surface recombination rate is directly proportional to D_{it}, which is related to the presence of the DBs and impurities present on the surface. One way to reduce D_{it} is the chemical saturation of DBs using passivation layer thin films such as Al_2O_3 and HfO_2, as shown in Figures 11.8 and 11.9, respectively. Elements such hydrogen and oxygen atoms in the passivation layer bind with the DBs so as to make these defect centers electrically inactive. The advantages of the chemical passivation are that it is simple, low cost to apply, and able to be carried out at room temperature.

Surface recombination rate can also be reduced by the field-effect passivation, which is achieved in two ways. The first is the passivation through the diffused surface. Through the additional doping atoms in the semiconductor, an electric field is formed near the surface, which repeals the minority charge carriers. The second way is the passivation through fixed charges in the dielectric layer. Some passivation materials have fixed positive or negative charges, which can influence the movement of charge carriers. When a dielectric material containing negative charges like Al_2O_3 is applied to p-type silicon, the high negative charge density in the passivation layer causes band bending downward of p-type c-Si, which in turn results in the repulsion of minority charge carriers and the reduction in the charge carrier recombination

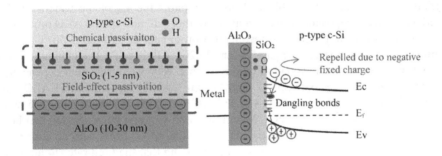

FIGURE 11.8 Mechanism and band diagram of the chemical and field-effect passivation of Al_2O_3/c-Si [22]. (Reprinted with permission from Elsevier.)

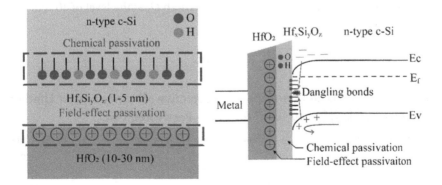

FIGURE 11.9 Mechanism and band diagram of the chemical and field-effect passivation of HfO_2/c-Si [22]. (Reprinted with permission from Elsevier.)

rate at the interface. In the same way, when a dielectric material containing inherent positive charges like HfO_2 is applied to n-type silicon, the high positive charge density at the interface repels the minority charge carriers.

11.2.2 Aluminum Oxide Thin Films

Al_2O_3 can be deposited by ALD using TMA precursor as aluminum source. There are two possible methods of ALD: thermal and PEALD. The former employs water or ozone as oxidant, whereas the latter employs plasma. The reactivity of PEALD can provide a better film quality with lower impurity levels [23]. Al_2O_3 deposition can also be achieved by other techniques, such as sputtering and plasma-enhanced chemical vapor deposition (PECVD). The advantages of ALD over PECVD are the excellent uniformity and conformity, which are achievable on large and non-flat substrates. ALD also allows the deposition of multilayer structures [23]. Table 11.2 compares τ_{eff} measured on 1.3 Ω-cm p-type floating zone (FZ)-Si wafers passivated by Al_2O_3 films deposited using spatial ALD, PECVD, and radio frequency (RF) magnetron sputtering [24]. Both spatial ALD and PECVD have the maximal surface recombination rate (S_{max}) of smaller than 10 cm s^{-1}, which clearly outperforms the sputtered Al_2O_3. However, the sputtered Al_2O_3 passivation layer shows S_{max} of 55.7 cm s^{-1}, which is still acceptable for the next generation of industrial and high-efficiency c-Si solar cells. Table 11.2 also compares the τ_{eff} of the wafers passivated using Al_2O_3 deposited by plasma-assisted, thermal, and spatial ALD processes. All the films are subjected to a post-deposition annealing at $(400 \pm 50)°C$ for 15 minutes to activate the surface passivation [25,26]. All three ALD techniques result in outstanding surface passivation quality with a very weak injection dependence over the injection range between 10^{13} and 10^{15} cm^{-3}. Al_2O_3 deposited by PEALD provides an effective lifetime of 4.8 m s at $\Delta n = 10^{15}$ cm^{-3}, which is better than that with "annealed" thermally grown SiO_2. Most importantly, it can be concluded that both traditional thermal ALD and spatial ALD provide Al_2O_3 films with an extremely high level of surface passivation. It is worth noticing that the high-rate (14 nm/min) spatial ALD can provide the excellent level of surface passivation, which is the same as the slow (<2 nm/min) thermal ALD [27]. Table 11.3 summarizes the precursors used and the fixed charge density (Q_f) of the Al_2O_3 films deposited by various techniques. It can be seen that all the films exhibit a high Q_f at the level of 10^{12} cm^{-2}.

TABLE 11.2

Deposition Methods and Corresponding Effective Minority Carrier Lifetime at the Injection Level of 10^{15} cm^{-3} for Al_2O_3-Passivated c-Si

Deposition Technique	Minority Carrier Lifetime (μs)	S_{max} (cm s^{-1})
Spatial or thermal ALD	2,000	7.25
PECVD	1,450	10
RF sputter	260	55.7
PEALD	4,800	3.02

TABLE 11.3

Deposition Methods of Al_2O_3 Films and Corresponding Fixed Negative Charge

Deposition Technique	Precursor	Q_f (cm^{-2})
Spatial or thermal ALD	$Al(CH_3)_3 + H_2O$	-2×10^{12}
PECVD	$Al(CH_3)_3 + N_2O + Ar$	-2.1×10^{12}
RF sputter	Al target $+ O_2$	-3×10^{12}
PEALD	$Al(CH_3)_3 + O_2$	-1×10^{12}
Sol–gel	$Al(C_4H_9O)_3$	-1.6×10^{12}
atmospheric pressure CVD	$C_9H_{21}AlO_3$	-3×10^{12}

11.2.3 Hafnium Oxide Thin Films

Hafnium oxide (HfO_2) thin films have attracted significant attention owing to their excellent properties, such as high dielectric constant (~20), high density (9.68 g cm^{-3}), large band gap (5.6~5.8 eV), and good thermodynamic stability. They have various applications. In the advanced semiconductor industry, HfO_2 thin films are used to replace SiO_2 as the gate dielectric in complementary metal-oxide-semiconductor devices (COMS) because of their high dielectric constant [28,29]. HfO_2 thin films are also used as protective coatings [30] due to their hardness and thermal stability. Additionally, the high refractive index of HfO_2 thin films makes it a potential candidate for antireflection coatings [31] and interference filters [32]. In recent years, the surface passivation of HfO_2 thin films, particularly on c-Si, has also been studied. The minority carrier lifetime of the HfO_2-passivated c-Si wafers is summarized in Table 11.4. In 2012, Jun Wang et al. [33] presented the surface passivation properties of a Si surface using a thin HfO_2 layer grown by ALD without further annealing.

TABLE 11.4

Deposition Methods and Corresponding Effective Minority Carrier Lifetime at the Injection Level of 10^{15} cm^{-3} for HfO_2-Passivated c-Si

Research Group	Substrate	Deposition Technique	Minority Carrier Lifetime (μs)
Wang et al. [33]	FZ n-Si 3–5 Ω cm	Thermal ALD	300
Lin et al. [34]	FZ p-Si 2.1 Ω cm	Thermal ALD	256 (p-Si)
	FZ n-Si 3.3 Ω cm		599 (n-Si)
Geng et al. [35]	p-Si 6 Ω cm	Thermal ALD	56.89
Gope et al. [37]	FZ n-Si 5 Ω cm	Thermal ALD	800
Cui et al. [38]	FZ p-Si 1 Ω cm	Thermal ALD	1,000 (p-Si)
	FZ n-Si 1 Ω cm		3,100 (n-Si)
Cheng et al. [39]	CZ n-Si 1.0 Ω cm	Thermal ALD	2,500
Zhang et al. [40]	CZ p-Si 30 Ω cm	PEALD	65.4
Gougam et al. [43]	FZ n-Si	Thermal ALD	5,000

They pointed out that as-grown HfO_2 is superior to as-grown Al_2O_3 for the passivation of Si surface. During the same year, F. Lin et al. [34] showed an effective surface recombination velocity of 55 cm s^{-1} and 24 cm s^{-1} on p-type and n-type c-Si due to high negative charges of HfO_2 thin films and low D_{it} of the Si/HfO_2 interface, respectively. In 2014, Huijuan Geng et al. [35] reported the advanced passivation using simple materials (Al_2O_3, HfO_2) and their compounds H(Hf)A(Al)O deposited by ALD. $I–V$ characteristics of Si solar cells fabricated with HfO_2 thin films indicate that the performance of cells is significantly improved. Meanwhile, Daniel K. Simon et al. [36] investigated a symmetrical passivation layer for n- and p-type Si using 20 nm Al_2O_3 combined with a thin HfO_2–SiO_2 interface. The effective surface recombination velocity is below 1 cm s^{-1} for both types of substrate doping. In 2015, Jhuma Gope et al. [37] studied HfO_2 thin films of different thicknesses that are deposited on n-type c-Si wafers at 300°C using thermal ALD process. They found that as-deposited HfO_2 thin film (~8 nm) shows better passivation with <100 cm s^{-1} S_{max} than the thicker films. The value of S_{max} (i.e., <100 cm s^{-1}) is reduced to ~20 cm/s when the best passivation sample is subjected to annealing at 400°C for 10 minutes. The capacitance–voltage ($C–V$) test shows that the D_{it} increases with an increase in film thickness, whereas its value decreases after annealing. In 2017, Jie Cui et al. [38] presented high level of surface passivation using 15 nm HfO_2 thin film deposited by ALD, which shows the value of S_{max} as low as 3.3 and 9.9 cm s^{-1} on n-type and p-type 1 Ω cm c-Si, respectively. Recently, Xuemei Cheng et al. [39] investigated different deposition parameters on the Si passivation by HfO_2 thin films. A S_{max} of 7.7 cm s^{-1} was obtained on float-zone n-type wafers. In addition, a significant improvement of Si passivation is observed after 100 hours light soaking. Xiao-Ying Zhang et al. [40] further reported the surface passivation of Si using HfO_2 thin films deposited by PEALD. The in situ oxygen remote plasma pretreatment and post-annealing in N_2 at 500°C were effective in reducing the trap density at Si/HfO_2 interface and improving the lifetime of the c-Si. In 2018, Evan Oudot et al. [41] studied hydrogen passivation of Si/SiO_2 interface using HfO_2 thin films deposited by ALD, which provides an excellent chemical passivation, adds negative charges located at interface, and thus associates with Si-O-Hf bonds. In the meantime, Jagannath Panigrahi et al. [42] enhanced the field-effect passivation of c-Si surface by introducing trap centers in a bilayer dielectric system consisting of HfO_2 and Al_2O_3. The S_{max} (~10 cm s^{-1}) is achieved at the intermediate bulk injection levels with thermal ALD HfO_2/Al_2O_3 bilayer system on n-type Si. In 2019, Adel B. Gougam et al. [43] reported the S_{max} value of 1.2 cms^{-1} using a 15-nm HfO_2 thin film grown on OH-terminated Si surface and annealing in air.

11.3 FABRICATION OF PASSIVATION THIN FILMS

11.3.1 EFFECT OF SURFACE MORPHOLOGY

The c-Si rear surface morphology will affect the deposition of the passivation films and rear reflectance. One way to achieve one-side texturing is to polish the rear of c-Si after the texturing process. The effects of the various c-Si rear surface morphologies are investigated by a standard double-sided texture process and further polishing processes for 10, 20, and 30 minutes for the c-Si rear surface, as shown in Figure 11.10 [44]. The typical textured surface morphology containing uniformly sized pyramids of around 5–8 µm (width of the base) is shown in Figure 11.10a. At the polishing time of 10 minutes, large and small pyramids start to coexist, as shown in Figure 11.10b. By increasing the polishing duration to 20 minutes, large pyramids disappear, leaving uniformly smaller and thinner ones, as shown in Figure 11.10c. Increasing the polishing duration to 30 minutes basically produces a mirror-like surface, although some pyramids are still observed, as shown in Figure 11.10d. These four samples A, B, C, and D show the four stages constituting a texturing cycle.

The injection level-dependent effective minority carrier lifetime and Auger lifetime for the four samples are shown in Figure 11.11a. The effective minority carrier lifetime is measured by Sinton WCT120, and the Auger carrier lifetime is calculated by the model proposed by Richter et al. [45]. The effective minority carrier lifetime values for samples A, B, C, and D are 174.4, 163.7, 200.1, and 225.3 µs at the injection level of 5×10^{15} cm^{-3}, respectively. The lifetime of the samples first decreases and then increases to its saturation value when the wafer is polished at increasing etching time. The reduction in lifetime for sample B can be associated with the nonuniform pyramidal structure consisting of many peaks and valleys of pyramids, which may cause film cracks. This damage traps charge carriers and reduces the lifetime. Another reason is a less efficient cleaning of the surface due to the presence of narrow valleys between pyramids. These factors can also account for the

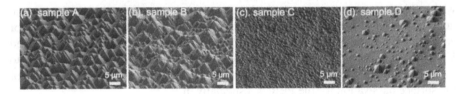

FIGURE 11.10 SEM images of rear side of the silicon wafers with further polishing duration for (a) 0, (b) 10, (c) 20, and (d) 30 minutes [44]. (Reprinted with permission from MDPI.)

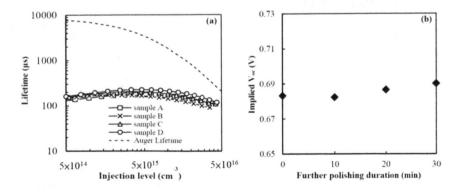

FIGURE 11.11 (a) Effective lifetime as a function of injection level; (b) implied V_{oc} for the Al$_2$O$_3$ films with different rear-side surface morphologies [44]. (Reprinted with permission from MDPI.)

better performance of samples C and D. The polishing not only leads to a flat surface but also removes texturing-induced defects, thus reducing the surface recombination rate. In addition, as both of the samples A and B have larger rear surface areas than the samples C and D, more defects are present at the Al_2O_3/c-Si interface, which act as the recombination centers. Sample D exhibits a slight increase in lifetime at an injection level of 10^{15} cm^{-3}, which is more apparent than that observed for the other samples due to the improved field-effect passivation on the polished surface. At a low injection level, very small amount of minority carriers are generated to charge the inversion layer, leading to a shorter minority carrier lifetime due to the asymmetry in the capture cross sections of electrons and holes at the Al_2O_3/Si interface [46]. At higher injection levels, the lifetime starts to decrease with the increasing Auger recombination. The implied V_{oc} values for the four samples, as shown in Figure 11.11b, are used for evaluating the maximum V_{oc} before the metallization process. The implied V_{oc} is given by the following equation [47]:

$$\tau_{eff}(\Delta n) = \frac{\Delta n(t)}{G - \Delta n(t)/\Delta t} \tag{11.3}$$

$$\text{Implied } V_{oc} = \frac{kt}{q} \ln\left(\frac{\Delta n(N_A + \Delta n)}{n_i^2}\right) \tag{11.4}$$

where G is the generation rate, Δn is the excess carrier concentration, Δt is the time taken in seconds, k is the Boltzmann constant, T is the temperature, N_A is the acceptor concentration of the wafer, and n_i is the intrinsic carrier concentration. The implied V_{oc} shows the same trend with the effective minority carrier lifetime with respect to the polishing time, and this can be helpful for the prediction of the performance of the surface passivation for the PERCs, especially in V_{oc}.

11.3.2 EFFECT OF POST-ANNEALING AMBIENT TEMPERATURE

11.3.2.1 Aluminum Oxide

Post-annealing is necessary for Al_2O_3 thin films to activate the passivation of c-Si, as the annealing changes the properties at the Al_2O_3/c-Si interface. The AlO_4 structure in Al_2O_3 is negatively charged, while AlO_6 is positively charged. Their ratio at the interface can be thus related to the field-effect passivation. Figure 11.12a shows the peak intensity ratio of AlO_4/AlO_6, which is evaluated using infrared spectral analysis, as a function of annealing temperature for different annealing gases [22,48]. With atmosphere annealing, the ratio increases from 2.45 to 3.54 with increasing temperature from 300°C to 600°C, and then it decreases to 2.9 as the temperature increases to 750°C. Similar trends can be seen in other annealing gases, but the ratios are lower than those of ATM annealing. The interfacial fixed charge density, Q_f, for the samples annealed in different gases and temperatures can be calculated using C–V measurement, and the result is shown in Figure 11.12b. For ATM annealing, Q_f increases from -4.19×10^{11} to -3.23×10^{12} cm^{-3} when the annealing temperature increases from 300°C to 600°C, and then it reduces to -1.34×10^{12} cm^{-3} at

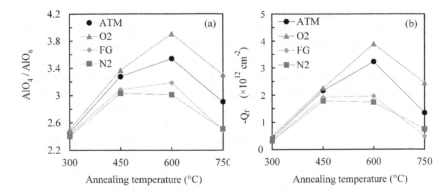

FIGURE 11.12 (a) AlO$_4$/AlO$_6$ ratio and (b) Q_f for the Al$_2$O$_3$ films as a function of post-annealing temperatures [22]. (Reprinted with permission from Elsevier.)

750°C. The forming gas (FG) and N$_2$ annealing give smaller Q_f values. The trend of Q_f confirms the prediction made from the IR result that ATM annealing leads to a higher fixed negative charge than the N$_2$ and FG annealing.

Figure 11.13a shows the Si-O-Si peak intensity as a function of temperature for different gases. The peak intensity of ATM annealing increases with the temperature, whereas the peak intensities of FG and N$_2$ annealing are smaller and only show slightly changes with the temperature. It is reported that the oxygen from the annealing gas can interchange with the oxygen atoms of Al$_2$O$_3$, producing mobile oxygen atoms that go deeper and repeat the interchange process [49]. Eventually, the oxygen propagation front can reach the Al$_2$O$_3$/Si interface. Therefore, during ATM annealing, the oxygen can reach the Al$_2$O$_3$/Si interface and enhance the thickness of the interfacial SiO$_x$. The other two annealing gases do not have a significant change in the interfacial oxide due to the absence of the oxygen. Furthermore, as the ATM annealing contains oxygen, the interfacial SiO$_x$ layer (consisting of mainly [SiO$_4$] tetrahedra) becomes thicker compared to FG and N$_2$ gas annealing processes.

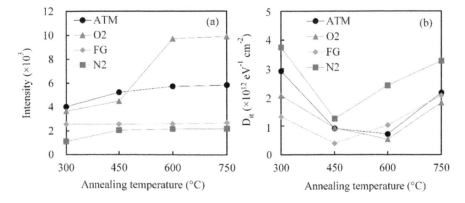

FIGURE 11.13 (a) Si-O-Si intensity and (b) D_{it} of the annealed Al$_2$O$_3$/Si samples as a function of annealing temperature [22]. (Reprinted with permission from Elsevier.)

More Al can be substituted in the Si site to form AlO_4 at the Al_2O_3/SiO_2 interface region [50]. Therefore, AlO_4/AlO_6 ratio is higher for the sample annealed in ATM process. This result predicts that ATM annealing can produce the highest amount of fixed negative oxide charges, as the fixed negative charge has been considered to originate from the AlO_4 at the interface [27,51]. D_{it} values can also be calculated from C–V measurement by Terman method [52], as shown in Figure 11.13b. The lowest D_{it} for ATM annealing is $7.21 \times 10^{11} eV^{-1} cm^{-2}$ at 600°C, whereas for FG and N_2 annealing, the lowest D_{it} values are, respectively, 3.98×10^{11} and $1.2 \times 10^{12} eV^{-1} cm^{-2}$ at 450°C. The oxygen in ATM and hydrogen in FG can bind with the DBs, so the D_{it} values of these two annealing gases are significantly lower than the value of N_2 annealing gas. Passivation by ATM annealing mostly depends on interfacial SiO_x quality. A higher temperature improves the density of this layer. In contrast, hydrogen passivation requires temperatures lower than 550°C to avoid dehydrogenation at interface [53]. The optimal temperature of ATM annealing is therefore higher than that of FG annealing. Overall, with decreasing D_{it}, hydrogen passivation provided by FG annealing is more effective than SiO_x passivation provided by ATM annealing.

The carrier lifetime values at the injection level of $1\times 10^{15} cm^{-3}$ are extracted for the purpose of comparison, as shown in Figure 11.14a. The corresponding Smax values are shown in Figure 11.14b. It can be seen that ATM annealing has a maximum lifetime value of 737.1 μs at 600°C, while the maximum lifetime values of FG and N_2 are observed at 450°C. Overall, among the aforementioned annealing gases, FG gives the best lifetime value of 933.8 μs. The highest lifetime value of the samples annealed in N_2 only reaches 497.2 μs. This indicates that annealing in gases containing passivating sources such as oxygen or hydrogen is strongly required as compared to annealing in inert gas.

To confirm the interfacial oxide thickness for the different annealing gases, the transmission electron microscopic (TEM) images of the Al_2O_3/Si samples are shown in Figure 11.15. An interfacial SiO_x layer between Al_2O_3 and c-Si can clearly be seen, and its thickness is indicated in parentheses. It can be seen that the interfacial oxide thickness insignificantly varies from 1.7 to 2 nm for FG and N_2 annealing.

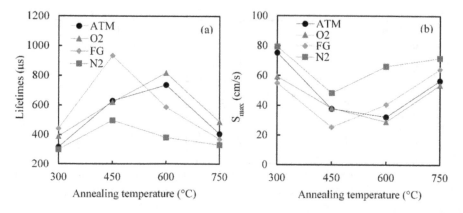

FIGURE 11.14   (a) Minority carrier lifetime and (b) S_{max} of the annealed Al_2O_3/Si samples as a function of annealing temperature [22]. (Reprinted with permission from Elsevier.)

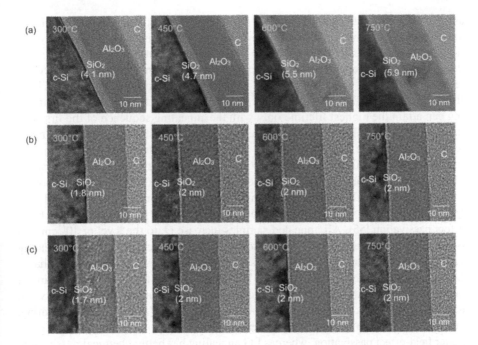

FIGURE 11.15 TEM images of the Al₂O₃/Si samples annealed in (a) ATM, (b) FG, and (c) N₂ at different temperatures [22]. (Reprinted with permission from Elsevier.)

However, the interfacial oxide thickness of the ATM-annealed samples monotonically increases from 4.1 to 5.9 nm when the temperature increases from 300°C to 750°C. As the silicon wafer oxidation in the post-annealing process requires oxygen atoms that react with silicon substrate surface to form Si-O bonds, this increased thickness confirms that oxygen in ATM can pass through the Al_2O_3 layer and reach the wafer surface to form SiO_x during the annealing process.

Table 11.5 shows the injection-level-dependent minority carrier lifetime values of the Al_2O_3/Si/Al_2O_3 samples without and with different annealing processes. Before annealing, the minority carrier lifetime value is low as below 100 μs over the whole injection-level range. The lifetime value greatly improves after the annealing process

TABLE 11.5

Q_f, D_{it}, Minority Carrier Lifetime, and Surface Recombination Rate for the Al_2O_3 Films without Annealing and with O_2 (600°C), FG (450°C), and Two-Step (O_{2+}FG) Annealing

	Q_f (cm⁻²)	D_{it} (eV⁻¹cm⁻²)	Lifetime (μs)	S_{max} (cm s⁻¹)
As-deposited	-3.18×10^{11}	3.59×10^{13}	97.1	224.6
O_2 (600°C)	-3.89×10^{12}	5.43×10^{11}	818.4	29
FG (450°C)	-1.9×10^{12}	3.98×10^{11}	933.76	25.4
Two-step	-2.35×10^{12}	3.05×10^{11}	1,097.51	21.6

as a consequence of chemical passivation and field-effect passivation brought by annealed Al_2O_3. However, the lifetime values are different in three annealing conditions, in which oxygen annealing has the lowest curve, FG annealing has the intermediate, and the two-step annealing has the highest. The lifetime values at the injection level of $3 \times 10^{15} cm^{-3}$ are found. The O_2-, FG-, and two-step-annealed samples have lifetime values of 818, 934, and 1,098 µs, respectively. Note that the two-step annealing can attain the highest lifetime only with the annealing sequence of the first step in O_2 and the second step in FG. The reverse sequence results in a lifetime similar to that of the sample with O_2 annealing alone. This might be because if FG annealing was performed first, the following O_2 annealing might cause dehydrogenation. Niwano et al. reported that for a wafer terminated by Si-H or Si-H_2 bonds, exposure to oxygen results in the replacement of hydrogen bonds with the Si-O bond [54]. The Q_f is $-3.2 \times 10^{-11} cm^{-2}$ for the as-deposited sample. Q_f at this level leads to weak field-effect passivation [55]. All annealed samples increase Q_f to the level of $10^{12} cm^{-2}$. It is seen that O_2 annealing gives the highest Q_f of $3.9 \times 10^{12} cm^{-2}$, two-step annealing gives the intermediate Q_f, and FG annealing gives the lowest Q_f. On the other hand, D_{it} value estimated by Terman method [56] is also shown to evaluate the chemical passivation. The as-deposited sample has a D_{it} of more than $10^{13} eV^{-1} cm^{-2}$. D_{it} value reduces to $5.4 \times 10^{11} eV^{-1} cm^{-2}$ for O_2 annealing, $3.7 \times 10^{11} eV^{-1} cm^{-2}$ for FG annealing, and $3.1 \times 10^{11} eV^{-1} cm^{-2}$ for two-step annealing. Thus, it is found that O_2 annealing has better field-effect passivation, whereas FG annealing has better chemical passivation. The former might be linked to the interfacial SiO_x growth. Unlike FG annealing, which is performed at a relatively low temperature and with the lack of oxygen, O_2 annealing is expected to have an improved SiO_x interfacial layer growth. This can increase the possibility of Al substitution for Si at the Al_2O_3/SiO_2 interface, which is regarded to be one of the possible origins of negative fixed charges [57]. Considering the two-step annealing, the intermediate Q_f is considered a combination of O_2 and FG annealing. However, its D_{it} value is lower than that of FG annealing. This is explained by the higher quality of interfacial oxide layer obtained during the first-step O_2 annealing. Some studies also reported that a denser SiO_x results in a better passivation [58]. The lower D_{it} in the two-step annealing sample can also be attributed to enhancing the hydrogenation process of silicon surface induced by hydrogen in Al_2O_3 film.

11.3.2.2 Hafnium Oxide

Post-annealing is an essential process affecting the properties and applications of HfO_2 thin films grown by ALD. Annealing temperature, annealing time, and annealing atmosphere are the important parameters for determining the properties of the HfO_2 thin films. Annealing temperature affects both chemical passivation and field-effect passivation processes. Figure 11.16 shows τ_{eff} and S_{max} for the samples at the injection level of $3 \times 10^{14} cm^{-3}$. At the beginning, as the temperature increases, the τ_{eff} also increases. When the annealing temperature further increases, the τ_{eff} decreases. There is an optimal annealing temperature for HfO_2 thin films. When the temperature increases from 400°C to 500°C, the τ_{eff} of the annealed HfO_2 thin film significantly increases. When the temperature is higher than 500°C, the τ_{eff} decreases, which might be due to the increasing microcrystalline fraction and grain boundaries in HfO_2 thin films [59]. The HfO_2 thin film annealed at 500°C has the

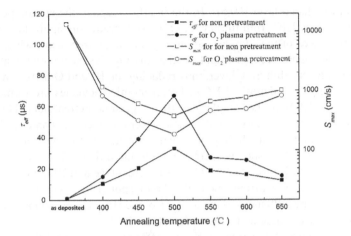

FIGURE 11.16 τ_{eff} and S_{max} of the HfO$_2$ thin films at the injection level of $3 \times 1,014\,\text{cm}^{-3}$ [40]. (Reprinted with permission from SpringerOpen.)

highest τ_{eff} of 67 μs, corresponding to a S_{max} value of 187 cm s^{-1}—this calculation was based on the τ_{eff} data at the injection level of $3 \times 10^{14}\,\text{cm}^{-3}$.

Figure 11.17 shows the elemental depth profiles of the HfO$_2$ thin films annealed at 500°C without and with O$_2$ plasma pretreatment and obtained by X-ray photoelectron spectroscopy (XPS). There are three regions found in this figure: in Region A, when the etching time is below 100 seconds, the atomic percentages of Hf and O are relatively uniform, corresponding to the ALD μc-HfO$_2$ layer. In Region B, the atomic percentages of Hf and O decrease when the etching time increases from 130 to 175 seconds, indicating that the O elements diffuse into the c-Si substrate,

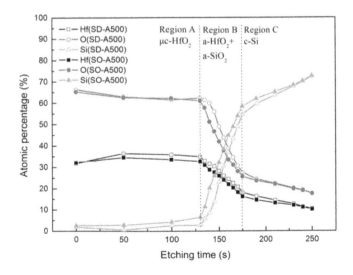

FIGURE 11.17 Elemental depth profiles of HfO$_2$ annealed at 500°C without and with O$_2$ plasma pretreatment versus etching time [40]. (Reprinted with permission from SpringerOpen.)

forming an interfacial layer [60,61]. In the interface region, the atomic percentages of Hf and O of the HfO_2 thin films with O_2 plasma pretreatment are lower than the atomic percentages of the sample without O_2 plasma pretreatment, while Si signal is reversed. A possible reason might be that the O_2 plasma pretreatment leads to the growth of a very thin SiO_2 layer, thus reducing the Hf and O diffusion resulting from the subsequently deposited HfO_2. Fewer atomic vacancies are formed by the diffusion of the HfO_2 on the sample with the O_2 plasma pretreatment. Thus, the O_2 plasma pretreatment can be expected to yield fewer interface traps and exhibit higher chemical passivation quality. In Region C, when the etching time increases above 175 seconds, the Si signal drastically increases up to more than 60%, corresponding to the surface of the c-Si substrate. The O atomic percentage and Hf atomic percentage in the c-Si substrates are attributable to the Ar ion sputtering effect. During the sputtering process of the XPS measurement, some of the Hf or O atoms may reside on the surface of silicon substrate and then be detected.

Figure 11.18 depicts the typical X-ray diffraction spectra of the as-deposited and annealed HfO_2 thin films. The as-deposited and low-temperature (400°C and 450°C)-annealed HfO_2 thin films are amorphous. For HfO_2 thin films annealed at higher temperatures (500°C, 550°C, and 600°C), diffraction peaks appear, indicating the formation of crystalline HfO_2. The peaks at $1/d = 0.319\,Å^{-1}$ and $0.354\,Å^{-1}$ correspond to (−111) and (111) planes of monoclinic phase (ICDD PDF#34-0104, space group P21/c), respectively. The peak at $1/d = 0.340\,Å^{-1}$ corresponds to (111) the plane of orthorhombic phase (ICDD PDF#21-0904, space group Pbcm). Other peaks near $1/d = 0.380\sim0.395\,Å^{-1}$ are (200), (020), and (002) planes of monoclinic and the (020) plane of orthorhombic phases. The monoclinic phase decreases, whereas orthorhombic phase increases when the annealing temperature increases. The orthorhombic HfO_2 predominantly shows the crystalline structure at higher annealing

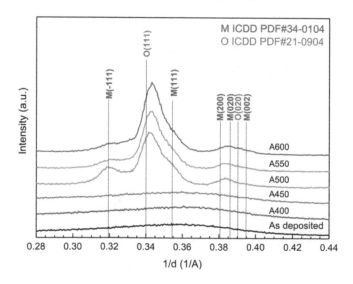

FIGURE 11.18 X-ray diffraction spectra of HfO_2 thin films annealed at different temperatures [59]. (Reprinted with permission from SpringerOpen.)

temperatures. However, the diffraction peaks of orthorhombic HfO_2 are observed at a lower $1/d$ (a smaller d-spacing) as compared to those in the ICDD PDF#21-0904. Furthermore, the shift of $1/d = 0.340\,\text{Å}^{-1}$ toward a higher value indicates that the d-spacing decreases with increasing annealing temperature.

Cross sectional images of the as-deposited and annealed HfO_2 thin films, as shown in Figure 11.19, were evaluated by a field-emission TEM (FE-TEM) for assessing the crystallization of HfO_2 thin films and HfO_2/Si interface. From the FE-TEM images, it can be predicted that the annealed HfO_2 thin films consist of three regions, namely, the HfO_2 layer, the Si substrate layer, and an interfacial layer with the thickness of 1–2 nm between HfO_2 and Si substrate. In Figure 11.19a, the as-deposited HfO_2 thin film is amorphous. However, a fraction of lattice arrangement with the d-spacing values of 2.82 and 3.12 Å is observed in the film annealing at 400°C, as shown in Figure 11.19b. The two d-spacing values are indexed to monoclinic HfO_2 (111) and

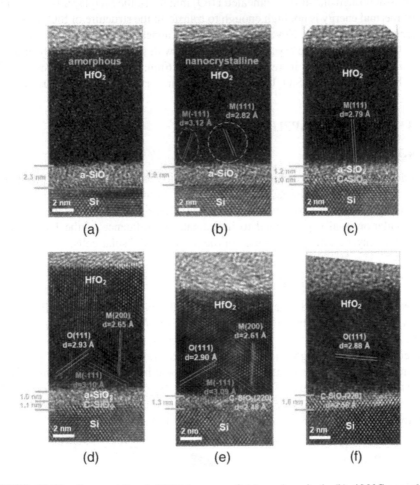

FIGURE 11.19 Cross-sectional TEM images of (a) as-deposited, (b) 400°C-annealed, (c) 450°C-annealed, (d) 500°C-annealed, (e) 550°C-annealed, and (f) 600°C-annealed HfO_2/Si [59]. (Reprinted with permission from SpringerOpen.)

monoclinic HfO$_2$ (–111) planes, respectively. As the annealing temperature increases from 400°C to 600°C, the crystal quality of HfO$_2$ film gradually enhances. As can be seen in Figure 11.19d and e, the main lattice arrangement consists of monoclinic HfO$_2$ (–111), monoclinic HfO$_2$ (200), and orthorhombic HfO$_2$ (111) when the annealing temperatures are at 500°C and 550°C. However, further increasing the annealing temperature to 600°C, the lattice structure of orthorhombic HfO$_2$ (111) still exists in the film, but the other two lattice arrangements gradually disappear. On the other hand, the d-spacing values of orthorhombic HfO$_2$ (111) planes for the 500°C-, 550°C-, and 600°C-annealed HfO$_2$ films are determined to be 2.93, 2.90, and 2.88 Å, respectively. The results agree well with the grazing incident X-ray diffraction (GIXRD) result that the orthorhombic HfO$_2$ (111) diffraction peak shifts toward the high angle direction when the annealing temperature increases from 500°C to 600°C. Another interesting phenomenon includes changing in the crystal structure and thickness of the SiO$_2$ layer. In the as-deposited and 400°C-annealed HfO$_2$ thin films, the SiO$_2$ layer is amorphous, as the thermal energy is not high enough to transform the structure of SiO$_2$ layer from amorphous to crystalline. Nevertheless, when the annealing temperature is increased from 450°C to 600°C, the crystalline SiO$_2$ layer (with the cubic SiO$_2$ (220) structure) appears and its thickness increases from 1.0 to 1.6 nm. In the 600°C-annealed HfO$_2$ thin film, the amorphous SiO$_2$ layer completely transforms to the cubic SiO$_2$ structure.

11.4 SOLAR CELL APPLICATIONS

11.4.1 PARAMETERS

In order to determine the effect of Al$_2$O$_3$ rear surface passivation layer on the performance of the p-type PERC solar cells, PC1D computer software is performed. The detailed simulation parameters are listed in Table 11.6 [48,62]. In the same way, the simulation is also performed to investigate the influence of the HfO$_2$ surface passivation layer on the performance of the n-type PERC solar cells.

11.4.2 DEVICE STRUCTURE

Figure 11.20 shows the diagram of the simulated p-type PERC solar cell. The rear passivation is achieved by SiN$_x$/Al$_2$O$_3$, whereas the front emitter is passivated by SiN$_x$. As the PERC solar cells have laser openings on the rear, which lead to unpassivated regions with high surface recombination rate. Therefore, the passivated region $S_{pass.}$ and the unpassivated region $S_{cont.}$ should be taken into account to determine the effective rear surface recombination rate ($S_{rear, eff}$) [63]. For the n-type PERC solar cell structure as shown in Figure 11.21, the rear passivation layer is SiN$_x$/HfO$_2$ and the front p-type emitter passivation layer is Al$_2$O$_3$.

11.4.3 PERFORMANCE

Figure 11.22 shows the simulated V_{oc} and conversion efficiency (η) as a function of the effective rear surface recombination rate ($S_{rear, eff}$) [48]. The $S_{rear, eff}$ values of 37.8–75.2 cm s^{-1} are obtained for the Al$_2$O$_3$ single-layer passivation. The corresponding

TABLE 11.6

PC1D Simulation Parameters for p-Type PERC Solar Cells with an Al$_2$O$_3$ Rear Passivation Layer, and for n-Type PERC Solar Cells with a HfO$_2$ Rear Passivation Layer

Parameter	Value p-Type PERC	n-Type PERC
Front surface texture depth	3 μm	3 μm
Exterior front reflectance	3%	3%
Emitter contact resistance	2.48 mΩ	2.48 mΩ
Base contact	1 μΩ	1 μΩ
Wafer thickness	200 μm	200 μm
Background doping	p-type 10^{16} cm^{-3}	n-type 10^{16} cm^{-3}
Emitter doping	n-type 10^{20} cm^{-3}	p-type 10^{20} cm^{-3}
Emitter diffusion depth	0.4 μm	0.4 μm
Bulk lifetime	862 μs	1000 μs
Emitter saturation current density	50 fA cm^{-2}	50 fA cm^{-2}
Effective rear surface recombination rate	Variable	Variable

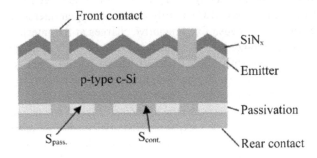

FIGURE 11.20 Structure of the p-type PERC solar cells with SiN$_x$/Al$_2$O$_3$ rear passivation layer [48]. (Reprinted with permission from Elsevier.)

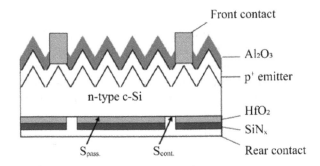

FIGURE 11.21 Structure of the n-type PERC solar cells with Al$_2$O$_3$ front passivation layer and HfO$_2$ rear passivation layer [62]. (Reprinted with permission from MDPI.)

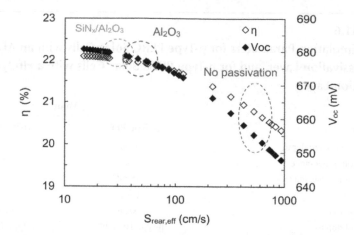

FIGURE 11.22 Open-circuit voltage and conversion efficiency of the p-type PERC solar cells as a function of effective surface recombination rate [48]. (Reprinted with permission from Elsevier.)

devices have V_{oc} of 675.4–678.5 mV and η of 21.82%–21.96%. For PERC solar cells with doubled-layer SiN_x/Al_2O_3, a $S_{rear, eff}$ of 27.1–55.6 cm s^{-1} is obtained, corresponding to V_{oc} of 676.9–679.7 mV and η of 21.9%–22.03%. It has been reported that the interface between silicon and Al_2O_3 mainly contains oxygen interstitials and aluminum vacancies that might cause fixed negative charges at the interface and provide field-effect passivation [64,65]. At the interface between Al_2O_3 and SiN_x, the oxygen vacancies will not significantly affect the electrical properties, since the SiN_x layer is used as a protection layer. Instead, the SiN_x film should have high density, and since the SiN_x layer is deposited with hydrogen-containing gases (e.g., SiH_4 and NH_3), the hydrogen atoms can move toward the silicon surface to further reduce the DBs during the deposition process. It is expected that a vacuum-type or plasma-assisted ALD can achieve a $S_{rear, eff}$ value closer to the ideal value when compared with the spatial ALD, but it will not be cost-effective to spend much more equipment cost and processing time in order to attain a slight improvement in device performance.

For the n-type PERC with rear HfO_2 and front Al_2O_3 passivation layers, the simulated V_{oc} and its efficiency as a function of base saturation current density (J_{0b}) are shown in Figure 11.23. J_{0b} values of 180–460 fA cm^{-2} are obtained for the single HfO_2 rear passivation layer, and the cell efficiency is around 22%. For PERC solar cells with an SiN_x/HfO_2 layer, a $S_{rear, eff}$ of 22 cm s^{-1} with a J_{0b} of 64 fA cm^{-2}, a V_{oc} of 689 mV, and an efficiency of 23.02% are obtained. This great improvement might be because the SiN_x layer can provide hydrogen atoms diffusing toward the substrate surface to further passivate the SiO_x/Si interface, similar to the case of the SiN_x/Al_2O_3 structure [64,66].

11.5   SUMMARY AND OUTLOOK

Looking back into the development of PERC solar cells, surface passivation is the most important technology, which effectively suppresses the charge carrier recombination. The rear morphology of c-Si can affect the performance of

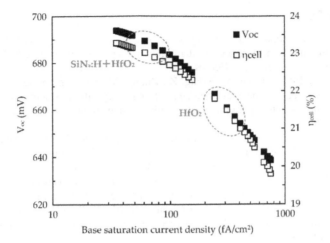

FIGURE 11.23 Open-circuit voltage and conversion efficiency of n-type PERC solar cells as a function of base saturation current density [62]. (Reprinted with permission from MDPI.)

PERC solar cells. Polished rear surface leads to superior field-effect passivation and re-absorption of long-wavelength light, as compared to the textured surfaces. Al_2O_3 and HfO_2 thin films are the suitable materials for passivating p-type and n-type c-Si, respectively, due to the field-effect passivation repelling the minority carrier from the surface and chemical passivation reducing the surface DBs. For Al_2O_3 passivation films, the negative fixed charge density can be positively correlated with AlO_4/AlO_6 intensity ratio. The optimal annealing temperature is around 600°C for Al_2O_3 annealed in oxygen-containing gases, whereas the annealing should be performed at a lower temperature of about 450°C for FG or N_2. In particular, oxygen annealing shows the highest fixed negative charge, high oxygen content, and intermediate hydrogen concentration at the Al_2O_3/c-Si surface, leading to the highest carrier lifetime value. For HfO_2 passivation thin films, annealing at 500°C effectively reduces the defect density at HfO_2/c-Si interface, with a mixed interfacial oxide consisting of a-HfO_2 and a-SiO_2. With the increase in annealing temperature, the reduction of a-HfO_2 component in the interfacial oxide layer results in poor interface properties. The optimization of the post-annealing process can control the crystalline structure and the interface properties of the HfO_2 films. For PERC solar cell applications, the passivation quality of the optimized Al_2O_3 as a rear passivation layer is close to that of the theoretical ideal passivation. The p-type PERC passivated by rear Al_2O_3 has a conversion efficiency of 22.1%. The n-type PERC passivated by front Al_2O_3 and rear HfO_2 exhibits high conversion efficiency of 23.02%. It is estimated that PERC solar cells will continue to compete with the traditional crystalline Si solar cell market in the next few years. Further improvement in PERC solar cell conversion efficiency can be achieved by very-high-quality c-Si substrate, optimizing emitter doping profile and reducing metal shading area. Efficiencies higher than 30% are possible for PERC-based multi-junction solar cells.

REFERENCES

1. Zhuang, Y.F., Zhong, S.H., Huang, Z.G., and Shen, W.Z. (2016) Versatile strategies for improving the performance of diamond wire sawn mc-Si solar cells. *Solar Energy Materials and Solar Cells*, **153**, 18–24.
2. Zhang, Y., Tao, J., Chen, Y., Xiong, Z., Zhong, M., Feng, Z., Yang, P., and Chu, J. (2016) A large-volume manufacturing of multi-crystalline silicon solar cells with 18.8% efficiency incorporating practical advanced technologies. *RSC Advances*, **6** (63), 58046–58054.
3. Cao, F., Chen, K., Zhang, J., Ye, X., Li, J., Zou, S., and Su, X. (2015) Next-generation multi-crystalline silicon solar cells: Diamond-wire sawing, nano-texture and high efficiency. *Solar Energy Materials and Solar Cells*, **141**, 132–138.
4. Liu, J., Yao, Y., Xiao, S., and Gu, X. (2018) Review of status developments of high-efficiency crystalline silicon solar cells. *Journal of Physics D: Applied Physics*, **51** (12), 123001.
5. Dullweber, T., and Schmidt, J. (2016) Industrial silicon solar cells applying the passivated emitter and rear cell (PERC) concept—A review. *IEEE Journal of Photovoltaics*, **6** (5), 1366–1381.
6. Ye, X., Zou, S., Chen, K., Li, J., Huang, J., Cao, F., Wang, X., Zhang, L., Wang, X.-F., Shen, M., and Su, X. (2014) 18.45%-Efficient multi-crystalline silicon solar cells with novel nanoscale pseudo-pyramid texture. *Advanced Functional Materials*, **24** (42), 6708–6716.
7. Hallam, B., Chen, D., Kim, M., Stefani, B., Hoex, B., Abbott, M., and Wenham, S. (2017) The role of hydrogenation and gettering in enhancing the efficiency of next-generation Si solar cells: An industrial perspective. *Physica Status Solidi A*, **214** (7), 1700305.
8. Armin, G.A., Matthew, B.B., Bram, H., and Thomas, M. (2012) Industrial silicon wafer solar cells – Status and trends. *Green*, **2** (4), 135–148.
9. Neuhaus, D.-H., and Münzer, A. (2007) Industrial silicon wafer solar cells. *Advances in OptoElectronics*, **2007**, 1–15.
10. Allen, T.G., Bullock, J., Yang, X., Javey, A., and De Wolf, S. (2019) Passivating contacts for crystalline silicon solar cells. *Nature Energy*, **4** (11), 914–928.
11. George, S.M. (2010) Atomic layer deposition: An overview. *Chemical Reviews*, **110** (1), 111–131.
12. Dingemans, G., and Kessels, W.M.M. (2012) Status and prospects of Al_2O_3-based surface passivation schemes for silicon solar cells. *Journal of Vacuum Science & Technology A: Vacuum, Surfaces, and Films*, **30** (4), 040802.
13. Levy, D.H., Jerr, S.R., and Carey, R.S. (2009) System for thin film deposition utilizing compensating forces. 2009/0217878, issued Sep. 3, 2009.
14. Granneman, E.H.A., and van Nooten, S.E. (2011) Apparatus and method for high-throughput atomic layer deposition, issued May 26, 2011.
15. Maas, D.J., van Someren, B., Lexmond, A.S., Spee, C.I.M.A., Duisterwinkel, A.E., and Vermeer, A.J.P.M. (2010) Apparatus and method for atomic layer deposition. 2010/024671, issued Mar. 4, 2010.
16. Barbos, C., Blanc-Pelissier, D., Fave, A., Botella, C., Regreny, P., Grenet, G., Blanquet, E., Crisci, A., and Lemiti, M. (2016) Al_2O_3 thin films deposited by thermal atomic layer deposition: Characterization for photovoltaic applications. *Thin Solid Films*, **617**, 108–113.
17. Batra, N., Gope, J., Vandana, Panigrahi, J., Singh, R., and Singh, P.K. (2015) Influence of deposition temperature of thermal ALD deposited Al_2O_3 films on silicon surface passivation. *AIP Advances*, **5** (6), 067113.
18. Potts, S.E., Dingemans, G., Lachaud, C., and Kessels, W.M.M. (2012) Plasma-enhanced and thermal atomic layer deposition of Al_2O_3 using dimethylaluminum isopropoxide, $[Al(CH_3)_2 (\mu\text{-}O^i Pr)]_2$, as an alternative aluminum precursor. *Journal of Vacuum Science & Technology A: Vacuum, Surfaces, and Films*, **30** (2), 021505.

19. van Hemmen, J.L., Heil, S.B.S., Klootwijk, J.H., Roozeboom, F., Hodson, C.J., van de Sanden, M.C.M., and Kessels, W.M.M. (2007) Plasma and thermal ALD of Al_2O_3 in a commercial 200 mm ALD reactor. *Journal of the Electrochemical Society*, **154** (7), G165.

20. Zhu, Z., Sippola, P., Lipsanen, H., Savin, H., and Merdes, S. (2018) Influence of plasma parameters on the properties of ultrathin Al_2O_3 films prepared by plasma enhanced atomic layer deposition below 100°C for moisture barrier applications. *Japanese Journal of Applied Physics*, **57** (12), 125502.

21. Hofmann, M. (2008) *Rear Surface Conditioning and Passivation for Locally Contacted Crystalline Silicon Solar Cells*, Verl. Dr. Hut, München.

22. Hsu, C.-H., Huang, C.-W., Cho, Y.-S., Wu, W.-Y., Wuu, D.-S., Zhang, X.-Y., Zhu, W.-Z., Lien, S.-Y., and Ye, C.-S. (2019) Efficiency improvement of PERC solar cell using an aluminum oxide passivation layer prepared via spatial atomic layer deposition and post-annealing. *Surface and Coatings Technology*, **358**, 968–975.

23. Langereis, E., Keijmel, J., van de Sanden, M.C.M., and Kessels, W.M.M. (2008) Surface chemistry of plasma-assisted atomic layer deposition of Al_2O_3 studied by infrared spectroscopy. *Applied Physics Letters*, **92** (23), 231904.

24. Schmidt, J., Werner, F., Veith-Wolf, B., Zielke, D., Bock, R., Brendel, R., Poodt, P., Roozeboom, F., Li, A., and Cuevas, A. (2010) Surface passivation of silicon solar cells using industrially relevant Al_2O_3 deposition techniques. *Photovoltaics International*, **10**, 52–57.

25. Lee, C.-Y., Deng, S., Zhang, T., Cui, X., Khoo, K.T., Kim, K., and Hoex, B. (2018) Evaluating the impact of thermal annealing on c-Si/Al_2O_3 interface: Correlating electronic properties to infrared absorption. *AIP Advances*, **8** (7), 075204.

26. Dingemans, G., Einsele, F., Beyer, W., van de Sanden, M.C.M., and Kessels, W.M.M. (2012) Influence of annealing and Al_2O_3 properties on the hydrogen-induced passivation of the Si/SiO_2 interface. *Journal of Applied Physics*, **111** (9), 093713.

27. Werner, F., Veith, B., Tiba, V., Poodt, P., Roozeboom, F., Brendel, R., and Schmidt, J. (2010) Very low surface recombination velocities on p- and n-type c-Si by ultrafast spatial atomic layer deposition of aluminum oxide. *Applied Physics Letters*, **97** (16), 162103.

28. Chi, X., Lan, X., Lu, C., Hong, H., Li, C., Chen, S., Lai, H., Huang, W., and Xu, J. (2016) An improvement of HfO_2/Ge interface by *in situ* remote N_2 plasma pretreatment for Ge MOS devices. *Materials Research Express*, **3** (3), 035012.

29. Singh, V., Sharma, S.K., Kumar, D., and Nahar, R.K. (2012) Study of rapid thermal annealing on ultra thin high-k HfO_2 films properties for nano scaled MOSFET technology. *Microelectronic Engineering*, **91**, 137–143.

30. Shimada, S., and Aketo, T. (2005) High-temperature oxidation at 1500° and 1600° of SiC/graphite coated with Sol-Gel-Derived HfO_2. *J American Ceramic Society*, **88** (4), 845–849.

31. Wang, Y., Lin, Z., Cheng, X., Xiao, H., Zhang, F., and Zou, S. (2004) Study of HfO_2 thin films prepared by electron beam evaporation. *Applied Surface Science*, **228** (1–4), 93–99.

32. Toledano-Luque, M., San Andrés, E., del Prado, A., Mártil, I., Lucía, M.L., González-Díaz, G., Martínez, F.L., Bohne, W., Röhrich, J., and Strub, E. (2007) High-pressure reactively sputtered HfO_2: Composition, morphology, and optical properties. *Journal of Applied Physics*, **102** (4), 044106.

33. Wang, J., Sadegh Mottaghian, S., and Farrokh Baroughi, M. (2012) Passivation properties of atomic-layer-deposited hafnium and aluminum oxides on Si surfaces. *IEEE Transactions on Electron Devices*, **59** (2), 342–348.

34. Lin, F., Hoex, B., Koh, Y.H., Lin, J.J., and Aberle, A.G. (2012) Low-temperature surface passivation of moderately doped crystalline silicon by atomic-layer-deposited hafnium oxide films. *Energy Procedia*, **15**, 84–90.

35. Geng, H., Lin, T., Letha, A.J., Hwang, H.-L., Kyznetsov, F.A., Smirnova, T.P., Saraev, A.A., and Kaichev, V.V. (2014) Advanced passivation techniques for Si solar cells with high-κ dielectric materials. *Applied Physics Letters*, **105** (12), 123905.
36. Simon, D.K., Jordan, P.M., Dirnstorfer, I., Benner, F., Richter, C., and Mikolajick, T. (2014) Symmetrical Al₂O₃-based passivation layers for p- and n-type silicon. *Solar Energy Materials and Solar Cells*, **131**, 72–76.
37. Gope, J., Vandana, Batra, N., Panigrahi, J., Singh, R., Maurya, K.K., Srivastava, R., and Singh, P.K. (2015) Silicon surface passivation using thin HfO₂ films by atomic layer deposition. *Applied Surface Science*, **357**, 635–642.
38. Cui, J., Wan, Y., Cui, Y., Chen, Y., Verlinden, P., and Cuevas, A. (2017) Highly effective electronic passivation of silicon surfaces by atomic layer deposited hafnium oxide. *Applied Physics Letters*, **110** (2), 021602.
39. Cheng, X., Repo, P., Halvard, H., Perros, A.P., Marstein, E.S., Di Sabatino, M., and Savin, H. (2017) Surface passivation properties of HfO₂ thin film on n-type crystalline Si. *IEEE Journal of Photovoltaics*, **7** (2), 479–485.
40. Zhang, X.-Y., Hsu, C.-H., Lien, S.-Y., Chen, S.-Y., Huang, W., Yang, C.-H., Kung, C.-Y., Zhu, W.-Z., Xiong, F.-B., and Meng, X.-G. (2017) Surface passivation of silicon using HfO₂ thin films deposited by remote plasma atomic layer deposition system. *Nanoscale Research Letters*, **12** (1), 324.
41. Oudot, E., Gros-Jean, M., Courouble, K., Bertin, F., Duru, R., Rochat, N., and Vallée, C. (2018) Hydrogen passivation of silicon/silicon oxide interface by atomic layer deposited hafnium oxide and impact of silicon oxide underlayer. *Journal of Vacuum Science & Technology A: Vacuum, Surfaces, and Films*, **36** (1), 01A116.
42. Panigrahi, J., Singh, R., and Singh, P.K. (2018) Enhanced field effect passivation of c-Si surface via introduction of trap centers: Case of hafnium and aluminium oxide bilayer films deposited by thermal ALD. *Solar Energy Materials and Solar Cells*, **188**, 219–227.
43. Gougam, A.B., Rajab, B., and Bin Afif, A. (2019) Investigation of c-Si surface passivation using thermal ALD deposited HfO₂ films. *Materials Science in Semiconductor Processing*, **95**, 42–47.
44. Kung, C.-Y., Yang, C.-H., Huang, C.-W., Lien, S.-Y., Zhu, W.-Z., Lin, H.-J., and Zhang, X.-Y. (2017) Performance improvement of high efficiency mono-crystalline silicon solar cells by modifying rear-side morphology. *Applied Sciences*, **7** (4), 410.
45. Richter, A., Glunz, S.W., Werner, F., Schmidt, J., and Cuevas, A. (2012) Improved quantitative description of Auger recombination in crystalline silicon. *Physical Review B*, **86** (16), 165202.
46. Lu, P.H., Wang, K., Lu, Z., Lennon, A.J., and Wenham, S.R. (2013) Anodic aluminum oxide passivation for silicon solar cells. *IEEE Journal of Photovoltaics*, **3** (1), 143–151.
47. Wang, H.-P., Li, A.-C., Lin, T.-Y., and He, J.-H. (2016) Concurrent improvement in optical and electrical characteristics by using inverted pyramidal array structures toward efficient Si heterojunction solar cells. *Nano Energy*, **23**, 1–6.
48. Hsu, C.-H., Huang, C.-W., Lai, J.-M., Chou, Y.-C., Cho, Y.-S., Zhang, S., Lien, S.-Y., Zhang, X.-Y., and Zhu, W.-Z. (2018) Effect of oxygen annealing on spatial atomic layer deposited aluminum oxide/silicon interface and on passivated emitter and rear contact solar cell performance. *Thin Solid Films*, **660**, 920–925.
49. da Rosa, E.B.O., Baumvol, I.J.R., Morais, J., de Almeida, R.M.C., Papaléo, R.M., and Stedile, F.C. (2002) Diffusion reaction of oxygen in aluminum oxide films on silicon. *Physical Review B*, **65** (12), 121303.
50. Kimoto, K., Matsui, Y., Nabatame, T., Yasuda, T., Mizoguchi, T., Tanaka, I., and Toriumi, A. (2003) Coordination and interface analysis of atomic-layer-deposition Al₂O₃ on Si(001) using energy-loss near-edge structures. *Applied Physics Letters*, **83** (21), 4306–4308.

51. Johnson, R.S., Lucovsky, G., and Baumvol, I. (2001) Physical and electrical properties of noncrystalline Al_2O_3 prepared by remote plasma enhanced chemical vapor deposition. *Journal of Vacuum Science & Technology A: Vbacuum, Surfaces, and Films*, **19** (4), 1353–1360.

52. Terman, L.M. (1962) An investigation of surface states at a silicon/silicon oxide interface employing metal-oxide-silicon diodes. *Solid-State Electronics*, **5** (5), 285–299.

53. Gupta, D., Vieregge, K., and Srikrishnan, K.V. (1992) Copper diffusion in amorphous thin films of 4% phosphorus-silcate glass and hydrogenated silicon nitride. *Applied Physics Letters*, **61** (18), 2178–2180.

54. Granneman, E.H.A., Kuznetsov, V.I., and Vermont, P. (2014) (Invited) Spatial ALD, deposition of Al_2O_3 films at throughputs exceeding 3000 wafers per hour. *ECS Transactions*, **61** (3), 3–16.

55. Deckers, J., Cornagliotti, E., Debucquoy, M., Gordon, I., Mertens, R., and Poortmans, J. (2014) Aluminum oxide-aluminum stacks for contact passivation in silicon solar cells. *Energy Procedia*, **55**, 656–664.

56. Li, M., Shin, H.-S., Jeong, K.-S., Oh, S.-K., Lee, H., Han, K., Lee, G.-W., and Lee, H.-D. (2014) Blistering induced degradation of thermal stability Al_2O_3 passivation layer in crystal Si solar cells. *JSTS:Journal of Semiconductor Technology and Science*, **14** (1), 53–60.

57. Ji, Y., Jiang, Y., Liu, H., Wang, L., Liu, D., Jiang, C., Fan, R., and Chen, D. (2013) Effects of thermal treatment on infrared optical properties of SiO_2 films on Si Substrates. *Thin Solid Films*, **545**, 111–115.

58. Hoex, B., Heil, S.B.S., Langereis, E., van de Sanden, M.C.M., and Kessels, W.M.M. (2006) Ultralow surface recombination of c-Si substrates passivated by plasma-assisted atomic layer deposited Al_2O_3. *Applied Physics Letters*, **89** (4), 042112.

59. Zhang, X.-Y., Hsu, C.-H., Lien, S.-Y., Wu, W.-Y., Ou, S.-L., Chen, S.-Y., Huang, W., Zhu, W.-Z., Xiong, F.-B., and Zhang, S. (2019) Temperature-dependent HfO_2/Si interface structural evolution and its mechanism. *Nanoscale Research Letters*, **14** (1), 83.

60. Fu, W.-E., Chang, C.-W., Chang, Y.-Q., Yao, C.-K., and Liao, J.-D. (2012) Reliability assessment of ultra-thin HfO_2 films deposited on silicon wafer. *Applied Surface Science*, **258** (22), 8974–8979.

61. Jiang, R., Xie, E., and Wang, Z. (2006) Interfacial chemical structure of HfO_2/Si film fabricated by sputtering. *Applied Physics Letters*, **89** (14), 142907.

62. Zhang, X.-Y., Hsu, C.-H., Cho, Y.-S., Lien, S.-Y., Zhu, W.-Z., Chen, S.-Y., Huang, W., Xie, L.-G., Chen, L.-D., Zou, X.-Y., and Huang, S.-X. (2017) Simulation and fabrication of HfO_2 thin films passivating Si from a numerical computer and remote plasma ALD. *Applied Sciences*, **7** (12), 1244.

63. Benick, J., Hoex, B., Dingemans, G., Kessels, W.M.M., Richter, A., Hermle, M., and Glunz, S.W. (2009) High-efficiency n-type silicon solar cells with front side boron emitter. *24th European Photovoltaic Solar Energy Conference*, 21–25 September 2009, 8 pages: Hamburg, Germany.

64. Simon, D.K., Jordan, P.M., Mikolajick, T., and Dirnstorfer, I. (2015) On the control of the fixed charge densities in Al_2O_3-based silicon surface passivation schemes. *ACS Applied Materials & Interfaces*, 7 (51), 28215–28222.

65. Zhao, Y., Zhou, C., Zhang, X., Zhang, P., Dou, Y., Wang, W., Cao, X., Wang, B., Tang, Y., and Zhou, S. (2013) Passivation mechanism of thermal atomic layer-deposited Al_2O_3 films on silicon at different annealing temperatures. *Nanoscale Research Letters*, **8** (1), 114.

66. Huang, H., Lv, J., Bao, Y., Xuan, R., Sun, S., Sneck, S., Li, S., Modanese, C., Savin, H., Wang, A., and Zhao, J. (2017) 20.8% industrial PERC solar cell: ALD Al_2O_3 rear surface passivation, efficiency loss mechanisms analysis and roadmap to 24%. *Solar Energy Materials and Solar Cells*, **161**, 14–30.

12 Organic Solar Cell

Shaohui Zheng
Southwest University

CONTENTS

12.1 Introduction 443
12.2 Classification of OSCs..444
12.3 Working Principles of OSCs..446
    12.3.1 Photon Absorption and Exciton Generation 447
    12.3.2 Exciton Diffusion and Dissociation...448
    12.3.3 Transport of Free Charge Carriers (Electron and Hole)...................448
12.4 Key Parameters of OSCs 449
12.5 Donor and Acceptor Materials 451
    12.5.1 Polymers 451
    12.5.2 Small Organic Molecules 452
12.6 Fullerene and Non-Fullerene Acceptors..454
    12.6.1 Fullerene Materials...455
    12.6.2 NFA ITIC and Its Derivatives 459
    12.6.3 NFA Y6 and Its Derivatives...465
References...471

12.1 INTRODUCTION

Nowadays, energy crisis is one of the most important matters for human beings due to global economic development, the rapid depletion of non-renewable fossil fuels, and the environmental issues such as serious pollution and greenhouse gas caused by the combustion products of fossil fuels [1]. This has inspired people to search for renewable and clean alternative energy sources.

Solar energy has great potential to satisfy the major energy demand of human society. Compared to other renewable energy sources, it is green, almost inexhaustible, and free of geographical restrictions. Therefore, the utilization of solar energy is extremely important for the sustainable development of human society.

So far, there are many ways to make use of solar energy. Among them, converting solar light into electricity is the most important way. Thus, different types of photovoltaic devices have been developed and improved. Photovoltaic devices are divided into three categories according to the types of materials: inorganic, organic, and organic–inorganic hybrids. Generally, the power conversion efficiency (PCE) of inorganic solar cells is higher than that of the others. However, the high cost and pollution to the environment appear during the production process of inorganic solar cells. In addition, poor flexibility and high weight also limit their usages. In order

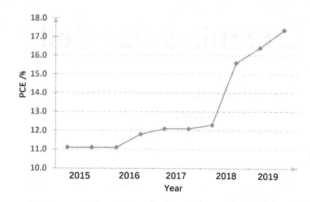

FIGURE 12.1   Development chart of PCE of the solar photovoltaic cells since 2015 [3,5–14].

to overcome these shortcomings, engineers and researchers have developed the next-generation solar cells: organic solar cells (OSCs) and organic–inorganic hybrid thin-film solar cells, including perovskites and dye-sensitized solar cells. OSCs have attracted much attention due to their large number of types of available organic materials, simple synthesis, lightweight, mechanical flexibility, and low cost.

The PCE of OSCs has increased from 1% to 17.35% over the years [2–4]. In particular, in the recent 5 years, the PCE of OSCs has been significantly improved, as shown in Figure 12.1. OSCs are expected to be commercialized in the coming future.

12.2   CLASSIFICATION OF OSCs

OSCs currently have two main disadvantages: low PCE and poor stability. Therefore, to overcome them, the research on OSCs mainly focuses on the development of materials of the organic active layer and buffer layer (electron/hole transfer layer), and the optimization of cell structures. The development of the active layer structure of OSCs has gone through three stages: single-layer, bi-layer heterojunction, and bulk heterojunction (BHJ) OSCs. On the other hand, OSCs can also be divided into single-junction and tandem OSCs according to whether they have a tandem structure.

Single-layer OSCs (Schottky cells) contain only one type of organic material in the active layer, which is sandwiched between a low work function (generally a metal material) and a high work function (generally an indium tin oxide (ITO)) material. Figure 2a shows a schematic diagram of a single-layer homojunction structure. The work function refers to the minimum energy required for an electron to escape from a metal surface. Under the light, electrons in the organic semiconductor are excited from the highest occupied molecular orbital (HOMO) to the lowest unoccupied molecular orbital (LUMO) or higher-level molecular orbitals, and then electron–hole pairs (called "excitons") are formed. Electrons and holes are then separated by a potential difference (a built-in electric field), and free electrons and holes are generated and later collected by a cathode and an anode, respectively. Consequently, the conversion from light to electricity is completed. In 1958, the first single-layer homojunction OSC with magnesium phthalocyanine (MgPc) as an active layer was

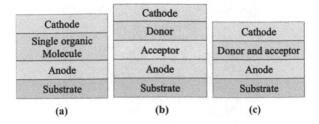

FIGURE 12.2 Schematic representation of typical structures used in OSCs, including single-layer homojunction solar cell (a), bi-layer heterojunction solar cell (b), and BHJ solar cell (c).

fabricated [15]. Its open-circuit voltage is 0.2 V; however, its PCE is extremely low. The improvements in the efficiency of the Schottky cell depend on the matching degree of different work functions and the selection of organic materials to achieve efficient exciton separation. Due to its low dielectric constant and large exciton binding energy, the efficiency of homojunction OSC is limited. However, because this type of OSC has some advantages such as a simple manufacturing process and good stability of the active layer, many researchers still commit to it. At present, the materials of the active layer of single-layer OSCs are generally donor–acceptor fused molecules. That is, the donor and acceptor are combined with the covalent bond. Therefore, a push–pull effect can be observed with this structure, and the aforementioned materials are involved in a wider range of light absorption and more efficient charge separations [16]. Today, the PCE of the single-layer OSCs with polymers and small molecules as active materials has reached 6.3% and 3.4% [17,18], respectively.

Bi-layer heterojunction solar cell was first fabricated in 1986 by Dr. Tang in Kodak Company, who revolutionarily utilized small-molecule donor (CuPC) and polymer acceptor materials to make a cell [4]. Its PCE is up to 1% and much higher than homojunction MgPC OSC. Compared to the homojunction OSCs, bi-layer structures greatly enhance the charge separation and reduce the charge recombination. The working principle of this structure is as follows: first, the built-in electric field is generated at the interface between donor and acceptor materials due to the potential difference of donor and acceptor. The electrons at or near the HOMO levels of donor material are excited to the LUMO or higher levels to form excitons under light. The excitons then diffuse to the interface between donor and acceptor, and are separated into electrons and holes by the built-in electric field at the interface between donor and acceptor. To end, the electrons and holes move to a cathode and an anode along with the acceptor and donor materials, respectively. And the electrical current is generated if the cathode and the anode are connected. With respect to homojunction OSC, bi-layer heterojunction structure revolutionarily improves the photoelectric conversion efficiency. Still, the exciton separation and the PCE of this type of OSCs are limited by the inadequate contact area between donor and acceptor. A schematic diagram of a bi-layer heterojunction OSC is shown in Figure 12.2b.

Aiming at the deficiencies of bi-layer heterojunction OSCs, BHJ OSCs with mixed, continuous, and osmotic structure of the donor and acceptor materials soon appear, which greatly increases the contact area between donor and acceptor. Subsequently, exciton separation at the interface is significantly enhanced, which

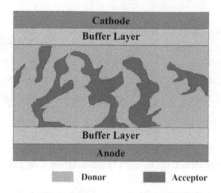

FIGURE 12.3   Bulk heterojunction OSC.

in turn improves PCE. An OSC with such a structure that forms a continuous and osmotic structure is called a "bulk heterojunction solar cell." At present, the BHJ structure is the mainstream of OSCs. The BHJ structure is shown in Figure 12.2c. A continuous and mutually infiltrated network formed by the donor and acceptor materials in the active layer is demonstrated in Figure 12.3.

Until now, only the structures of single-junction OSC are discussed. Similar to the tandem structure of the inorganic solar cell, the multi-junction, i.e., tandem structure of OSCs, is also an important research field in recent years. This structure has attracted widespread attention of many researchers. Remarkably, Chen group at Nankai University has successfully improved the PCE of OSC up to 17.3% with tandem structure [19]. On the one hand, tandem cells can achieve the superposition of open-circuit voltages. On the other hand, the selection of complementary active layer materials can achieve excellent utilization of different wavelength bands in the visible and near-infrared regions (NIRs). For tandem OSCs, the principles of material selection of the active layer are extremely important. Researchers have to consider the matching of energy gaps of different active layers to obtain a wide range of light absorption in the visible and NIRs; the intermediate transport layer, the hole transport layer, and the electron transport layer in the tandem structure should be as transparent as possible in the visible and NIRs to reduce the light absorption. The schematic diagram of a tandem structure is shown in Figure 12.4.

12.3   WORKING PRINCIPLES OF OSCs

How OSCs work is shown in Figure 12.5. Basically, the working principle of OSC can be simply divided into three steps. First, an electron is excited from the HOMO (or near orbitals) to the LUMO (or near orbitals) when a photon is absorbed, and an exciton (electron–hole pair) is formed. Second, excitons diffuse to the interface between donor and acceptor, and then separate into an electron and a hole under the built-in electric field. Third, the electron and the hole move to cathode and anode, respectively.

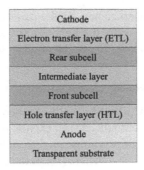

FIGURE 12.4 The structure of tandem solar cells.

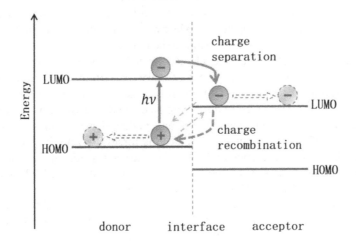

FIGURE 12.5 Schematic representation of the working principle of OSCs.

12.3.1 PHOTON ABSORPTION AND EXCITON GENERATION

The active layer of an OSC is mainly composed of a donor and an acceptor material. An electron is excited from the HOMO or near molecular orbital of the donor and acceptor material to the LUMO or the virtual orbitals at higher energy level upon the absorption of a photon. A hole is then formed in the HOMO energy level corresponding to the excited electron. Due to the low dielectric constant of organic semiconductors, the delocalized electrons and holes cannot be directly generated after absorbing photons. This electron–hole pair with a certain binding energy (0.1–1 eV) [20] is also called an "exciton," and the energy is called the "exciton binding energy." The choice of materials will greatly affect the absorption of photons and exciton binding energy, thus affecting the generation of photocurrent. Note that an organic material with a narrow band gap can achieve the coverage of the absorption spectrum in the NIR: the main energy of sunlight is distributed in the visible and NIRs. At the same time, the relative complementation of the absorption spectra of the donor and acceptor materials should be considered so that the photons can be absorbed enough.

Therefore, external quantum efficiency (EQE) can be improved. For the tandem structure, it is necessary to consider the light absorption range of the front and rear subcells and the intermediate layer to gain an efficient absorption of photons.

12.3.2 Exciton Diffusion and Dissociation

Next, the thermally induced diffusion of the formed excitons will happen. One important concept has to be introduced, i.e., exciton diffusion length (EDL). EDL is defined as the distance that the excitons can diffuse before they decay. Generally, EDLs in OSC are about tens of nanometers [19]. If the size of the microstructure unit of donor or acceptor is longer than EDL, the excitons will decay and cannot diffuse to the interface between donor and acceptor. So charge separations cannot happen. In BHJ structure, donor and acceptor materials form a continuous and osmotic network, which critically increases the contact area between donor and acceptor with respect to the bi-layer heterojunction. Meanwhile, the small domain size in BHJ structure facilitates the exciton diffusion to the interface.

At the donor–acceptor interface, electron transfer may happen from an excited donor (exciton) to an acceptor. There are two important parameters affecting charge separations in this process. One is the exciton binding energy (E_b), which is defined as the energy required to dissociate an excited electron–hole pair into free charge carriers and has important effects on whether electron transfer can happen at the interface [21–23]. The other is the driving force. Commonly, for heterojunction OSCs based on fullerenes and their derivatives (as acceptors), the efficient charge separation of excitons at the interface requires that the LUMO energy-level difference (driving force) [24] between donor and acceptor should be above 0.3 eV [25]. However, the driving force in non-fullerene acceptor (NFA)-based OSCs could be much smaller (<0.1 eV) [24]. At present, there is no reasonable and widely accepted explanation about the difference in driving force between fullerene- and NFA-based solar cells. This is one of the important research directions in the field of OSCs [16].

12.3.3 Transport of Free Charge Carriers (Electron and Hole)

After charge transfer at the interface between donor and acceptor, the electron and the hole may travel to two electrodes along the electron and hole "channels" in the acceptor and donor materials, respectively. (Of course, they may recombine too.) Then, the free electron and the hole are extracted by two electrodes, and thus, light is converted into electrical energy.

The separation and transport processes of electrons and holes are very sensitive to the micromorphology of donor and acceptor. For BHJ OSCs, the poor mutual solubility between donor and acceptor materials, and the aggregation of donor or acceptor material will reduce the area of interface, and are therefore unfavorable for charge separation. Only an appropriate phase separation and interpenetrating network are beneficial for charge separation and transport. A large domain size due to unnecessary bulky phase separation may cause electrons and holes to recombine rather than to diffuse to the electrodes. In addition, a small domain size will be not beneficial to the diffusion of electrons or holes as well. Finally, the impurity in a BHJ OCS and its vicinity cause the recombination of electrons and holes, thus affecting the transport of electrons and holes and subsequent charge extraction process too.

12.4 KEY PARAMETERS OF OSCs

The critical parameters that influence the PCEs of organic photovoltaic materials include EQE, internal quantum efficiency (IQE), open-circuit voltage (V_{oc}), short-circuit current density (J_{sc}), and fill factor (FF).

EQE: It is defined as the ratio of extracted free electrons to the incident photons of an OSC, or the ratio of electrical output power to incident optical power of an OSC.

IQE: It is defined as the ratio of the number of extracted free electrons to the number of photons that enter an OSC (excluding reflection and transmission of light).

V_{oc}**:** The open-circuit voltage of an OSC is defined as the voltage at which the current density output is 0, as shown in Figure 12.6. It is one of the most important parameters of an organic photovoltaic device.

The open-circuit voltage of an OSC is determined by many factors such as the strength of incident light, the types of materials of donor and acceptor, electron/hole transfer layer, electrode materials, temperature, morphology of active layers, and fabrication conditions [26]. However, one may utilize the following simple equation to quickly estimate V_{oc} of a BHJ OSC [27]:

$$V_{oc} = \left(\frac{1}{e}\right)\left(\left|E^{D}_{HOMO}\right| - \left|E^{A}_{LUMO}\right|\right) - 0.3\,V \qquad (12.1)$$

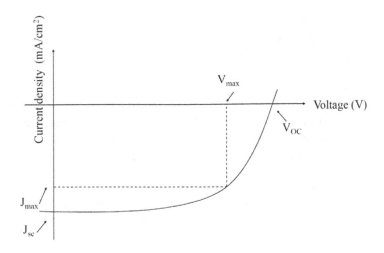

FIGURE 12.6 Schematic J–V curve of an OSC under illumination.

where e means the electron charge, ε^A_{LUMO} is the LUMO energy of acceptor, and ε^D_{HOMO} is the HOMO energy of donor. The open-circuit voltage of a BHJ OSC depends on the energy-level matching of donor and acceptor materials [28].

J_{sc}: Short-circuit current (J_{sc}) represents the current through an OSC when the voltage is zero (short-circuited), as shown in Figure 12.6. J_{sc} is affected by many factors such as the absorption range of light, the microstructure of the active layer material, the contact between the active layer material and electron/hole transfer layer, the interfaces between buffer layers and two electrodes, and so on. All these factors affect the processes of photoexcitation and charge transfer and extraction. There are some important principles: the absorption wavelength of the materials in the active layer should be within the visible and NIRs; donor and acceptor materials complement each other when absorbing photons; finally, appropriate microstructure, phase separation size, and contact form are critical for obtaining a higher short-circuit current too.

FF: As shown in Figure 12.6, FF is the ratio of the product of the current (J_{max}) and voltage (V_{max}) at the point of a J–V curve with the maximum output power to the product of J_{sc} and V_{oc}.

$$FF = \frac{J_{max} \times V_{max}}{J_{sc} \times V_{oc}} \tag{12.2}$$

Here, J_{max} and V_{max} are the current density and the voltage at the point of maximum output power, respectively. J_{sc} and V_{oc} represent the short-circuit current and open-circuit voltage of an OSC, respectively.

PCE: It can be defined as the ratio of the electrical output power of an OSC to the incident light power, which is expressed as follows:

$$PCE = \frac{J_{sc} \times V_{oc} \times FF}{P_{in}} \tag{12.3}$$

where J_{sc}, V_{oc}, FF, and P_{in} denote the short-circuit current, open-circuit voltage, FF, and the incident light power, respectively [29].

The aforementioned key parameters of OSCs are not isolated and the interplay among them should be carefully considered. The aggregation of molecules (or the micromorphology of materials) is one of the important factors affecting the diffusion of excitons and charge transfer processes. Poor stacking of the active layer material (or poor crystallinity controlled by the molecular structure) will hinder exciton diffusion and subsequent electron/hole migration to two electrodes. In contrast, the material with excellent stackability and crystallinity forms a large number of aggregations which affect phase separation and exciton diffusion either. Therefore, the selection of donor/acceptor materials is critical to the efficiency of an OSC. The molecular structure and the property determine the accumulation of materials. Of course, during fabrication processes, heat treatment and the use of certain additives also affect the microstructure of an OSC. In summary, the synthesis of new materials and the appropriate selection of donor/acceptor materials have been frontier research fields of OSCs. At present, through the careful selection of active layer materials and the optimization of organic photovoltaic devices, it has been successful to obtain a PCE of more than 17% for single-junction OSCs [2,3,30].

12.5 DONOR AND ACCEPTOR MATERIALS

12.5.1 POLYMERS

A polymer is a chemical compound formed through the polymerization of many repeating organic monomers. The number of repeating units is called the "degree of polymerization." Generally, a polymer with a small degree of polymerization and small molecular mass is called as an "oligomer," and a polymer with a high degree of polymerization and large molecular mass is called a "high polymer."

Polymers are commonly used in OSCs. Certainly, polymer-based OCSs have some drawbacks such as batch-to-batch variation, difficult exciton diffusion, and low mobility of charge carriers. However, polymer-based OSCs are an important research field because they are cheap, easy to produce, and flexible, and can harvest photons efficiently in the visible and NIRs. Furthermore, they can be solution-processed when making a polymer OSC with a roll-to-roll manufacturing process [31]. Polymers can be used as donor and/or acceptor in OSCs. All-polymer-based OSCs have been synthesized and fabricated, and the highest PCE is currently up to 11% [32]. Some of the commonly used polymers in OSCs are listed below.

P3HT: This polymer has some advantages such as simple structure, easy manufacturing, and relatively low cost. The structure of P3HT is shown in Figure 12.7a. It is often used as a donor material. P3HT has a relatively wide band gap, and the wavelength of absorption mainly ranges from 300 to 600 nm. The energy levels of LUMO and HOMO are about −3.0 and −5.0 eV, respectively. In terms of morphology, P3HT exhibits promising crystalline properties after annealing [33].

PTB7-Th: This is a popular polymer donor material. PTB7-Th has a narrow band gap, and its light absorption peak is around 800 nm. The LUMO and HOMO energy levels of PTB7-Th are −3.64 and −5.22 eV, respectively. PTB7-Th also shows a certain crystallinity. Its structure is shown in Figure 12.7b [34].

PDBT-T1: This polymer, often used as a donor material, has a medium band gap, and its wavelength of absorption is around 700 nm. It has a deeper HOMO energy level (−5.36 eV) and LUMO energy level (−3.43 eV). It has good crystallinity and exhibits good phase separation when mixing with other organic acceptor material. The structure of PDBT-T1 is shown in Figure 12.7c [35].

PBDTTT-C-T: This polymer donor material is an analogue of polymer PTB7-Th. Compared to PTB7-Th, its repeating unit lacks a fluorine atom, and some branches of the backbone are different either. PBDTTT-C-T has higher HOMO and LUMO energy values of −5.11 and −3.25 eV, respectively.

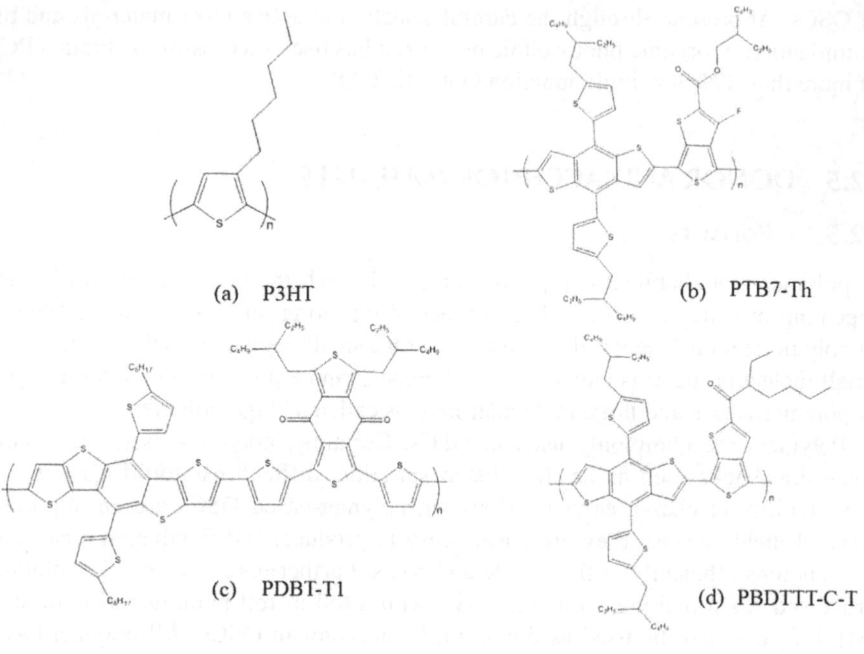

FIGURE 12.7 Molecular structures of typical polymers: (a) P3HT, (b) PTB7-Th, (c) PDBT-T1, and (d) PBDTTT-C-T. "n" denotes the degree of polymerization.

It has a similar light absorption range from 500 to 800 nm like PTB7-Th. Its crystallinity is similar to PTB7-Th as well. The structure of PBDTTT-C-T is shown in Figure 12.7d [36].

In summary, polymer OSCs, including all-polymer OSCs, have a brilliant future because of their unique merits. With the quick development of new organic synthesis method and the presence of new building blocks, polymer OSCs with high PCE (>10%) may be commercialized in 10 years.

12.5.2 Small Organic Molecules

A small organic molecule refers to an organic molecule with a well-defined structure and small molecular mass.

Small organic molecules have been widely used in OSCs because of their advantages such as batch-to-batch reproducibility, crystallinity, easy synthesis, purification, and modification [37–39]. Many small organic molecules have been synthesized and fabricated into photovoltaic devices. To date, the most promising small organic molecules in OSCs are the types of donor–acceptor fused material. Such designed small organic molecules can promote the charge transfer within the molecules due to the "push–pull" effect. In addition, the energy level of frontier molecular orbitals and the range of absorption in the visible and NIRs can be modified through this method [40]. Generally, to synthesize the donor–acceptor fused small organic molecule, there are two typical combinations of donor (D) and acceptor (A): A-D-A and D-A-D. Some typical examples are listed as follows.

1. A-D-A small molecule

There are many A-D-A types of small organic molecules. Here, we select a series of simple organic molecules, named DCCnT, to demonstrate the characteristics of A-D-A type of molecules, as shown in Figure 12.8a. DCCnT (where n is 1–4, which means the number of thiophene rings) is built up with dicyanovinyl as an acceptor unit, thiophene as a donor unit, and cyclohexene as linkage. This series of molecules is a representation of oligomers because the number of repeating units of these molecules is limited to 4 [38,41,42]. In addition, they can also be easily modified by adding different functional groups to improve their performance. These typical A-D-A small molecules have been used as donor materials. For instance, DCC3T-Me is a new small molecule by adding methyl groups on the thiophene ring of DCC3T. The PCE of DCC3T-Me/C_{60} OSC can further be increased to 4.4% [43].

2. D-A-D small organic molecule

Similarly, one D-A-D type of small organic molecule, named TPA-BT-HTT, is given here as one example, as shown in Figure 12.8b. This donor material has a star-shaped configuration. This molecule is synthesized with triphenylamine (TPA) as the core and donor unit, benzothiadiazole as the bridge and acceptor unit, and oligothiophene as another donor unit. The TPA-BT-HTT/$PC_{61}BM$ OSC has a PCE of 4.3%. This star-shaped molecule is also solution-processable [44].

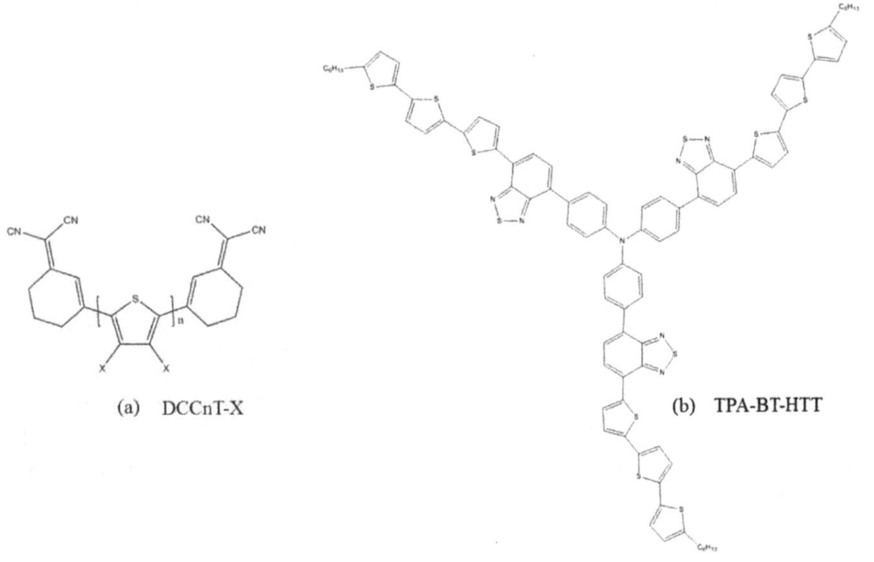

(a) DCCnT-X

(b) TPA-BT-HTT

FIGURE 12.8 Molecular structures of A-D-A type of DCCnT-X (a) and D-A-D type of TPA-BT-HTT (b).

TABLE 12.1

Selected OSCs Fabricated with Small Organic Molecules and Polymer and Their Key Parameters

Small Molecule (Donor)	Small Molecule (Acceptor)	V_{oc} (V)	J_{sc} (mA·cm^{-2})	FF (%)	PCE (%)	References
DF-PCIC	ITIC	0.910	15.66	72.00	10.14	[45]
BDT-2t-ID	PC$_{71}$BM	0.970	14.20	50.00	6.90	[46]
H11	IDIC	0.977	15.21	65.46	9.73	[47]
DRTB-T	IC-C6IDT-IC	0.980	14.25	65.00	9.08	[37]
Polymer (Donor)	**Small Molecule (Acceptor)**	V_{oc} (V)	J_{sc} (mA·cm^{-2})	FF (%)	PCE (%)	References
PBDB-T	NCBDT	0.839	20.33	71.00	12.12	[48]
FTAZ	IDIC	0.840	21.40	67.50	12.14	[49]
PBDT-S-2TC	ITIC	0.960	16.40	64.30	10.12	[50]
FTAZ	INIC3	0.852	19.68	68.50	11.50	[51]
P2F-EHp	Y6	0.810	26.68	74.11	16.02	[52]
PM6	Y6	0.820	25.20	76.10	15.70	[53]
Polymer (Donor)	**Polymer (Acceptor)**	V_{oc} (V)	J_{sc} (mA·cm^{-2})	FF (%)	PCE (%)	References
PBDB-T	N2200	0.800	11.84	54.13	5.15	[54]
PBDB-T	PNDI-2T-TR(5)	0.860	14.34	58.67	7.23	[54]
PTB7-Th	PTbPDI	0.65	10.97	43.00	3.14	[34]
PTB7-Th	PTTbPDI	0.68	8.75	46.00	2.81	[34]
Small Molecule (Donor)	**Polymer (Acceptor)**	V_{oc} (V)	J_{sc} (mA·cm^{-2})	FF (%)	PCE (%)	References
DTSi(FBTTh2Cy)2	N2200	0.78	5.65	60.00	2.68	[55]
DTGe(FBTTh2Cy)2	N2200	0.78	6.08	63.30	3.04	[55]

V_{oc}, open-circuit voltage; J_{sc}, short-circuit current density; FF, fill factor; PCE, power conversion efficiency.

The selected OSCs and their key parameters are listed in Table 12.1.

All molecular structures of donor and acceptor materials mentioned in Table 12.1 are listed in Figures 12.9 and 12.10, respectively.

12.6   FULLERENE AND NON-FULLERENE ACCEPTORS

In the history of OSCs, fullerene materials played a prevailing role. After 2015, since the appearance of high-performance NFAs such as ITIC and Y6 [53,56], the PCE of OSCs has been significantly improved. Therefore, it is worth discussing both fullerene and NFAs in this chapter.

Fullerene and its derivatives have been widely used as acceptor materials in OSCs because of their LUMO energy levels, excellency as electron acceptors, and high isotropic electron mobility. However, fullerene and its derivatives show significant disadvantages such as high price, weak absorption in the visible region, poor stability (easy dimerization), and difficulty in adjusting energy levels through structural changes. Therefore, in recent years, high-performance non-fullerene materials have become the focus of OSCs. Specifically, the single-junction OSCs based on small molecular NFA Y6 have achieved a breakthrough in PCE of more than 17% [2,57,58].

12.6.1 FULLERENE MATERIALS

Ever since the first fullerene (C_{60}) was synthesized in 1985 through laser evaporation of graphite by Richard Smalley, Harry Kroto, and Robert Curl, it has attracted the broad interest of global researchers [59]. Because the molecular structure of C_{60} is similar to that of football, it is also called "footballene." In addition, it was inspired by the spherical dome shell structure of the Montreal World Expo in Canada designed by the architect Buckminster Fuller, so C_{60} is also called "Buckminster fullerene" or "buckyball."

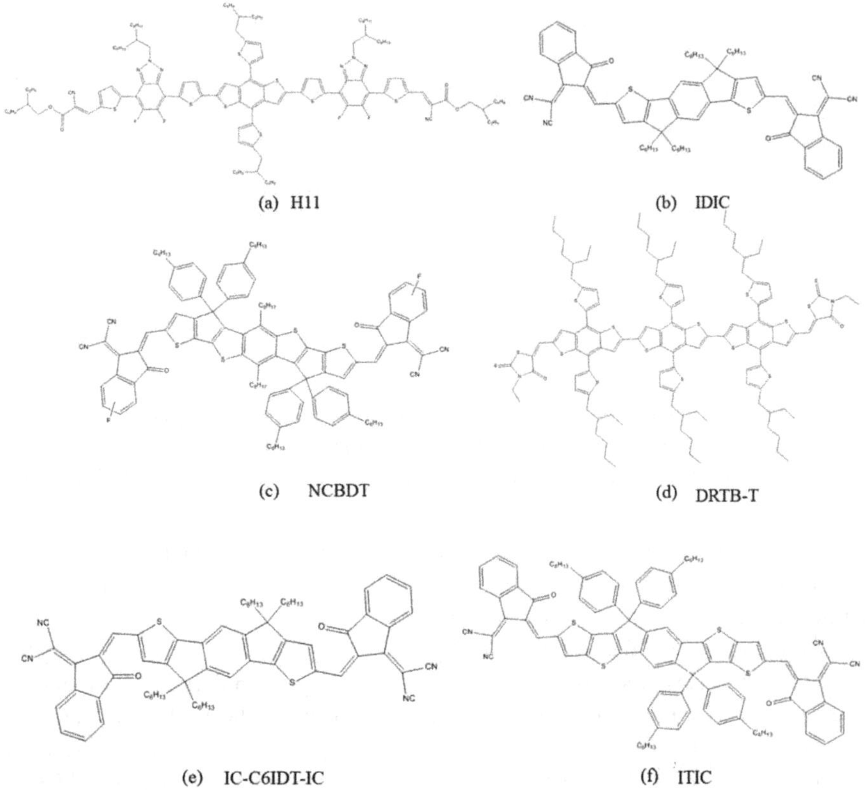

(a) H11   (b) IDIC

(c) NCBDT   (d) DRTB-T

(e) IC-C6IDT-IC   (f) ITIC

FIGURE 12.9 Molecular structures of small organic molecules mentioned in Table 12.1. (a) H11, (b) IDIC, (c) NCBDT, (d) DRTB-T, (e) IC-C6IDT-IC, (f) ITIC, (g) INIC3, (h) BDT-2t-ID, (i) Y6, (j) PC71BM, (k) DF-PCIC, (l) DTSi(FBTTh2Cy)2/DTGe(FBTTh2Cy)2.

(Continued)

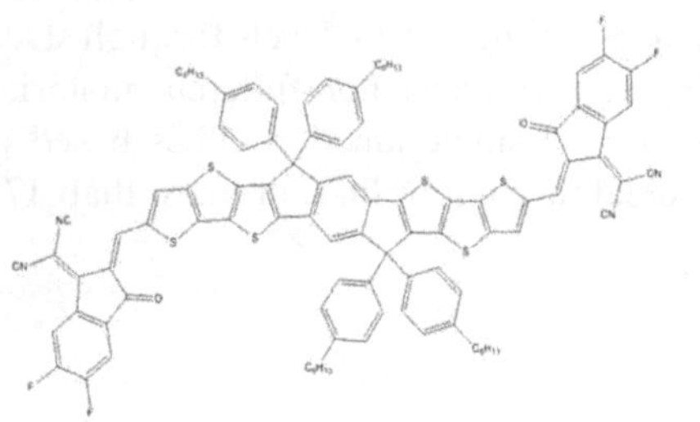

(g)   INIC3

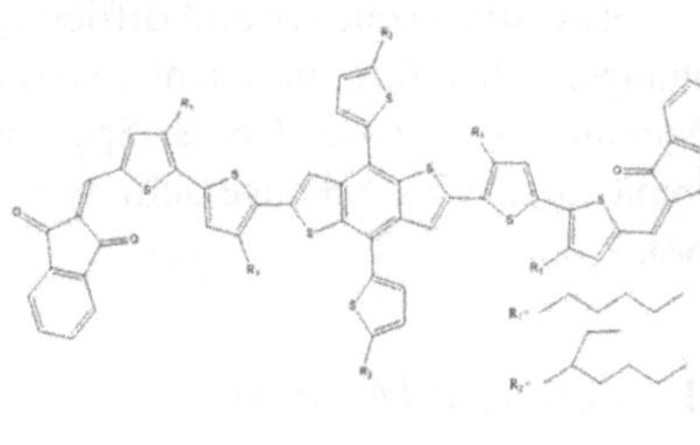

(h)   BDT-2t-ID

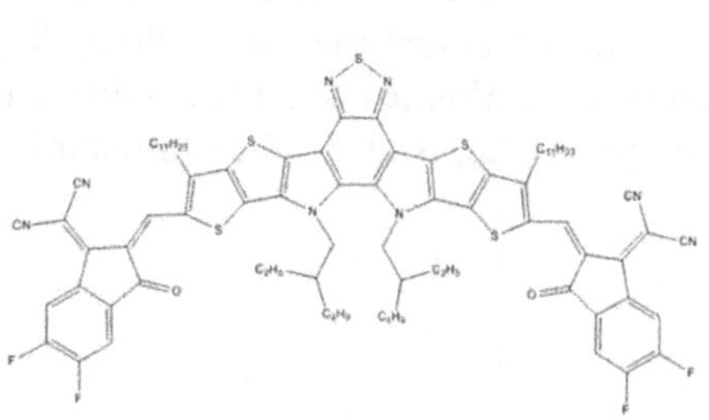

(i)   Y6

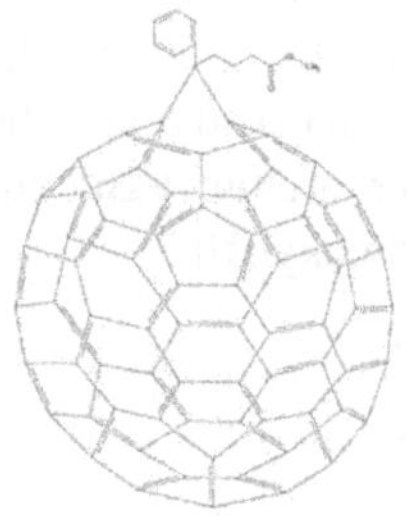

(j)   PC71BM

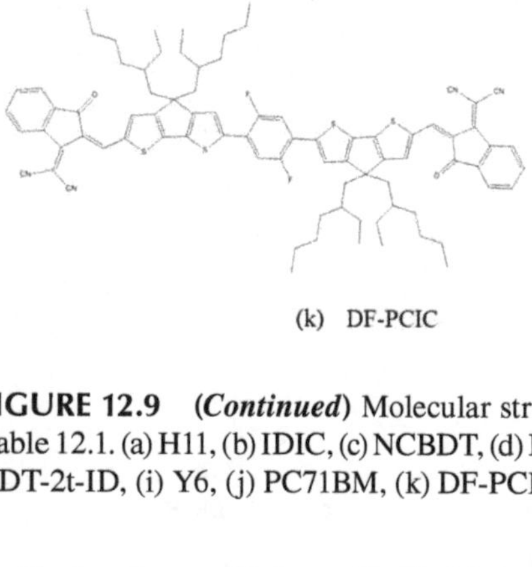

X=Si/Ge

(k)   DF-PCIC

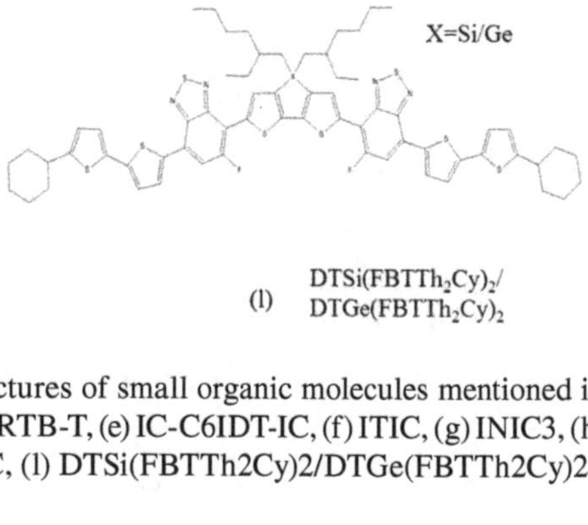

(l)   DTSi(FBTTh₂Cy)₂/
DTGe(FBTTh₂Cy)₂

FIGURE 12.9 (*Continued*) Molecular structures of small organic molecules mentioned in Table 12.1. (a) H11, (b) IDIC, (c) NCBDT, (d) DRTB-T, (e) IC-C6IDT-IC, (f) ITIC, (g) INIC3, (h) BDT-2t-ID, (i) Y6, (j) PC71BM, (k) DF-PCIC, (l) DTSi(FBTTh2Cy)2/DTGe(FBTTh2Cy)2.

C_{60} has been widely applied in biomedicine, catalysts, superconducting materials, organic photovoltaic, and other fields. The methods of making C_{60} include chemical synthesis, arc discharge, and laser heating [60].

C_{60} is composed of 20 hexagons and 12 pentagons. With a large cavity, C_{60} may have some metal particles or small molecules inside to change its properties. Ever since the appearance of C_{60}, more fullerene molecules have been synthesized. Besides C_{60}, C_{70} (consisting of 25 hexagons), $PC_{61}BM$ ([6,6]-phenyl-C_{61}-butyric acid methyl ester), and $PC_{71}BM$ ([6,6]-phenyl-C_{71}-butyric acid methyl ester) are also widely used in OSCs.

The chemical structures of two representative fullerenes, namely, C_{60} and its derivative $PC_{61}BM$, are shown in Figure 12.11.

C_{60} and C_{70} are regarded as excellent electron acceptor materials in OSCs due to their conjugated spherical and ellipsoidal structures, respectively. In order to improve their solubility, functionalized C_{60} and C_{70} have been synthesized. Side chains like

alkoxy group have been added to C_{60} and C_{70} balls to obtain the fullerene derivatives $PC_{61}BM$ and $PC_{71}BM$, and their solubility is improved. In 1992, Sariciftci et al.
discovered that electrons in excited states of the donor could be quickly injected into
C_{60} because the surface of spherical C_{60} is a large conjugated structure, which allows
the electrons to be delocalized very well and stabilized [61,62]. This pioneering work
inspired many researchers. Through the continuous efforts of global researchers, the
efficiency of OSCs with fullerenes as electron acceptors has exceeded 12% [63,64].

However, fullerenes have the following inherent disadvantages:

First, it is difficult to modify and adjust the band gap and LUMO energy of
fullerene. Functional groups (side chains) can be added to C_{60} and C_{70} to
improve their solubility; however, the energy level of frontier molecular

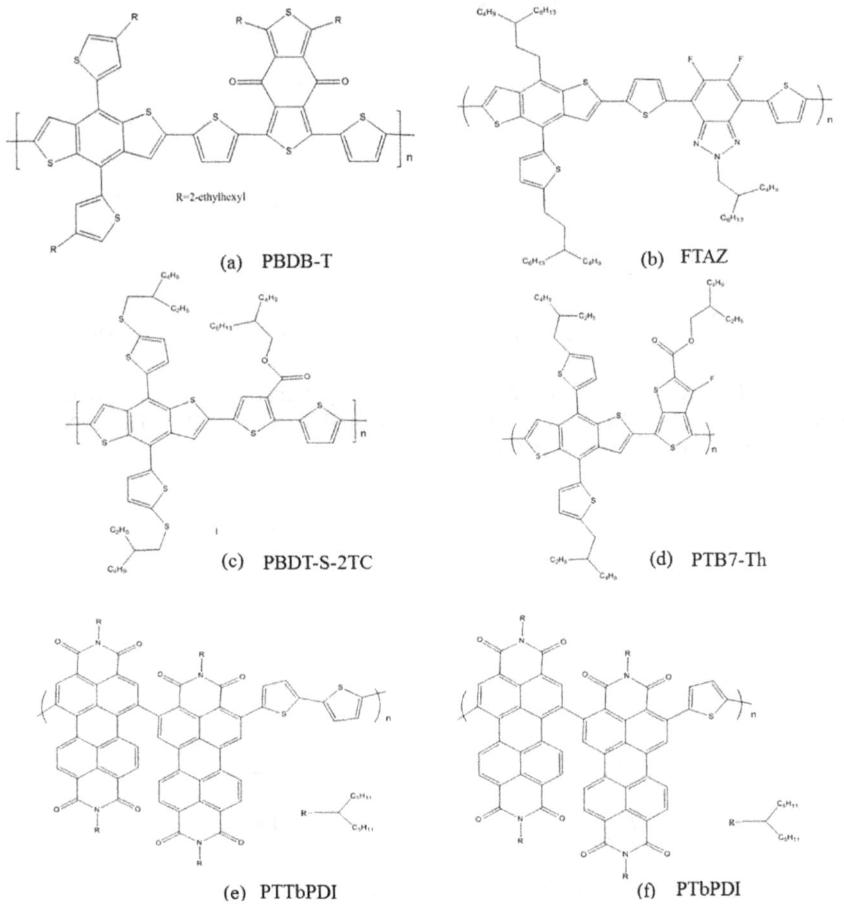

(a) PBDB-T

(b) FTAZ

(c) PBDT-S-2TC

(d) PTB7-Th

(e) PTTbPDI

(f) PTbPDI

FIGURE 12.10 Molecular structures of polymers mentioned in Table 12.1. (a) PBDB-T, (b)
FTAZ, (c) PBDT-S-2TC, (d) PTB7-Th, (e) PTTbPDI, (f) PTbPDI, (g) N2200, (h) PNDI-2T-
TR(x), (i) PM6, and (j) P2F-EHp.

(Continued)

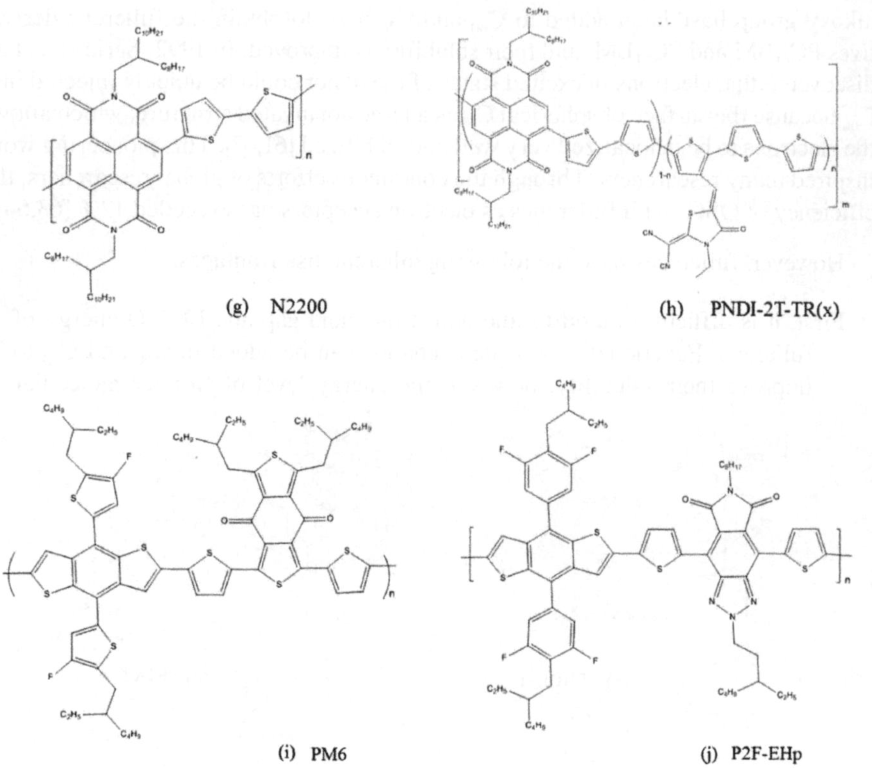

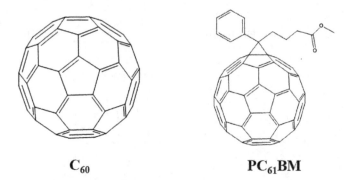

(g) N2200

(h) PNDI-2T-TR(x)

(i) PM6

(j) P2F-EHp

FIGURE 12.10 (*Continued*) Molecular structures of polymers mentioned in Table 12.1. (a) PBDB-T, (b) FTAZ, (c) PBDT-S-2TC, (d) PTB7-Th, (e) PTTbPDI, (f) PTbPDI, (g) N2200, (h) PNDI-2T-TR(x), (i) PM6, and (j) P2F-EHp.

C_{60} $PC_{61}BM$

FIGURE 12.11 The chemical structures of C_{60} and $PC_{61}BM$.

orbitals and light absorption ability cannot be adjusted through this method. Therefore, the selected donor materials have to match the energy level of fullerene acceptor in order to obtain the high open-circuit voltage. This greatly limits the choice of donor materials. Experiments have shown that the PCE of single-junction photovoltaic devices based on fullerene acceptor is limited [63]. Second, as mentioned above, the absorption of OSCs based on fullerene acceptor is limited in the visible and NIRs. The reason for the weak absorption of fullerene is attributed to the fact that the C_{60} and C_{70} are spherical and ellipsoidal structures with high symmetry, respectively. This directly leads to the failure of improving J_{sc} and PCE. Furthermore, fullerenes are easy to aggregate and dimerize due to their molecular structures, affecting the effective separation of charges and stability of OSCs. Finally, the voltage loss of fullerene-based OSC is generally much higher than that of silicon-based and perovskite solar cells, resulting in low open-circuit voltages.

The shortcomings of fullerenes as electron acceptors listed above have prevented the further improvement of the PCE of OSCs in the last two decades. Under this situation, it is very necessary for researchers to find new NFAs to overcome the problems of fullerene acceptors.

Compared to fullerene acceptors, NFAs have the characteristics such as tunable energy levels and absorption range, better solubility, and lower energy loss [17,29]. To date, a large number of NFA materials have been successfully synthesized and applied, achieving the highest PCE over 17% [3,30,37–39]. Among them, two NFA materials, namely, ITIC and Y6, are particularly noteworthy.

12.6.2 NFA ITIC AND ITS DERIVATIVES

In 2015, Dr. Xiaowei Zhan and his research group first synthesized a new type of non-fullerene small-molecule ITIC at the Institute of Chemistry, Chinese Academy of Sciences. The OSC (donor: PTB7-TH; acceptor: ITIC) was further fabricated and a new record PCE of 6.80% was obtained for the fullerene-free OSCs [56]. ITIC has an A-D-A structure, as shown in Figure 12.12. One can see that the backbone of ITIC is composed of a bulky seven-ring-fused core as donor unit, namely, indacenodithieno[3,2-b]thiophene (IT), and two terminal groups, i.e., 2-(3-oxo-2,3-dihydroinden-1-ylidene)malononitrile (IC) as acceptor units. Moreover, four 4-hexylphenyl groups are added as branched chains to increase the solubility and non-planarity.

As a promising NFA, ITIC has excellent photoelectric properties. It has a wide and strong absorption, ranging from 500 to 750 nm. And the HOMO and LUMO energies of this molecule, measured using cyclic voltammetry (CV), are −5.48 and −3.83 eV, respectively [56]. Both values are higher than those of popular fullerene C_{60} (experimental HOMO: −6.20 and LUMO: −3.70 eV) [65]. Since ITIC was synthesized, many kinds of research around this new material have been performed, including donor matching, molecular modification of ITIC, and device optimization [41–43].

Donor matching: By considering the HOMO/LUMO energy level and complementarity of absorption, polymer donors are first selected to pair with ITIC donor because they are of great robustness, tunable with different functional groups, stable, and flexible [66]. In 2016, Li et al. selected two polymer donors named J50 and J51

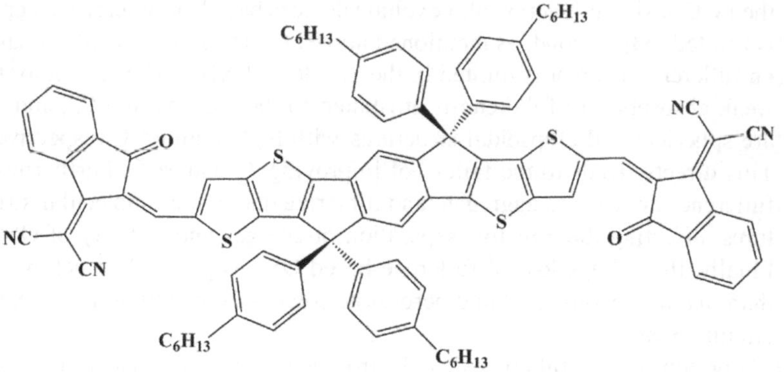

FIGURE 12.12 The molecular structure of ITIC.

to fabricate the devices with ITIC NFA. The difference between J50 and J51 is the fluorination of the backbone of the polymer. A high PCE of 9.26% was achieved by choosing fluorinated J51 as a medium band gap polymer donor and ITIC as a low band gap acceptor [67]. This OSC has excellent complementary absorption, a high short-circuit current density (J_{sc}) of 16.47 mA·cm^{-2}, and a high open-circuit voltage (V_{oc}) of 0.82 V. In contrast, J50 without fluorine substitutions has a modest PCE of 4.8%. This study indicates the importance of fluorination of the backbone of polymer donor materials. Furthermore, a series of J51 derivatives such as J52 with branched alkyl, J60 with branched alkylthio, and J61 with linear alkylthio were synthesized, and the OSCs based on them were tested by the same research group [68]. The champion of these OSCs is J61/ITIC, which exhibits the PCE of 9.53% with an enhanced J_{sc} of 17.43 mA·cm^{-2} and a high V_{oc} of 0.89 V. Very recently, a novel polymer donor material named PBTA-PSF, with benzo(1,2-b:4,5-b')dithiophene (BDT) derivative containing fluorine and sulfur atoms, was successfully synthesized [69]. This copolymer PBTA-PSF showed a low HOMO energy level and complementary absorption with the narrow-bandgap n-type small-molecule ITIC. As a result, the OSC based on PBTA-PSF:ITIC demonstrated a high PCE of 13.91%, which is the highest record of OSCs with ITIC as an acceptor material. The photovoltaic performance parameters of the OSCs based on polymer: ITIC are listed in Table 12.2. In addition, some typical polymer donor materials are shown in Figure 12.13.

Modification of ITIC: Since the synthesis of ITIC in 2015, numerous ITIC derivatives have been designed and synthesized. The modification of ITIC molecule includes the design of a new core ring, adding halogen elements at end groups, and the changes of the side chain. These modifications aim to adjust the energy level and light absorption range of ITIC derivatives, and improve the morphology of blend film, so as to achieve more excellent photovoltaic performance.

The m-ITIC, in which four hexyl groups of ITIC are at the meta-position of the benzene ring, was designed by Li group [70], as shown in Figure 12.14. Compared to the prototype ITIC, m-ITIC shows a higher absorption coefficient and electron mobility. Accordingly, the photovoltaic devices with m-ITIC as an electron acceptor demonstrated a more remarkable performance than ITIC-based devices. Li et al.

TABLE 12.2
The Photovoltaic Performance Parameters of the OSCs Based on Polymer: ITIC

Device	V_{oc} (V)	J_{sc} (mA cm^{-2})	FF	PCE (%)	References
J50: ITIC	0.71	12.93	0.53	4.80	[67]
J51: ITIC	0.82	16.47	0.69	9.26	[67]
J52: ITIC	0.73	13.11	0.58	5.51	[68]
J60: ITIC	0.91	16.33	0.60	8.97	[68]
J61: ITIC	0.89	17.43	0.61	9.53	[68]
PBTA-PSF: ITIC	1.01	18.51	0.74	13.91	[69]

further fabricated the device based on J61:m-ITIC and obtained a higher PCE of 11.77%. It is noticeably higher relative to the OSC based on J61:ITIC (10.57%) [70]. In addition, Li and coworkers synthesized a new donor material named J91. By matching with m-ITIC as electron acceptor, they obtained the PCE of 11.63% [71].

Another side-chain modification of ITIC is the substitution of benzene by thiophene, which was first synthesized by Zhan group and named ITIC-Th [72]. Due to the σ-inductive effect of thienyl side chains, ITIC-Th exhibits a lower energy level: HOMO is −5.66 eV and LUMO is −3.93 eV. When matching with a wide band gap polymer donor PDBT-T1, a PCE value of 9.3% was achieved. In 2017, on the basis of ITIC-Th, Zhan et al. synthesized a new fluorinated molecule named ITIC-Th1 [73]. The introduction of fluorine atoms increases the intramolecular electron push–pull effects and reduces the bandgap, which enhances light harvesting and consequently increases J_{sc}. Therefore, the OSC device with a wide bandgap polymer FTAZ as donor and ITIC-Th1 as acceptor exhibits a PCE of 12.1%. The molecular structures of ITIC-Th and ITIC-Th1 are presented in Figure 12.15.

It is also feasible to introduce a linear alkyl side chain. Han et al. synthesized a new molecule named C8-ITIC by substituting side phenylalkyl chain with a linear eight-carbon alkyl chain. This new molecule has a reduced optical gap and shows better solubility. Han group also synthesized a fluorinated polymer donor material PFBDB-T and achieved an impressive PCE of 13.2% when matching with C8-ITIC acceptor [74].

The introduction of halogen atoms at end groups is another successful strategy to improve the performance of ITIC. Hydrogen atoms of terminal groups are replaced with halogen atoms. Halogen atom is electron deficient. The introduction of halogen can effectively enhance the intramolecular charge transfer effect, redshift absorption spectrum, further reduce the energy level, and influence the morphology of thin film [17]. In addition, the substitutions with different numbers (two or four) of fluorine or chlorine atoms also have different effects. Researchers have synthesized IT-2F, IT-2Cl, IT-4F, and IT-4Cl, as shown in Figure 12.16. Currently, the PCE of fluorinated or chlorinated ITIC-based OSCs has exceeded 13.0% [48,49,75]. More details are given as follows.

In 2017, Hou et al. first synthesized IT-4F and fabricated a device with a novel copolymer PBDB-T-SF as a donor. The measured V_{oc}, J_{sc}, and FF are 0.88 V, 20.5 mA

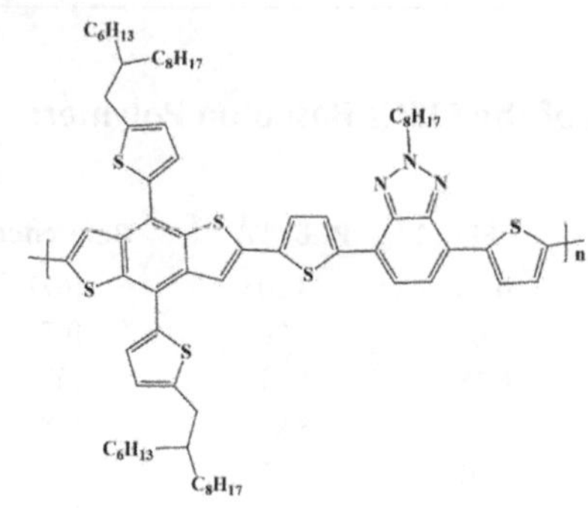

J50

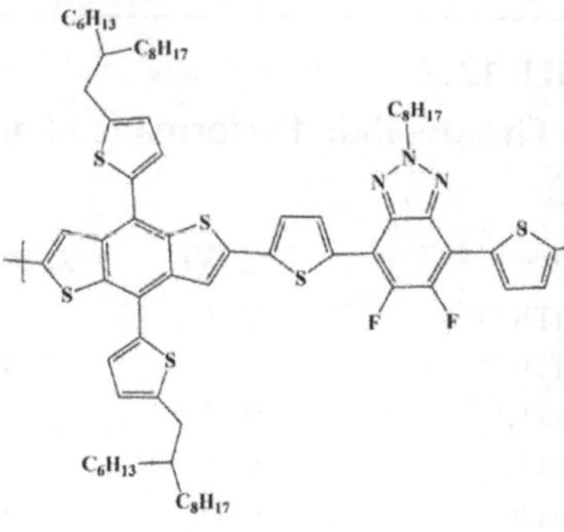

J51

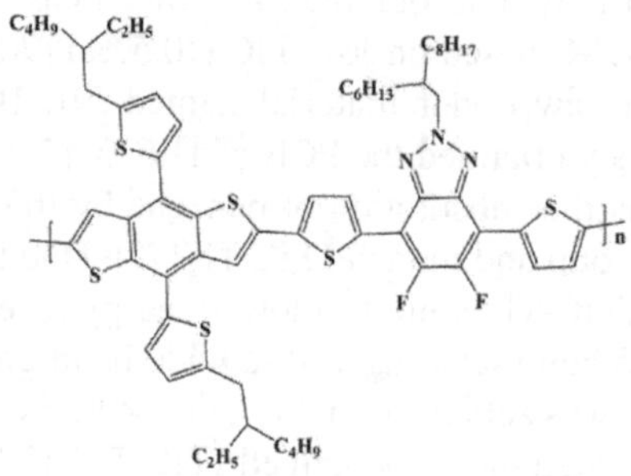

J52

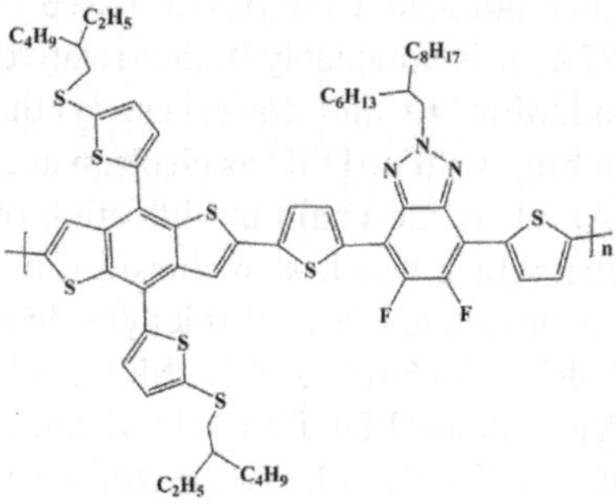

J60

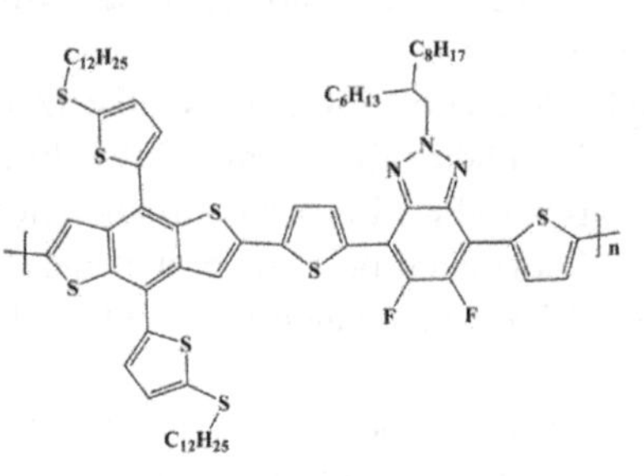

J61

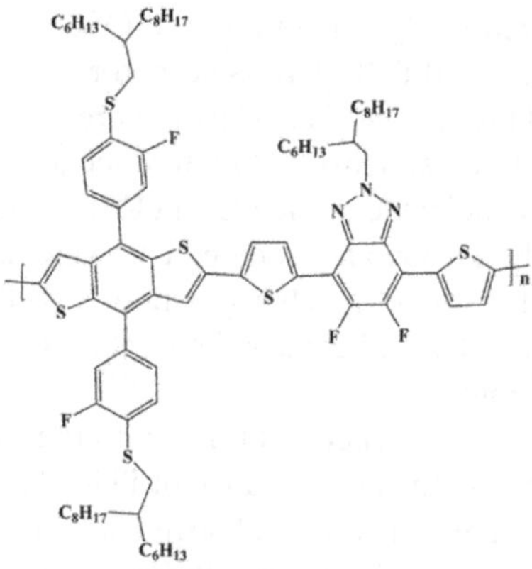

PBTA-PSF

FIGURE 12.13 The molecular structures of polymer donor materials.

cm^{-2}, and 72%, respectively. And this OSC has a record-breaking PCE of 13.1% [76]. Later, the record was further improved. The PCEs of OSCs with the polymers P2 and PBDB-T-2Cl as donors are increased to 14.2% and 14.29%, respectively [49,51]. In 2018, Hou et al. designed and synthesized chlorinated ITIC: IT-2Cl and IT-4Cl

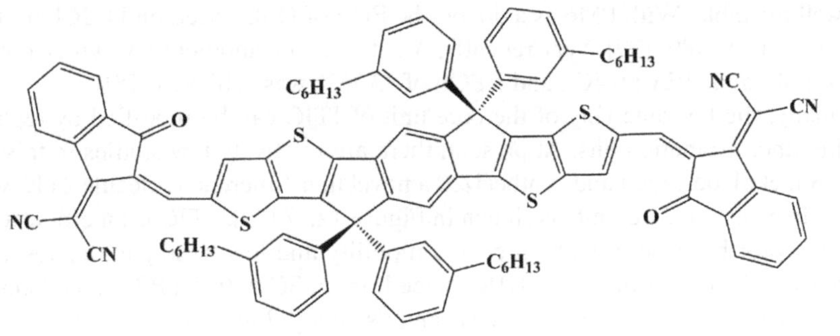

FIGURE 12.14 Molecular structure of m-ITIC.

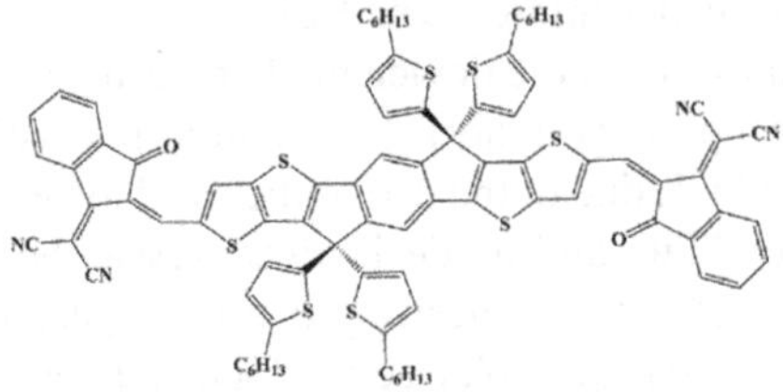

ITIC-Th 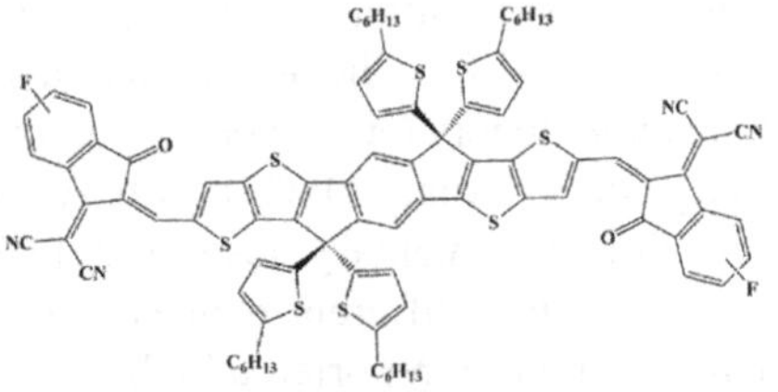   ITIC-Th1

FIGURE 12.15 Molecular structures of ITIC-Th and ITIC-Th1.

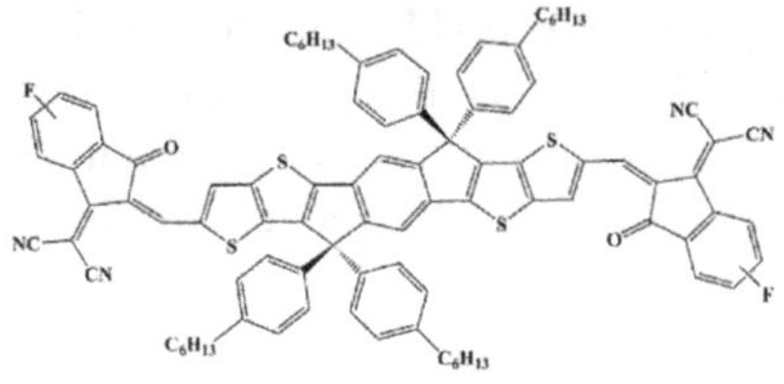

IT-2F 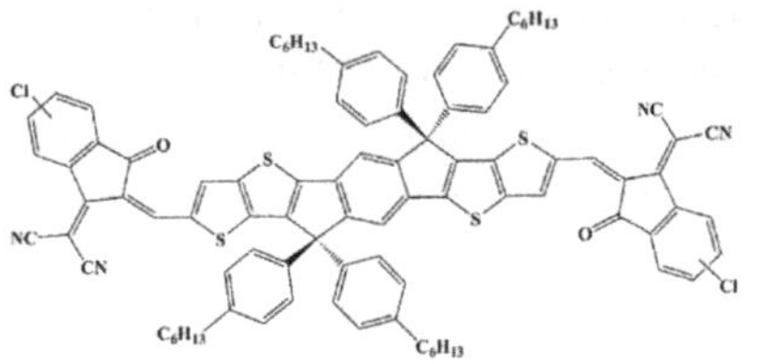   IT-2Cl

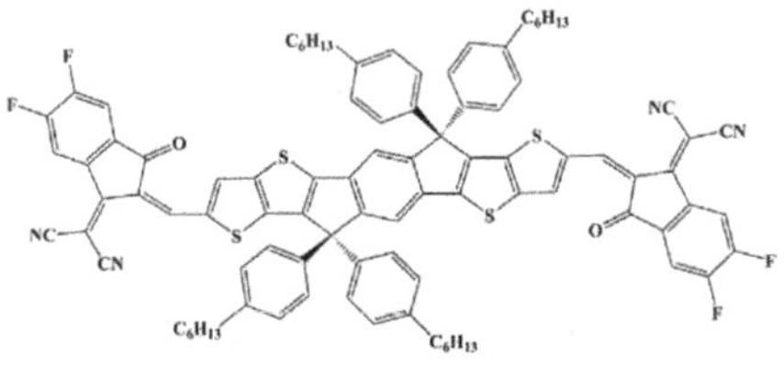

IT-4F 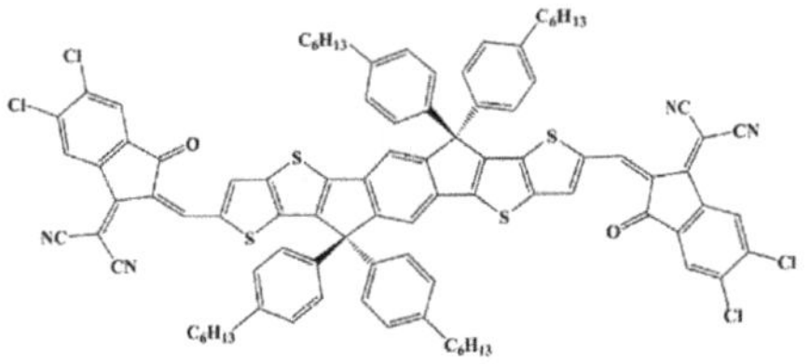   IT-4Cl

FIGURE 12.16 Molecular structures of IT-2F, IT-2Cl, IT-4F, and IT-4Cl.

[77]. Because the C–Cl bond has a large local dipole moment, it enhances the intermolecular charge transfer effect and the molecular packing, thereby expanding absorption range in the visible region and lowering the energy level of frontier molecular orbitals. Both IT-2Cl and IT-4Cl have similar chemical structures and

are well miscible. With PM6 as a donor, the PCE of OSC based on IT-2Cl or IT-4Cl NFA is up to 14.18% [77]. Very recently, An et al. made another OSC with a polymer donor PM7 and NFA IT-4Cl, and a PCE of 13.76% was achieved [78].

Finally, the benzene ring of the core unit of ITIC can be modified by replacing it with other aromatic units. At present, there are relatively few studies in this area. Only Shi et al. designed and synthesized a novel non-fullerene molecule 4TIC with a bithiophene as the core unit, as shown in Figure 12.17 [79]. 4TIC is an efficient electron acceptor because of its high charge mobility and low band gap. Its absorption range extends to 900 nm in the NIR. Indeed, the OSC with PTB7-Th as donor and 4TIC as acceptor has a PCE of 10.43%, and its energy loss is low to 0.33 eV.

The key photovoltaic parameters of OSCs with ITIC derivatives as electron acceptors are summarized in Table 12.3. Clearly, the three strategies of modification of ITIC to improve its performance are valid and successful. Of course, it is also important to find suitable donor materials to further improve the PCE.

Ternary OSCs: The definition of a ternary OSC is the device made by using two compatible donors (or acceptors) and one acceptor (or donor) [67]. Comparing with binary OSCs, ternary OSCs may have higher PCE due to the use of the third materials with complementary absorption [81]. For ITIC and its derivatives, researchers have performed different recipes of ternary OSC to further improve their performance. Sun et al. reported a high-performance ternary solar cell, which was fabricated with two NFAs (SdiPBI-Se and ITIC-Th) and a polymer donor (PDBT-T1). The PCE of this ternary OSC is 10.27% [82]. Later, the same group fabricated another ternary OSC with a polymer donor (PDBT-T1) and two acceptor materials (PC$_{71}$BM and ITIC-Th) in 2017, and the PCE was increased to 10.48% [81]. Hou et al. obtained 14.18% high efficiency by combining two compatible NFAs (IT-2Cl and IT-4Cl) with the copolymer donor PM6 [77]. Note that the third component is usually used to improve the blend morphology of the active layer and harvest more photons. The photovoltaic parameters of ternary OSCs based on ITIC derivatives are listed in Table 12.4.

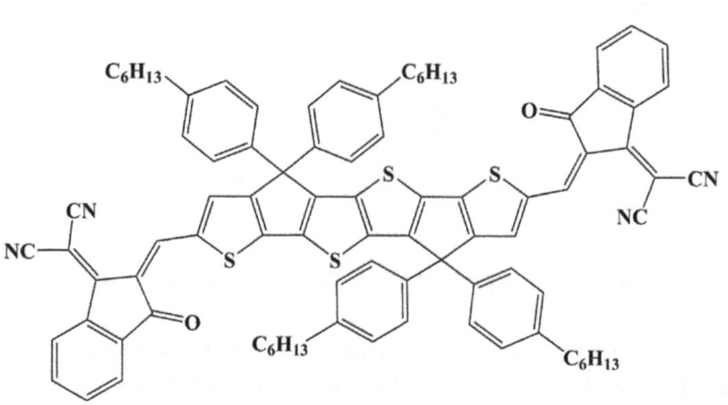

FIGURE 12.17 Molecular structure of 4TIC.

TABLE 12.3
The Photovoltaic Performance Parameters of the OSCs Based on Polymer: ITIC Derivatives

Device	V_{oc} (V)	J_{sc} (mA cm^{-2})	FF	PCE (%)	References
J61:m-ITIC	0.91	18.31	0.71	11.77	[70]
J91:m-ITIC	0.98	18.03	0.66	11.63	[71]
PDBT-T1:ITIC-Th	0.88	16.24	0.67	9.60	[72]
FTAZ:ITIC-Th1	0.85	19.33	0.74	12.10	[73]
PFBDB-T:C8-ITIC	0.94	19.6	0.72	13.20	[74]
PBDB-T-SF:IT-4F	0.88	20.50	0.72	13.10	[76]
PBDB-T-2Cl:IT-4F	0.89	20.58	0.78	14.29	[75]
P2:IT-4F	0.90	20.73	0.76	14.20	[80]
PM7:IT-4Cl	0.87	20.53	0.77	13.76	[78]
PTB7-Th:4TIC	0.78	18.8	0.72	10.43	[79]

TABLE 12.4
The Photovoltaic Performance Parameters of the Ternary OSCs

Device	V_{oc} (V)	J_{sc} (mA cm^{-2})	FF	PCE (%)	References
PDBT-T1: SdiPBI-Se: ITIC-Th	0.93	15.37	0.70	10.27	[82]
PDBT-T1: PC$_{71}$BM :ITIC-Th	0.93	15.54	0.71	10.48	[81]
PSTZ:ITIC:IDIC	0.95	17.40	0.67	11.10	[83]
PM6:IT-4Cl:IT-2Cl	0.84	22.03	0.76	14.18	[77]
PBT1-C:ICBA:ITIC-2Cl	0.90	19.7	0.77	13.4	[84]

12.6.3 NFA Y6 AND ITS DERIVATIVES

High-performance NFAs develop very quickly. ITIC and its derivatives are based on an A-D-A-type molecular framework [77]. To further enhance light harvesting in the NIR, a fused DAD (donor–acceptor–donor) moiety was introduced to the core of A-D-A NFA in order to improve the intramolecular electron push–pull effect and lower the optical band gaps [85]. This strategy is first proposed and implemented by Prof. Yingping Zou, who along with coworkers developed a high-performance A-DAD-A-type NFA, named Y6, as shown in Figure 12.18. This molecule contains a fused DAD core, i.e., dithienothiophen[3.2-b]-pyrrolobenzothiadiazole, (TPBT), and two terminal groups 2-(5,6-difluoro-3-oxo-2,3-dihydro-1H-inden-1-ylidene) malononitrile (2FIC) as acceptor units. The PCE of the first single-junction OSC with Y6 as acceptor (a polymer PM6 as the donor) is already up to 15.7% [53]. This is a breakthrough. Until now, the maximum PCE for a single-junction OSC based on Y6 has reached 17.35%, which is certified by NREL (National Renewable Energy Laboratory) [3].

As shown in Figure 12.18 and given in Table 12.5, A-DAD-A-type NFAs can be divided into two types according to the differently fused DAD cores: benzo[b]

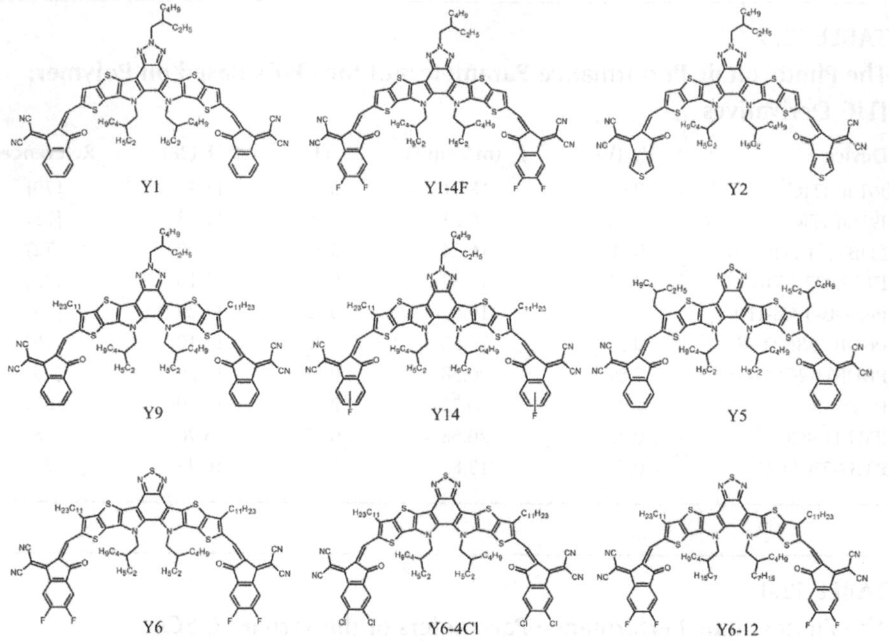

FIGURE 12.18 Molecular structures of A-DAD-A-type NFAs.

[1,2,3]triazole and 2,1,3-benzothiadiazole. The former includes Y1, Y1–4F, Y2, Y9, and Y14, in which Y1 can be regarded as the prototype. The latter includes Y5, Y6, Y6-4Cl, and Y6–12. Similarly, Y5 is the prototype. Please note that PM6 is also named PBDB-T-F.

As mentioned before, introducing halogen atoms (F, Cl) into the end group of the donor–acceptor fused NFAs is one of the successful strategies to enhance the absorption in the visible region and lower energy gap [83]. Meanwhile, introducing halogen atoms influences the molecular interaction behavior, thereby affecting the morphology of the active layer [77]. Indeed, compared to both prototypes Y1 and Y5, the V_{oc} of OSCs with halogenated NFAs (Y1–4F, Y14, Y6, and Y6-4Cl) slightly decreases, while J_{sc} and FF increase noticeably, and the PCEs are improved [85–87].

Side-chain engineering is another effective strategy for manipulating energy levels and morphology and improving the photovoltaic performance of devices based on these NFAs. For instance, Y6–12 has a longer alkyl side chains (2-butyloctyl) than Y6 (2-ethylhexyl), which improves crystallinity. In addition, a higher electron mobility was observed due to the enhanced lamellar stacking in in-plane direction [87]. Likewise, an undecyl chain was added to thienylthiophene in Y9 prototype with respect to Y1 prototype. With the same donor, despite the slight increase in V_{oc} and J_{sc} of OSC which is based on Y9, the effect of additional undecyl chain on intramolecular interactions resulted in a significant reduction of FF. Even after the optimization of morphology with 1-chloronaphthalene, the PCE of Y9-based OSC was still slightly lower than that of Y1-based OSC [88].

TABLE 12.5

Optoelectronic Properties and OSC Device Parameters of A-DAD-A Type NFAs

Acceptor	λ_{max} (nm)	LUMO (eV)	HOMO (eV)	Gap (eV)	Donor	V_{oc} (V)	J_{sc} (mA cm^{-2})	FF (%)	PCE (%)	References
Y1	802	−3.95	−5.45	1.50	PBDB-T	0.87	21.68	70.12	13.42	[85]
					PM6	0.92	12.9	55.50	7.10	[85]
Y1-4F	851	−4.11	−5.56	1.45	PBDB-T	0.74	22.7	57.4	9.90	[85]
					PM6	0.83	25.2	68.5	14.80	[85]
Y2	827	−4.04	−5.43	1.39	PBDB-T	0.82	23.12	70.80	13.40	[85]
Y9	808	−3.78	−5.59	1.81	PBDB-T	0.90	23.28	63	13.26	[88]
Y14	853	−4.01	−5.56	1.55	PBDB-T	0.80	26.15	71.48	14.92	[89]
					PM6	0.87	16.83	58.24	8.49	[89]
Y5	783	−3.87	−5.55	1.68	PBDB-T	0.87	22.7	70.2	14.0	[90]
					PM6	0.94	13.2	59.6	7.5	[85]
Y6	821	−4.10	−5.65	1.55	PBDB-T	0.72	25.0	62.1	11.2	[85]
					PM6	0.84	24.7	76.7	15.9	[85]
Y6-4Cl	839	−4.12	−5.68	1.56	PM6	0.87	25.2	73.7	16.5	[86]
Y6–12	-	−4.06	−5.68	1.62	PM6	0.85	25.0	74	16.4	[87]

Note that PBDB-T-F is also named PM6.

Except for the strategies mentioned above, Zou et al. replaced the end-group INIC (2-(3-oxo-2,3-dihydro-1-H-indene 2,1ylidene)malononitrile) of Y1 with the stronger electron-deficient INTC (2-(6-oxo-5,6-dihydro-4H-cyclopenta[c]thiophen-4-ylidene) malononitrile). The new molecule is named Y2 molecule [85]. Through this way, the molecular energy levels and absorption spectrum of A-DAD-A-type of molecule can be effectively tuned by manipulating the intramolecular electron push–pull effects. The enhancement of intramolecular push–pull effects can decrease the energy gap and further increase the absorption in the NIR. However, the lowered LUMO energy level (improving driving force) reduces the open-circuit voltage. This is a dilemma. So the modification of Y1 by changing end acceptor units does not significantly improve the photovoltaic performance.

Except for the molecular modifications of NFA Y6 to improve PCE, another effective strategy is to find better donor materials to match [91,92]. The chemical structures of selected donors and key parameters of OSCs with these donors and Y6 as acceptor are presented in Figure 12.19 and Table 12.6, respectively.

For most of the classical and promising polymer donors, they generally have D–π–A backbone, which consists of donor (D), π bridge, and acceptor (A) units [93]. The absorption spectra, energy levels, crystallinity, and charge mobility of D–π–A types of polymers can be modulated by changing the building blocks [94]. So far, there are three popular methods to optimize polymer donor: extending conjugated length, changing side chains, and introducing fluorine or chlorine atom into building blocks. Until now, for Y6 NFA, many polymer donors have been modified and tested to pack high-performance OSCs. With the introduction of the chlorine and fluorine atoms to the prototype PBDB-T, there are two new derivatives: PBDB-T-Cl and PM6. The latter as a polymer donor is fabricated with Y6 acceptor, and the OSC has the best photovoltaic performance (PCE = 17%) until now [58,95]. Similarly, with respect to PE61, the introduction of fluorine atom also significantly improves the performance of PE62-based OSC. Next, changing side chains is another effective strategy. With different side chains (–H and –OCH$_3$), the OSCs based on polymers PE31 and PE32 have very different FF and J_{sc}. For PE62 and PE63 (with/without sulfur atom on side chain), the OSCs based on them have small differences of J_{sc}, V_{oc}, and FF. Finally, the combination of three strategies is often applied. For PE2 and PE4, changing side chains, introducing halogen atoms, and extending conjugated length are all applied. Likewise, the combination of introducing a halogen atom and changing side chain is applied for J51-Cl and J52-FS [52,91,96].

Besides polymers, the donors of small organic molecules have been applied to fabricate with Y6 to make high-performance all-small-molecule (ASM) OSC. Compared to polymers, small molecules have merits such as high purity, well-defined molecular structures, and high reproducibility in OSCs [29]. Lu et al. developed a small-molecule donor BTR-Cl, as shown in Figure 12.19. The BTR-Cl:Y6 OSC has liquid crystalline morphology in the active layer, leading to a PCE of 13.6% [97]. Wei et al. designed and synthesized a small-molecule donor, named ZR1. The single-crystal structural analyses of ZR1 reveal that it has an explicit molecular planarity and compact intermolecular packing. The ASM OSC of ZR1/Y6 achieves the recorded PCE of 14.34% by optimizing the hierarchical morphologies [98].

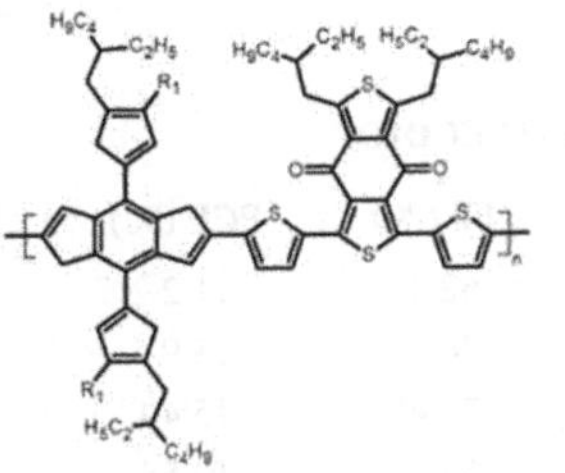

PBDB-T R_1 = -H
PBDB-T-F R_1 = -F
PBDB-T-Cl R_1 = -Cl

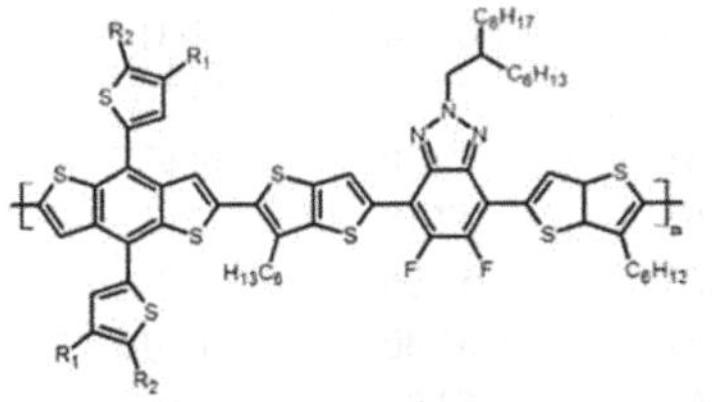

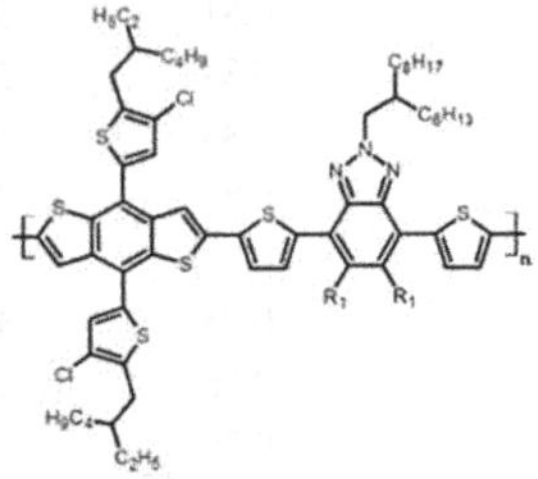

PE2 R_1 = -F, R_2 = $-S-CH_2CH(C_2H_5)C_4H_9$
PE4 R_1 = -Cl, R_2 = $-CH_2CH(C_2H_5)C_4H_9$

PE31 R_1 = -H
PE32 R_1 = $-OCH_3$

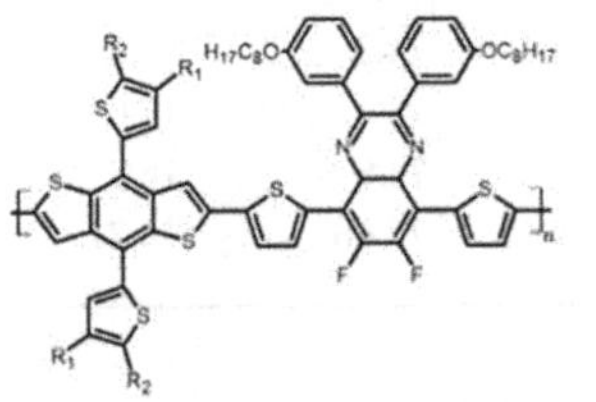

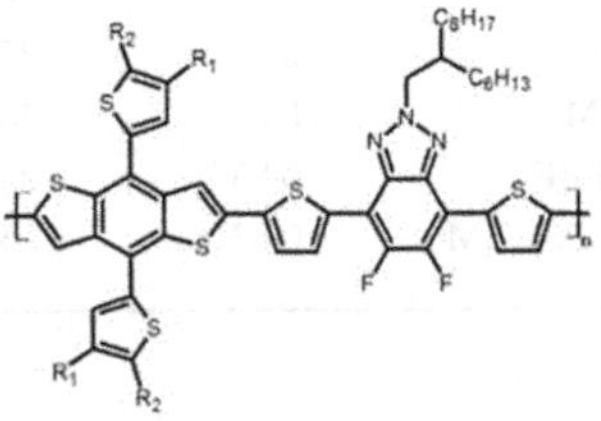

PE61 R_1 = -H, R_2 = $-CH_2CH(C_2H_5)C_4H_9$
PE62 R_1 = -F, R_2 = $-CH_2CH(C_2H_5)C_4H_9$
PE63 R_1 = -F, R_2 = $-S-CH_2CH(C_2H_5)C_4H_9$

J52-FS R_1 = -F, R_2 = $-S-CH_2CH(C_2H_5)C_4H_9$
J52-Cl R_1 = -Cl, R_2 = $-CH_2CH(C_2H_5)C_4H_9$

BTR-Cl

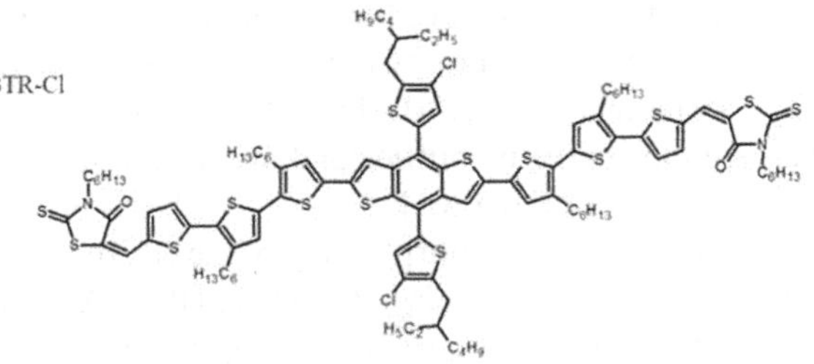

ZR1

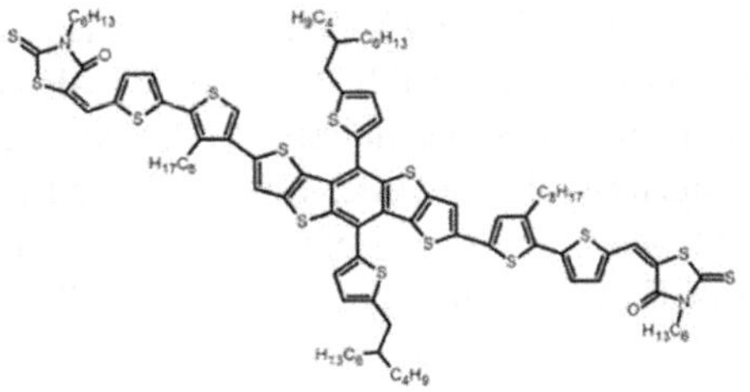

FIGURE 12.19 Chemical structures of polymer and small-molecule donor matching with Y6 acceptor. Note that PBDB-T-F is also named PM6.

TABLE 12.6

Photovoltaic Parameters of OSCs with Y6 as Acceptor

Active Layer	V_{oc} (V)	J_{sc} mA.cm^{-2}	FF (%)	PCE (%)	References
PBDB-T:Y6	0.72	25.0	62.1	11.2	[85]
PM6:Y6	0.84	24.7	76.7	15.9	[85]
PBDB-T-Cl:Y6	0.87	24.98	71.42	15.49	[87]
J52-FS:Y6	0.81	22.92	57	10.58	[96]
J52-Cl:Y6	0.84	23.18	61.44	12.31	[52]
PE2:Y6	0.83	23.24	70	13.50	[96]
PE4:Y6	0.83	21.68	75.20	14.02	[52]
PE31:Y6	0.80	20.63	47.18	7.62	[52]
PE32:Y6	0.75	17.53	51.95	7.31	[52]
PE61:Y6	0.66	23.41	55.30	8.61	[91]
PE62:Y6	0.78	24.64	62.22	12.02	[91]
PE63:Y6	0.83	24.68	63.74	13.10	[91]
BTR-Cl:Y6	0.86	24.17	65.5	13.61	[97]
ZR1:Y6	0.86	24.34	68.44	14.34	[90]
PM6:Y6:PC$_{61}$BM	0.84	25.1	74.3	16.5	[102]
PM6:Y6:PC$_{71}$BM	0.84	26.0	78	17.0	[58]
PM6:Y6:3TP3T-4F	0.85	25.9	74.9	16.7	[103]
PBDB-T-Cl:Y6:PC$_{71}$BM	0.87	25.44	75.66	16.71	[87]

Multiple-compound blending in the active layers is an effective approach to integrate the advantages of different donors or acceptors [99]. With the addition of fullerene derivatives or ITIC derivatives, the PCEs of highly efficient ternary OSCs have been up to 17.0% [58]. Through adding the third component, the homogeneous film morphology and the π–π stacking pattern of the host binary blend are maintained, and the phase purity can be increased [100]. In addition, the third component like PCBM can also increase the charge mobilities and absorption in the visible region, leading to the increase in J_{sc} and FF [87,101].

In OSCs, the molecular packing and morphology of the active layer have a significant impact on exciton separation and charge transport [104]. A counterexample is A-DAD-A BTPT-4F NFA synthesized by Lei et al. [30]. The PCE of single-junction P2F-EHp: BTPT-4F OSC is only 1.09%. The main reason for the poor photovoltaic performance of device based on P2F-EHp: BTPT-4F is that large aggregates or grains in this system should be detrimental to the efficient charge separation at the donor–acceptor interfaces. Meanwhile, Lei et al. used dibenzylether (DBE) as the solvent additive to optimize the morphology of the active layer. As the DBE ratio is adjusted, the PCE of the OSC device based on P2F-EHp: BTPTT-4F(Y6) is increased from 11.14% to 16.02% [30].

In summary, the addition of additives and the modification of donor and acceptor can adjust the morphology of the active layer, tune the energy level, and improve the photoelectric performance of OSC devices. However, the morphology of the active layer is very sensitive to the composition of each component.

At present, the successful modifications based on PM6/Y6 are some small changes under the condition of maintaining its original configuration. In the development of new A-DAD-A NFAs, the adjustment of active layer morphology is still a big challenge.

REFERENCES

1. M. Notarianni, J. Liu, K. Vernon, N. Motta, Synthesis and applications of carbon nano-materials for energy generation and storage, *Beilstein Journal of Nanotechnology* 7 (2016) 149–196.
2. X. Xu, K. Feng, Z. Bi, W. Ma, G. Zhang, Q. Peng, Single-junction polymer solar cells with 16.35% efficiency enabled by a platinum(II) complexation strategy, *Advanced Materials* 0 (2019) 1901872.
3. M.A. Green, E.D. Dunlop, D.H. Levi, J. Hohl-Ebinger, M. Yoshita, A.W.Y. Ho-Baillie, Solar cell efficiency tables (version 55), *Progress in Photovoltaics* 28 (2020) 3–15.
4. C.W. Tang, 2-Layer organic photovoltaic cell, *Applied Physics Letters* 48 (1986) 183–185.
5. M.A. Green, K. Emery, Y. Hishikawa, W. Warta, E.D. Dunlop, Solar cell efficiency tables (version 46), *Progress in Photovoltaics* 23 (2015) 805–812.
6. M.A. Green, K. Emery, Y. Hishikawa, W. Warta, E.D. Dunlop, Solar cell efficiency tables (version 48), *Progress in Photovoltaics* 24 (2016) 905–913.
7. M.A. Green, K. Emery, Y. Hishikawa, W. Warta, E.D. Dunlop, Solar cell efficiency tables (version 47), *Progress in Photovoltaics* 24 (2016) 3–11.
8. M.A. Green, Corrigendum to solar cell efficiency tables (version 49), *Progress in Photovoltaics* 25 (2017) 333–334.
9. M.A. Green, K. Emery, Y. Hishikawa, W. Warta, E.D. Dunlop, D.H. Levi, A.W.Y. Ho-Baillie, Solar cell efficiency tables (version 49), *Progress in Photovoltaics* 25 (2017) 3–13.
10. M.A. Green, Y. Hishikawa, W. Warta, E.D. Dunlop, D.H. Levi, J. Hohl-Ebinger, A.W.Y. Ho-Baillie, Solar cell efficiency tables (version 50), *Progress in Photovoltaics* 25 (2017) 668–676.
11. M.A. Green, Y. Hishikawa, E.D. Dunlop, D.H. Levi, J. Hohl-Ebinger, A.W.Y. Ho-Baillie, Solar cell efficiency tables (version 52), *Progress in Photovoltaics* 26 (2018) 427–436.
12. M.A. Green, Y. Hishikawa, E.D. Dunlop, D.H. Levi, J. Hohl-Ebinger, A.W.Y. Ho-Baillie, Solar cell efficiency tables (version 51), *Progress in Photovoltaics* 26 (2018) 3–12.
13. M.A. Green, E.D. Dunlop, D.H. Levi, J. Hohl-Ebinger, M. Yoshita, A.W.Y. Ho-Baillie, Solar cell efficiency tables (version 54), *Progress in Photovoltaics* 27 (2019) 565–575.
14. M.A. Green, Y. Hishikawa, E.D. Dunlop, D.H. Levi, J. Hohl-Ebinger, M. Yoshita, A.W.Y. Ho-Baillie, Solar cell efficiency tables (version 53), *Progress in Photovoltaics* 27 (2019) 3–12.
15. D. Kearns, M. Calvin, Photovoltaic effect and photoconductivity in laminated organic systems, *Journal of Chemical Physics* 29 (1958) 950–951.
16. J. Hou, O. Inganas, R.H. Friend, F. Gao, Organic solar cells based on non-fullerene acceptors, *Nature Materials* 17 (2018) 119–128.
17. G. Feng, J. Li, Y. He, W. Zheng, J. Wang, C. Li, Z. Tang, A. Osvet, N. Li, C.J. Brabec, Y. Yi, H. Yan, W. Li, Thermal-driven phase separation of double-cable polymers enables efficient single-component organic solar cells, *Joule* 3 (2019) 1765–1781.
18. S. Lucas, T. Leydecker, P. Samori, E. Mena-Osteritz, P. Bauerle, Covalently linked donor-acceptor dyad for efficient single material organic solar cells, *Chemical Communications (Cambridge, England)* 55 (2019) 14202–14205.

19. L. Meng, Y. Zhang, X. Wan, C. Li, X. Zhang, Y. Wang, X. Ke, Z. Xiao, L. Ding, R. Xia, H.-L. Yip, Y. Cao, Y. Chen, Organic and solution-processed tandem solar cells with 17.3% efficiency, *Science* 361 (2018) 1094–1098.

20. M. Knupfer, Exciton binding energies in organic semiconductors, *Applied Physics a-Materials Science and Processing* 77 (2003) 623–626.

21. R.S. Bhatta, M. Tsige, Chain length and torsional dependence of exciton binding energies in P3HT and PTB7 conjugated polymers: A first-principles study, *Polymer* 55 (2014) 2667–2672.

22. K. Hummer, P. Puschnig, S. Sagmeister, C. Ambrosch-Draxl, Ab-initio study on the exciton binding energies in organic semiconductors, *Modern Physics Letters B* 20 (2006) 261–280.

23. P. Puschnig, C. Ambrosch-Draxl, Excitons in organic semiconductors, *Cr Physics* 10 (2009) 504–513.

24. W. Zhao, D. Qian, S. Zhang, S. Li, O. Inganäs, F. Gao, J. Hou, Fullerene-free polymer solar cells with over 11% efficiency and excellent thermal stability, *Advanced Materials* 28 (2016) 4734–4739.

25. S. Li, Z. Zhang, M. Shi, C.Z. Li, H. Chen, Molecular electron acceptors for efficient fullerene-free organic solar cells, *Physical Chemistry Chemical Physics: PCCP* 19 (2017) 3440–3458.

26. N.K. Elumalai, A. Uddin, Open circuit voltage of organic solar cells: An in-depth review, *Energy and Environmental Science* 9 (2016) 391–410.

27. M.C. Scharber, D. Mühlbacher, M. Koppe, P. Denk, C. Waldauf, A.J. Heeger, C.J. Brabec, Design rules for donors in bulk-heterojunction solar cells: Towards 10% energy-conversion efficiency, *Advanced Materials* 18 (2006) 789–794.

28. J. Liu, S. Chen, D. Qian, B. Gautam, G. Yang, J. Zhao, J. Bergqvist, F. Zhang, W. Ma, H. Ade, O. Inganäs, K. Gundogdu, F. Gao, H. Yan, Fast charge separation in a non-fullerene organic solar cell with a small driving force, *Nature Energy* 1 (2016) 16089.

29. Y. Chen, X. Wan, G. Long, High performance photovoltaic applications using solution-processed small molecules, *Accounts of Chemical Research* 46 (2013) 2645–2655.

30. B. Fan, D. Zhang, M. Li, W. Zhong, Z. Zeng, L. Ying, F. Huang, Y. Cao, Achieving over 16% efficiency for single-junction organic solar cells, *Science China Chemistry* 62 (2019) 746–752.

31. A.J. Heeger, 25th anniversary article: Bulk heterojunction solar cells: Understanding the mechanism of operation, *Advanced Materials* 26 (2014) 10–28.

32. Z. Li, L. Ying, P. Zhu, W. Zhong, N. Li, F. Liu, F. Huang, Y. Cao, A generic green solvent concept boosting the power conversion efficiency of all-polymer solar cells to 11%, *Energy and Environmental Science* 12 (2019) 157–163.

33. S. Li, W. Liu, M. Shi, J. Mai, T.-K. Lau, J. Wan, X. Lu, C.-Z. Li, H. Chen, A spirobifluorene and diketopyrrolopyrrole moieties based non-fullerene acceptor for efficient and thermally stable polymer solar cells with high open-circuit voltage, *Energy and Environmental Science* 9 (2016) 604–610.

34. R. Lenaerts, T. Cardeynaels, I. Sudakov, J. Kesters, P. Verstappen, J. Manca, B. Champagne, L. Lutsen, D. Vanderzande, K. Vandewal, E. Goovaerts, W. Maes, All-polymer solar cells based on photostable bis (perylene diimide) acceptor polymers, *Solar Energy Materials and Solar Cells* 196 (2019) 178–184.

35. L. Huo, T. Liu, X. Sun, Y. Cai, A.J. Heeger, Y. Sun, Single-junction organic solar cells based on a novel wide-bandgap polymer with efficiency of 9.7%, *Advanced Materials* 27 (2015) 2938–2944.

36. L. Huo, S. Zhang, X. Guo, F. Xu, Y. Li, J. Hou, Replacing alkoxy groups with alkylthienyl groups: A feasible approach to improve the properties of photovoltaic polymers, *Angewandte Chemie International Edition* 50 (2011) 9697–9702.

37. L. Yang, S. Zhang, C. He, J. Zhang, H. Yao, Y. Yang, Y. Zhang, W. Zhao, J. Hou, New wide band gap donor for efficient fullerene-free all-small-molecule organic solar cells, *Journal of the American Chemical Society* 139 (2017) 1958–1966.

38. R. Fitzner, E. Reinold, A. Mishra, E. Mena-Osteritz, H. Ziehlke, C. Körner, K. Leo, M. Riede, M. Weil, O. Tsaryova, A. Weiß, C. Uhrich, M. Pfeiffer, P. Bäuerle, Dicyanovinyl-substituted oligothiophenes: Structure-property relationships and application in vacuum-processed small molecule organic solar cells, *Advanced Functional Materials* 21 (2011) 897–910.

39. Y. Chen, C. Li, P. Zhang, Y. Li, X. Yang, L. Chen, Y. Tu, Solution-processable tetra-zine and oligothiophene based linear A–D–A small molecules: Synthesis, hierarchical structure and photovoltaic properties, *Organic Electronics* 14 (2013) 1424–1434.

40. H.-I. Je, J. Hong, H.-J. Kwon, N.Y. Kim, C.E. Park, S.-K. Kwon, T.K. An, Y.-H. Kim, End-group tuning of DTBDT-based small molecules for organic photovoltaics, *Dyes and Pigments* 157 (2018) 93–100.

41. K. Schulze, C. Uhrich, R. Schueppel, K. Leo, M. Pfeiffer, E. Brier, E. Reinold, P. Baeuerle, Efficient vacuum-deposited organic solar cells based on a new low-bandgap oligothiophene and fullerene C-60, *Advanced Materials* 18 (2006) 2872.

42. R. Fitzner, E. Mena-Osteritz, A. Mishra, G. Schulz, E. Reinold, M. Weil, C. Korner, H. Ziehlke, C. Elschner, K. Leo, M. Riede, M. Pfeiffer, C. Uhrich, P. Bauerle, Correlation of pi-conjugated oligomer structure with film morphology and organic solar cell performance, *Journal of the American Chemical Society* 134 (2012) 11064–11067.

43. R. Fitzner, E. Mena-Osteritz, K. Walzer, M. Pfeiffer, P. Bäuerle, A-D-A-type oligothio-phenes for small molecule organic solar cells: Extending the π-system by introduction of ring-locked double bonds, *Advanced Functional Materials* 25 (2015) 1845–1856.

44. H. Shang, H. Fan, Y. Liu, W. Hu, Y. Li, X. Zhan, A solution-processable star-shaped molecule for high-performance organic solar cells, *Advanced Materials* 23 (2011) 1554–1557.

45. S. Li, L. Zhan, F. Liu, J. Ren, M. Shi, C.Z. Li, T.P. Russell, H. Chen, An unfused-core-based nonfullerene acceptor enables high-efficiency organic solar cells with excellent morphological stability at high temperatures, *Advanced Materials* 30 (2018) 1705208.

46. H. Komiyama, T. To, S. Furukawa, Y. Hidaka, W. Shin, T. Ichikawa, R. Arai, T. Yasuda, Oligothiophene-indandione-linked narrow-band gap molecules: Impact of pi-conjugated chain length on photovoltaic performance, *ACS Applied Materials and Interfaces* 10 (2018) 11083–11093.

47. H. Bin, Y. Yang, Z.G. Zhang, L. Ye, M. Ghasemi, S. Chen, Y. Zhang, C. Zhang, C. Sun, L. Xue, C. Yang, H. Ade, Y. Li, 9.73% efficiency nonfullerene all organic small mol-ecule solar cells with absorption-complementary donor and acceptor, *Journal of the American Chemical Society* 139 (2017) 5085–5094.

48. B. Kan, J. Zhang, F. Liu, X. Wan, C. Li, X. Ke, Y. Wang, H. Feng, Y. Zhang, G. Long, R.H. Friend, A.A. Bakulin, Y. Chen, Fine-tuning the energy levels of a nonfullerene small-molecule acceptor to achieve a high short-circuit current and a power conversion efficiency over 12% in organic solar cells, *Advanced Materials* 30 (2018) 1704904.

49. Y. Lin, F. Zhao, S.K.K. Prasad, J.-D. Chen, W. Cai, Q. Zhang, K. Chen, Y. Wu, W. Ma, F. Gao, J.-X. Tang, C. Wang, W. You, J.M. Hodgkiss, X. Zhan, Balanced partnership between donor and acceptor components in nonfullerene organic solar cells with >12% efficiency, *Advanced Materials* 30 (2018) 1706363.

50. Y. An, X. Liao, L. Chen, J. Yin, Q. Ai, Q. Xie, B. Huang, F. Liu, A.K.Y. Jen, Y. Chen, Nonhalogen solvent-processed asymmetric wide-bandgap polymers for nonfullerene organic solar cells with over 10% efficiency, *Advanced Functional Materials* 28 (2018) 1706517.

51. S. Dai, F. Zhao, Q. Zhang, T.-K. Lau, T. Li, K. Liu, Q. Ling, C. Wang, X. Lu, W. You, X. Zhan, Fused nonacyclic electron acceptors for efficient polymer solar cells, *Journal of the American Chemical Society* 139 (2017) 1336–1343.

52. A. Tang, Q. Zhang, M. Du, G. Li, Y. Geng, J. Zhang, Z. Wei, X. Sun, E. Zhou, Molecular engineering of D-pi-A copolymers based on 4,8-bis(4-chlorothiophen-2-yl)benzo 1,2-b:4,5-b' dithiophene (BDT-T-Cl) for high-performance fullerene-free organic solar cells, *Macromolecules* 52 (2019) 6227–6233.

53. J. Yuan, Y. Zhang, L. Zhou, G. Zhang, H.-L. Yip, T.-K. Lau, X. Lu, C. Zhu, H. Peng, P.A. Johnson, M. Leclerc, Y. Cao, J. Ulanski, Y. Li, Y. Zou, Single-junction organic solar cell with over 15% efficiency using fused-ring acceptor with electron-deficient core, *Joule* 3 (2019) 1140–1151.

54. D. Chen, J. Yao, L. Chen, J. Yin, R. Lv, B. Huang, S. Liu, Z.G. Zhang, C. Yang, Y. Chen, Y. Li, Dye-incorporated polynaphthalenediimide acceptor for additive-free high-performance all-polymer solar cells, *Angewandte Chemie* 57 (2018) 4580–4584.

55. D. Han, T. Kumari, S. Jung, Y. An, C. Yang, A comparative investigation of cyclohexyl-end-capped versus hexyl-end-capped small-molecule donors on small donor/polymer acceptor junction solar cells, *Solar RRL* 2 (2018) 1800009.

56. Y. Lin, J. Wang, Z.-G. Zhang, H. Bai, Y. Li, D. Zhu, X. Zhan, An electron acceptor challenging fullerenes for efficient polymer solar cells, *Advanced Materials* 27 (2015) 1170–1174.

57. D. Deng, Y. Zhang, J. Zhang, Z. Wang, L. Zhu, J. Fang, B. Xia, Z. Wang, K. Lu, W. Ma, Z. Wei, Fluorination-enabled optimal morphology leads to over 11% efficiency for inverted small-molecule organic solar cells, *Nature Communications* 7 (2016) 13740.

58. Y. Lin, B. Adilbekova, Y. Firdaus, E. Yengel, H. Faber, M. Sajjad, X. Zheng, E. Yarali, A. Seitkhan, O.M. Bakr, A. El-Labban, U. Schwingenschlogl, V. Tung, I. McCulloch, F. Laquai, T.D. Anthopoulos, 17% efficient organic solar cells based on liquid exfoliated WS2 as a replacement for PEDOT:PSS, *Advanced Materials* (2019). DOI: 10.1002/adma.201902965.

59. H.W. Kroto, J.R. Heath, S.C. O'Brien, R.F. Curl, R.E. Smalley, C_{60}: Buckminsterfullerene, *Nature* 318 (1985) 162–163.

60. A.A. Bogdanov, D. Deininger, G.A. Dyuzhev, Development prospects of the commercial production of fullerenes, *Technical Physics* 45 (2000) 521–527.

61. N.S. Sariciftci, L. Smilowitz, A.J. Heeger, F. Wudl, Photoinduced electron transfer from a conducting polymer to buckminsterfullerene, *Science* 258 (1992) 1474.

62. N.S. Sariciftci, D. Braun, C. Zhang, V.I. Srdanov, A.J. Heeger, G. Stucky, F. Wudl, Semiconducting polymer-buckminsterfullerene heterojunctions: Diodes, photodiodes, and photovoltaic cells, *Applied Physics Letters* 62 (1993) 585–587.

63. J. Zhao, Y. Li, G. Yang, K. Jiang, H. Lin, H. Ade, W. Ma, H. Yan, Efficient organic solar cells processed from hydrocarbon solvents, *Nature Energy* 1 (2016) 15027.

64. S.M. Ryno, M.K. Ravva, X. Chen, H. Li, J.-L. Brédas, Molecular understanding of fullerene: Electron donor interactions in organic solar cells, *Advanced Energy Materials* 7 (2017) 1601370.

65. K.L. Mutolo, E.I. Mayo, B.P. Rand, S.R. Forrest, M.E. Thompson, Enhanced open-circuit voltage in subphthalocyanine/C_{60} organic photovoltaic cells, *Journal of the American Chemical Society* 128 (2006) 8108–8109.

66. H. Peng, X. Sun, W. Weng, X. Fang, 5- energy harvesting based on polymer, in: H. Peng, X. Sun, W. Weng, X. Fang (Eds.), *Polymer Materials for Energy and Electronic Applications*, Academic Press, Cambridge, MA (2017), pp. 151–196.

67. L. Gao, Z.-G. Zhang, H. Bin, L. Xue, Y. Yang, C. Wang, F. Liu, T.P. Russell, Y. Li, High-efficiency nonfullerene polymer solar cells with medium bandgap polymer donor and narrow bandgap organic semiconductor acceptor, *Advanced Materials* 28 (2016) 8288–8295.
68. H. Bin, Z.-G. Zhang, L. Gao, S. Chen, L. Zhong, L. Xue, C. Yang, Y. Li, Non-fullerene polymer solar cells based on alkylthio and fluorine substituted 2D-conjugated polymers reach 9.5% efficiency, *Journal of the American Chemical Society* 138 (2016) 4657–4664.
69. X. Li, G. Huang, N. Zheng, Y. Li, X. Kang, S. Qiao, H. Jiang, W. Chen, R. Yang, High-efficiency polymer solar cells over 13.9% with a high V_{oc} beyond 1.0 V by synergistic effect of fluorine and sulfur, *Solar RRL* 3 (2019) 1900005.
70. Y. Yang, Z.-G. Zhang, H. Bin, S. Chen, L. Gao, L. Xue, C. Yang, Y. Li, Side-chain isomerization on an n-type organic semiconductor ITIC acceptor makes 11.77% high efficiency polymer solar cells, *Journal of the American Chemical Society* 138 (2016) 15011–15018.
71. L. Xue, Y. Yang, J. Xu, C. Zhang, H. Bin, Z.-G. Zhang, B. Qiu, X. Li, C. Sun, L. Gao, J. Yao, X. Chen, Y. Yang, M. Xiao, Y. Li, Side chain engineering on medium band-gap copolymers to suppress triplet formation for high-efficiency polymer solar cells, *Advanced Materials* 29 (2017) 1703344.
72. Y. Lin, F. Zhao, Q. He, L. Huo, Y. Wu, T.C. Parker, W. Ma, Y. Sun, C. Wang, D. Zhu, A.J. Heeger, S.R. Marder, X. Zhan, High-performance electron acceptor with thienyl side chains for organic photovoltaics, *Journal of the American Chemical Society* 138 (2016) 4955–4961.
73. F. Zhao, S. Dai, Y. Wu, Q. Zhang, J. Wang, L. Jiang, Q. Ling, Z. Wei, W. Ma, W. You, C. Wang, X. Zhan, Single-junction binary-blend nonfullerene polymer solar cells with 12.1% efficiency, *Advanced Materials* 29 (2017) 1700144.
74. Z. Fei, F.D. Eisner, X. Jiao, M. Azzouzi, J.A. Röhr, Y. Han, M. Shahid, A.S.R. Chesman, C.D. Easton, C.R. McNeill, T.D. Anthopoulos, J. Nelson, M. Heeney, An alkylated indacenodithieno 3,2-b thiophene-based nonfullerene acceptor with high crystallinity exhibiting single junction solar cell efficiencies greater than 13% with low voltage losses, *Advanced Materials* 30 (2018) 1705209.
75. Y. Zhang, H. Yao, S. Zhang, Y. Qin, J. Zhang, L. Yang, W. Li, Z. Wei, F. Gao, J. Hou, Fluorination vs. chlorination: A case study on high performance organic photovoltaic materials, *Science China Chemistry* 61 (2018) 1328–1337.
76. W. Zhao, S. Li, H. Yao, S. Zhang, Y. Zhang, B. Yang, J. Hou, Molecular optimization enables over 13% efficiency in organic solar cells, *Journal of the American Chemical Society* 139 (2017) 7148–7151.
77. H. Zhang, H. Yao, J. Hou, J. Zhu, J. Zhang, W. Li, R. Yu, B. Gao, S. Zhang, J. Hou, Over 14% efficiency in organic solar cells enabled by chlorinated nonfullerene small-molecule acceptors, *Advanced Materials* 30 (2018) 1800613.
78. C. Jiao, C. Pang, Q. An, Nonfullerene organic photovoltaic cells exhibiting 13.76% efficiency by employing upside-down solvent vapor annealing, *International Journal of Energy Research* 43 (2019) 8716–8724.
79. X. Shi, L. Zuo, S.B. Jo, K. Gao, F. Lin, F. Liu, A.K.Y. Jen, Design of a highly crystalline low-band gap fused-ring electron acceptor for high-efficiency solar cells with low energy loss, *Chemistry of Materials* 29 (2017) 8369–8376.
80. S. Li, L. Ye, W. Zhao, H. Yan, B. Yang, D. Liu, W. Li, H. Ade, J. Hou, A wide band gap polymer with a deep highest occupied molecular orbital level enables 14.2% efficiency in polymer solar cells, *Journal of the American Chemical Society* 140 (2018) 7159–7167.

81. T. Liu, X. Xue, L. Huo, X. Sun, Q. An, F. Zhang, T.P. Russell, F. Liu, Y. Sun, Highly efficient parallel-like ternary organic solar cells, *Chemistry of Materials* 29 (2017) 2914–2920.

82. T. Liu, Y. Guo, Y. Yi, L. Huo, X. Xue, X. Sun, H. Fu, W. Xiong, D. Meng, Z. Wang, F. Liu, T.P. Russell, Y. Sun, Ternary organic solar cells based on two compatible non-fullerene acceptors with power conversion efficiency >10%, *Advanced Materials* 28 (2016) 10008–10015.

83. W. Su, Q. Fan, X. Guo, X. Meng, Z. Bi, W. Ma, M. Zhang, Y. Li, Two compatible non-fullerene acceptors with similar structures as alloy for efficient ternary polymer solar cells, *Nano Energy* 38 (2017) 510–517.

84. Y. Xie, F. Yang, Y. Li, M.A. Uddin, P. Bi, B. Fan, Y. Cai, X. Hao, H.Y. Woo, W. Li, F. Liu, Y. Sun, Morphology control enables efficient ternary organic solar cells, *Advanced Materials* 30 (2018) 1803045.

85. J. Yuan, T. Huang, P. Cheng, Y. Zou, H. Zhang, J.L. Yang, S.-Y. Chang, Z. Zhang, W. Huang, R. Wang, D. Meng, F. Gao, Y. Yang, Enabling low voltage losses and high photocurrent in fullerene-free organic photovoltaics, *Nature Communications* 10 (2019). DOI: 10.1038/s41467-41019-08386-41469.

86. Y. Cui, H. Yao, J. Zhang, T. Zhang, Y. Wang, L. Hong, K. Xian, B. Xu, S. Zhang, J. Peng, Z. Wei, F. Gao, J. Hou, Over 16% efficiency organic photovoltaic cells enabled by a chlorinated acceptor with increased open-circuit voltages, *Nature Communications* (2019). DOI: 10.1038/s41467-41019-10351-41465.

87. L. Hong, H. Yao, Z. Wu, Y. Cui, T. Zhang, Y. Xu, R. Yu, Q. Liao, B. Gao, K. Xian, H.Y. Woo, Z. Ge, J. Hou, Eco-compatible solvent-processed organic photovoltaic cells with over 16% efficiency, *Advanced Materials* (2019). DOI: 10.1002/adma.201903441.

88. M. Luo, L. Zhou, J. Yuan, C. Zhu, F. Cai, J. Hai, Y. Zou, A new non-fullerene acceptor based on the heptacyclic benzotriazole unit for efficient organic solar cells, *Journal of Energy Chemistry* 42 (2020) 169–173.

89. M. Luo, C. Zhao, J. Yuan, J. Hai, F. Cai, Y. Hu, H. Peng, Y. Bai, Z.a. Tan, Y. Zou, Semitransparent solar cells with over 12% efficiency based on a new low bandgap fluorinated small molecule acceptor, *Materials Chemistry Frontiers* 3 (2019) 2483–2490.

90. J. Yuan, Y. Zhang, L. Zhou, C. Zhang, T.-K. Lau, G. Zhang, X. Lu, H.-L. Yip, S.K. So, S. Beaupre, M. Mainville, P.A. Johnson, M. Leclerc, H. Chen, H. Peng, Y. Li, Y. Zou, Fused benzothiadiazole: A building block for n-type organic acceptor to achieve high-performance organic solar cells, *Advanced Materials* (2019). DOI: 10.1002/adma.201807577.

91. T. Yan, W. Song, J. Huang, R. Peng, L. Huang, Z. Ge, 16.67% rigid and 14.06% flexible organic solar cells enabled by ternary heterojunction strategy, *Advanced Materials* (2019). DOI: 10.1002/adma.201902210.

92. L. Duan, Y. Zhang, H. Yi, F. Haque, R. Deng, H. Guan, Y. Zou, A. Uddin, Trade-off between exciton dissociation and carrier recombination and dielectric properties in Y6-sensitized nonfullerene ternary organic solar cells, *Energy Technology* (2019). DOI: 10.1002/ente.201900924.

93. H.S. Vogelbaum, G. Sauve, Recently developed high-efficiency organic photoactive materials for printable photovoltaic cells: A mini review, *Synthetic Metals* 223 (2017) 107–121.

94. G. Zhang, J. Zhao, P.C.Y. Chow, K. Jiang, J. Zhang, Z. Zhu, J. Zhang, F. Huang, H. Yan, Nonfullerene acceptor molecules for bulk heterojunction organic solar cells, *Chemical Reviews* 118 (2018) 3447–3507.

95. J. Gao, J. Wang, Q. An, X. Ma, Z. Hu, C. Xu, X. Zhang, F. Zhang, Over 16.7% efficiency of ternary organic photovoltaics by employing extra $PC_{71}BM$ as morphology regulator, *Science China-Chemistry* (2019). DOI: 10.1007/s11426-11019-19634-11425.

96. Y. Chen, Y. Geng, A. Tang, X. Wang, Y. Sun, E. Zhou, Changing the pi-bridge from thiophene to thieno 3,2-b thiophene for the D-pi-A type polymer enables high performance fullerene-free organic solar cells, *Chemical Communications* 55 (2019) 6708–6710.

97. H. Chen, D. Hu, Q. Yang, J. Gao, J. Fu, K. Yang, H. He, S. Chen, Z. Kan, T. Duan, C. Yang, J. Ouyang, Z. Xiao, K. Sun, S. Lu, All-small-molecule organic solar cells with an ordered liquid crystalline donor, *Joule* 3 (2019) 3034–3047.

98. R. Zhou, Z. Jiang, C. Yang, J. Yu, J. Feng, M.A. Adil, D. Deng, W. Zou, J. Zhang, K. Lu, W. Ma, F. Gao, Z. Wei, All-small-molecule organic solar cells with over 14% efficiency by optimizing hierarchical morphologies, *Nature Communications* 10 (2019). DOI: 10.1038/s41467-41019-13292-41461.

99. Z. Zhou, S. Xu, J. Song, Y. Jin, Q. Yue, Y. Qian, F. Liu, F. Zhang, X. Zhu, High-efficiency small-molecule ternary solar cells with a hierarchical morphology enabled by synergizing fullerene and non-fullerene acceptors, *Nature Energy* 3 (2018) 952–959.

100. W. Zhao, S. Li, S. Zhang, X. Liu, J. Hou, Ternary polymer solar cells based on two acceptors and one donor for achieving 12.2% efficiency, *Advanced Materials* 29 (2017) 7.

101. M.-A. Pan, T.-K. Lau, Y. Tang, Y.-C. Wu, T. Liu, K. Li, M.-C. Chen, X. Lu, W. Ma, C. Zhan, 16.7%-efficiency ternary blended organic photovoltaic cells with PCBM as the acceptor additive to increase the open-circuit voltage and phase purity, *Journal of Materials Chemistry A* 7 (2019) 20713–20722.

102. R. Yu, H. Yao, Y. Cui, L. Hong, C. He, J. Hou, Improved charge transport and reduced nonradiative energy loss enable over 16% efficiency in ternary polymer solar cells, *Advanced Materials* (2019). DOI: 10.1002/adma.201902302.

103. J. Song, C. Li, L. Zhu, J. Guo, J. Xu, X. Zhang, K. Weng, K. Zhang, J. Min, X. Hao, Y. Zhang, F. Liu, Y. Sun, Ternary organic solar cells with efficiency >16.5% based on two compatible nonfullerene acceptors, *Advanced Materials* (2019). DOI: 10.1002/adma.201905645.

104. C.J. Takacs, Y. Sun, G.C. Welch, L.A. Perez, X. Liu, W. Wen, G.C. Bazan, A.J. Heeger, Solar cell efficiency, self-assembly, and dipole-dipole interactions of isomorphic narrow-band-gap molecules, *Journal of the American Chemical Society* 134 (2012) 16597–16606.

13 High-Performance Electrolytes for Batteries

Yixiang Ou
Beijing Academy of Science and technology

CONTENTS

13.1 Introduction 479
13.2 Advances in Electrolytes for LIBs.. 483
    13.2.1 Development of Electrolytes for LIBs 483
    13.2.2 Inorganic Solid Electrolytes (ISEs) 484
    13.2.3 Solid Polymer Electrolytes (SPEs)... 485
    13.2.4 Organic–Inorganic Hybrid Composite Electrolytes 487
13.3 Advances in Electrolytes for SOFC... 488
    13.3.1 Typical Electrolytes for SOFC... 492
    13.3.2 Deposition of YSZ Thin-Film Electrolytes 492
    13.3.3 Microstructure of YSZ Thin Films 495
    13.3.4 Electrical Properties of YSZ Thin Films... 498
13.4 High-Performance New-Type Electrolytes for SOFC 500
13.5 Conclusions... 503
Acknowledgments... 503
References... 503

13.1 INTRODUCTION

In face of massive consumption of fossil fuels along with the deteriorating environmental problems, clean energy sources in terms of power generation and stored energy devices have been the main choices to replace gasoline for automotive applications, especially for pure electric vehicles. Nuclear power generation, as one of the high-efficient clean energy sources, is extensively thought to solve the shortage of electricity to some extent. Battery, one of the promising electrochemical devices composed of electrodes (anode and cathode) and electrolytes (liquid and solid), directly converts chemical energy to electricity. Since the lead-acid battery was invented in 1836, their commercial applications were extensively promoted later. However, due to low energy density, the lead-acid battery is indeed a collection of cells, each of which contains two electrodes immersed in a strong solution of sulfuric acid. Figure 13.1 reveals the schematic diagram of the lead-acid battery. It is reported that producing one lead-acid battery threw off three kilograms of lead emissions, by one estimate. Hence, the heavy metal pollution, especially lead, always results in

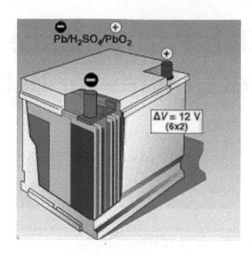

FIGURE 13.1   Schematic diagram of lead-acid battery.

environment problem. Besides, a strong solution of sulfuric acid causes the damage
to the movement of the lead-acid batteries.

Afterward, dry cells appeared in terms of Zn-Mn, Mg-Mn systems, etc., where
the mushy electrolytes were widely used free of leaks problem as compared with
liquid electrolytes of the lead-acid battery. However, the energy density is still
very limited. At the earlier of the 1910s, the commercial production of iron-nickel
rechargeable battery begun. The appearance of rechargeable batteries significantly
promotes extensive applications in every field of daily life. For instance, in Zn-Mn
dry cell, the copper is connected with carbon rod as anode electrode. Meanwhile,
MnO_2 and carbon are coated on the surface of the carbon rod to eliminate the electrode polarization. The cathode is Zn tube. Mushy electrolytes are made up of Nh_4Cl
and $ZnCl_2$. Figure 13.2 shows the schematic diagram of Zn-Mn dry cells.

Based on the requirements of high energy density and much more durable battery,
lithium ion batteries (LIBs) are developed by the movement of Li ions between anode
and cathode. A membrane is set between anode and cathode to prevent shortage.

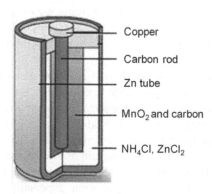

FIGURE 13.2   Schematic diagram of Zn-Mn dry cells.

LIBs are widely used as a promising electrical energy storage for electronic products and electric vehicles thanks to high specific energy density, fast charge/discharge, and long lifetime [1,2]. Figure 13.3 exhibits the schematic diagram of LIBs.

LIBs can be fabricated in various shapes according to special demand in electronics, such as button, flat rectangle, cylinder, and rectangle. Besides, the lightweight of LIBs also reduces the weight of electronic devices, which promotes wide applications. Electrolytes are composed of solutes ($LiClO_4$, $LiPF_6$, $LiBF_4$, etc.) and organic solvents (diethyl ether, vinyl carbonate, propylene carbonate, diethyl carbonate, etc.). It is noted that when LIBs are in charge, organic solvents often make a damage to the structure of graphite, thus causing desquamation. In addition, solid electrolyte interphase formed on the surface of graphite cathode results in electrode passivation. Organic solvents are also prone to cause safety issues because they are flammable and explosive. Actually, in order to enhance power density, the thickness of the membrane is reduced as thin as possible. Thus, the thin membrane leads to a high risk of shortage and even explosions in severe conditions. Therefore, the high-performance solid electrolytes are developed in terms of high ion conductivity at room temperature and high safety.

Fuel cells are the same as the normal batteries, which are composed of anode, cathode, and electrolyte. Figure 13.4 shows the schematic diagram of fuel cells. According to the different electrolyte materials, fuel cells are divided into five types, namely, alkaline fuel cell (AFC), phosphoric acid fuel cell (PAFC), proton exchange membrane fuel cell (PEMFC), molten carbonate fuel cell (MCFC), and solid oxide fuel cell (SOFC). AFC and PEMFC can be used as a portable power source.

It is noteworthy that AFC has a short life if it is used in atmospheres. PAFC could be built as a small power station of 100 kW. SOFC and MCFC have to work at a high

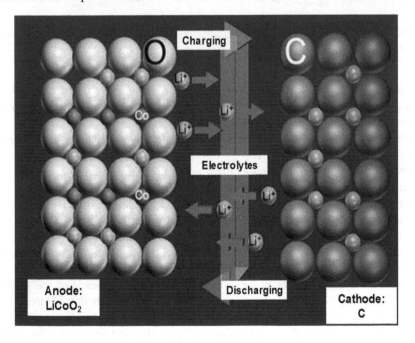

FIGURE 13.3 Schematic diagram of LIBs.

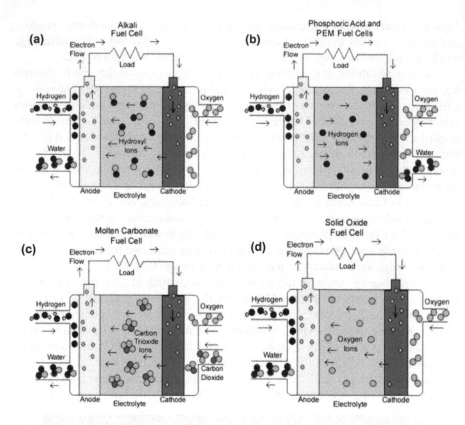

FIGURE 13.4   Schematic diagram of fuel cells: (a) AFC, (b) PAFC and PEMFC (c) MCFC, and (d) SOFC.

temperature of 500°C–1,000°C. PAFC is often working at 20°C–80°C. SOFC and MCFC are usually used in the large power stations of MW. Figure 13.5 presents the schematic diagram of the work temperature of fuel cells.

SOFC and hydrogen fuel cells are the typical fuel cells, which are generators that convert directly chemical energy into electrical energy. SOFC is a new way of power generation, which exhibits great potential in the solution of energy crisis and environmental pollution, and gets wide attention in distributed generation. Also, advances in SOFC make it come true that natural gas can be directly used as fuel without reforming to convert chemical energy into electrical energy. The efficiency of SOFC depends on the ionic conductivity of solid electrolytes.

Proton exchange membrane (PEM) hydrogen cells are promising power batteries widely used in the pure electric vehicle as a power source. In PEM hydrogen cells, electron in a hydrogen atom is separated out under the influence of the platinum catalyst when hydrogen fuel is delivered to anode plate (cathode). Hydrogen ions losing electron (protons) can get through the PEM to reach the fuel cell cathode plate (anode). However, the electron does not pass through the PEM. The electron passes only through an external circuit to the fuel cell cathode plate to generate an

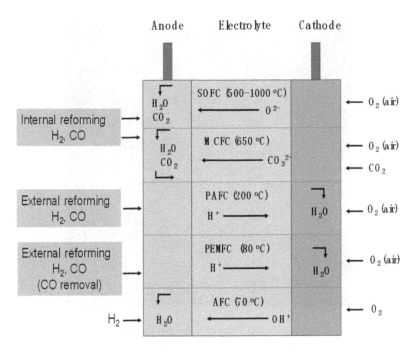

FIGURE 13.5 Schematic diagram of work temperature of fuel cells.

electric current in the external circuit. When the electrons reach the cathode plate, they recombine with oxygen and hydrogen ions to form water. Figure 13.6 presents the schematic diagram of PEM of hydrogen fuel cells.

One of the greatest challenges for LIBs and SOFC is unquestionably to obtain sufficient conductivity at a lower operation temperature. The ionic conductivity of electrolyte material is greatly influenced by the chemical composition and microstructure, which plays an important role in the performance of batteries. Hence, in this chapter, we introduce electrolyte material designs and controllable preparation processes used for LIBs and SOFCs.

13.2 ADVANCES IN ELECTROLYTES FOR LIBs

13.2.1 DEVELOPMENT OF ELECTROLYTES FOR LIBs

Research and development of LIBs began since the early 1980s [1], while the first commercial applications appeared in 1990 [2]. A large number of applications of LIBs are used for electronic products and electric vehicles due to high specific energy density, high working voltage, low self-discharge rate, fast charge/discharge, long lifetime, and no memory effect [3–5]. However, it is noted that volatile and flammable liquid organic solvents widely used as electrolyte solutions in LIBs are prone to cause safety problems during cycling due to the dendrite growth of metallic Li anode [6]. Hence, the utilization of solid electrolytes is expected to enhance the safety of LIBs by the almost complete elimination of the growth of Li dendrites [7–9].

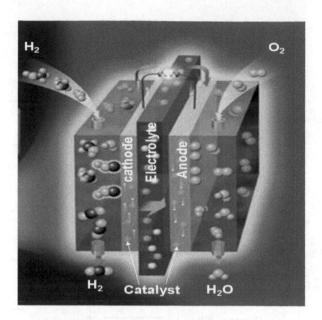

FIGURE 13.6 Schematic diagram of PEM of hydrogen fuel cells.

Additionally, liquid organic solvents will result in irreversible capacity losses due to the formation of stable solid electrolyte interphase to hinder the increase in cycle life and limit the temperature window, which pose severe safety concerns on LIBs [10]. All solid-state electrolyte (SSE) materials have been attracted increasing interests, which mainly include inorganic solid electrolytes (ISEs), solid polymer electrolytes (SPEs), and organic–inorganic hybrid composite electrolytes [6].

13.2.2 INORGANIC SOLID ELECTROLYTES (ISEs)

Among all SSE materials, ISEs are classified into oxide-based, sulfide-based, etc [11–16]. Figure 13.7 shows the crystal structure of parent garnet-like $Li_5La_3M_2O_{12}$ [15]. It is found that lithium ion transference number of ISEs is almost unity and its ionic conductivity is almost comparable to that of organic liquid electrolyte [17]. Also, there are still two main challenges for achieving high-performance ISEs [18,19]: The first one is how to create favorable solid–solid interface between electrode and electrolyte, and the second one is how to obtain high ionic conductivity at room temperature (e.g., above 10^{-3} S cm^{-1}).

Due to their high ionic conductivity and adequate mechanical features for lamination, sulfide composites have attracted an increasing attention as solid electrolytes used in all-solid-state batteries. Their smaller electronegativity and binding energy to Li ions and bigger atomic radius provide high ionic conductivity and make them more attractive for practical applications. In recent years, noticeable efforts have been made to develop high-performance sulfide SSEs. The improvement of the ionic conductivity of LISICON-type SSEs focuses on the replacement of oxide by sulfur in the framework, which is referred to as thio-LISICON [10]. Since the radius of

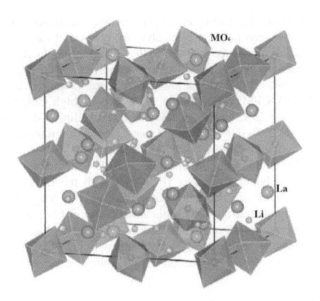

FIGURE 13.7 Crystal structure of parent garnet-like $Li_5La_3M_2O_{12}$ [15]. (Reproduced with permission with Elsevier publisher.)

S^{2-} is higher than that of O^{2-}, this substitution can significantly enlarge the size of Li^+ transport bottlenecks. Besides, S^{2-} has better polarization capability than O^{2-}, thus weakening the interaction between skeleton and Li^+ ions. Therefore, thio-based materials can achieve high ionic conductivity. It is reported that $Li_{3.25}Ge_{0.25}P_{0.75}S_4$ possesses the high conductivity of 2.2×10^{-3} S cm^{-1} at room temperature, high electrochemical stability, and no phase transition up to 500°C [20].

As is known, most sulfide SSEs are derived from the Li-P-S systems. Among these different sulfide materials, the LGPS family has shown superior ionic conductivity (up to 2.5×10^{-2} S cm^{-1} at room temperature), even higher than the conventional liquid electrolytes [21]. Thus, sulfide-based all solid-state LIBs have a great potential in electrode/electrolyte synthesis and cell fabrication methods.

13.2.3 Solid Polymer Electrolytes (SPEs)

SPEs are the promising electrolyte materials, which are extensively used in electrochemical devices, especially polymer LIBs. They also exhibit potential applications in flexible and wearable devices. Generally, SPEs are composed of polymer host as solid matrix along with alkali metal salt without the addition of organic liquid solvents [22]. SPEs have no leakage of electrolytes, low flammability, good flexibility, and safety as compared to the conventional liquid electrolytes. It is noteworthy that the stable contact between the electrode and the electrolyte greatly enhances the interfacial impedance due to the strong adhesive property on the surface of electrodes [23,24]. Besides, SPE process is flexible. In addition, SPEs of LIBs are required for good mechanical strength, excellent flame retardancy, superior thermal stability, high ionic conductivity at ambient temperature, and wide electrochemical window [25–28]. However, at a higher temperature, the chemical reaction will occur at the interfaces between polymer electrolyte materials and electrodes. Thus, the increased interfacial impedance will lead to the deterioration of LIBs.

The research and development of all solid-state polymer electrolyte materials aim to enhance high ionic conductivity at low temperature. As suggested by Yuan et al. [13], the flexible PAN-PEO copolymer of SPEs shows a high ionic conductivity of 6.79×10^{-4} S cm^{-1} at 25°C and an electrochemical stability of 4.8 V vs. Li$^+$/Li, as shown in Figure 13.8. Besides, it is noted that Li-dendrite growth in the charging process of Li batteries can be effectively inhibited by PAN [14].

Kim et al. [29] demonstrated a shape-deformable and thermally stable plastic crystal composite polymer electrolyte (PC-CPE), with the high ionic conductivity of 1.02×10^{-3} S cm^{-1} at room temperature only slightly lower than that of the carbonate-based liquid electrolyte, which is attributed to its high diffusivity, plasticity, solvating power, and well-interconnected ion-conductive channels, as shown in Figure 13.9. Meanwhile, the electrochemical window of the PC-CPE up to 5.0 V vs. Li$^+$/Li, indicates the potential application to high-voltage batteries. Also, due to its

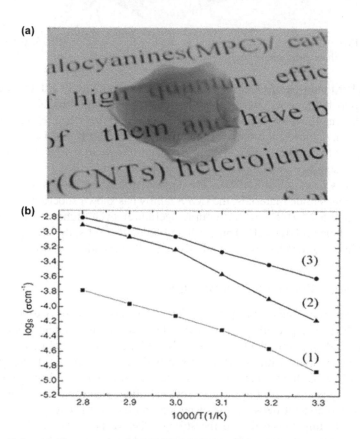

FIGURE 13.8 (a) Photograph of PAN-PEOSPE film; (b) Arrhenius plot of PAN-PEO copolymer SPE film doped with (1) 0 wt%, (2) 10 wt%, and (3) 20 wt% of PEO (Mn¼ 3,000,000) [13]. (Reproduced with permission with Elsevier publisher.)

FIGURE 13.9 Structural characterization of PC-CPE [29]. (Reproduced with permission from Royal Society of Chemistry (RSC).)

unique chemical/structural peculiarity, PC-CPE significantly improves the mechanical flexibility and thermal stability. Notably, the cell incorporating the self-standing PC-CPE delivered the stable charge/discharge behavior without suffering from safety problems, even after exposure to thermal shock condition [6].

13.2.4 ORGANIC–INORGANIC HYBRID COMPOSITE ELECTROLYTES

Organic–inorganic hybrid composite electrolytes integrate the merits of organic polymers and inorganic ceramics, as shown in Figure 13.10. The improvements in mechanical properties, ionic conductivity, and interfacial stability of the polymer electrolytes are achieved by using inorganic materials filled into a polymer substrate. The polymer electrolytes are generally prepared by dispersing inorganic fillers, such as inert ceramic fillers (Al_2O_3, SiO_2, TiO_2, etc.), ferroelectric ceramic fillers ($BaTiO_3$, $PbTiO_3$, and $LiNbO_3$), carbon nanotubes (CNTs), and fast ionic conductors, into the polymer matrix [6,30].

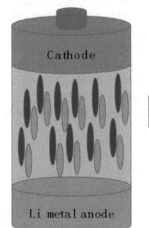

Organic–inorganic
composite solid electrolytes:
1. High ionic conductivity
2. Superior mechanical properties
3. High flexible
4. Non flammable
5. Stability and safety

FIGURE 13.10 Characteristics of organic–inorganic composite solid electrolytes.

In organic–inorganic hybrid composite electrolytes, the fillers provide a support matrix to retain an overall solid structure, even at high temperature. It is known that the ionic conductivity of the composite electrolytes increases with the decreasing particle size because the nanoparticles reduce the crystallinity of the polymer–salt system. Another reason for the enhanced ionic conductivity is the percolating interfacial effect: Anions adsorb on the surface of fillers, then break up the ion pair, and lead to increased interfacial ionic conductivity. Thus, the composite electrolytes are considered as high energy and safe rechargeable LIB materials.

It has been found that nanosized ceramic powders incorporated into the polymer-based electrolytes not only act as solid plasticizers to inhibit crystallization kinetics, but also show the effects on solvent and cationic mobility. As suggested by Sun et al. [31], the addition of ferroelectric materials such as $BaTiO_3$, $PbTiO_3$, and $LiNbO_3$ into the PEO-Li_x polymer electrolyte significantly enhances the ionic conductivity. This phenomenon is related to the association of anions with lithium cations and the spontaneous polarization of ferroelectric ceramic particles, due to their particular crystal structure. A combination of rutile and barium titanate decreases the interfacial resistance between the lithium electrode and the composite polymer electrolyte. The composition of the electrolyte can be optimized by a proper choice of the type and morphology of ferroelectric ceramics. Moreover, all the electrolytes studied show the decomposition potentials higher than 4 V vs. Li/Li$^+$. This might lead to the development of a polymer electrolyte having a true solid-state configuration and thus good mechanical properties, combined with high conductivity and low interfacial resistance with a lithium metal electrode. Hence, it is believed that the ferroelectric materials and the PEO-Li_x polymer hybrid composite electrolytes are suitable candidates for the practical applications [6].

ISEs, like fast ionic conductors, have presented many potential advantages, such as no electrolyte leakage, high electrochemical stability, nonflammability, and high thermal stability [6]. Organic–inorganic hybrid composite electrolytes are composed of ISEs and flexible polymer materials that can synergistically combine the beneficial properties of both glass ceramics and polymers [31]. Kobayashi et al. [32] proposed a composite concept in which a ceramic electrolyte is placed at the positive electrode interface and a polymer electrolyte at the negative electrode interface [33]. Ethylene oxide co-2-(2-methoxyethoxy) ethyl ether-LiBF4 polymer film was placed between (Li, La)TiO$_3$ and Li metal, and showed relatively high lithium ion conductivity, typically 10^{-3} S cm^{-1} at 22°C. Moreover, the all-solid-state battery [LiMn$_2$O$_4$/(Li, La)TiO$_3$/dry polymer/Li] showed good cycle characteristics at 60°C (Figure 13.11). It is demonstrated that this new composite method should be a promising electrolyte applied in solid-state batteries based on lithium metal electrode.

13.3 ADVANCES IN ELECTROLYTES FOR SOFC

SOFC is one of the advanced electrochemical devices that convert chemical energy into electrical energy, which is a kind of clean energy expected for the substitution of fossil fuels. It is noted that the conversion efficiency is temperature dependent; i.e., the transport of oxygen anions in an electrolyte material is highly influenced by temperature [34,35]. Hence, these devices have to work at high temperature up to

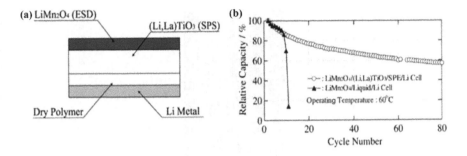

FIGURE 13.11 (a) Composite-type all-solid-state lithium secondary battery; (b) charge/discharge cycle characteristics of composite-type all-solid-state battery and liquid battery. Operation condition: 4.3 V/3.0 V cutoff, 60°C [32]. (Reproduced with permission with Elsevier publisher.)

800°C–1,000°C in order to obtain high ionic conductivity. Moreover, the development of high-performance electrolyte material is an effective way to improve the conversion efficiency of SOFC.

Yttria-stabilized zirconia (YSZ), strontium and magnesium-doped lanthanum gallate (LSGM), $Ce_{0.8}Sm_{0.2}O_{2-\delta}$ (SDC), $Gd_2Zr_2O_7$, gadolinium- or samarium-doped ceria (CGO or CSO), etc., have been widely used as electrolytes in SOFC [36]. Among them, YSZ has been widely used as an electrolyte material in SOFCs due to its high ionic conductivity and chemical stability in both oxidizing and reducing conditions [35]. High values of the oxygen conductivity at high temperatures are attributed to oxygen vacancies in doping ZrO_2 with Y_2O_3 structure. Generally, these devices have to work at a high intermediate temperature in order to obtain the sufficient conductivity. Accordingly, the follow-on problems are high costs, including manufacture and maintenance at high temperature, and poor durability [37]. In order to achieve the desired properties, a decrease in thickness and grain size in nanometer scale of electrolytes was found to effectively enhance the ionic conductivity and reduce the operation temperature.

It is found by Garcia-Barriocanal et al. [38] that a high lateral ionic conductivity of eight orders of magnitude enhancement near the room temperature was obtained in YSZ/strontium titanate epitaxial heterostructures. Figure 13.12 shows a schematic diagram of the structure of STO/YSZ/STO superlattice thin films. More detailed information can be found in Ref. [38]. The coherent interface between STO and YSZ reveals the epitaxial growth of YSZ along with STO substrate. With the decrease in thickness of YSZ layer in $[STO_{10\,nm}/YSZ_{X\,nm}/STO_{10\,nm}]$ trilayer electrolyte materials, a distinct increase in temperature dependence of ionic conductivity and a decrease in activation energy are obtained, respectively, due to the contributions of interface structure.

Advances in novel electrolyte materials and preparation process are expected to reduce the operating temperature of SOFC. The multilayers, nanofibers, and doped electrolyte materials have been extensively investigated in terms of the modulation of phase and surface/interface structure to enhance the ionic conductivity at a low temperature. Besides, thin-film electrolytes with a lower ohmic loss for SOFC are obtained, thus gaining increasing interests in scientific and industrial fields due to compensation

STO YSZ STO

FIGURE 13.12 Schematic diagram of the structure of STO/YSZ/STO superlattice thin films.

for the performance degradation encountered in lowering operating temperature [39–41]. The thickness of YSZ electrolytes varies from several nanometers in micro-SOFC to hundreds micrometers in industrial SOFCs. Tape casting is the most common method to produce thick YSZ electrolytes with different shapes according to actual applications. Then, YSZ sizing agents are sintered into YSZ electrolytes. However, the surface feature and microstructure greatly depend on the sintering process.

Two typical sputtering processes, namely, spraying process and radio frequency (RF) magnetron sputtering, are usually used to fabricate the thin films. It is noted that RF magnetron sputtering is the flexible process, which is widely used for the preparation of high-quality YSZ electrolytes without sintering. The improvement in density and surface morphology promoted the increase in ionic conductivity [42,43]. It is reported that YSZ thin films with a thickness of less than 60 nm showed an enhanced ionic conductivity, which is attributed to increasing YSZ/MgO interfacial conductivity with decreasing film thickness [44]. Nanostructured YSZ thin films with the increased number of grain boundaries exhibited high ionic conductivity [45]. Similar results show that the conductivity of nanocrystalline doped ceria electrolyte increases as the grain size decreases, suggesting an unblocking effect of homophase interfaces [46]. Therefore, the modulation of surface and the interface of electrolytes in terms of grain size, density, and morphology have been demonstrated to be essential in order to enhance the electrochemical performance of electrolyte materials for SOFCs and gas sensors. Figure 13.13 shows the cross-sectional morphologies of high-resolution transmission electron microscopic images and selected area electron diffraction (SAED) of highly textured YSZ thin films deposited on MgO substrate.

As compared with bulk electrolytes, thin-film electrolytes have been synthesized at relatively low sintering temperature or without sintering by RF magnetron sputtering [47], spray pyrolysis [45,48], pulsed laser deposition [44], plasma spraying [49], electron-beam evaporation [50], sol–gel [51], etc. Among them, RF magnetron sputtering is a flexible and efficient process for the fabrication of thin-film electrolytes [47,51].

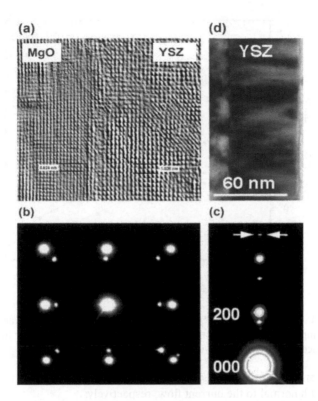

FIGURE 13.13 Lattice structure and orientation of YSZ films deposited on MgO substrate: (a) Cross-sectional TEM morphologies of YSZ/thin films; (b, c) SAED patterns; (d) the lattice mismatch between film and substrate resulting in strain contours [44]. (Reproduced with permission with Elsevier publisher.)

Moreover, the microstructure of thin films can be tailored by controlling plasma characteristics via deposition parameters [52,53].

The enhanced ad-atom mobility benefited from the increased number of energetic ions, or the impact energy of the deposited species results in smooth and dense thin film with improved adhesion and electrical properties by changing sputtering power, substrate bias, temperature, work pressure, and the distance between substrate and target [54–56]. Depositions at increased temperature occur, and the grain growth results from grain coarsening by the coalescence of small islands [57]. In addition, the increased ad-atom mobility promotes the increased defects and re-nucleation rates induced by ion bombardment [58]. Figure 13.14 shows the resistance of YSZ thin films as a function of their thickness determined at different temperatures.

The thickness (d) of the thin film has a close correlation with the resistance, which affects the cross-sectional area of the electrode, which is explained by the following equation [44]:

$$R = \frac{1}{\sigma} \frac{a}{bd}.$$

(13.1)

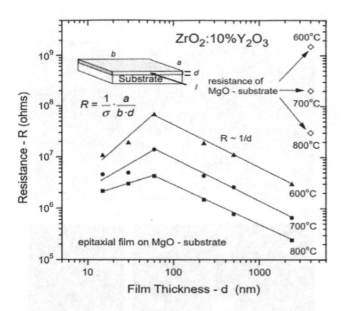

FIGURE 13.14 The resistance of YSZ thin films as a function of their thickness determined at different temperatures. The resistance of MgO substrate is presented for the comparison [44]. (Reproduced with permission with Elsevier publisher.)

where σ, a, and b are the conductivity, the distance between electrodes, and the electrode width normal to the current flow, respectively.

13.3.1 TYPICAL ELECTROLYTES FOR SOFC

Ionic conductivity of the electrolyte materials has a strong dependence on materials' design and preparation process. YSZ is the typical electrolyte material that can improve the ionic conductivity by controlling the surface/interface structure. A package of anode- or cathode-supported YSZ electrolyte stacks are assembled into SOFC (Figure 13.15). The thickness of YSZ electrolytes is always increased from tens of nanometers to hundreds of micrometers depending on the actual applications. As usual, thick YSZ coatings are fabricated using tape casting. However, the electrolytes used in micro-SOFC are prepared by the sputtering technology.

The black tape obtained is plastic and easy to handle. Thus, the black tapes will produce different shapes, such round, quadrate, and strip, depending on the applications [59]. Figure 13.16 presents the photograph of an anode-supported cell with 16 cm^2 (4 cm × 4 cm) active area.

13.3.2 DEPOSITION OF YSZ THIN-FILM ELECTROLYTES

RF magnetron sputtering has an advantage in forming high-quality YSZ thin films in terms of the enhanced surface features and microstructure by the precise control of deposition conditions. Figure 13.17 shows the schematic drawing of RF magnetron sputtering system for the depositions of thin-film electrolytes. Three plasma guns mainly

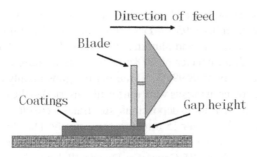

FIGURE 13.15 Schematic diagram of casting for YSZ electrolyte tapes.

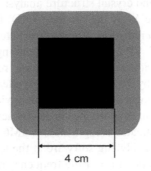

FIGURE 13.16 Photograph of an anode-supported cell with 16 cm² (4 × 4 cm) active area.

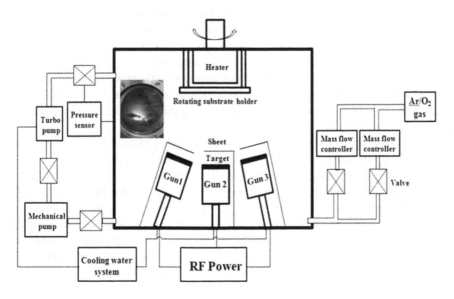

FIGURE 13.17 Schematic drawing of RF magnetron sputtering system for the depositions of thin-film electrolytes.

containing targets, magnetrons, and cooling routes in targets were uniformly inserted in the middle of the chamber ($395 \times 403 \times 401$ mm). YSZ thin films of 696 nm thickness were deposited on quartz and alumina substrates ($15 \times 12 \times 1$ mm) by sputtering 8 at% YSZ target with a diameter of 50.8 mm in a gas mixture of Ar (30 sccm) and O_2 (8 sccm) with the purity of 99.999%. Three plasma guns mainly containing targets, magnetrons, cooling routes in targets were uniformly inserted in the middle of the chamber ($395 \times 403 \times 401$ mm). Prior to depositions, substrates were ultrasonically cleaned in denatured acetone and alcohol for 15 minutes, respectively. The chamber was pumped down to a base pressure less than 1.0×10^{-5} Pa. YSZ target was sputtered at an RF power of 50–150 W using Seren R301 RF Generator. During all depositions, work pressure was maintained at 0.67–4.0 Pa. No external substrate heating and substrate bias were used for all depositions. As-deposited samples were annealed in air at 600°C for 2 hours.

The phase identification and crystal structure analysis of YSZ thin films were carried out using a D8-Advance X-ray diffraction (XRD) from 27° to 65° in the θ–2θ configuration with the Cu $K\alpha$ radiation. The microstructure of the coatings was examined by a ZEISS SUPRA-55 VP field emission scanning electron microscope and a JEOL JEM-2100F transmission electron microscope, respectively. Parallel Pt electrode stripes were deposited by the traditional DC sputtering on the film surface used as electrodes for the measurement of electrical conductivity. Ag paste was then applied to cover the Pt pads in order to improve electrode contact and fix Ag wires. The electrical conductivity was investigated using a Zahner IM6 electrochemical workstation in combination with Z-view analyzing software at the temperature of 400°C–650°C with an increment of 25°C in dry air over the frequency range from 8 MHz to 0.1 Hz.

YSZ thin films were deposited at an RF power of 50–150 W and at a work pressure of 0.67–4.0 Pa by RF magnetron sputtering without external substrate heating and bias. The optimal distance between substrate and target is fixed at 30 mm during all depositions by assuring substrates immersed in plasma flux. Thus, the plasma-enhanced deposition process was developed in order to efficiently fabricate YSZ thin films with controllable microstructures and properties. The increase in RF power resulted in increasing input energy for the target sputtering during depositions. Consequently, the increasing average number of energetic ions led to increasing ion bombardment on film growth. The increased deposition rate (as shown in Table 13.1) is attributed to the increased amount of sputtering target species (neutrals and ions) arriving at substrates per unit of time.

An appropriate increase in work pressure promotes plasma discharge to obtain a high density of plasma. However, at higher work pressure, the density of reactive gas in the chamber leading to the increased fraction of neutral flux in the deposition flux with a wider angular distribution due to more collisions resulted in the increased atomic shadowing [54,58,59]. Correspondingly, due to the low kinetic energy of arriving ions and the substrate temperature induced by ion bombardment, the surface diffusion is limited. Actually, RF power and work pressure showed competitive impacts on the microstructure and properties of thin films. Liu et al. [56] reported Sm^{3+} and Nd^{3+} co-doped CeO_2 (SNDC) thin films deposited at 100–300 W and 2–15 Pa by RF magnetron sputtering. The films deposited at low RF power (100 W) and low work pressure (2 Pa), or high low RF power (300 W) and low work pressure (15 Pa), showed the lower electrical conductivity due to intergranular voids

TABLE 13.1

Deposition Parameters, Deposition Rate, Mechanical Properties, and Electrical Conductivity at 650°C of YSZ Films Deposited on Quartz Substrate by RF Magnetron Sputtering

RF Power (W)	Power Density of Target (W cm⁻²)	Work Pressure (Pa)	Deposition Rate (nm min⁻¹)	Electrical Conductivity (S cm⁻¹)	Hardness	Young's Modulus (GPa)	H/E*	H³/E*²
50	2.47	4.0	5.54	0.0101	7.8	175	0.0446	0.0155
100	4.94	4.0	7.21	0.0191	11.3	228	0.0495	0.0278
150	7.41	4.0	9.27	0.0145	9.2	191	0.0482	0.0213
100	2.47	0.67	5.76	0.0118	8.6	184	0.0467	0.0188
100	2.47	1.33	6.12	0.0124	10.6	218	0.0486	0.0250

and rough surface. Therefore, based on plasma surface interactions, it is essential to design depositions for high-quality thin-film electrolyte used for SOFCs and gas sensors by controlling deposition conditions.

13.3.3 Microstructure of YSZ Thin Films

Figure 13.18 shows XRD patterns of YSZ thin films deposited on quartz and alumina substrates at various RF power and work pressure conditions by RF magnetron sputtering. (111), (200), (220), and (222) peaks were detected in all XRD patterns,

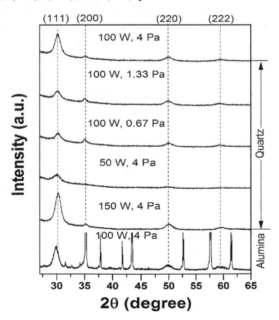

FIGURE 13.18 XRD patterns of YSZ thin films deposited on quartz and alumina substrate at various RF power and work pressure conditions by RF magnetron sputtering.

indicating a single phase with a cubic fluorite structure in YSZ thin films. Under the deposition conditions of the lower RF power and work pressure, random (111), (200), and (220) crystallographic plane orientations were obtained due to the competition between impacts of surface energy and strain energy [55,58]. With an increase in RF power and work pressure, the increasing intensity of (111) reflection shows the tendency of (111) preferred orientation for the growth of YSZ thin films due to the lowest strain energy $(U_{200} > U_{220} > U_{111})$ [60–62].

Figure 13.19 shows the cross-sectional and top-view SEM micrographs of YSZ thin films deposited at an RF power of 50–150 W and a work pressure of 4.0 Pa by RF magnetron sputtering. The microstructure evolution with the changing of sputtering power and work pressure can be explained by the structure zone diagram (SZD) [52,57]. At 50 W, the thin film shows porous columnar grains with the grain size of about 100 nm comprising the equiaxed grains with the in-plane grain size of about 30 nm (Figure 13.19b).

Lager columnar intergranular gaps and rough surface morphology are clearly detected, which are attributed to the limited surface diffusion, resulting in the aggregation of nano-grains (Zone 1), as shown in Figure 13.19a. With increasing RF power to 100 W, dense equiaxed grains and smooth surface morphology were gradually obtained, while the grain size decreased to 36 nm (Zone T) (Figure 13.19c, d). Further increasing RF power to 150 W, the increased energetic ions promoted the continuous grain growth to form underdense nano-columnar with a grain size of 78 nm (Zone 2) (Figure 13.19e, f). The increased kinetic energy of arriving ions induced by ion bombardment resulted in the rapid growth. However, the nature of inhomogeneous distribution of RF power in sputtering led to the low plasma density and deposition flux [59]. Although the features of microstructure and morphology are strongly influenced by substrate temperature [52], the ad-atom mobility are insufficiently driven by the increased substrate temperature of 100–200°C induced by ion bombardment in these cases. Thus, the limited surface diffusion resulted in the appearance of columnar intergranular gaps.

Figure 13.19g, h shows the top-view SEM micrographs of YSZ thin films deposited at an RF power of 100 W and a work pressure of 0.67 and 1.33 Pa by RF magnetron sputtering. The films deposited at 0.67 Pa show the feature of porously packed fibrous grains. The intergranular gaps and rough surface morphology are detected, which are similar features to those of the films deposited at 50 W and 4.0 Pa (Figure 13.19a). With an increase in work pressure to 1.33 Pa, the films become relatively smooth with the gradual disappearance of intergranular gaps and rough surface morphology. Meanwhile, the aggregation of nano-grains disappeared instead of the irregular nano-grains. Further increasing work pressure to 4.0 Pa, as shown in Figure 13.19c, d, dense and smooth YSZ thin films together with the formation of equiaxed grain are obtained. The increased work pressure resulted in the transformation from underdense packed grains to densely equiaxed grains, but still in Zone T with a wide distribution of grain size.

In film depositions, the microstructure and properties of the films are influenced by competitive factors such as work pressure, sputtering power, target distance to substrate, and substrate temperature. In consideration of the limitation of the pressure controller (max 30 mTorr) for this equipment, the deposition conditions of 100 W and 4.0 Pa are the optimized processes for YSZ thin films without substrate heating

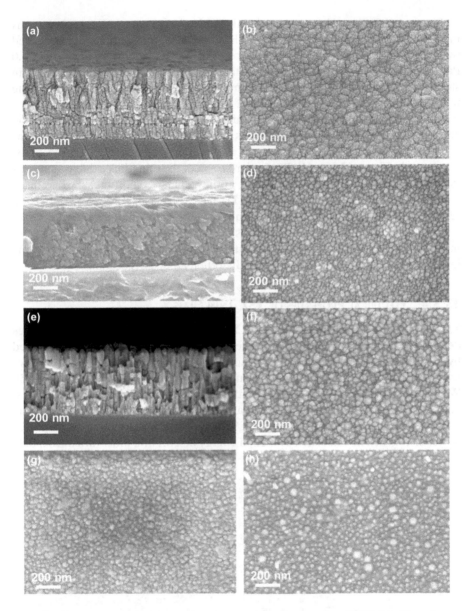

FIGURE 13.19 Cross-sectional and top-view SEM micrographs of YSZ thin films depos-ited at an RF power of 50–150 W at 4.0 Pa and a work pressure of 0.67–4.0 Pa at 100 W by RF magnetron sputtering. (a, e) 50 W and 4.0 Pa; (b, c) 100 W and 4.0 Pa; (d, e) 150 W and 4.0 Pa; (g) 100 W and 0.67 Pa; (h) 100 W and 1.33Pa.

and bias. Based on current results above and SZD, it can be inferred that the further increasing work pressure most likely results in the growth of nano-grains, leading to the intergranular gaps and rough surface morphology. Similar results were found in the SNDC thin films deposited by RF magnetron sputtering at 300 W and 15 Pa [56].

Figure 13.20 shows the cross-sectional TEM micrographs with SAED patterns inserted and high-resolution TEM (HRTEM) images of YSZ thin films deposited by RF magnetron sputtering at a target power of 100 W and a work pressure of 4.0 Pa. As shown in Figure 13.20a, 696-nm-thick films exhibit fully dense nano-grain structure with fine grain size and smooth morphology as well as dense film/substrate interface, which agreed with the results of Figure 13.19.

No intergranular gaps and voids are detected. SAED patterns with bright and continuous rings are further identified as a single cubic structure with reflections of (111), (200), (220), and (222), as well as good crystallinity of YSZ thin films. As shown in Figure 13.20b, HRTEM micrograph of the thin films exhibits dominating (111)-orientated grain growth, dense grain boundary, and film/substrate interface. The measured lattice spacing from (111) and (220) planes d is 2.868 and 1.726 Å, respectively. The smallest grain size is 9 nm along the growth direction. The corrugated fringes reveal a large amount of dislocations inside crystalline grains, resulting in the localized strain fields in the lattice that accounted for the build-up of the internal stress, which led to slight lattice distortion around dislocations.

13.3.4 ELECTRICAL PROPERTIES OF YSZ THIN FILMS

The in-plane electrical conductivity of YSZ thin films with various thicknesses deposited on quartz and alumina substrates at various RF powers was characterized by a two-probe method. Figure 13.21 shows Arrhenius plots of electrical conductivity for YSZ thin films deposited on quartz substrate at an RF power of 50–150 W and at a work pressure of 4.0 Pa. With an increase in RF power, the electrical conductivity of YSZ thin films exhibits an initial increase followed by a decrease. The highest electrical conductivity of 0.0191 S cm^{-1} at 650°C was observed at an RF power of 100 W. The calculated activation energy of YSZ thin films varied from 0.88

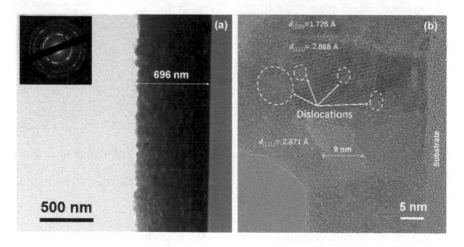

FIGURE 13.20 Cross-sectional TEM micrographs with SAED patterns inserted of YSZ thin films deposited by RF magnetron sputtering at a target power of 100 W and a work pressure of 4.0 Pa.

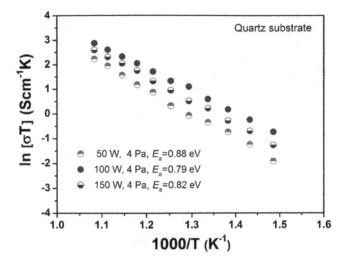

FIGURE 13.21 Arrhenius plots of electrical conductivity for YSZ thin films deposited on quartz substrate at an RF power of 50–150 W and at a work pressure of 4.0 Pa.

to 0.79 eV, corresponding to the similar tendency as that of electrical conductivity as a function of RF power.

Figure 13.22 shows the impedance spectra measured at 400°C–600°C for YSZ thin films deposited on alumina substrate at an RF power of 100 W and a work pressure of 4.0 Pa. It is found that the impedance spectra are composed of a semicircular arc at the high frequency and a tail at the low frequency. Thus, the inserted equivalent circuit, including two R-CPE elements connected in series, the conduction through thin films and electrode polarization were utilized to fit impedance spectra. Figure 13.23 shows the Arrhenius plots of electrical conductivity for YSZ thin films deposited at 100 W and 4.0 Pa.

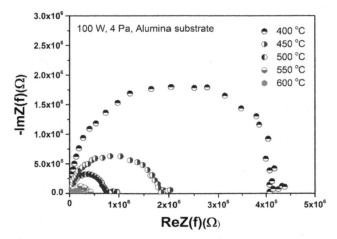

FIGURE 13.22 Impedance spectra measured at 400–600°C for YSZ thin films deposited on alumina substrate at an RF power of 100 W and at a work pressure of 4.0 Pa.

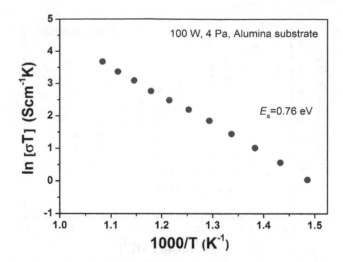

FIGURE 13.23 Arrhenius plots of electrical conductivity for YSZ thin films deposited on alumina substrate at an RF power of 100 W and at a work pressure of 4.0 Pa.

The variation in electrical conductivity is influenced by the microstructural evolution from porous columnar grains to dense nano-grains to porous nano-columnar grains that resulted from the changes in plasma discharge conditions. The enhanced electrical conductivity is attributed to the increased density and refined grains. Moreover, the increased number of dense grain boundary and the lattice distortion induced by the large number of dislocations acted as the rapid pathway for oxygen ion diffusion. In addition, the increase in work pressure from 0.67 to 4.0 Pa resulted in changes in electrical conductivity measured at 650°C from 0.0118 to 0.0191 S cm^{-1}, as shown in Table 13.1. The improvement in electrical conductivity is attributed to the microstructural evolution from irregular nano-grains to equiaxed nano-grains, along with the increased density.

13.4 HIGH-PERFORMANCE NEW-TYPE ELECTROLYTES FOR SOFC

Expect for the typical YSZ electrolyte materials, and new-type electrolytes, such as 8YSZ/Al$_2$O$_3$ and SDC/Al$_2$O$_3$ multilayer, SNDC composite materials are fabricated for electrolytes of SOFC due to their high ionic conductivity. Figure 13.24 shows the schematic diagram of the impedance spectroscopy setup to measure electrical conductivity and to determine the microstructure of YSZ/Al$_2$O$_3$ multilayer films [63]. The interfacial structure of YSZ and Al$_2$O$_3$ layers provides highly efficient channels for the oxygen diffusion. Hence, the lateral electrical conductivity is measured in this case.

Figure 13.25 shows the Arrhenius plot of the total electrical conductivity of 8YSZ/Al$_2$O$_3$ multilayer film compared with that of traditional 8YSZ ceramic [63]. It is found that the ionic conductivity of 8YSZ/Al$_2$O$_3$ multilayer thin film is 3.5 times more than that of traditional 8YSZ ceramic, along with a slightly lower activation

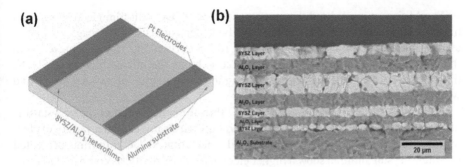

FIGURE 13.24 (a) Schematic diagram of the impedance spectroscopy setup to measure electrical conductivity and (b) microstructure of YSZ/Al2O3 multilayer films [63]. (Reproduced with permission with Elsevier publisher.)

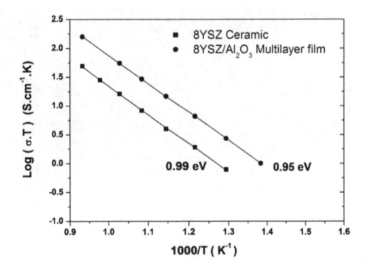

FIGURE 13.25 Arrhenius plot of the total electrical conductivity of 8YSZ/Al$_2$O$_3$ multilayer film compared with that of traditional 8YSZ ceramic [63]. (Reproduced with permission with Elsevier publisher.)

energy. Besides, grain boundaries in the 8YSZ layers in multilayer films were obtained, which were perpendicular to the applied electric field direction to effectively reduce charge carriers' mobility.

Hence, the heterostructure interface structure formed by nano-grain and amorphous layers shows a great enhancement of ionic conductivity than that of 8YSZ ceramic. It is thought that the improvement of ionic conductivity is attributed to the increased tensile stress in 8YSZ layer due to the difference in thermal expansion coefficients between the 8YSZ/Al$_2$O$_3$ layers, as discussed previously in the literature for nanometric films, and also due to differential shrinkage between YSZ and Al$_2$O$_3$ layers.

In addition, it is reported that nanocrystalline SNDC thin-film electrolytes with high in-plane electrical conductivities have been deposited using RF magnetron sputtering. High electrical conductivity of 9×10^{-3} S cm^{-1} at 500°C was achieved with the optimal sputtering process. Figure 13.26 exhibits the microstructure of SNDC thin-film electrolytes deposited by RF magnetron sputtering at various power and working pressure conditions.

Meanwhile, it is also found that SNDC thin films deposited on polycrystalline alumina, single-crystal alumina, quartz, and bulk substrates exhibit a variety of electrical conductivities. Highly (111) preferred orientation SNDC thin films deposited on single crystal alumina show a higher electrical conductivity, as compared to films with random orientation on the polycrystalline alumina substrate. It also indicates that the electrical conductivity of the SNDC film is higher than that of the corresponding bulk due to the lower activation energy. Figure 13.27 shows the electrical conductivity of SNDC thin films deposited on various substrates.

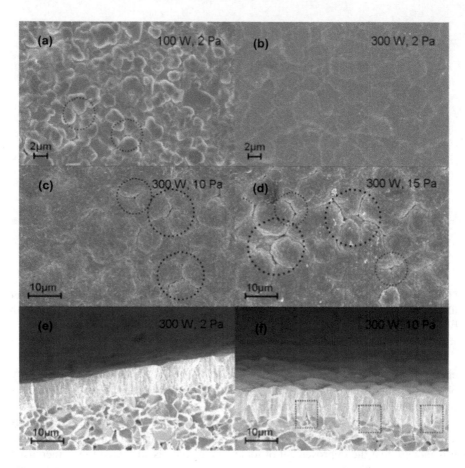

FIGURE 13.26 Microstructure of SNDC thin-film electrolytes deposited by RF magnetron sputtering at various power conditions of 100–300 W and at working pressure conditions of 2–15 Pa [56]. (Reproduced with permission with Elsevier publisher.)

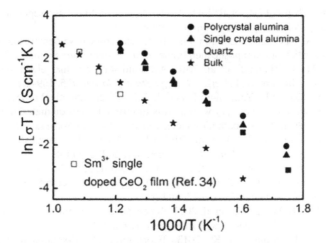

FIGURE 13.27 Electrical conductivity of SNDC thin films deposited on various substrates [56]. (Reproduced with permission with Elsevier publisher.)

13.5 CONCLUSIONS

In recent years, the development of all SSE materials has been a research hotspot in LIBs due to the elimination of safety issues caused by dendrite growth of Li. The challenge of LIBs is how to reduce interface impedance and improve ionic conductivity at room temperature. Much more efforts have been made, such as the synthesis and application of ISEs, SPEs, and organic–inorganic hybrid composite electrolytes. As for SOFCs, novel material designs using advanced technologies have been applied to synthesize high-performance electrolyte materials by improving the surface/interface properties to enhance ionic conductivity (above 10^{-3} S cm^{-1}) at a low temperature (lower than 400°C). However, above all, there is still a long way to go.

ACKNOWLEDGMENTS

The author appreciates Dr. Haoqi Wang and Qian Zhou for their contributions for the partial writing of Sections 13.2 and 13.3 in this chapter. This work is also supported by China Postdoctoral Science Foundation (No: 2016M600085) and 2020 talent plan granted by Beijing Academy Science and Technology.

REFERENCES

1. Li M., Lu J., Chen Z., et al. 2018. 30 Years of Lithium-Ion Batteries. *Advanced Materials* 30: 1800561–85.
2. Yoshio M., Brodd R.J., Kozawa A. *Lithium-ion Batteries*, Springer-Verlag, New York, 2009.
3. Li S., Zhao Y., Shi X., et al. 2012. Effect of sulfolane on the performance of lithium bis (oxalato) borate-based electrolytes for advanced lithium ion batteries. *Electrochimica Acta* 65: 221–7.

4. Xiang H., Chen J., Li Z., et al. 2011. An inorganic membrane as a separator for lithium-ion battery. *Journal of Power Sources* 196: 8651–5.
5. Huang X.S., Hitt J. 2013. Lithium ion battery separators: development and performance characterization of a composite membrane. *Journal Membrane Science* 425–426: 163–8.
6. Yue L., Ma J., Zhang J., et al. 2016. All solid-state polymer electrolytes for high-performance lithium ion batteries. *Energy Storage Material* 5: 139–64.
7. Amiki Y., Sagane F., Yamamoto K., et al. 2013. Electrochemical properties of an all-solid-state lithium-ion battery with an in-situ form edelectrode material grown from a lithium conductive glass ceramics sheet. *Journal of Power Sources* 241: 583–8.
8. Ahn C.W., Choi J.J., Ryu J., et al. 2014. Electrochemical properties of $Li_7La_3Zr_2O_{12}$-based solid state battery. *Journal of Power Sources* 272: 554–8.
9. Kobayashi E., Plashnitsa L.S., Doi T., et al. 2010. Electrochemical Properties of Li symmetric solid-state cell with NASICON-type solid electrolyte and electrodes. *Electrochemica Communication* 12: 894–6.
10. Cao C., Li Z.B, Wang X.L., et al. 2014. Recent advances in inorganic solid electrolytes for lithium batteries. *Frontiers in Energy Research* 2: 1–10.
11. Kotobuki M., Kanamura K. 2013. Fabrication of all-solid-state battery using $Li_5La_3Ta_2O_{12}$ ceramic electrolyte. *Ceramics International* 39: 6481–7.
12. Knauth P. 2009. Inorganic solid Li ion conductors: an overview. *Solid State Ionics* 180: 911–6.
13. Yuan F., Chen H.Z., Yang H.Y., et al. 2005. PAN-PEO solid polymer electrolytes with high ionic conductivity. *Materials Chemical Physics* 89: 390–4.
14. Huang H., Chen L., Huang X., et al. 1992. Studies on PAN-based lithium salt complex. *Electrochimica Acta* 37: 1671–3.
15. Takada K. 2009. Electrolytes: solid oxide. *Enc. Electrochemica Power Sources* 5: 328–36.
16. Kamaya N., Homma K., Yamakawa Y., et al. 2011. A lithium super ionic conductor. *Nature Materials* 10: 682–6.
17. Kotobuki M., Kanamura K. 2013. Fabrication of all-solid-state battery using $Li_5La_3Ta_2O_{12}$ ceramic electrolyte. *Ceramics International* 39: 6481–7.
18. Ohta N., Takada K., Zhang L., et al. 2006. Enhancement of the high-rate capability of solid-state lithium batteries by nanoscale interfacial modification. *Advanced Materials* 18: 2226–9.
19. Sakuda A., Hayashi A., Ohtomo T., et al. 2011. All-solid-state lithium secondary batteries using $LiCoO_2$ particles with pulsed laser deposition coatings of Li_2S-P_2S_5 solid electrolytes. *Journal Power Sources* 196: 6735–41.
20. Kanno R., Murayama M., et al. 2001. Lithium ionic conductor thio-LISICON: the Li_2S-GeS_2-P_2S_5 system. *Journal of the Electrochemical Society* 148(7): A742–6.
21. Kato Y., Hori S., Saito T., et al. 2016. High-power all-solid-state batteries using sulfide superionic conductors. *Nature Energy* 1(4): 16030–6.
22. Ibrahim S., Yassin M.M., Ahmad R., et al. 2011. Effects of various $LiPF_6$ salt concentrations on PEO-based solid polymer electrolytes. *Ionics* 17: 399–405.
23. Ahmad S. 2009. Polymer electrolytes: characteristics and peculiarities. *Ionics* 15: 309–21.
24. Meyer W.H. 1998. Polymer electrolytes for lithium-ion batteries. *Advanced Materials* 10(6): 439–48.
25. Fergus J.W. 2010. Ceramic and polymeric solid electrolytes for lithium-ion batteries. *Journal of Power Sources* 195: 4554–69.
26. Song J.Y., Wang Y.Y., Wan C.C. 1999. Review of gel-type polymer electrolytes for lithium-ion batteries. *Journal of Power Sources* 77: 183–97.
27. Ramesh S., Ng H.M. 2011. An investigation on PAN-PVC-LiTFSI based polymer Electrolytes system. *Solid State Ionics* 192: 2–5.

28. Zhang Z., Fang S. 2000. Novel network polymer electrolytes based on polysiloxane with internal plasticizer. *Electrochimica Acta* 45: 2131–8.
29. Kim S.H, Choi K.H., Cho S.J., et al. 2014. A shape-deformable and thermally stable solid-state electrolyte based on a plastic crystal composite polymer electrolyte for flexible/safer lithium-ion batteries. *Journal of Materials Chemistry. A* 2(28): 10854–61.
30. Zhang D., Xu X., Wang Z., et al. 2019. Recent progress of organic-inorganic composite solid electrolytes for all-solid-state lithium batteries. *Chemistry-A European Journal* 04461. doi: 10.1002/chem.201904461.
31. Sun H.Y., Takeda Y., Imanishi N., et al. 2000. Ferroelectric materials as a ceramic filler in solid composite polyethylene oxide-based electrolytes. *Journal of the Electrochemical Society* 147 (7): 2462–7.
32. Kobayashi Y., Miyashiro H., Takeuchi T., et al. 2002. All-solid-state lithium secondary battery with ceramic polymer composite electrolyte. *Solid State Ionics* 152–153: 137–42.
33. Cho J., Liu M., et al. 1997. Preparation and electrochemical properties of glass-polymer composite electrolytes for lithium batteries. *Electrochimica Acta* 42: 1481–8.
34. Minh N.Q. 1993. Ceramic fuel cells. *Journal American Ceramics Society* 76: 563–88.
35. Maskell W.C. 2000. Progress in the development of zirconia gas sensors. *Solid State Ionics* 134: 43–50.
36. Stambouli A.B., Traversa E. 2002. Solid oxide fuel cells (SOFCs): a review of an environmentally clean and efficient source of energy. *Renewable & Sustainable Energy Reviews* 6: 433–55.
37. Wachsman E.D., Lee K.T. 2011. Lowering the temperature of solid oxide fuel cells. *Science* 334: 935–9.
38. Garcia-Barriocanal J., Rivera-Calzada A., Varela M., et al. 2008. Colossal ionic conductivity at interfaces of epitaxial $ZrO_2:Y_2O_3/SrTiO_3$ heterostructures. *Science* 321: 676–80.
39. Kim K.J., Park B.H., Kim S.J., et al. 2015. Micro solid oxide fuel cell fabricated on porous stainless steel: a new strategy for enhanced thermal cycling ability. *Scientific Reports* 6: 22443–51.
40. Beckel D., Bieberle-Hütter A., Harvey A., et al. 2007. Thin films for micro solid oxide fuel cells. *Journal of Power Sources* 173: 325–45.
41. El-Toni A.M., Yamaguchi T., Shimizu S., et al. 2008. Development of a dense electrolyte thin film by the ink-jet printing technique for a porous LSM substrate. *Journal of the American Ceramics Society* 91: 346–9.
42. Riess I., Braunshtein D., Tannhauser D. S. 1981. Density and ionic conductivity of sintered $(CeO_2)_{0.82}(GdO_{1.5})_{0.18}$. *Journal of the American Ceramics Society* 64: 479–85.
43. Gibson I.R., Dransfield G.P., Irvine J.T.S. 1998. Sinterability of commercial 8 mol% yttria-stabilized zirconia powders and the effect of sintered density on the ionic conductivity. *Journal of Materials Science* 33: 4297–305.
44. Kosacki I., Rouleau C.M., Becher P.F., et al. 2005. Nanoscale effects on the ionic conductivity in highly textured YSZ thin films. *Solid State Ionics* 176: 1319–26.
45. Garcia-Sanchez M.F., Peña J., Ortiz A., et al. 2008. Nanostructured YSZ thin films for solid oxide fuel cells deposited by ultrasonic spray pyrolysis. *Solid State Ionics* 179: 243–9.
46. Bellino M.G., Lamas D.G., Walsöe de Reca N.E. 2006. A mechanism for the fast ionic transport in nanostructured oxide-ion solid electrolytes. *Advanced Materials* 18: 3005–9.
47. Yao L., Liu W., Ou G., et al. 2015. Enhanced ionic conductivity in magnetron-sputtered $Ce_{0.8}Sm_{0.2}O_2$-δ/Al_2O_3 multilayers. *Electrochimica Acta* 158: 196–201.
48. Perednis D., Gauckler L.J. 2004. Solid oxide fuel cells with electrolytes prepared via spray pyrolysis. *Solid State Ionics* 166: 229–39.

49. Lang M., Henne R., Schaper S., et al. 2001. Development and characterization of vacuum plasma sprayed thin film solid oxide fuel cells. *Journal of Thermal Spray Technology* 10: 618–25.

50. Hartmanova M., Thurzo I., Jergel M., et al. 1998. Characterization of yttria-stabilized zirconia thin films deposited by electron beam evaporation on silicon substrates. *Journal of Materials Science* 33: 969–75.

51. Gaudon M., Laberty-Robert C., Ansart F., et al. 2006. Thick YSZ films prepared via a modified sol-gel route: thickness control (8–80 µm). *Journal of the European Ceramics Society* 26: 3153–60.

52. Thornton J.A. 1974. Influence of apparatus geometry and deposition conditions on the structure and topography of thick sputtered coatings. *Journal of Vacuum Science & Technology* 11: 666–70.

53. Thompson C.V., Carel, R. 1995. Texture development in polycrystalline thin films. *Materials Science and Engineering: B* 32: 211–9.

54. Zheng B.C., Meng D., Che H. L., et al. 2015. On the pressure effect in energetic deposition of Cu thin films by modulated pulsed power magnetron sputtering: A global plasma model and experiments. *Journal of Applied Physics* 117: 203–302.

55. Ou Y.X., Lin J., Tong S., et al. 2016. Structure, adhesion and corrosion behavior of CrN/TiN superlattice coatings deposited by the combined deep oscillation magnetron sputtering and pulsed dc magnetron sputtering. *Surface and Coatings Technology* 293: 21–7.

56. Liu W., Li B., Liu H., et al. 2011. Fabrication of Sm^{3+} and Nd^{3+} co-doped CeO_2 thin-film electrolytes by radio frequency magnetron sputtering. *Electrochimica Acta* 56: 8329–33.

57. Anders A. 2010. A structure zone diagram including plasma-based deposition and ion etching. *Thin Solid Films* 518: 4087–90.

58. Petrov I., Barna P.B., Hultman L., et al. 2003. Microstructural evolution during film growth. *Journal of Vacuum Science & Technology A* 21: S117–28.

59. Timurkutluk B., Celika S., Ucar E. 2019. Influence of doctor blade gap on the properties of tape cast NiO/YSZ anode supports for solid oxide fuel cells. *Ceramics International* 45: 3192–8.

60. Lieberman M.A., Lichtenberg A.J. *Principles of Plasma Discharge and Materials Processing*, Second edition, John Wiley, New York, 2005.

61. Pelleg J., Zevin L.Z., Lungo S., et al. 1991. Reactive-sputter-deposited TiN films on glass substrates. *Thin Solid Films* 197: 117–28.

62. Ou Y.X., Ouyang X.P., Liao B., et al. 2020. Hard yet tough CrN/Si_3N_4 multilayer coatings deposited by the combined deep oscillation magnetron sputtering and pulsed dc magnetron sputtering. *Applied Surface Science* 502: 144168–77.

63. Antunes F.C., Goulart C.A., Andreeta M. R. B., de Souza D.P. 2018. YSZ/Al_2O_3 multilayer thick films deposited by spin coating using ceramic suspensions on Al2O3 polycrystalline substrate, *Materials Science and Engineering: B* 228: 60–6.

Index

A

AAO, 246, 247, 258–259
ABO_3 perovskite, 4
absolute power, 248
acid-doped, 219
activation energy, 251
activation loss, 244
active layer, 444
A-D-A, 453
adatoms, 130, 131, 141, 145, 147, 152, 154
additive selection rule, 34
additives, 18, 212
adeciduate mechanical features, 484
adhesion, 272, 491
agglomeration, 252
air, 322, 324, 333, 334, 336, 338–340, 343, 345, 347, 350, 365–367, 369
ALD, 245, 253, 254, 258, 263, 269, 272, 416–419, 422, 423, 430, 431, 436
aliovalent-doped ceria, 249
alkaline fuel cell (AFC), 481
all-inorganic halide perovskite, 23
alloy model, 402–404
all solid-state electrolyte materials, 484
AlN, 126, 128–132, 134–137, 139, 141–161
Al_2O_3, 417–429, 434–437
aluminum oxide, 417, 420
$Al_xGa_{1-x}N$, 126, 128, 129, 160
amphoteric membrane, 207
angled fins, 361–363
anion exchange membrane, 207
anode/anodes, 243, 252–253, 258, 289, 290
anode or cathode-supported YSZ electrolytes stacks, 492
area-specific resistance, 245
area utilization, 263, 272
aromatic polymers, 216
array, 268
Arrhenius plots of electrical conductivity, 498
aspect ratio, 257
atomic layer deposition, 416, 417
atomic peening, 273

B

background pressure, 248
back surface field, 414
band structure, 285, 306
barium carbonate, 252
batteries, 286, 287, 289, 290, 292, 295

BCY, 252
biaxial strain, 147, 157
bi-layer heterojunction, 444
blue energy, 113–117
boundary, 298, 299
BSF, 414–416
buckling, 263, 265, 273–276
buckling stress, 275–276
buffer layer, 248
bulk electrolytes, 490
bulk heterojunction, 444
butane, 248, 253
BYZ, 249, 258, 272
BZCY, 252

C

C_{60}, 455
capillary, 293
carbon decomposition, 253
carbon nanotubes, 487
catalytic activity, 286, 253, 295–296
catalytic burner, 248
cathode, 243, 258
cathode reaction kinetics, 252
cation exchange membrane, 207
cell configuration, 242, 246
ceramic fillers, 487
$Ce_{0.8}Sm_{0.2}O_{2-\delta}$ (SDC), 489
charge carrier ions, 203
charge carrier mobility, 12
charge-discharge, 206
charge/discharge, 483
charge neutrality, 250
charge transfer, 288, 299, 301, 302, 396–397
charge transport, 13
charging/charging process, 321, 322, 332, 334, 336, 339, 340, 351
chemical compatibility, 255
chemical energy, 241, 482
chemical passivation, 420, 421, 424, 430, 432, 437
chemical solution deposition, 258
chemical stability, 249, 251
chemisorption, 254
circular membrane, 263, 265, 267, 274, 275–276
coagulating bath, 226
co-firing process, 248
combinatorial etching, 263
combined energy, 364–368

complementary metal-oxide-semiconductor devices, 423
composite electrolyte, 251
composite membrane, 221
composite polymer electrolyte (PC-CPE), 486
COMS, 423
concentration loss, 244
conduction, 321, 344, 351, 361
conductivity, 319, 320, 322, 323, 325, 328, 330, 332, 333, 344, 351, 353, 354, 356, 361, 370
conformal deposition, 254
contact resistance, 295, 297
convection, 321, 337, 338, 341, 344, 350, 352, 353, 356, 359–362, 370
conversion, 286, 287, 289, 291
conversion efficiency, 414, 416, 434, 436, 437, 488
corrugated membrane, 267, 274
co-sputtering, 253
Coulombic efficiency, 289
crater patterns, 268
critical stress, 274
crystallinity, 130–132, 134, 135, 139, 141, 142, 146, 149, 150, 152–154, 157, 160, 161, 253
crystal structure of halide perovskite, 5
crystal structure/symmetry, 249, 285
c-Si, 414–416, 419–424, 426, 428, 432, 436
$CsPbI_3$, 17
$CsPbI_3$ has four phases, 27
cubic fluorite, 250
current collector, 248, 259
current density, 245
current leakage, 242
current output, 260
current-voltage curve, 244
curtain, 417
CVD, 296

D

D-A-D, 453
dangling bonds (DBs), 286, 420, 421, 428, 436, 437
decoupled, 201
deepultraviolet, 128, 129
deep UV, 125, 128, 129, 159
deformation, 269, 272
deintercalation, 289
delamination, 275
demixing, 228
dendrite growth, 289, 483
dielectric-to-conductor, 93–97
dielectric-to-dielectric, 86–93
Differential Scanning Calorimetry (DSC), 325–328, 333
diffraction pattern, 136, 149, 154

diffusion coefficient, 304
diffusion length L and lifetime, 12
dimethylamine, 29
discharging/discharging process, 321, 322, 332, 334, 336, 339, 340
dislocation behavior, 136, 140, 141, 150, 151, 154, 161
dislocations, 247
dispersing inorganic fillers, 487
D_{it}, 420, 424, 427–430
domain boundaries, 298, 299
Donnan exclusion, 210
dopant concentration, 250
DRIE, 263, 267, 269
driving force, 448
dry cells, 480

E

effective lifetime, 419, 422, 425
effective mass, 12
effective rear surface recombination rate, 434
efficiency, 206–207
electrical energy, 241, 482
electrical energy storage, 481
electrical leakage, 247
electrical properties, 491
electrical transport spectroscopy, 302, 303
electric conductivity, 245
electrochemical, 200, 285–287, 289, 291–304, 306
electrochemical approach, 286, 291
electrochemical devices, 287, 291, 304
electrochemical on-chip, 285–306
electrochemical performance, 298
electrochemical reaction, 243, 267
electrochemical stability, 488
electrochemical vapor deposition, 242
electrochemical window, 485–486
electrode cross section area, 491
electrolysis, 287
electrolyte/electrolytes, 202, 243, 245, 257, 481
electrolyte leakage, 488
electrolyte materials, 481
electronic, 285, 286, 305
electronic shortening, 251
electron transport layer, 386–394
ELOG, 126, 154, 156, 161
encapsulation, 331, 348
energy density, 481
energy minimization, 272
energy storage, 318
energy storage technique, 199
enthalpy, 319–322
epilayer, 127, 128, 130, 131, 139, 141, 142, 144, 146, 147, 152, 154, 156–161
epitaxial, 275

epitaxial heterostructures, 489
epitaxial lateral overgrowth, 126, 128, 146, 147, 150, 154
etching, 246, 255, 263, 269
etching pit density (EPD), 142, 152, 157, 159, 161, 162
ethylenediamine (EDA), 32
ethylenediamine-pyrocatechol, 246
eutectic, 318–320, 330
excimer laser, 254
exciton, 375, 444
exciton binding energy, 447
exciton diffusion length, 447
exergoeconomic factor, 367, 368, 370
exergy, 366–370
external circuit, 241, 244
external quantum efficiency, 449
external reformer, 241
extrinsic stress, 272

F

fast ionic conductors, 487
FCC, 250
ferroelectric ceramic fillers, 487
FG, 427–430, 437
field-effect passivation, 420, 421, 426, 437
fill factor, 449
first working CsPbI$_3$ device, 27
fixed charge density, 422, 426, 437
flat sapphire substrate (FSS), 142, 143, 146–149, 161
flexible batteries, 56–68
flexible supercapacitors, 46, 48, 49, 58, 70
floating zone (FZ), 422, 423
fluorite type, 249
formamidinium lead iodide, 17
fracture toughness, 243
free-standing, 255–257, 263–267, 267, 271, 274–276
fuel cells, 481
fuel oxidation, 253
fullerene, 455
full width at half maximum (FWHM), 130–132, 134–136, 139, 146, 147, 149, 154–156, 158–161

G

gadolinium- or samarium-doped ceria (CGO or CSO), 489
galvanostatic charge/discharge, 289, 290
GaN, 126–128, 160
gas diffusion, 243
gas exchange, 252
gas leakage, 242
GDC, 247, 251, 258

Gd$_2$Zr$_2$O$_7$, 489
Gibbs free energy, 287, 288, 295, 297
glass ceramic, 257
good chemical and mechanical stability, 205
graft, 215
grain boundaries, 247
grain size, 247
graphene, 296
graphene oxides, 217
graphitic, 292

H

hafnium oxide, 420
halide perovskite photovoltaics, 1–36
heat capacity, 320–322, 326, 343
heat energy, 254
heat exchanger, 322, 364
heat flux, 343, 353, 354, 360
HER, 287, 296–299, 302
heterogeneous catalysis, 300
heterophase, 298, 299
HfO$_2$, 420–423, 430–437
HI addition, 28
high ion conductivity, 204
high-performance ISEs, 484
high-performance sulfide solid-state electrolytes, 484
high polymer, 451
high working voltage, 483
hole transport layer, 376, 394–396
HOMO, 444
homophase interfaces, 490
HPbI$_3$ being lead source, 28
hybrid membrane, 212
hydrocarbon, 253
hydrocarbon membranes, 222
hydrofluoric acid, 257
hydrogen fuel cells, 483
hydrogen oxidation reaction, 253
hydroxyl groups, 418, 221

I

ideal voltage, 244
immersion-precipitation induced phase separation, 226
incompressible, 356
individual factors, 297
injector, 417
inorganic, 319, 320, 326
inorganic solid electrolytes (ISEs), 484
in-plane, 285, 286, 305
in-plane electrical conductivity, 498
in-situ, 300, 305
instability issue of MA cation, 16
intercalation, 288–290, 305

interconnect, 259
interface defect density, 420
interface engineering, 386
interfacial conductivity, 490
interfacial diffusion, 242
interfacial layer, 290
interfacial resistance, 488
interlayer, 252
internal quantum efficiency (IQE), 128, 129, 152,
         153, 158–161, 449
internal reforming, 248
interstitial implantation, 273
intrinsic defect, 14
intrinsic stress, 272
intrinsic thermal stability of halide perovskite
         materials, 15–18
inverted structure, 377
ion bombardment, 491
ion exchange capacity, 205
ion exchange membrane, 203
ionic conductivity, 245, 249, 251, 253, 483
ionic transportation, 250
ionization, 253
ion selectivity, 223
irreversible capacity losses, 484
isovalent-cation stabilized bismuth
         oxides, 249
ITIC, 459

K

key component, 201
KOH etching, 269

L

lager columnar intergranular gaps, 496
laminar, 356
lamination, 259
large-scale energy storage system, 199
latent heat, 318, 320, 322, 326, 328, 332, 344,
         350, 358
lateral ionic conductivity, 489
lattice distortion, 273
layer-by-layer self-assembly, 211
layered materials, 285, 286, 289
lead-acid battery, 479
levelized cost, 366
ligand, 254
lightemitting diode, 129
Li-P-S systems, 485
liquid organic solvents, 484
lithium ion batteries (LIBs), 480, 481
lithography, 247, 258, 267
low cost, 205
low defect density, 152
lower dimensional perovskites, 11

low self-discharge rate, 483
low vanadium ion permeation, 204
LSTN, 259
LUMO, 444

M

MA- and Br-free composition, 23
magnetron sputtering, 497
MAPbI$_3$, 7
mass transport, 252
material degradation, 242
maximal surface recombination rate, 422
MEA, 246, 255, 257, 259, 263, 269, 272
mechanical robustness, 248
mechanical stability, 250, 255, 263, 266, 267,
         269, 272, 274, 276
mechanism, 286, 289, 291, 292, 295, 300,
         302, 306
mechanistic study, 285–306
melting, 319–322, 325–328, 332–334, 337, 338,
         345, 350, 352, 354–356, 358–361, 363,
         367, 369, 370
melting point, 325, 326, 328, 332–334
membrane, 200
membrane-edge support, 263
membrane fracture, 263
membrane resistivity, 206
MEMS, 245, 255, 258, 276
mesh/meshing, 345–347, 353, 356, 360
metal catalysis, 172
metal catalyst, 253
metallic grids, 248, 271, 276
metalorganic chemical vapor deposition
         (MOCVD), 126–128, 130, 133, 142,
         152, 160
methane, 253
methanol, 258
micro/nano power source, 109
micro-phase separation, 222
micro-reactor, 291, 292, 294–300, 306
minority carrier, 416, 422, 423, 425, 426,
         429, 437
mixed cation and mixed halogen system, 23
"mixed cation" approach, 20
mobility, 250, 273, 286, 297
modification materials, 208
molten carbonate fuel cell (MCFC), 481
monoclinic, 250
monocrystalline silicon, 414
monolayer MoS$_2$, 294–298
morphological degradation, 252
morphology, 14, 253
MoS$_2$, 286, 287–291, 293–299, 305
MQW, 126, 153, 158, 160
multiple quantum well, 126
mushy electrolytes, 480

N

Nafion, 207
nanoenergy, 78–79
nanoionics, 246
nano-patterned sapphire substrate (NPSS), 130,
    142–152, 153–157, 160, 161
nanoscale thin film, 246
nanosphere lithography, 268
nanotubular array, 269
N-containing groups, 217
NH_3, 126, 130
nickel grid, 271
nickel supporting grid, 263
nitrides, 286
non-flammability, 488
non-fluorinated membranes, 216
non-solvent, 225
η, 434
Nyquist curves, 301

O

octahedral factor, 5
octahedral sites, 250
OER, 288, 300–302
OH, 418, 424
ohmic loss, 244–245, 489
Ohm's law, 245
oligomer, 451
on-chip, 291, 292, 295, 298
open circuit voltage (OCV), 245, 251, 252, 258,
    259, 449
operating temperature/operation temperature,
    245, 249, 252, 263, 264, 489
optical, 286, 291, 294, 299, 305
optical absorption coefficient, 7
optimized E_g, 4
optoelectronic, 285
organic, 319, 320, 326
organic-inorganic hybrid composite electrolytes,
    484, 487
organic liquid electrolyte, 484
organic solar cell/organic solar cells, 374, 444
organic solution, 259
overpotential, 287–289, 292, 296, 297, 299, 300
oxidant, 243
oxidation process, 241
oxides, 286
oxygen ion, 243, 246, 249, 252
oxygen reduction reaction (ORR), 252, 292, 293
oxygen vacancies, 250

P

paraffin, 319, 345
partially-fluorinated membranes, 214

passivated emitter and rear contact (PERC),
    414–416, 434–437
passivated region, 434
passivation, 413–437
payback period, 365, 370
$PC_{61}BM$, 456
$PC_{71}BM$, 456
PCM-based TES, 318, 320, 321, 323, 330–333,
    336, 341, 344, 345, 350, 353, 356, 357,
    364–370
Pd, 253
PEALD, 418, 419, 422–424
perfluorinated sulfonic acid polymer, 207
performance of batteries, 483
perovskite, 4, 251
phase change materials (PCMs), 318–322, 326,
    330, 332, 341, 343, 347, 361, 364,
    369, 370
phase change period, 318, 320, 340
phase change temperature, 321, 343
phase invention method, 223
phase stability of all-inorganic $CsPbI_3$ perovskite,
    23–35
phase stability of $FAPbI_3$ perovskite, 18–23
phase transformation, 272
phase transition, 328, 332, 348
phosphoric acid fuel cell (PAFC), 481
photoluminescence, 127, 153
photovoltaic or solar cell, 2
physicochemical parameters, 291, 302
pinholes, 246, 257, 267
PL, 153, 158, 160
plasma discharge, 494
plasma enhanced ALD, 418
plasma-enhanced chemical vapor deposition
    (PECVD), 422, 423
plasma etching, 257
plasma guns, 494
plasmonic effect, 379
PLD, 251, 253, 259, 263, 264, 272
plume temperature, 254
poisson ratio, 272, 276
polarization, 247, 252
poly(diallyldimethylammonium chloride), 211
poly(ether ether ketone), 216
poly(ether sulfone), 226
poly(ethylene–tetrafluoroethylene), 215
poly (methyl methacrylate) (PMMA), 293,
    295, 303
poly(vinyl pyrrolidone), 226
polyaniline, 210
polybenzimidazole, 218
polycrystalline, 275
polydopamine, 218
polyethylene glycol, 224
polyethyleneimine, 210
polyimide, 221

polymer, 451
polymer-lean phase, 226
polymer matrix, 487
polymer-rich phase, 226
polymer-salt system, 488
polypyrrole, 210, 228
polytetrafluoroethylene, 213, 214
polyvinylidene fluoride, 213
porogen, 224, 229
porosity, 247
porous electrode, 241
porous membranes, 223–230
porous nickel, 247
porous substrates, 247, 255, 258
positively charged groups, 228
post-annealing, 424, 426, 427, 429, 437
post-buckling, 276
potential difference, 444
powder-based processes, 242, 245
power conversion efficiency, 2, 450
power density, 247, 248, 251, 252, 259, 268
power generation, 318, 330, 331, 364–367
power output, 248, 261, 272
precursor, 250, 254, 419, 423
pretreatment, 213
principal stress, 266
proton/protons, 203, 243, 245
proton conductivity, 249, 251
proton exchange membrane fuel cell
          (PEMFC), 481
Pt, 252, 258, 271
pulse duration, 254
purge, 254
PVD, 253
pyridine groups, 220
pyridinic, 292

Q

Q_f, 422, 423, 426, 427, 429
quantum dots (QDs), 32

R

radio frequency magnetron sputtering, 490
Raman shift, 148, 156
rate-limiting reaction, 252
reactant, 243, 254
reaction area, 243
rechargeable battery, 480
recombination, 414, 416, 419–421, 424, 426, 436
redox flow batteries, 200
redox reactions, 202
relative product cost, 367, 368, 370
renewable energy, 199
residual stress, 142, 144, 147, 148, 157, 161, 255,
          263, 264, 265, 272–273, 276

RF sputtering, 246
Ru, 253

S

scaling up, 266, 277
Schottky cell, 444
S_{cont}, 434
screen printing, 259
selected area electron diffraction, 490
self-limiting, 416, 418
self-powered sensors, 109–113
semiconductor device, 97–104
semiconductor-to-conductor, 101–104
semiconductor-to-dielectric, 100–101
Shockley-Read-Hall, 419
short-circuit, 263
short-circuit current density, 449
sieving action, 223
silicon bulk micromachining, 246
silicon micromachining processes, 247, 255
silicon nitride, 255, 416
silicon oxide, 420
silicon wafers, 246, 255, 257
single cell structure, 201
single-junction, 444
single-layer, 444
sintering process, 242, 246, 258, 490
sintering temperature, 490
SiN_x, 416, 434–436
SiO_2, 422, 423, 428, 430, 432, 434, 437
skin layer, 226
SL, 130, 141, 142, 144–146, 150, 152, 153,
          156–158, 160, 162
small grain size, 32
small organic molecule, 452
S_{max}, 422, 424, 428–431
solar cell, 414–416, 419, 422, 424, 434–437
solar energy and photovoltaic, 1–4
sol-gel process/sol-gel technique, 208, 253
solid oxide fuel cell (SOFC), 481
solid polymer electrolytes (SPEs), 484
solid-state configuration, 488
solution recasting, 212–213
solvent anneal, 14
solvent-template method, 224–225
solvent treatment, 228
space sensors, 168
spalling, 275
S_{pass}, 434
spatial ALD, 417, 422, 436
specific capacity, 289
specific energy density, 481
specific heat, 322, 323, 328, 330, 332, 334, 344
spectrum of the sun light, 3
Spider-web pattern, 271
spin coating, 253

sponge-like pores, 228
spot size, 254
spray pyrolysis, 253
sputtering, 246, 252, 259, 269, 272
square, 263, 266
$S_{rear,eff}$, 434, 436
SRH, 419, 420
stability of halide perovskites, 15–35
stabilize black $CsPbI_3$, 26
stainless steel, 247, 259
storage, 286, 287, 289, 304–306
strain, 34
stress absorber, 266
stress concentration, 263, 265
stress distribution, 266
stress relaxation, 273
strontium and magnesium-doped lanthanum
    gallate (LSGM), 489
sulfide-based all solid-state lithium batteries, 485
sulfide composites, 484
sulfonated poly(ether ether ketone), 216
sulfonation, 216
sulfonation degree, 216
sulfonic acid groups, 216
summary, 35–36
super-cooling, 320, 321
superlattice (SL), 130, 141, 142, 144–146, 150,
    152, 153, 156–158, 160, 162, 287
superlattice thin films, 489
supporting electrolyte, 202
supporting substrate, 246, 255
support layer, 226
surface energy, 272
surface exchange, 252
surface functionalization, 20
surface/interface structure, 492
surface modification, 210
surface morphology, 496
surface passivation, 414, 419, 422, 423, 426,
    434, 436
surface recombination rate, 419, 420, 426, 429,
    434–436
surface roughness, 247, 259
surface structure, 285
sustainability, 364, 368–370
swelling ratio, 206

T

Tafel slopes, 295
tandem OSC, 444
tape-casting process, 248, 259
tapered edge support, 264–267
target to the substrate distance, 254
TDD, 126, 127
TEM, 131, 134, 136, 138, 139, 149, 150, 152,
    155–157, 161

template, 127, 128, 152, 153, 158, 159, 161, 162
tensile strength, 206
tensile stress, 272
TES, 318, 321, 324–326, 330, 333, 336, 341, 343,
    353, 366, 369, 370
tetrahedral sites, 250
tetravalent cation, 249
theoretical maximum efficiencies, 7
thermal ALD, 417–419, 422–424
thermal cycling, 242, 248, 260, 269
thermal energy, 317–370
thermal energy storage, 318
thermal expansion coefficient, 257, 273
thermal shock condition, 487
thermal stability, 253, 259, 269, 274, 321, 488
thermal stress, 269, 273–274
thermodynamic/thermodynamics, 244,
    367–370
thermodynamic absorptions, 286
thermo-mechanical integrity, 271
thin film deposition method, 242, 245, 253
thin-film electrolytes, 490
T-history, 328, 329
threading dislocation density (TDD), 126, 127
TMA, $Al(CH_3)_3$, 418
tolerance factor, 35
τ_{eff}, 419, 422, 430, 431
transitionmetal carbides, 286
transition metal dichalcogenides, 286
transmission electron microscope, 490
transmission electron microscopy, 131, 149
triboelectric nanogenerator (TENG), 79–85
trimethylaluminum, 418
triple phase boundary, 252
trivalent dopant cation, 251
tunability of bandgap, 8
turbulent dissipation rate, 342
turbulent energy, 342
2D materials, 286–290, 295, 298, 299, 305, 306

U

ultimate tensile strength, 276
ultra-violet oxidation, 250
unpassivated region, 434

V

vanadium ion permeability, 205
vanadium redox flow battery, 203
van der Waals gaps, 286
vapor condensation, 269
vehicle and Grotthuss mechanism, 204
V/III ratio, 130–134, 136, 139, 142, 152, 160, 161
V_{oc}, 425, 426, 434, 436
voltage drop, 244
volumetric change, 251

W

water, 318, 320, 336, 339, 340, 343, 364
water splitting, 287
water uptake, 206
Weibull statistics, 274
work function, 444
wrinkles, 265, 266

X

X-ray diffraction, 131, 494

Y

Y6, 465
YDC, 251, 269
yellow hexagonal non-perovskite, 17
Young's modulus, 272, 273, 276
Yttria-stabilized zirconia (YSZ), 242, 246,
          247–251, 257–260, 263, 265, 266, 269,
          271–277, 489